9th International Symposium on Electromagnetic Processing of Materials (EPM 2018)

IOP Conference Series: Materials Science and Engineering
Volume 424

Hyogo, Japan
14 – 18 October 2018

ISBN: 978-1-5108-7335-3
ISSN: 1757-8981

Printed from e-media with permission by:

Curran Associates, Inc.
57 Morehouse Lane
Red Hook, NY 12571

Some format issues inherent in the e-media version may also appear in this print version.

Printed by Curran Associates, Inc. (2018)

For permission requests, please contact the Institute of Physics at the address below.

Institute of Physics
Dirac House, Temple Back
Bristol BS1 6BE UK

Phone: 44 1 17 929 7481
Fax: 44 1 17 920 0979

techtracking@iop.org

Additional copies of this publication are available from:

Curran Associates, Inc.
57 Morehouse Lane
Red Hook, NY 12571 USA
Phone: 845-758-0400
Fax: 845-758-2633
Email: curran@proceedings.com
Web: www.proceedings.com

Table of contents

Volume 424

9th International Symposium on Electromagnetic Processing of Materials (EPM 2018) 14–18 October 2018, Hyogo, Japan

Accepted papers received: 29 August 2018
Published online: 13 October 2018

Preface

Papers

IOP Conf. Series: Materials Science and Engineering **424** (2018) 011001 doi:10.1088/1757-899X/424/1/011001

PREFACE

The current volume represents contributed papers of the proceedings of the 9th international symposium on Electromagnetic Processing of Materials, Yumebutai International Conference Center, Awaji island, Hyogo, Japan on October 14 – 18, 2018.

The series of EPM symposiums was initiated in 1994 by Prof. S. Asai (Japan) and Prof. M. Garnier (France). It takes place every three years alternating between Europe and Asia (Nagoya 1994, Paris 1997, Nagoya 2000, Lyon 2003, Sendai 2006, Dresden 2009, Beijing 2012, Cannes 2015). This is the 9th EPM symposium. There are more than 140 participants and more 100 presentations from industry and academia coming from 15 countries.

Electromagnetic Processing of Materials (EPM) has been widely applied for improving properties of materials and it is also expected to be extended to various materials processes due to unique advantages of electromagnetic interactions. This symposium is traditionally open for any kind of materials processing involving electric and / or magnetic fields. It brings together leading academics, researchers and engineers from industry, and commercial software developers.

Hideyuki YASUDA

On behalf of the EPM2018 Organizing Committee

Organizing committee

Hideyuki Yasuda	Kyoto University
Hiroshi Harada	Nippon Steel & Sumitomo Metal Corporation
Noriyuki Hirota	National Institute for Materials Science
Kazuhiko Iwai	Hokkaido University
Yuji Miki	JFE Steel Corporation
Tomoyuki Kakeshita	Osaka University & Fukui University of Technology
Keiichiro Kashimura	Chubu University
Toshiyuki Kozuka	Kumamoto University
Akitoshi Matsui	JFE Steel Corporation
Takashi Miyake	Kobe Steel
Kohei Morishita	Kyushu University
Kazuki Morita	University of Tokyo
Hideyuki Ohtsuka	National Institute for Materials Science
Nobuhiro Okada	Nippon Steel & Sumitomo Metal Corporation
Tomoki Shibata	Daido Castings
Shin-ichi Shimasaki	National Institute of Technology, Kagawa College
Eiichi Takeuchi	Kinki Polytechnic College Kyoto
Sadahiro Tsurekawa	Kumamoto University
Kazuyuki Ueno	Iwate University
Kazumi Yamamoto	Daido Steel
Noboru Yoshikawa	Tohoku University

Organized by

一般社団法人
日本鉄鋼協会

The Iron and Steel Institute of Japan (ISIJ)

Co-organized by

The 19th Committees, Japan Society for the Promotion of Science

Supported by

Hyogo International Association

The Japan World Exposition 1970 Commemorative Fund

Tsutomu Nakauchi Foundation

ISIJ Kansai Regional Office

Hyuga Memorial Grant for International Conference

Japan Superconductor Technology, Inc. (JASTEC)

IOP Conf. Series: Materials Science and Engineering **424** (2018) 011002 doi:10.1088/1757-899X/424/1/011002

Peer review statement

All papers published in this volume of *IOP Conference Series: Materials Science and Engineering* have been peer reviewed through processes administered by the proceedings Editors. Reviews were conducted by expert referees to the professional and scientific standards expected of a proceedings journal published by IOP Publishing.

Improvement of the properties of light metal matrix Micro/Nanocomposite Materials: myth or reality?

M. Garrido, L. Davoust, R. Daudin, L. Salvo, Y. Fautrelle

Grenoble Polytechnic Institute/CNRS, SIMAP Laboratory, BP 75, 38402 Saint Martin d'Heres cedex, France

Corresponding author: yves.fautrelle@simap.grenoble-inp.fr

Abstract
The use of materials reinforced by micro-nanoparticles (MMC) is more and more widespread. According to recent studies, the addition of micro-particles in solid materials (polymers, concrete or metal alloys) gives them interesting properties in terms of thermal conductivity, electrical conductivity and mechanical resistance. These properties open new perspectives in terms of environmental consequences. They allow for example to lighten structures while improving their mechanical characteristics, to increase the performance of electrical conductors by reducing energy losses. We provide an example related to light metal alloys widely used in automotive industry and aeronautics. However, the introduction of particles presents several difficulties. The efficiency of particle addition at liquid state depends on many factors, like their dimension, the type of sedimentation forces and the chemical composition of both liquid alloys and particles:
(i) at the metallurgical level, the particles are often very reactive and may modify the composition and the micro-structure of the alloy, which can be sometimes beneficial and sometimes damaging,
(ii) in terms of scale-up and development of actual processes, the introduction of micro- and especially nanoparticles in liquid phase is a very difficult operation to achieve. Many types of devices are proposed but none is really satisfactory, especially regarding the homogeneity of the materials.
We present some application cases related to magnesium alloys that perfectly illustrate the advantages and disadvantages of MMCs.

Key words : metallic-matrix composites, nano-particules, dispersion of particules, magnesium alloys, electromagnetic stirring

The majority of modern technologies for the production and processing of materials utilize diverse methods to tailor and optimise their properties using suitable volume or surface treatments. It has been recently shown that nano-particles dispersed in a metal could lead to superior properties of materials for electro-technical or mechanical applications. The impact expected from those materials is of major importance, namely (i) substantial enhancement of material properties, e.g. shaping of magnetic hysteresis loops, decrease of magnetic coercivity, increase of yield strength, ductility, surface hardness of metallic sheets, extension of lifetime, increase of operating temperature ranges, and/or (ii) new microstructures via processing of e.g. polymers in electric fields, including controlled dispersion of fillers, and/or (iii) reduction of costs. The dispersion of ceramic particles of micro and nanometric size in metals is a promising method to potentially provide composites with enhanced mechanical properties. Materials reinforced with ceramics, also called ceramic metal matrix composites (MMCs), present a tradeoff in properties due to the direct relation of stiffness and/or strength to the increase of volume fraction of reinforcement particles. For example, studies carried out by Nikhilesh Chawla and Yu-Lin Shen[1] on Al-Cu-Mg (2080) reinforced with SiC particles showed that ductility can decrease significantly when higher percentages of particles are incorporated. Their studies also showed that at same volume fraction the reinforced material presented a better ductility when smaller particles were used. The spatial distribution of the particles plays an important role in the improvement of the mechanical properties of composite materials, and significant improvements in tensile strength and hardening rate may obtained when the particles were well dispersed [1,2].

Various methods are used in order to elaborate composite materials. Elaboration of MMCs faces several major issues like introduction and dispersion of particles, chemical reaction with the alloy elements etc. The introduction of particles into a matrix remains a difficult problem. This is why various solutions are used. For example, the production of so-called ODS steel is achieved at solid state from a mixture of steel powders and particles[3]. The mixture is then compacted and sintered by rolling. Introduction of particles may also be achieved by injection into the liquid matrix. The latter operation is not always efficient since the crossing of the liquid free surface by particles is often difficult[4]. As far as nanoparticles are concerned, an additional issue comes from the trend of those particles to form clusters which are very difficult to de-agglomerate. Furthermore, elaboration of MMCs at liquid state requires an efficient dispersion of particles which conditions the performances of the material. That problem increases when the scale of the process becomes larger and larger. Accordingly, efficient stirring devices are necessary. They may use mechanical stirring, electromagnetic one and/or ultra-sound devices which are efficient to de-agglomerate nano-clusters [5-7].

The electromagnetic stirring provided by a travelling magnetic field (TMF) applied during melting of metals generates a flow that can be controlled in direction and intensity[6]. In principle, particles of different sizes and compositions can be accurately dispersed by controlling the metal flow. The control of the flow can be achieved by modifying the current applied to the electromagnet. However, for micro-particles both gravitational and electromagnetic sedimentation forces lower the ability of such device to produce a satisfactory dispersion. This was shown by experiments dedicated to mixing SiC macro-particles under the influence of TMF in pure magnesium and industrial magnesium alloy AZ91[6].

When industrial alloys are used, the particles may react with the various components. For example, Magnesium alloy AZ91 inoculated with Al_2O_3 particles presented an increase of the number of Al_8Mn_5 intermetallic particles in the alloy which suggest that Al_2O_3 particles served as nucleants for these intermetallic compounds[4]. Moreover, the Al_2O_3 particles react with the matrix material to form magnesium oxide MgO and other intermetallics precipitates like Al_8Mn_5. It turns out that refinement of the sample was obtained thanks to the presence of the bi-product MgO, which acts as a good grain refiner, and not Al_2O_3 particles directly. In that case mechanical tests showed that the presence of Al_2O_3 particles did not improve the hardness nor the tensile strength of the inoculated alloy.

Other tests were performed with a second industrial magnesium alloy: Magnesium ELEKTRON 21. Mg EK21 is a high strength fully heat treatable magnesium based alloy designed to be used for lightweight structural applications subjected to high temperature. Its excellent mechanical properties make this light alloy widely used in automotive and aerospace applications. The composition of magnesium EK21 contains rare earths such as neodymium and gadolinium which provide high strength at high temperatures and also a great potential for precipitation strengthening. Zirconium is also present in its composition and serves as grain refiner, which improves its mechanical properties at ambient temperature and enhances its corrosion resistance. Dispersion-strengthening using ceramic nanoparticles which are theoretically stable at high temperatures, holds a high potential as a method to improve high-temperature properties of magnesium alloys. The reinforcement of metal matrices using ceramic nanoparticles effectively dispersed can be achieved by different mechanisms, such as Orowan pinning of dislocations or by restricting grain boundary sliding[8]. For instance, the dispersion of SiC nanoparticles in pure magnesium using ultrasounds stirring system was carried out by Dieringa[9] produced the improvement of its creep resistance. This feature was related to Orowan strengthening and dislocation generation. Liu *et al*[10] demonstrated that dispersion of nanoparticles of aluminium nitride (AlN) into nanocrystalline aluminium enhanced its elastic modulus and hardness. One of the particularities presented by AlN particles is its lattice parameter which is similar to the one presented by pure magnesium. This feature makes this particle a potential grain refiner for magnesium matrices. Different authors demonstrated that improvement of mechanical properties of particulate reinforced composites highly depends on the good dispersion of the particles within the matrix material[11]. We have carried out a microstructural and mechanical analysis of magnesium EK21-based samples in which nano- particles of AlN were electromagnetically dispersed using a travelling magnetic field. Among the different advantages offered by electromagnetic stirring, the contact-less interaction with the melted material, is of utmost importance in order to avoid melt contamination. Moreover, the sedimentation effects on nano-particles are weak. The capacity to monitor the direction of the flow and therefore the particle distribution patterns as well as the capacity to maintain the stirring until the end of the solidification stand also as decisive assets.

The addition of AlN nanoparticles to magnesium EK21 produces measurable changes in the microstructure and mechanical properties of the material. The first effect observed is an unexpected increase in grain size with respect to the pure material processed under the same casting conditions. The reinforced material presents an average grain size 200% bigger than the average grain size of pure magnesium EK21. The study of the equilibrium reactions of AlN with pure magnesium shows that an amount equivalent to 0.3% in weight of aluminium is released into the magnesium matrix under the casting conditions applied during the experiments. The aluminium dissolves in the melted magnesium and tends to react with the available elements in the alloy such as zirconium which is used in this alloy as grain refiner. The reaction between zirconium and aluminium produces different compounds such as AlZr. The lattice mismatch between zirconium and magnesium is reported to be less than 1%[12] , whereas the lattice mismatch between AlZr and magnesium is approximately 4% which suggest that this particle is not well adapted as a grain refiner of magnesium matrices. The formation of this particle is suspected to reduce the grain refinement effect of zirconium and consequently to be responsible for the increase of the grain size observed. Aluminium was also found in the intermetallic phase rich in Gadolinium and Neodymium. The high difference of electronegativity between aluminium and rare earths leads to a high tendency to form Al-RE compounds. Aluminium compounds containing rare earths such as Al_3Zr have been related to the enhancement of creep strength on different alloys such as Mg-Li-Al[13]. The formation of $AlNd_2$ compounds was predicted by thermodynamic calculations and confirmed during XRD analysis of the materials produced. The presence of aluminium in the intermetallic phase can be also related to the presence of AlN nanoparticles rejected to grain boundaries during solidification. The presence of AlN nanoparticles in grain boundaries has been detected showing that despite of the partial dissolution of the particles during material processing, most of the nanoparticles survived. The presence of particles in the intergranular space could have contributed to the reinforcement of the material related to intergranular reinforcement. Signs of interaction between particles and dislocations, presented as stress variations at the end of the tests, were observed during the compression tests of the samples containing nanoparticles. Finally, the compression tests put in evidence the decrease of the dynamic recovery in the materials reinforced with nanoparticles. This could be a possible evidence of the reduction of annihilation of dislocations by the presence of the particles which could have acted as obstacles for the moving dislocations during recovery. The effects produced in the microstructure and creep resistance by the incorporation of AlN nanoparticles into magnesium EK21 matrices has been assessed. The following conclusions can be drawn: The creep threshold of the samples containing nanoparticles increased with respect to the pure alloy, and the flow stress rose up by 20 % to 50% which supports other studies found in literature.

In conclusions, the development of liquid phase MMCs with good particle distribution homogeneity is a real challenge, especially for large size processes. This partly explains why the performances of these materials are contrasted. In addition to the difficulty of injecting the particles into a liquid metal, the addition of particles does not always lead to any improvement in the performance of the material, as our experiments showed. Finally, as far as industrial alloys are concerned, the particles react chemically with the elements of the alloy and cause unexpected effects.

Acknowledgements
The research leading to these results has been carried out under EXOMET project under the European Community's Seventh Framework Programme, Contract no. FP7-NMP3-LA-2012-280421. ESA with Wim Sillekens, GPM2 team.

References

1. B. N. Chawla and Y. Shen, Adv. Eng. Mater. 3-6 (2001), 357–370
2. D. Zhang, K. Sugio, K. Sakai, H. Fukushima, and O. Yanagisawa, Met. Mater. Trans., 48-2 (2007), 171–177
3. K.D. Zilnyk, H.R.Z. Sandim, R.E. Bolmaro, R. Lindau, A. Möslang, A. Kostka, D. Raabe, Journal of Nuclear Materials 448 (2014), 33–42
4. M. Garrido, Dispersion d'origine électromagnétique de micro/nano particules dans un alliage métallique solidifié, PhD thesis, Université Grenoble-Alpes (2017)
5. Z-Y Fan, International Materials Reviews, 47-2 (2002), 49-85
6. V. Bojarevics, K. Pericleous, M. Garrido, Y. Fautrelle, and L. Davoust, Magnetohydrodynamics, 51-3 (2015), 453-460
7. D. G. Eskin, in Processing of materials by external fields, D. Eskin and J. Mi eds., Springer Verlag ,New York (2018)
8. T. H. Courtney, Mechanical Behavior of Materials: Second Edition, Waveland Pr Inc (2005)
9. H. Dieringa, J. Mater. Sci., 46-2 (2011), 289–306
10. Y. Q. Liu, H. T. Cong, W. Wang, C. H. Sun, and H. M. Cheng, Mater. Sci. Eng. A, 505 no. 1–2 (2009), 151–156
11. I. M. R. Banerjee, Ceramic Nanocomposites, ed. by R. Banerjee and I. Manna, Woodhead Publishing Series, Elsevier (2013)
12. G. Klosch, B. McKay, and P. Schumacher, in Essential Readings in Magnesium Technology, J. Wiley & Sons (2014)
13. M. Avedesian and B. Hugh, ASM Specialty Handbook: Magnesium and Magnesium Alloys. 1999.

Electromagnetic sensor just below CC mold by using magnetic transformation of steel

H. Harada[1], M. Nagashima[2], T.Konno[2], M.Yamana[2] and T. Toh[3]

[1] Nippon Steel & Sumitomo Metal Corporation (Process Research Labs., Futtsu, Japan)
[2] Nippon Steel & Sumitomo Metal Corporation (Hirohata Steel Works, Himeji, Japan)
[3] Nippon Steel & Sumikin Technology Co.Ltd (Resource and Process Solution Div., Futtsu, Japan)

Corresponding author: harada.248.hiroshi@jp.nssmc.com

Abstract
Electromagnetic sensor to detect the unbalanced flow in CC strand by using the magnetic transformation just below CC mold has been proposed, noticing that the narrow face of slab has been cooled at beneath of mold and surface temperature rapidly decreases to Curie temperature (Tc). Laboratory experiments have been performed to verify the principle of the proposed sensor, which is composed of primary and secondary solenoid coil. The obtained experimental results showed that the relationship between the surface temperature of slab and induced electric voltage could be explained by Curie-Weiss law, when AC magnetic field has been imposed on the slab surface. Moreover, plant tests have been conducted by installing the sensor just below the narrow face of mold. It can be found that the proposed sensor have potential to measure surface temperature under severe conditions and detect the unbalanced flow in CC strand.

Key words : magnetic transformation, Curie temperature, continuous casting, unbalanced flow, solidification

Introduction

Molten steel flow in CC mold largely affects the cast slab quality and the productivity of CC. In CC mold, unsymmetrical discharged flow from submerged entry nozzle causes unbalanced flow in CC strand and local remelting of solidified shell at the impinging points of the discharged flow, as shown in **Fig.1**. Additionally, the unbalanced flow leads to the entrapment of mold powder at meniscus and the inner defects due to deeply transported Ar bubbles. Therefore, suitable control of the discharged flow and the entire flow pattern in CC strand are necessary. At the same time, the detection of the unbalanced flow becomes important in order to escape from serious casting troubles and slab defects. Several detection methods of the unbalanced flow in CC mold have been reported [1] [2]. However, it is impossible to directly detect the unbalanced flow in CC strand. Harada et al. [3] proposed electromagnetic sensor just below CC mold by utilizing magnetic transformation of steel, noticing that strong cooling beneath the mold causes magnetic transformation of steel. In this study, noticing that the unbalanced flow forms the large differences of surface temperature of slab just below the mold, the electromagnetically detecting methods of unbalanced flow in CC strand have been developed.

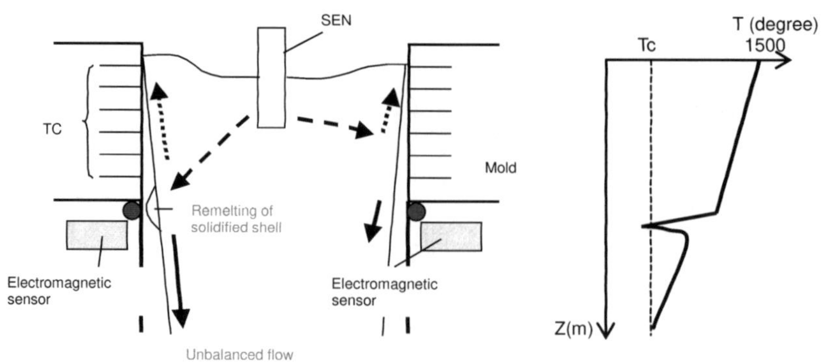

Fig.1: Schematic view of unbalanced flow in CC strand and surface temperature change during casting.

The principle of electromagnetic sensor and laboratory experiment for verification

Fig.2 shows the principle of the proposed electromagnetic sensor [3]. When AC magnetic field is applied normal to the slab surface, the distribution of magnetic field in the slab is dependent of magnetism of slab surface. When surface temperature (Ts) is much higher than Tc, the distribution of magnetic field line is the same as the one in vacuum. On the contrary, when Ts approaches to and pass through Tc, the applied magnetic field is distorted by the magnetism of slab surface, in other words, slab surface temperature. By measuring the induced electric voltage due to the change of imposed magnetic field line, surface temperature can be estimated. The sensor is composed of primary coil and secondary coil. A set of coils is mounted into the cylinder made of stainless steel. AC electric current is supplied to primary coil by using a constant current power amplifier, AC magnetic field is imposed normal to the slab surface. The noise in secondary coil is minimized with low pass filter. Moreover, only the signal with same frequency as imposed electric current, is elucidated by using Lock in amplifier.

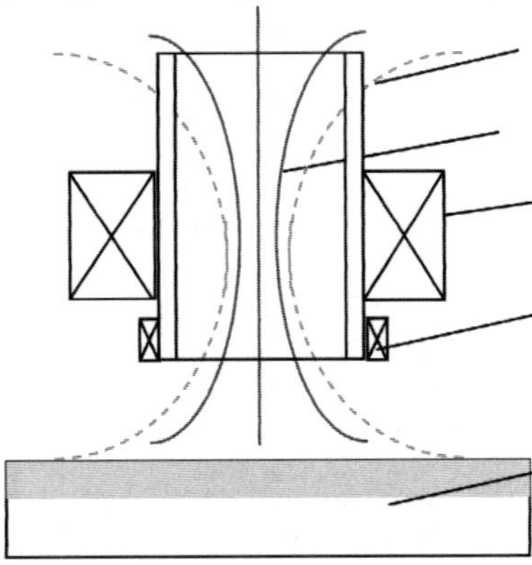

Fig.2: Schematic view of electromagnetic sensor and the change of magnetic field line due to magnetism of slab surface.

Laboratory experiment has been performed to clarify the relationship between sensor signals and surface temperature of slab. Slab samples are inserted into a heating furnace and heated into 1350 degree. A thermocouple is inserted into the slab, in which the position is 1mm from the surface of slab. After extraction of the heated slab, aforementioned sensor is set on the heated samples and the relationship between induced electric voltage of secondary coils and surface temperature of slab has been investigated. **Fig.3** shows the typical example of the obtained results. It can be found that the induced electric voltage is almost constant in high temperature and low temperature region. On the contrary, the electric voltage changes with temperature, when surface temperature approaches to Curie temperature. When magnetic field is applied to paramagnet whose temperature is changing with time, the change of magnetic flux and the induced electric potential of the coil is governed by Curie-Weiss law and Faraday law of induction as expressed by equations (1)-(4). After transformation of these equations, the relationship between electric potential and temperature of paramagnet can be finally obtained by using integral equation (5). The calculated results by using equation (5) and experimental results are plotted on **Fig.4**. It can be found that experimental results show good agreement with the calculated results based on Curie-Weiss law and that the proposed electromagnetic thermometer is the sensor based on magnetic transformation of steel.

$$V = -n\frac{d\emptyset}{dt} \tag{1}$$

$$\emptyset = B \cdot S \tag{2}$$

$$\mathrm{B} = \mu_0(1+\chi)H \tag{3}$$

$$\chi = \frac{c}{T - T_c} \tag{4}$$

$$\int Vdt = \int \frac{\mu_0 nSHc}{(T - T_c)^2} dT \tag{5}$$

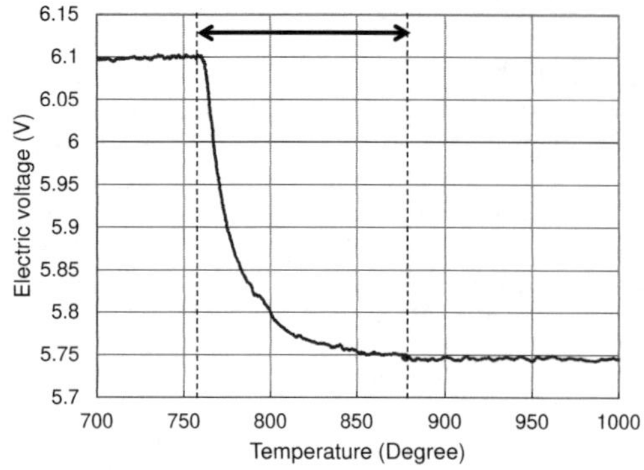

Fig. 3: Relationship between slab surface temperature and induced electric voltage of secondary coil.

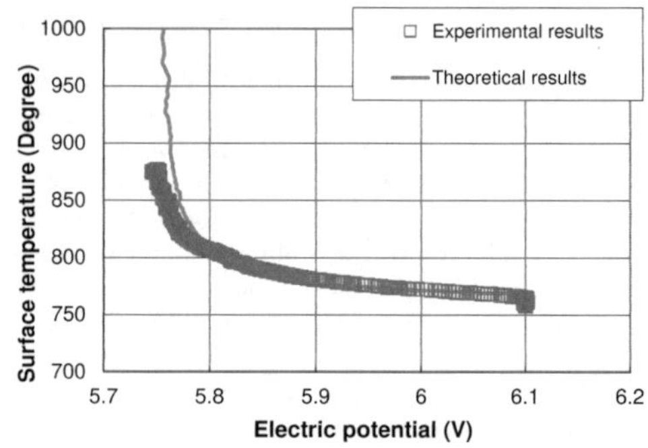

Fig.4: Comparison of experimental results with theoretical results based on Curie-Weiss law, about the relationship between slab surface temperature and induced electric voltage.

Plant test for verification of the proposed sensor

A pair of sensor was installed under the respective narrow face of mold at bending type of continuous casting machine and plant tests have been performed to investigate the effects of casting conditions on the sensor signals. **Table 1** shows the casting conditions. **Fig.5** shows the time-change of the measured surface temperature by using sensor, when casting speed changes in sequential casting. It can be found that surface temperature of narrow face of slab changes with casting speed and surface temperature always changes under the constant casting speed conditions. The difference in surface temperature measured by both sensors have been compared with the difference in the fluid flow velocity at solidification front at both narrow face of slab. The fluid flow velocity at solidification front can be estimated from the measured value of inclination angle of columnar dendrites and solidification speed by using experimentally obtained formula by Okano et al. [4] The difference in thus obtained fluid flow velocity at both narrow face has been compared with the difference in the measured surface temperature of both sensors. The obtained results are plotted in **Fig.6**. The good correlation between the measured temperature difference and the estimated fluid flow difference can be recognized. This means that the temperature differences increases with the fluid flow difference, in other words, the magnitude of the unbalanced flow. As a result, it can be found that the proposed sensor have potential to detect the unbalanced flow in CC strand by measuring surface temperature of cast slab just below CC mold.

Table1 Casting conditions of plant test.

Slab width	1000 ~ 1800mm
Casting speed	0.75 ~ 1.2m/min.
Steel grade	Middle Carbon Al-killed steel, Low Carbon Al-killed steel, IF steel
Sensor position	(i)Vertical position: just below the narrow face of mold (ii)Distance between sensor and narrow face of slab : 30mm

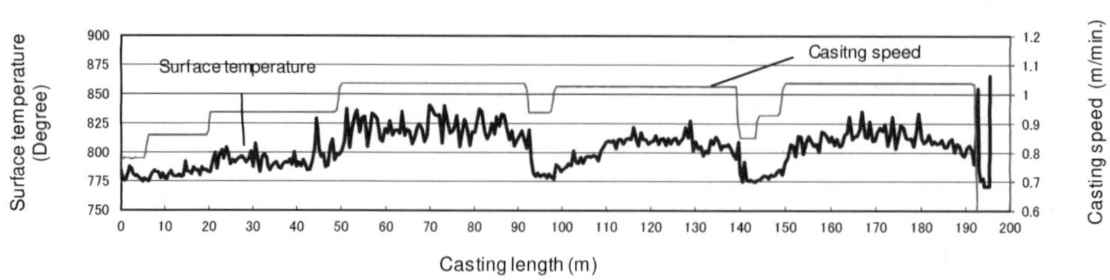

Fig.5: Successive change of surface temperature of slab with casting speed.

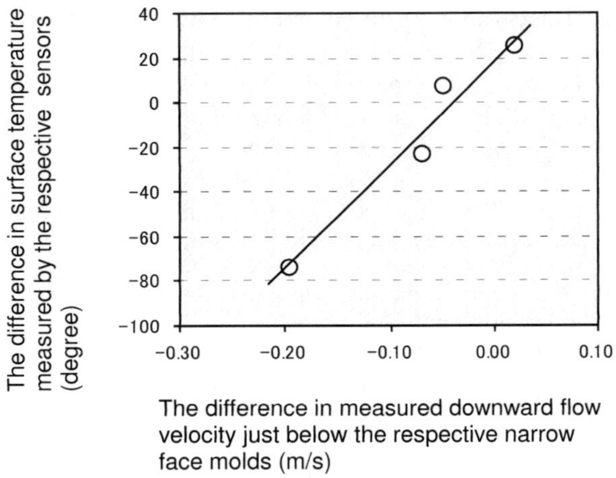

Fig.6: Correlation between temperature difference and fluid flow velocity difference.

Conclusions

Electromagnetic sensor to detect the unbalanced flow in CC strand by using the magnetic transformation just below CC mold has been proposed. The obtained results are summarized as follows:

(1) Noticing that the narrow face of slab has been cooled below Curie temperature at the beneath of mold, the sensing method to measure the change of imposed AC magnetic field due to slab surface temperature, has been constructed.

(2) Laboratory experiments have been performed to verify the principle of the proposed sensor which is composed of primary and secondary solenoid coil. The obtained experimental results showed that the relationship between the surface temperature of slab and induced electric voltage could be explained by Curie-Weiss law, when AC magnetic field has been imposed on the slab surface.

(3) Plant tests have been performed by installing the sensor under the respective narrow face mold to show that the proposed sensor could measure surface temperature under severe conditions and detect the unbalanced flow in CC strand.

Nomenclature

B: magnetic flux density, *c*: constant, *H*: magnetic field, *n*: turn number of electric coil, *S*: area, t: time, *T*: temperature, *Tc*: Curie temperature, *Ts*: Surface temperature of slab, *V*: electric voltage , ϕ: magnetic flux, χ: magnetic susceptibility, μ_0: magnetic permeability

References

1. M.Ohji (1994), 153_{rd} and 154_{th} Nishiyama Memorial Lecture, ISIJ, 1.
2. T.Ishii, M.Tanaka, A.Uehara, H.Kimura and N.Tsutsumi (1994), Proc. Int. Symposium on Electromagnetic Processing of Materials, Nagoya, ISIJ, 396.
3. H.Harada, M.Nagashima and T.Toh (2015), Proc. of 8th Int. Symposium on ElectromagneticProcessing of Materials, Canne.
4. S.Okano, T.Nishimura, H.Ohi and T.Chino: Tetsu-to-Hagane, 61(1975), 2982.

Real-time control of the mould flow in a model of continuous casting in frame of the TOMOCON project

T. Wondrak[1]*, U. Hampel[1], M. Ratajczak[1], I. Glavinic[1], F. Stefani[1], S. Eckert[1], D. van der Plas[2], P. Pennerstorfer[3], I. Muttakin[4], M. Soleimani[4], S. Abouelazayem[5], J. Hlava[5], A. Blishchik[6], S. Kenjeres[6]

[1] Helmholtz-Zentrum Dresden - Rossendorf, Bautzner Landstraße 400, Dresden, Germany
[2] Tata Steel, Wenckebachstraat 1, Velsen-Noord, The Netherlands
[3] Primetals Technologies Austria, Turmstrasse 44, Linz, Austria
[4] University of Bath, Claverton Down, Bath, United Kingdom
[5] Technical University of Liberec, Studentská 1402/2, Liberec, Czech Republic
[6] Delft University of Technology, van der Maasweg 9, Delft, The Netherlands

Corresponding author: t.wondrak@hzdr.de.

Abstract
In continuous casting, the flow structure of the liquid steel in the mould and the two-phase distribution in the submerged entry nozzle (SEN) are crucial for the quality of the produced steel. In order to effectively control the flow in the mould by electromagnetic brakes (EMBrs) and the injection of gas into the SEN, even a rough knowledge of the flow structure would be very helpful. In the framework of the TOMOCON project, the contactless inductive flow tomography (CIFT) and the mutual inductance tomography (MIT) will be integrated into a control loop for slab casters. This control loop will be developed and implemented at the Mini-LIMMCAST facility, which is available at the Helmholtz-Zentrum Dresden – Rossendorf. In this paper a short overview of this project will be given.

Key words: continuous casting, electromagnetic brake, inductive measurement techniques, flow control

Introduction

Continuous casting is widely used in the steel industry for the production of billets, blooms and slabs. Typically, the liquid steel flows from a tundish through a submerged entry nozzle (SEN) into the mould. A stopper rod or a sliding gate controls the flow rate of the melt. The walls of the mould are made of copper and are constantly cooled with water so that a shell solidifies at the walls. The partly solidified strand exits the mould at its bottom and is cooled further until it is solidified completely. For most steel grades, argon gas is injected into the SEN in order to prevent nozzle clogging and to catch inclusions.

The flow in the upper part of the mould affects the formation of inclusions and surface defects that affect the quality of the cast product [1]. For instance, the flow should enable the transport of bubbles towards the free surface to avoid inclusions in the solidified steel. In addition, too high velocities at the meniscus can lead to slag entrapment. On the other hand, too low velocities may lead to premature freezing of the steel at the meniscus, promoting surface defects. Thus, the flow pattern and flow velocity at the meniscus must be controlled within tight limits. Therefore, it is desirable to control the flow in the upper part of the mould during casting. Typically, electromagnetic stirrers (EMS) or electromagnetic brakes (EMBrs) are used for this purpose [2]. Ideally, the control loop of those actuators should take the current flow structure in the mould into account. However, up to now this is not feasible due to the lack of appropriate measurement techniques. Only a few local measurement techniques are available for specific plant trials because of the high temperatures of about 1500°C and the opaqueness of the liquid steel. Among them are the nail board dip test, the strain gauge method, mould flow control (MFC) sensors from AMPEA [3] or the evaluation of the oscillation mark shape [4]. Only recently, a new technique was proposed, based on temperature measurements in the meniscus area of the mould [5].

In the last decade two promising contactless measurement techniques were developed: contactless inductive flow tomography (CIFT) and mutual inductance tomography (MIT) which have the potential to be incorporated in a real-time process control loop. CIFT is able to reconstruct the flow structure in the mould by applying a magnetic field to the mould and measuring the flow induced perturbances of that field outside the mould. Based on these measurements, the velocity field is reconstructed by solving the corresponding linear inverse problem [6-9]. MIT is able to detect the conductivity distribution in one cross section of a pipe [10-14]. The sensor consists of an array of coils which are arranged around the pipe. From the measured mutual inductances between these coils the gas/liquid distribution is reconstructed by solving the corresponding non-linear inverse problem. It has been applied for imaging the distribution of liquid metal and argon gas in the SEN of a small model [8] as well as of a pilot [11] and a full size plant [12]. In the framework of the EU project TOMOCON (smart TOMOgraphic sensors for advanced industrial process CONtrol) a new control mechanism based on these measurement techniques will be developed and demonstrated using the Mini-LIMMCAST facility, which is available at Helmoltz-Zentrum Dresden – Rossendorf [15]. It is a model of a slab caster and is operated at room temperature with the eutectic alloy GaInSn. It will be equipped with both measurement techniques and an EMBr. In this paper, a short overview about the project will be given. After a short description of the Mini-LIMMCAST facility, the two tomographic measurement techniques and a preliminary draft of the control loop are presented.

Demonstrator setup

Fig. 1 shows a schematic and a photo of the Mini-LIMMCAST [15]. Here, instead of liquid steel, the eutectic alloy GaInSn is used whose material properties are described in [16]. The melt flows from the tundish via the SEN made of acrylic glass into the mould with a cross section of 140 x 35 mm^2. The position of the stopper rod controls the flow rate in the SEN. Additionally, argon gas is injected into the SEN via the tip of the stopper rod. The liquid flows from the bottom of the mould into a catchment tank from where it is pumped back into the tundish. The model will be operated in continuous mode allowing measurement times of several hours. For modelling the influence of an electromagnetic brake (EMBr) on the flow, a DC magnet is attached to the mould that produces a transverse magnetic field with a maximum field strength of 0.31 T. The pole faces of the magnet cover the wide side of the mould completely. The model has following actuating variables: the flow rate of the liquid metal by the position of the stopper rod, the flow rate of the argon gas and the strength of the EMBr. Additionally, the new inductive measurement techniques will be validated using ultrasound Doppler velocimetry measurements.

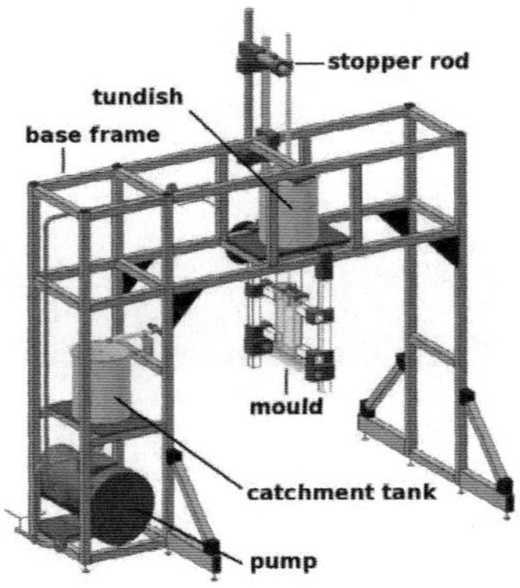

(a)

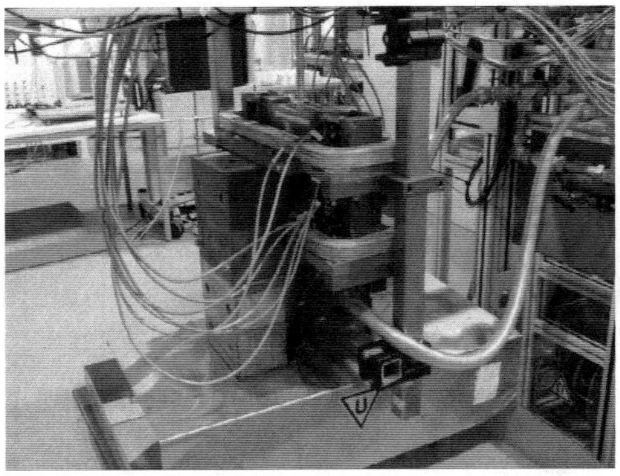

(b)

Fig 1.: Schematic (a) and photo (b) of the Mini-LIMMCAST facility

Contactless inductive flow tomography

Fig. 2a shows a schematic of the sensor arrangement for CIFT. Two rectangular coils generate the primary magnetic field of about 1 mT. Seven magnetic field sensors along each narrow face of the mould record the secondary field, which has a maximum strength of about 200 nT. Fig. 2b shows the reconstructed velocity for one instant in time without an electromagnetic brake. The typical double roll flow structure can be clearly identified. One challenge of this technique is the accurate measurement of the secondary magnetic field in the presence of the ferromagnetic pole shoes of the EMBr. These ferromagnetic parts possibly affect the secondary magnetic field and this alternation is not accounted for in the CIFT formulation. It was shown numerically that for the present sensor arrangement the impact of the pole shoes on the secondary magnetic field is negligible. Additionally, gradiometric induction coils are used with an alternating primary magnetic field [9]. The typical excitation frequency is in the range of a few Hz so that the skin effect is negligible and the magnetic field can penetrate all the liquid. Induction coils have the advantage that they are insensitive to all static magnetic fields. While in previous work the strength of the EMBr was constant during the entire measurement, the effects of a varying strength of the EMBr have to be investigated. The time resolution of the magnetic field measurement and the subsequent flow reconstruction will be about 1 Hz. For the integration of CIFT into a control loop, a real time reconstruction algorithm has to be developed.

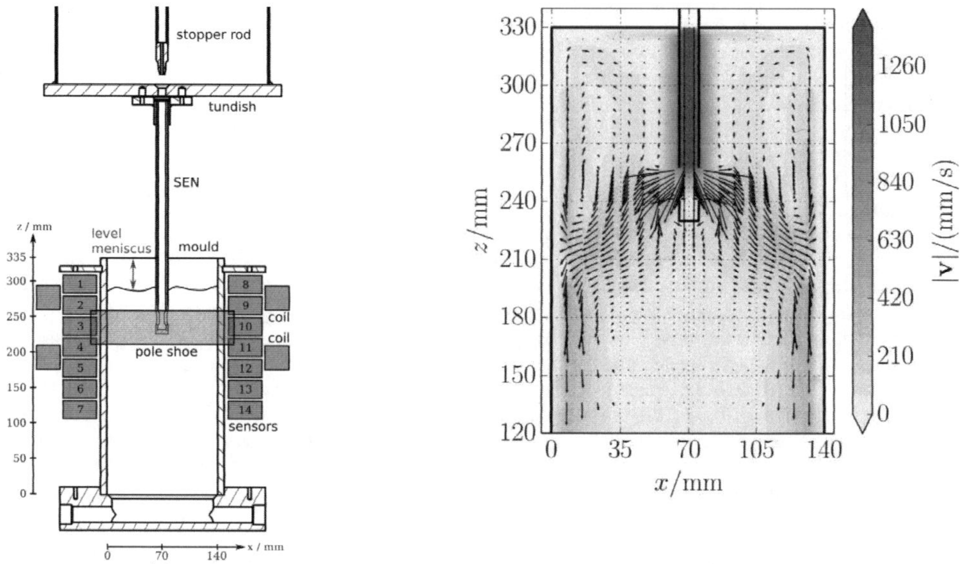

(a) (b)

Fig 2.: Schematic of the sensor arrangement for CIFT and one flow reconstruction without EMBr

Mutual inductance tomography

MIT is able to reconstruct the conductivity distribution in one cross section of the SEN, and thereby to distinguish between liquid metal and argon in case of a two-phase flow. Existing systems typically use an array of 8 coils around the SEN and operate with a frame rate of 20 to 40 frames per second [8,14]. After measuring the mutual inductances between the 8 coils, the non-linear inverse problem has to be solved. Fig. 3 shows a schematic sketch of an MIT sensor and the reconstructed liquid metal distribution at the Mini-LIMMCAST facility [8,14]. These experiments showed that the outer surface of the liquid metal strand in the SEN was reliably detected. However, it was difficult to identify bubbles inside the liquid jet. Additionally, the high velocity of about 1.4 m/s of the melt requires a high frame rate to visualise bubbles of the size of 5 mm. A proposed solution could be the combination of the electrical capacitance tomography (ECT) with MIT. The MIT part will be operated with low frequencies for imaging the interior of the liquid metal jet and ECT will be used to detect the outer shape of the jet [18].

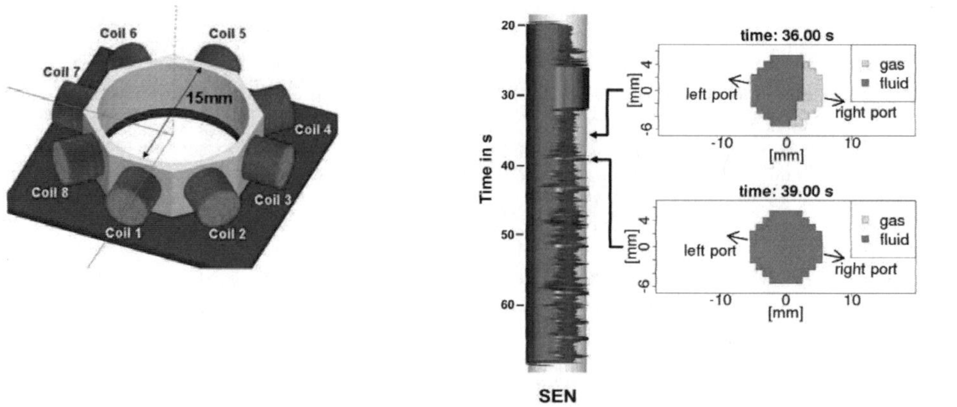

(a) (b)

Fig 3.: Schematic (a) of an MIT sensor and reconstructed argon/GaInSn distribution (b) in the SEN at the Mini-LIMMCAST facility (see [8])

Control loop

In order to combine both tomographic measurement systems with the available actuating variables of the demonstrator, a control loop is proposed which is shown in Fig. 4. The challenge is the design and implementation of the controller. Advanced control designs from the class of model-based control (MBC) strategies are most likely to cope successfully with this challenge. In the first step, a variety of numerical flow simulations have to be carried out in order to identify the special features of the flow which are suitable for the control mechanism. The numerical simulation will be validated by experiments at the Mini-LIMMCAST using both tomographic sensors as well as ultrasound Doppler velocimetry measurements and X-ray radiographic measurements for two-phase flows. These results will form the basis for the design of the controller. A specific challenge for the control loop may be that process adaptations should not occur too often, as these may result in initial instabilities.

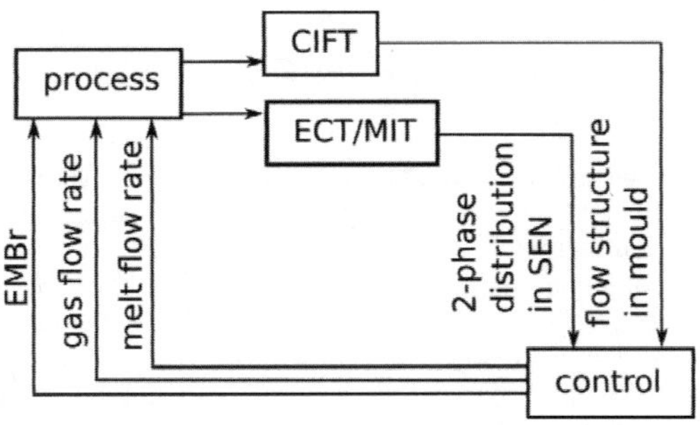

Fig 4: Proposed control loop

Conclusions

Both measurement techniques, CIFT and MIT, allow the measurement of the current state of the flow regime in the SEN as well as in the mould for the first time. Additionally, both tomographic techniques have the potential to be applied robustly in the harsh environment of an industrial caster. The detailed knowledge of the flow enables to implement a control mechanism, which is capable to properly react in a timely manner on changes in the flow in order to achieve sufficient product quality. However, the development of the control loop is a challenging task due to the complexity of the flow. In the TOMOCON project, a first control loop will be designed and will be implemented at a model of a continuous caster to show the feasibility of this approach.

Acknowledgment

Financial support from the EU in the framework of the Marie Skłodowska-Curie grant agreement No 764902 (TOMOCON) is acknowledged.

References

1. L.F. Zhang, B.G. Thomas, Iron and Steel Institute of Japan International 43 (2003), 271 -291
2. P. Dauby, Revue de Métallurgie 109 (2012), 113-136
3. B.G. Thomas et al., Iron and Steel Institute of Japan International 41 (2001), 1262-1271
4. K. Cukierski, B.G. Thomas, Metallurgical and Materials Transactions B 39 (2008), 94-107
5. M. Seden et al, Proceedings 9th ECCC, Vienna, 26-29th June 2017, 67- 75
6. F. Stefani, T. Gundrum, G. Gerbeth, Physical Review E 70 (2004), 056306
7. T. Wondrak et al., Measurement Science & Technology 21 (2010), 045402
8. T. Wondrak et al., Metallurgical and Materials Transactions B 42 (2011), 1201-1210
9. M. Ratajczak et al., Philosophical Transactions of the Royal Society A 374 (2016), 20150330
10. R. Binns et al., Measurement Science & Technology 12 (2001), 1132-1138
11. X. Ma et al., IEEE Sensors Journal 5 (2005), 224-232
12. J. van Oord et al., Proc. 2nd Steelsim Conference, Graz, 12-14 September 2008, 179-184
13. X. Ma et al., Measurement Science & Technology 17 (2006), 111-118
14. N Terzija et al., Flow Measurement and Instrumentation 22 (2011), 10-16
15. K. Timmel et al., Iron and Steel Institute of Japan International 50 (2010), 1134-1145
16. Y. Plevachuck et al., Journal of Chemical and Engineering Data 59 (2014), 757-763
17. K. Timmel et al., Metallurgical and Materials Transactions B 46 (2015), 700-710
18. T. Wondrak, M. Soleimani, Measurement Science & Technology 28 (2017), 064001

Application of modulated calorimetry to the liquid metals using electromagnetic levitation and static magnetic field

O. Budenkova[1], M. Milgravis[2], Ch. Garnier[1], A. Gagnoud[1], Y. Delannoy[1], S. Semenov[1], P. Chometon[3], S. Rivoirard[3], M. Alamir[4], J. Etay[1]

[1] Univ. Grenoble Alpes, CNRS, Grenoble INP, SIMAP, 38000 Grenoble, France
[2] Institute of Physics, University of Latvia, Riga, LV-2169, Latvia
[3] Univ. Grenoble Alpes, CNRS, Institute Néél, 38000 Grenoble, France
[4] Univ. Grenoble Alpes, CNRS, GIPSA-Lab, Grenoble INP, 38000 Grenoble, France

Corresponding author : olga.budenkova@simap.grenoble-inp.fr

Abstract
Measurement of the thermophysical properties of liquid metals is challenging because of their high chemical activity and high temperatures. The electromagnetic levitation allows one to hold the electrically conductive liquid sample containerless in an inert atmosphere in thermal equilibrium while measurements on the sample can be taken in a non-contact way followed by extraction of some thermophysical properties. Yet, the electromagnetic forces within the skin layer inside the sample cause convective flow of the liquid thus disabling the data extraction. A static magnetic field imposed over a sample is known to damp the convective flow. With these ideas, an experimental set-up with a DC magnetic field directed perpendicular to the gravity vector was constructed and first experiments were performed with liquid Ni and some other materials. In most of the experiments the instability of the levitated sample during a slow variation of the DC magnetic field was observed which is reported in the present article.

Key words: modulated calorimetry, electromagnetic levitation, DC magnetic field, liquid metals, thermophysical properties

Introduction

Measuring of thermophysical properties of the most liquid metals and their alloys is a challenging problem, related to their high melting temperatures and chemical reactivities. The latter implies that it is preferably to keep the liquid sample under an inert atmoshpere and without any contact with a container, i.e. in levitation. A good electrical conductivity of the liquid metals allows one to use the electromagnetic levitation in a field created by a high frequency (HF) monophase AC current which circulates in an inductor (e.m. levitator). Simultaneously, due to the Joule heating, the sample can be heated and kept at different temperatures in thermal equilibrium. Further, a modulated calorymetry can be applied which consists in perturbing the thermal equilibrium of the sample by a modulated heating. The latter can be done either using the same e.m. levitator with a specially adapted modulation of a HF current [1] or with use of the laser heating [2]. The "thermal response" of the sample (i.e. the temperature of the latter in a certain zone) to the perturbation is registered and analysed. In a case of a purely conductive heat transfer inside the sample and with use of an appropriate modulation of the heating it is possible to extract the data about the specific heat C_p of the sample and its thermal conductivity κ. Such procedure has been succefully applied to various samples under microgravity conditions on the International Space Station [1] where the electromagnetic levitation is required, in fact, only to stabilize the sample in space, but there is no need to overcome the gravity. That makes possible of a special adaptation of a geometry of the e.m. levitator to minimize the convective flow in the sample which nevertheless remains non-zero.

The possibility of the application of the same measurement procedure in on-ground condition is highly attractive, yet related to many technical difficulties. In fact, the electromagnetic forces required to keep the liquid sample in levitation create strong convection inside the droplet [3] thus making the data processing adapted to the coductive heat transfer invalid and data extraction impossible. However, a static magnetic field (DC field) can be used to damp the convective flow. This idea was realized in [4]-[7] with a DC field parallel to the gravity vector, i.e. directed along the axis of symmetry of the e.m. levitator (with neglecting the helicity of the latter). The intensity of the DC magnetic field necessary to damp any oscillation of the droplet's shape was found to be higher than 3T. The mesurement procedure developed in [4-6] is based on a modulated laser heating of the sample in its equatorial zone and mesurement of the temperature of the pole of the sample. In principle, such procedure requires preliminary knowledge of the optimal frequency of the modulated heating as well as the size of the equatorial and polar zones within which the temperature is supposed to be uniform.

IOP Conf. Series: Materials Science and Engineering **424** (2018) 012004 doi:10.1088/1757-899X/424/1/012004

A mesurement protocol, proposed in [8], is similar to the one used at International Space Station and does not require additional equipement (laser) thus it is simpler than the one described above since the heating of the sample is produced by the same e.m. inductor. Another advantage of the proposed protocol is the use of a heating in a form of a white noise with a fixed amplitude instead of its sinusoidal modulation with a predefined frequency. Further data processing based on the theory of the dynamic systems allows one to define all the parameters, namely, C_p, κ, and total emissivity ε provided the power input into the sample is known.

Experimental set-up and measurement procedure
The scheme of the experimental set-up is presented briefly below and can be found elsewhere [9]. A Helmholtz coil (superconducting magnet) and a double wall water-cooled steel chamber are the two main parts of the set-up (Fig. 1). The chamber contains water cooled electromagnetic inductor which is mounted over a base together with a mirror allowing for the observation of the lateral side of the sample. The top of the chamber is covered with a transparent quartz lid through which the sample and the mirror are visbile. Two pyrometers are mounted above the Helmoltz coil and measure the temperature of the polar and equatorial zone of the sample. Another mirror mounted close to the pyrometers allows us to registered the image of the sample in two projections due to its reflection in the lateral mirror. Before the measuring procedure the chamber containig the levitator and the sample is pumped to obtain a low vacuum and is filled with the argon after that.

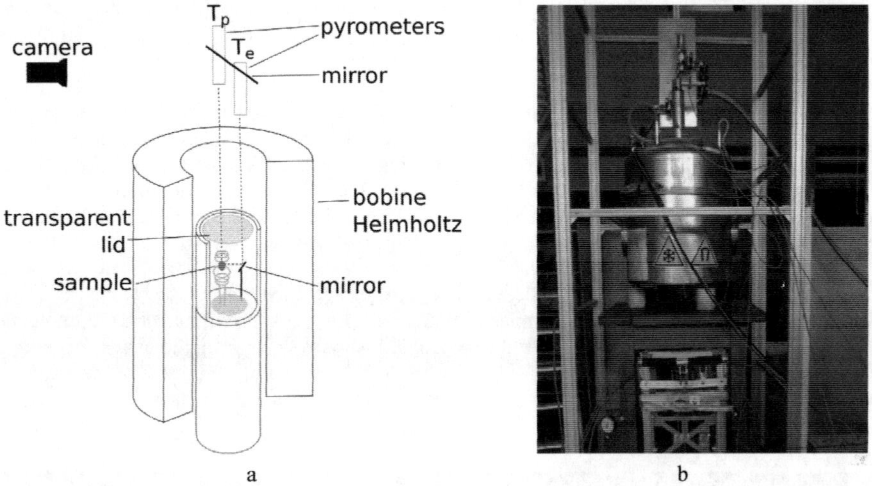

Fig. 1: The scheme of the experimental set-up (a) and the photo of the Helmholtz coil with pyrometers mounted above it and a water-cooled chamber inside (b)

The variation of the control parameters during the measurement procedure is shown schematically in Fig. 2, where V is a voltage commanded on the generator to be supplied to the e.m. levitator and B is the variation of the intensity of the DC field inside the supermagnet. At the beginning of the experiment the sample is placed at the bottom of the electromagnetic levitator. At the time t_0 we switch on and increase the voltage supply to the levitator up to a predefined fixed value as fast as possible to make the solid sample levitate and further to melt it already in levitation. The electric current supplied to the Helmoltz bobine is switched on when the sample is liquid, i.e. after the Curie temperature is attained that can be seen duering the experiment due to temperature measurement. The speed at which the intensity of the DC magnetic field increases during the period from t_2 to t_3 is limited by techical conditions. When a required value of a DC field is reached, the modulation of the voltage is performed with a lower (m1) and higher (m2) amplitude. In Fig. 2 only two series of the modulation are shown, but in experiment we can perform more. After the perturbations are done, the magnetic DC field is directed to zero (from time t_2 to t_3) and when $B=0$, the voltage on the e.m. levitator decreases till zero. The temperature of the sample goes down, the latter solidifies and descend back to the bottom of the e.m. levitator. A refractive ceramics at the bottom of the levitator protects it from being melted by the sample which reamins rather hot. At this stage the sample can be contaminated due to its interaction with ceramics, but the measurements are already performed.

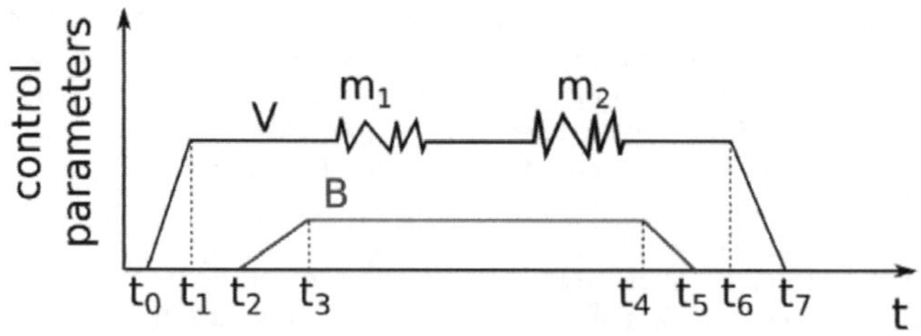

Fig. 2: Protocol for the variation of the control parameters during the measuring procedure with V the voltage supplied to the electromagnetic levitator and B the intensity of the DC field.

Using this (or similar) protocols, we performed experiments with the samples of Ni and a Ni-based high entropy alloy. In some experiments the perturbation of the samples were performed with varios intensity of the DC field in order to observe its effect on the thermal response of the sample, the most strong field which we used was $B=1.2T$. Generally it was found that with a more intense DC field the difference between the temperatures measured at the equatorial and the polar zone strongly decreases. This can be explained by damping of the convective flow which otherwise creates vortices inside the liquid thus promoting and supporting in the sample the existence of the areas with the strong temperture difference.

Yet, due to visualisation of the samples we have strong grounds to believe that non-zero flow remained in the sample, the latter was strongly rotated and shape variation were equally observed. This observation is in agreement with [4-6] where the damping of such effects was reached at B>3T. The Helmolhtz coil which we used, in principle, allowed us to generate the DC field up to 5T but we found the sample extremely unstable when the slow variation of the DC magnetic field was performed in order to reach a preset value. Thes osciallation and displacement of the droplet become more intense with a higher value of B although the speed at which it was varied was kept fixed. An example of these observations is given in Fig.3 with the time interval between the images about 0.1. The image A is due to the observation of the sample from the top (i.e. after 1 reflection in the mirror mounted outside the Helmholtz coil) and the image B is a lateral image of the top part of the sample formed after reflections in two mirrors: the one close to the electromagnetic levitator and another one outside the Helmholtz coil. A droplet contour on the image A is shown with a dashed line. A bright area outside the contour is strong reflection from the ceramics at the bottom of the electromagnetic levitator. The variation of the intensity of the image B can be related with the temperature distribution but this has to be confirmed.

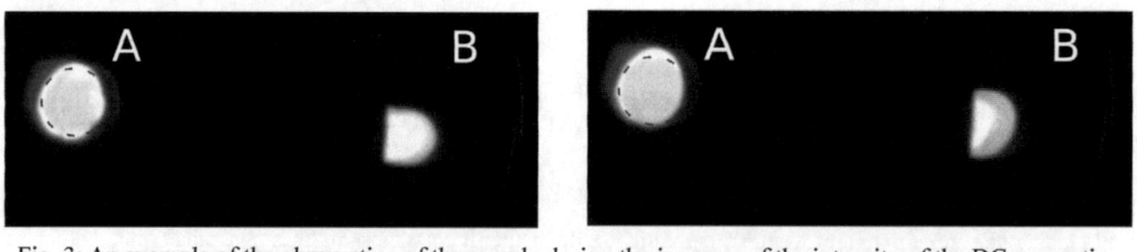

Fig. 3: An example of the observation of the sample during the increase of the intensity of the DC magnetic field from $B=1T$ to $B=1.2T$, the time interval between the two images is about 0.1s. The image A is due to the observation of the sample from the top (after 1 reflection in the mirror mounted outside the Helmholtz coil) and the image B is a lateral image of the top part of the sample formed after reflections in two mirrors: the one close to the electromagnetic levitator and another one outside the Helmholtz coil.

These instabilities did not allow us to perform the measurements with a higher DC magnetic field because of the risk to loose the sample that, because of uncontrolled trajectory, could stick to a levitator and melt it.

Conclusion

The effect of the DC magnetic field perpendicular to the gravity vector during the experiments on the application of the modulated calorimetry to the metallic sample in levitation was observed. It was found that a fixed value of the DC did provide the damping effect, but its slow variations required to obtain its preset value caused strong instabilities in droplet motions.

Acknowledgment
The work is supported by the CNES under Material Sciences Program and by the ESA via THERMOPROP project (AO-1999-022).

References

1. R. Wunderlich and H. Fecht, Measurement Science and Technology, **16** (2005) 402-416
2. T.Tsukada, H. Fukuyama, H. Kobatake, Int.J. Heat Mass Transfer, **50** (2007) 3054-3061
3. V. Bojarevics, K. Pericleous, ISIJ International, **43** (2003), 890–898
4. H. Fukuyama, H. Kobatake, K. Takahashi, I. Minato, T. Tsukada, S. Awaji, Meas. Sci. Tech., **18** (2007) 2059 – 2066
5. M. Watanabe, M. Adachi, H. Fukuyama, J. Mater. Sci, **52** (2017) 9850-9858
6. M. Watanabe, J. Takano, M. Adachi, M. Uchikoshi, H. Fukuyama, J. Chem. Thermodynamics, **121** (2018) 142-152
7. H. Yasuda, I. Ohnaka, Yu. Ninomiya, R. Ishii, S. Fujita, K. Kishio, J. Crystal Growth, **260** (2004) 475–485
8. P. Schetelat, J. Etay, Heat and Mass Transfer, **47** (2010) 759-769
9. J. Etay et al., Magnetohydrodynamics, **53** (2017) 3-12

Experimental study on the behaviour of the submerged jet in a cold liquid metal model for continuous casting of round blooms under the influence of rotating magnetic fields

D. Schurmann, B. Willers, S. Eckert

Helmholtz-Zentrum Dresden-Rossendorf e.V. (HZDR), Bautzner Landstraße 400, 01328 Dresden, Germany

Corresponding author: d.schurmann@hzdr.de

Abstract
Experimental results were obtained at the Mini-LIMMCAST facility [1], an experimental setup to model continuous casting of steel operated continuously as a loop with the cold liquid metal alloy GaInSn. The setup consists of a round mould with an inner diameter of 80 mm, representing a mould used in industrial bloom casting in a scale of 1:3. A mould-electromagnetic-stirrer (M-EMS) generated a rotating magnetic field (RMF) with different magnetic flux densities. Velocity profiles inside the mould were measured by Ultrasound Doppler Velocimetry (UDV). The experiments revealed an unexpected behaviour of the submerged jet: for medium magnetic field strengths of the RMF it can be observed, that the jet stops circling and stays at one position in the mould, bearing the risk of breakout due to continuous impingement of superheated material to one position of the solidified shell.

Key words: Liquid Metal Model, Ultrasound Doppler Velocimetry (UDV), Continuous Casting, Rotating Magnetic Field (RMF)

Introduction

In the continuous casting process of round blooms or billets Rotating Magnetic Fields (RMF) applied by electromagnetic-stirrers (EMS) are widely used to promote equiaxed solidification and therefore an improved quality of the cast product. In a recent study [2] we showed that intensive stirring results in high horizontal velocities in the vicinity of the meniscus and creates swirl structures at the free surface which might be responsible for bringing inclusions from the slag layer into the solidification zone. In this study we investigate the vertical velocities that develop in the mould caused by the jet which emerges from the submerged entry nozzle (SEN) under the influence of different strengths of the rotating magnetic field.

Experimental Setup

The LIMMCAST experimental facilities at HZDR are used to model the continuous casting process in low temperature liquid metal experiments [1, 3]. The facilities allow investigations of several aspects of the continuous casting process and focus mainly on the determination of the flow structures that develop in the continuous casting mould. The high versatility of the facilities allow measurements in different casting geometries, also under the influence of magnetic fields.

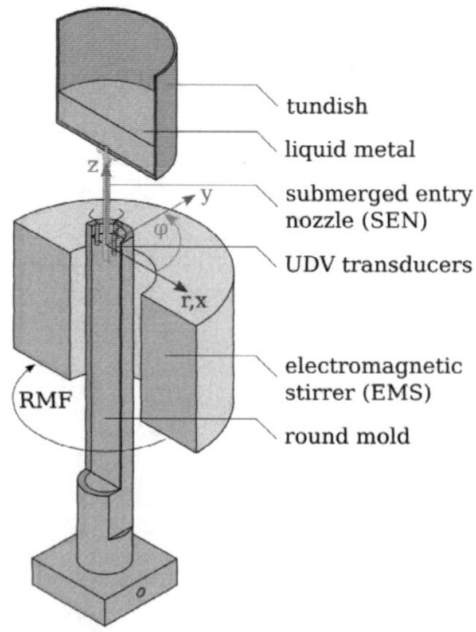

Fig. 1: Schematic view of the setup.

Table 1: Some physical properties of liquid steel at 1500 °C and GaInSn and water at 20 °C. The values for liquid steel are exemplary values from [4], [5], GaInSn values are from measurements [6] and water from [7].

		Liquid Steel	GaInSn	Water
density	kg / m³	~ 7 000	6 353	988
dyn. viscosity	mPa s	~ 2 ... 18	2.1	1
el. conductivity	1 / Ω m	~ 0.833×10^6	3.29×10^6	<50

The study presented in this work is performed at the Mini-LIMMCAST facility, which is composed of a liquid metal loop, operated at room temperature with GaInSn. Some physical properties of the model fluid are compared to those of liquid steel and water in Table 1. From the low electrical conductivity of water compared to that of steel it can be concluded that the use of a liquid metal as a model fluid is compulsory when the influence of magnetic fields are of interest.

Fig. 1 shows a schematic view of the setup. The liquid metal is pumped continuously from a storage vessel into the tundish, where the level is kept constant, by controlling the rotation speed of the pump. Driven by gravity, the liquid metal flows through the SEN into the model of mould and strand, over a dam to set a constant level of the meniscus inside the mould and finally back into the storage vessel. The PMMA model of mould and strand represent an industrial configuration in a scale of 1:3 for continuous casting of round blooms. It has an inner diameter of 80 mm and a length of approx. 800 mm. The submerged entry nozzle (SEN) has a single vertical outlet with a diameter of 10 mm. A mould electromagnetic stirrer, positioned between $-380\ \mathrm{mm} < z < -70\ \mathrm{mm}$, is used to apply a rotating magnetic field (RMF) with a frequency of $f = 2.5$ Hz and a magnetic flux density of up to $B_0 = 20$ mT to the set up. The coordinate system for the measurements is located in the centre of the cylindrical model with the z-axis pointing to the top and $z = 0$ located on the free surface in the model mould.

Measurements of the vertical velocity component w are performed by means of Ultrasound Doppler Velocimetry (UDV) by a DOP 3010 (Signal Processing S.A., Savigny, Switzerland, [8]) with up to ten individual ultrasound transducers. The transducers are positioned at $z = -10$ mm in direct contact with the liquid metal, every 90° at radial positions of $r = 30$ mm.

Results

A tempo-spatial contour plot of a three minute long measurement of the vertical velocity component w from below the meniscus down to $z = -300$ mm in the mould at four sensor locations on the radius $r = 30$ mm is shown in Fig. 2. The magnetic flux density B_0 of the RMF has been increased two times, at $t = 62$ s from $B_0 = 10$ mT to 15 mT, and finally at $t = 125$ s to $B_0 = 20$ mT.

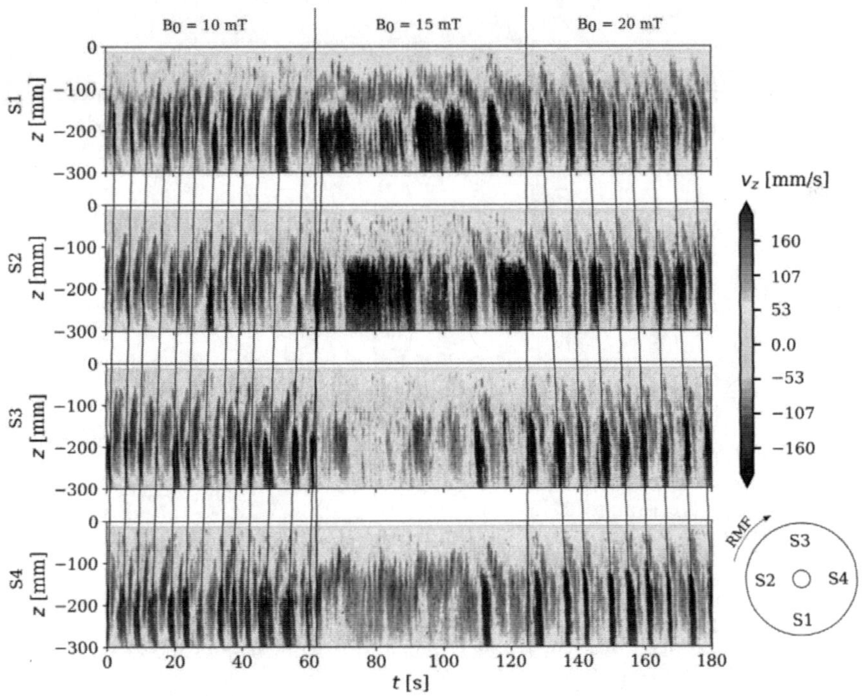

Fig. 2: Time evolution of the vertical velocity component measured by four sensors in the mould at $r = 30$ mm under the influence of three different magnetic flux densities B_0.

In the first part of the measurement, where a small magnetic flux density of $B_0 = 10$ mT is applied ($t < 62$ s), a periodic oscillation between positive and negative velocities can be observed at each sensor. A negative velocity is detected when the submerged jet, which emerges from the SEN, passes the measurement line of the ultrasound transducer. A positive velocity represents an upward flow to the free surface. The velocities are shifted in time between the sensors, which indicates a circling movement from one sensor to the other. The tilted black lines in the diagram show that the jet passes the sensors in a counter-clockwise direction, which is opposite to the rotating direction of the RMF as applied here to the setup.

At a medium magnetic flux density of 15 mT (62 s $< t <$ 125 s) the periodic oscillation between upward and downward flow cannot be observed any more. Negative velocities which represent the presence of the jet are primarily measured by sensor S1 and S2, while the upward flow is recorded by the sensors at the opposite site of the mould. This means that the jet stops its circling motion through the mould and stays mainly at one position. The magnetic flux density at which the circling of the jet stops is referred to as critical magnetic flux density $B_{0,crit}$.

At higher magnetic flux densities of the RMF ($B_0 = 20$ mT, $t > 125$ s) a behaviour similar to the lowest RMF can be observed: a periodic change of positive and negative velocities at each sensor and a time-shift from sensor to sensor, indicating the circling of the jet inside the mould. In contrast to the first part, at $B_0 = 20$ mT the circling direction of the jet is clockwise, which is in the same direction in which the RMF is applied.

To quantify the circling behaviour of the jet, measurements of the vertical velocity component w in the mould with six different magnetic flux densities from $B_0 = 2.4$ mT to 19.8 mT are performed. A discrete fast-Fourier-transformation (FFT) is applied to the velocity data which is measured along the three dimensions time t, angular position of the UDV transducers φ and measurement depth z at discrete values (k, l, m):

$$w = w_{klm}(z_k, \varphi_l, t_m) \qquad k = 1 \ldots K, \qquad l = 1 \ldots L, \qquad m = 1 \ldots M \tag{1}$$

The 3d-FFT is then defined as

$$W_{\chi\psi f} = \sum_{k=1}^{K} \sum_{l=1}^{L} \sum_{m=1}^{M} w_{klm}\, e^{-j2\pi(\chi k/K + \psi l/L + fm/M)} \tag{2}$$

and yields the three frequencies χ, ψ, f, which are related to the three dimensions z, φ, t of the measurement. Extracting the frequency f, where the absolute value of the FFT is maximal, allows the quantification of the jet's circling frequency and direction.

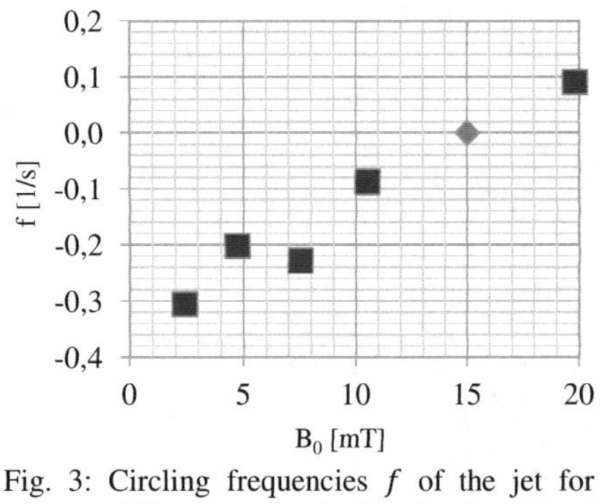

Fig. 3: Circling frequencies f of the jet for different magnetic flux densities B_0 of the RMF.

The extracted frequencies for the measurements with variable B_0 are shown in Fig. 3. Three situations can be distinguished: for magnetic flux densities which are smaller than the critical value $B_{0,crit}$, negative circling frequencies can be observed, which corresponds to a circling of the jet in opposite direction than the applied RMF. The closer B_0 is to to the critical magnetic flux density, the lower the circling frequency of the jet. As mentioned earlier, at the critical value $B_{0,crit} = 15$ mT no significant circling of the jet can be observed, which corresponds to a circling frequency of $f = 0$. For increasing magnetic flux densities a positive circling frequency, corresponding to a circling of the jet in the same direction as the applied RMF is found. Frequencies between -0.3 Hz $\leq f \leq 0.1$ Hz were found.

Discussion

By measuring the tangential velocity in this configuration it was shown in [2], that with increasing magnetic flux density of the RMF the rotation of the fluid in the strand intensifies.

When we assume that the jet is disturbed and enters the rotating fluid column a bit off centre, it will be deformed in the direction of the main rotational flow. The fluid that leaves the jet is responsible for a momentum transport and therefore a propelling of the jet. If the kinetic energy of the deformed jet is higher than that of the rotating column, the jet is able to propel itself against the direction of the cylindrical flow. This is the case for a low RMF or a high casting speed. When the magnetic flux density of the RMF is increased towards $B_{0,crit}$, the kinetic energies of jet and rotating fluid column are similar and the momentum transport of the deformed jet is not high enough to propel the jet against the fluid column, resulting in a stop of the jet circling. If the RMF is above the critical magnetic flux density, the jet's momentum transport is not sufficient to propel itself against the rotating fluid column, but gets slowly transported in the direction of the rotating fluid column. Therefore, the frequencies which could be extracted from the FFT in Fig. 3 are below those of the rotating fluid column (c.f. [2]) and much lower than the frequency of the applied RMF.

The situation when the jet stops circling at $B_0 = B_{0,crit}$ can be dangerous for the real process, since the superheated material brought to one side of the mould by the jet would remelt and weaken the solidified strand shell. Therefore, the shell might not be able to withstand the high pressure of the liquid steel in the mould and break, causing an expensive stop of the production. Therefore we suggest to avoid the operation of a continuous casting strand with the application of a RMF close to the critical magnetic flux density of the particular system. It is very likely, that this critical parameter is dependent on the casting velocity and the geometric parameters of mould and SEN. With a higher casting velocity the kinetic energy of the jet increases, which would move the critical magnetic flux density to higher values. In contrast, a lower casting speed and therefore a lower kinetic energy of the jet will probably move the critical magnetic flux density to lower values. Therefore, when a reduction of the casting velocity occurs, e.g. at the end of the casting process, when less liquid steel is in the tundish, the magnetic flux density of the RMF has to be adapted accordingly, to avoid the critical value for that casting speed.

Conclusions

Measurements of the vertical velocities in a cold liquid metal model for the continuous casting of round blooms with applied electromagnetic stirring have been performed at the Mini-LIMMCAST facility of HZDR by means of the Ultrasound Doppler Velocimetry.

The measurements revealed several interesting phenomena related to the behaviour of the submerged jet under the influence of rotating magnetic fields, which have not been described before:

1. The jet circles in the mould.
2. The jet stops circling at a critical magnetic flux density $B_{0,crit}$, which should be avoided in the casting process to prevent a possible breakout of liquid steel out of the strand.
3. The jet circulation direction depends on the magnetic flux density of the RMF:
    a. $B_0 < B_{0,crit}$: The jet circles in opposite direction than the RMF.
    b. $B_0 > B_{0,crit}$: The jet circles in the same direction as the RMF.
4. The circling frequency of the jet is much lower than the rotational frequency of the RMF.

References

1. K. Timmel, S. Eckert, G. Gerbeth, F. Stefani, and T. Wondrak, 'Experimental Modeling of the Continuous Casting Process of Steel Using Low Melting Point Metal Alloys - the LIMMCAST Program', *ISIJ Int.*, vol. 50, no. 8, pp. 1134–1141, 2010.
2. B. Willers, M. Barna, J. Reiter, and S. Eckert, 'Experimental Investigations of Rotary Electromagnetic Mould Stirring in Continuous Casting Using a Cold Liquid Metal Model', *ISIJ Int.*, vol. 57, no. 3, pp. 468–477, 2017.
3. K. Timmel, C. Kratzsch, A. Asad, D. Schurmann, R. Schwarze, and S. Eckert, 'Experimental and Numerical Modeling of Fluid Flow Processes in Continuous Casting: Results from the LIMMCAST-Project', *IOP Conf. Ser. Mater. Sci. Eng.*, vol. 228, p. 012019, Jul. 2017.
4. C. Y. Ho and T. K. Chu, 'Electrical resistivity and thermal conductivity of nine selected AISI stainless steels', DTIC Document, 1977.
5. M. Korolczuk-Hejnak, P. Migas, and W. Ślęzak, 'Determination of the liquid steel viscosity curves using a high temperature rheometer', *J. Phys. Conf. Ser.*, vol. 602, p. 012037, Apr. 2015.
6. Y. Plevachuk, V. Sklyarchuk, S. Eckert, G. Gerbeth, and R. Novakovic, 'Thermophysical Properties of the Liquid Ga–In–Sn Eutectic Alloy', *J. Chem. Eng. Data*, vol. 59, no. 3, pp. 757–763, Mar. 2014.
7. VDI, Ed., *VDI-Wärmeatlas*, 11th ed. Berlin, Heidelberg: Springer, 2013.
8. S. A. Signal Processing, 'DOP3000 series User's manual'.

Experimental and numerical investigation on particle-induced liquid metal flow using Lorentz force velocimetry

Ch. Karcher[1], Z. Lyu[1], Th. Boeck[1], N. Tran[2], U. Lüdtke[2]

[1] Institute of Thermodynamics and Fluid Mechanics, Technische Universität Ilmenau
P.O. Box 100565, D-98684 Ilmenau, Germany

[2] Electrothermal Energy Conversion Group, Technische Universität Ilmenau
P.O. Box 100565, D-98684 Ilmenau, Germany

Corresponding author: christian.karcher@tu-ilmenau.de

Abstract
The dynamics of electrically non-conducting particles in liquid meal flow are of interest in a number of metallurgical processes. For instance, during continuous casting of steel, proper removal of slag particles interacting with mold flow is crucial for the quality of the product. However, as such flows are very complex, analyzing single effects are almost impossible. Therefore, we investigate experimentally a simplified configuration where spherical particles of known size are pulled in the vertical direction with a controlled speed at a given position through a liquid metal column initially at rest. As a test melt we use the alloy GaInSn in eutectic composition which is liquid at room temperature. The displacement flow induced by the particle movement is measured using Lorentz force velocimetry. This method is based on recording the flow-induced force acting on an externally arranged permanent magnet. In the present paper we extend earlier work by analyzing not only the drag force in the direction of the particle movement but also the lift force acting in the horizontal direction. Moreover, we use an improved signal processing routine that allows us to perform measurements at higher particle Reynolds numbers. The experimental results are in both quantitatively and qualitatively agreement with the predictions of numerical simulations using the commercial code ANSYS/FLUENT.

Key words: liquid metal, Lorentz force, particle-induced flow

Introduction

Lorentz force velocimetry (LFV) is a non-contact electromagnetic flow measurement technique in electrically conducting fluids like liquid metal melts [1]. The measurement principle is based on the interactions of a moving melt with a magnetic field, i.e. on the effects of magnetohydrodynamics (MHD) [2]. When the melt enters into a magnetic field, it will slightly bend the field lines in the direction of its movement. As a result, electrical eddy currents are induced. These eddy currents interact with the magnetic field generating flow-braking Lorentz forces within the melt. Physically, these Lorentz forces are the restoring forces of the bent magnetic field lines when considered to behave as elastic strings. The magnitude F_L of the Lorentz force is proportional to the melt velocity u according to the relation

$$F_L \sim \sigma u B^2 V, \qquad (1)$$

where σ is the electrical conductivity of the melt, B the magnetic flux density and V a characteristic volume corresponding to the volume penetrated by the magnetic field. In LFV, we measure the counter force which is acting on the magnet system that generates the field and which is pointing in the direction of the melt flow. The respective measuring device is called Lorentz force flowmeter (LFF) which basically consists of a magnet system and an attached force sensor. In case the flowrate of the melt is of interest, the magnet system consists of an array of permanent magnets in such a way that the magnetic field penetrates the entire cross-section of the flow. On the other hand, if local velocities shall be measured, the magnetic system consists of a single tiny permanent magnet. As the LFF is arranged outside of the melt domain, LFV can be considered to be a non-contact electromagnetic method.

In this paper we apply this technique to the case when melt flow is induced by spherical solid and non-conducting particles of known size pulled with a constant speed u through a liquid metal column initially at rest. Using this simplified configuration we are able to investigate purely the displacement flow induced by the rising particles. This model experiment shall be the first step on the way of applying LFV to more complex cases like detection of gas bubbles or particle-flow in metallurgical applications. We extend previous work [3], [4] by not only recording the drag force induced by the displacement flow but also the lift force acting in the horizontal direction. Furthermore, we describe a new data processing scheme that allows us to perform measurements at higher particle Reynolds numbers.

The present study is organized as follows. First, we describe the improved experimental set-up and explain the new features of the data processing method employed. Secondly, we present some selected results obtained using these new techniques. Next we compare the experimental findings with the predictions of numerical simulations. Finally, we shall give a short summary of the main conclusions.

Experimental set-up

The experimental set-up is shown in Fig. 1. It consists of a vessel made of plexiglass of size 60 × 60 × 400 mm^3 having a wall thickness of 8 mm. The vessel is filled with liquid metal GaInSn which is liquid at room temperature. The spherical particle is made of plastic, i.e. it is electrically non-conducting, and has diameter of d = 6 mm. The particle is fixed on a thin fishing line of thickness 0.1 mm in order to fix its lateral position at a distance of 10 mm away from the left side-wall. During an experimental run the fishing line is pulled through the top and bottom holes of the vessel. The pulling velocity of the sphere is controlled by an additional linear driver, which provides speeds in the range of 0 to 200 mm/s. The highest speed corresponds to a particle Reynolds number of Re = 3600, where Re = ud/ν, u is the pulling speed and ν is the kinematic viscosity of the melt.

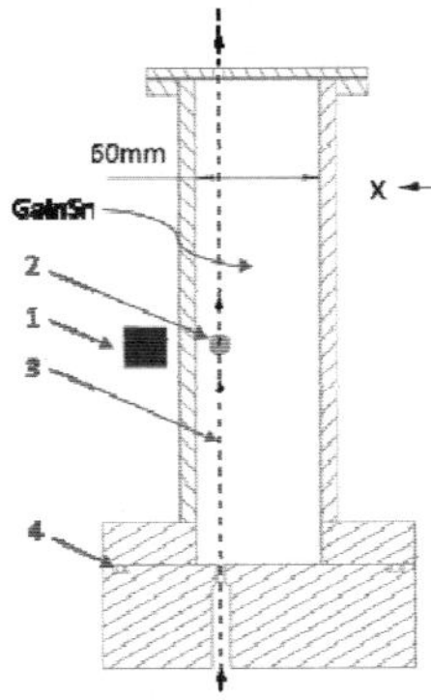

Fig. 1: Sketch of the experimental set-up (1 LFV, 2 particle, 3 fishing line, 4 sealing).

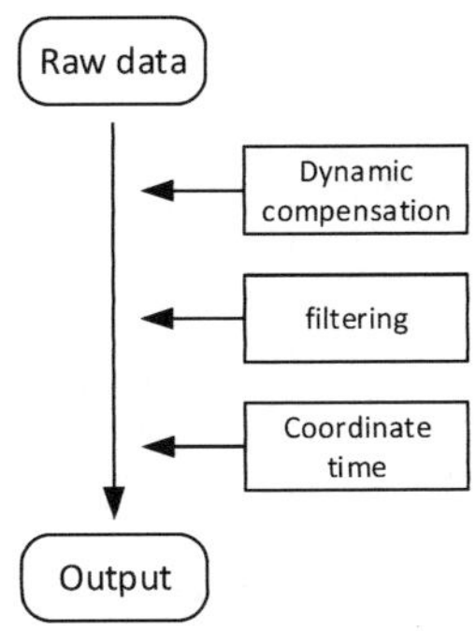

Fig. 2: Flow chart of the data processing.

Our LFV consists of a 12 × 12 × 12 mm^3 permanent magnet (NdFeB 42) which is installed in 1 mm distance from the outer wall of the vessel. It is arranged in such a way that the magnetic field lines are pointing into the horizontal direction. The magnetic flux density of the magnet is about B = 575 mT at its surface. However, this value deceases rapidly with increasing distance from the surface. Within the melt domain, we measure values of B = 98 mT at distance 9 mmm corresponding to the inner vessel wall, and B = 10 mT at distance 19 mm, i.e. the position of the fishing line. Hence, according to Eq. (1) we expect magnitudes of the Lorentz forces in the range of micronewtons. Such tiny forces can only be measures by special sensors. We use an Interference-Optical-Force-Sensor (IOFS) to measure either the x- or z-component, corresponding to the lift or the drag component of the Lorentz force induced by the displacement flow of the melt around the particle, respectively. The sensor measures the displacement of an elastic body under the load to the force by interferometric means and converts the signal into a voltage in the range of nanovolts. Finally, using the calibration curve of the sensor, the voltage is converted into a certain force value.

The flow chart of our data processing is shown in Fig. 2. Our improved procedure can be described as follows. The dynamic characterization of the IOFS force sensor corresponds to the frequency response of the second order transfer function. Using the given eigenfrequencies of the sensor of 16 Hz when arranging in the in z-direction and of 14 Hz when arranged in the x-direction, respectively, we define the transfer function G(s) and then use the inverse function $G(s)^{-1}$ to compensate the measurement data. Thus, we eliminate the dynamic effects of IOFS sensor. Afterwards, a low pass filter is applied to denoise the signal. Additionally, a synchronized Ultrasonic Doppler Velocimetry (UDV) measurement is applied to provide the accurate time when the particle is passing the magnet. We record it as the reference zero-time for the LFV measurement. Using this sensor and data processing we can conduct measurement up to particle Reynolds number of 2000. By that, we can extend the range of Re by a factor of about 4.

Numerical simulation

For the numerical simulation we use the commercial CFD software ANSYS/FUENT and its MHD extension. Here we modify a numerical scheme developed for the simulation of Argon bubbles rising in a GaInSn column initially at rest. Details of the scheme are given elsewhere [6], [7]. Here, we replace the free-slip velocity condition at the bubble interface by a no-slip condition at the surface of the rigid particle. The code solves the incompressible Navier-Stokes equations that are extended by the Lorentz force density $\mathbf{f}_L$ defined by the relation

$$\mathbf{f}_L = \mathbf{j} \times \mathbf{B}. \tag{2}$$

IOP Conf. Series: Materials Science and Engineering **424** (2018) 012006 doi:10.1088/1757-899X/424/1/012006

Here, **j** denotes the eddy current density. It can be calculated via Ohm's law given by

$$\mathbf{j} = \sigma[-\nabla\Phi + \mathbf{u} \times \mathbf{B}], \tag{3}$$

where Φ the electrical potential, and u is the velocity vector. The set of equations is closed by calculating Φ upon applying the charge conservation law, i.e.

$$\nabla \cdot \mathbf{j} = 0 \tag{4}$$

to Eq. (3). Finally, for **B** we take an analytical representation for a cubic permanent magnet [7]. Hence, we solve the MHD equations in the limit of small magnetic Reynolds number when the effect of the induced magnetic field can be neglected. Vessel walls and sphere are taken as electrically non-conducting. Fluid properties are those of GaInSn.

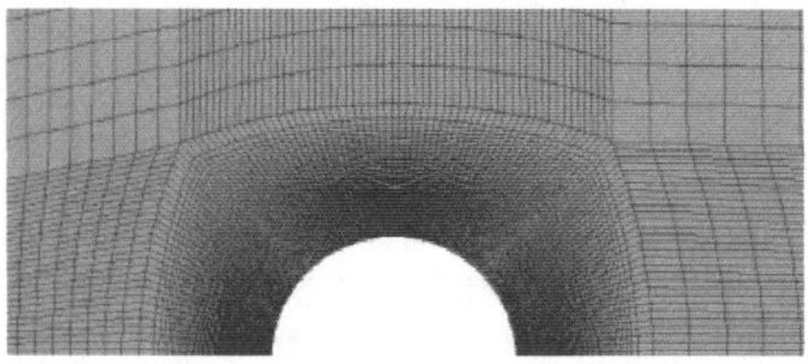

Fig. 3: Mesh around sphere for Re = 300.

The simulations are carried out as follows. The sphere is fixed in space and vessel, melt, and magnet all of which are moved with a constant velocity -u downwards. In order to resolve the boundary layer with at least 10 nodes, a Reynolds number dependent mesh is used around the sphere, see Fig. 3. The code is validated for the pure hydrodynamic case of flow around a sphere. Here we compare our calculated drag coefficient with experimental data [8]. In each case errors are less than 8%.

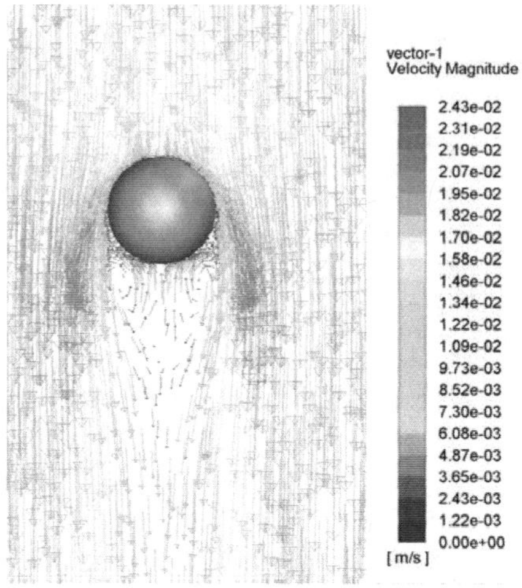

Fig. 4: MHD flow around sphere at Re = 300.

As an example, Fig. 4 shows the flow field around the sphere at a Reynolds number of Re = 300. The plot refers to reference time zero, i.e. when the magnet on the left wall is at the same height with the sphere. We observe that boundary layer separation occurs at about 80° on the right hand side. This value is in excellent agreement with purely hydrodynamical experiments. However, on the left hand side, separation occurs at about 90°. We attribute this stabilizing effect to the presence of both, a near wall on the left and the MHD interactions.

Results and comparison

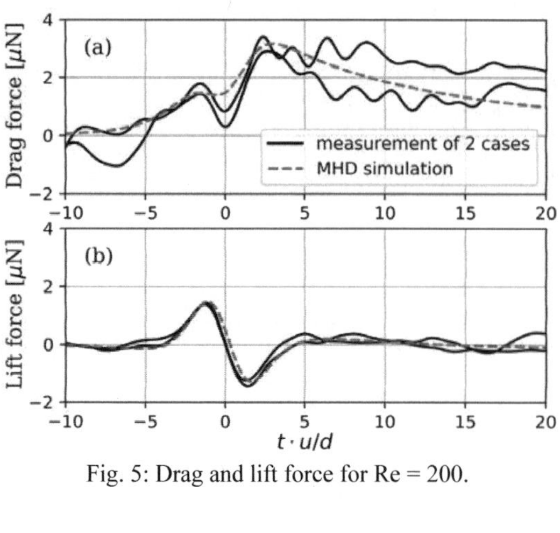

Fig. 5: Drag and lift force for Re = 200.

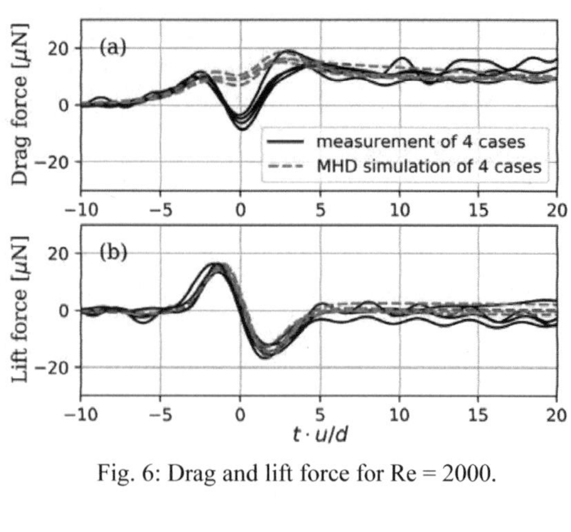

Fig. 6: Drag and lift force for Re = 2000.

Figs. 5 and 6 show some selected results of our experiments and the supporting numerical simulations. Here, Fig. 5 refers to a Reynolds number of Re = 200. In this case the flow is steady. The upper graph (a) shows the measured drag force (black curves) together with the predictions of the simulation as a function the normalized position t·u/d. Here, t·u/d = 0 refers to reference time zero. The lower graph (b) shows the measured lift force and the respective results of the simulation. As expected from Eq. (1), the forces are in the range of some micronewtons. We observe that in the experiments, the drag force can only be measured within an accuracy given by the two black curves, i.e. the repeatability of the experiment is limited at relatively small Reynolds numbers. However, the overall agreement with the results of the simulation is remarkably good, especially in the downstream area t·u/d > 5. In the region -5 < t·u/d < 5, i.e. when the magnet is close to the sphere, the signals look more complex. We first find a slight increase of the drag when the sphere approaches the magnet, then a drop to almost zero at reference time zero and another sharper increase when the sphere leaves the magnet. The simulations predict a continuously increasing rag in this region. We may attribute this difference to both the restrictions of the experimental set-up in measuring such tiny forces and the assumptions made in the simulations. However, we can clearly see from Fig. 5 (b) that the agreement of experiments and simulations is excellent with respect to the lift force. In the region -5 < t·u/d < 5 we observe first an increase reaching a maximum at about t·u/d = -1.5, then a decrease to zero at t·u/d = 0, then a further decrease to a minimum with negative lift, and finally a relaxation to zero in the wake region. Moreover, we can observe that the repeatability of the experiment concerning lift force is also excellent. Hence, in possible applications of analyzing particle dynamics using LFV, recording the lift force is favorable.

Fig. 6 shows the results for a Reynolds number of Re = 2000. In this case the flow is transient. Again, the upper graph shows the measured and the calculated drag forces while the lower graph shows the respective lift forces. As the flow is transient, we have added the results of 4 numerical runs with slightly different initial conditions. We observe that the repeatability of the experiments are much better than in the low-Reynolds number flow as the forces are significantly higher. Moreover, the overall agreement between the experimental findings and the predictions of the simulations become likewise much better. However, in the region -5 < t·u/d < 5 the measurement and the calculation differ again, as already observed in the case Re = 200. In the experiments we find even negative drag forces at zero position while the simulations predict positive values. More experimental runs and simulations are needed to explain this difference. However, as for the low-Re case, the agreement between experimental findings and numerical prediction is again excellent. We can conclude that the lift force is more reliable for particle detection in the entire range of Reynolds numbers investigated.

Summary and conclusions

In this study we have investigated experimentally and numerically the application of Lorentz force velocimetry for analyzing particle dynamics in liquid metal flow. We consider the simplified case when non-conducting spherical particles are pulled at a prescribed speed from bottom to top of the test melt GaInSn initially at rest. We measure and compute both the electromagnetically induced drag and the lift force that is exerted by the displacement flow generated by the moving particle on a small permanent magnet arranged outside the melt. We observe an excellent agreement of the experimental findings and the prediction of the simulations when analyzing the lift force. In contrary, we observe some major difference in the drag force within the region where the magnet field strongly interact with the displacement flow, i.e. when particle and magnet are at the same height. We may attribute these differences to imperfections in both the experimental set-up and the numerical method.

Acknowledgment

The authors acknowledge support from Deutsche Forschungsgemeinschaft (DFG) within the Research Training Group on Lorentz force velocimetry and Lorentz force eddy current testing under grant GRK1567.

References

1. A. Thess, E. Votyakov, Y. Kolesnikov, Physical Review Letters, 96(16) (2006), 164501-164520.
2. P. A. Davidson. An introduction to Magnetohydrodynamics. Cambridge University Press, (2001).
3. Z. Lyu, Ch. Karcher, Magnetohydrodynamics, 53(1) (2017), 67-78.
4. Z. Lyu, Th. Boeck, Ch. Karcher, A. Thess, Magnetohydrodynamics, 53(4) (2017), 653–666.
5. N. Tran, T. Boeck, U. Lüdtke, Z. Lyu, Ch. Karcher, Magnetohydrodynamics, 53(4) (2017), 619-632.
6. N. Tran, U. Lüdtke, H. Schwanbeck. Proc. CADFEM ANSYS Simulation Conference, 2017, Koblenz, Germany.
7. P. Furlani. Permanent magnet and electromechanical devices: materials, analysis and application, Acad. press, (2001).
8. H. Schlichting, Boundary-Layer Theory, McGraw-Hill, (1979).

Visualisation of the large scale circulation in Rayleigh-Bénard convection using contactless inductive flow tomography

T. Wondrak[1], F. Stefani[1], V. Galindo[1], S. Eckert[1]

[1] Helmholtz-Zentrum Dresden - Rossendorf, Bautzner Landstraße 400, Dresden, Germany

Corresponding author : t.wondrak@hzdr.de

Abstract
Rayleigh-Bénard (RB) convection plays an important role in geo- and astrophysics as well as in many metallurgical applications. At sufficiently high values of the Rayleigh number, a large scale circulation (LSC) is formed whose dynamics had turned out to be surprisingly rich. In this paper, the applicability of the contactless inductive flow tomography (CIFT) for the detection of the torsional mode of the LSC is investigated. CIFT enables the three-dimensional reconstruction of flow structures in liquid metals by applying one or more magnetic fields and measuring the flow induced perturbations of those fields outside the melt. Additionally, preliminary measurements of the flow induced magnetic field with a similar sensor arrangement will be presented.

Key words: flow measurement techniques, Rayleigh-Bénard convection

Introduction

Thermally driven flows of liquid metals play an important role for instance in the production of mono-crystalline silicon using the Czochralski (Cz) crystal growth method. The measurement of the velocity of the liquid silicon is a challenging task because the liquid is opaque and has a temperature of about 1500 °C. Therefore, non-contacting methods for the determination of the flow are needed. The contactless inductive flow tomography (CIFT) [1-2] is able to reconstruct the three dimensional structure of the flow by applying primary magnetic fields to the melt and by measuring the flow induced magnetic field perturbation outside the fluid volume. Based on these measurements, the flow field is reconstructed by solving a linear inverse problem. Appropriate regularization methods, similar to Tikhonov regularisation in combination with the L-curve technique, are necessary to mitigate the non-uniqueness problem.

In order to investigate the applicability of CIFT for Cz crystal growth, an existing modified Rayleigh-Bénard (RB) setup had been equipped with one pair of excitation coils and 20 magnetic field sensors for CIFT at one plane at mid-height of the cylinder [3]. One of the challenges for CIFT is the reliable measurement of the flow induced magnetic field which is about 4-5 orders smaller than the applied magnetic field. The reason for this is the low velocity in the order of 1 cm/s. In the previous study [3], the flow induced magnetic field was robustly measured for a variety of temperature differences between 2.3 K and 80.8 K over a period of 8 hours. It was demonstrated that dynamical features of the large scale circulation (LSC), like azimuthal rotations and cessations, as well as typical frequencies of the torsional/sloshing mode are well detectable. However, due to the lack of information over the height of the cylinder, it was not possible to distinguish between torsional modes [4] and sloshing modes [5,6]. Therefore, a new sensor arrangement has to be found that enables a robust reconstruction while using as few as sensors possible. The procedure is as follows: in the first step, the secondary field is computed outside the vessel for a numerically simulated time-dependent flow. In the second step, the flow is reconstructed using different sensor arrangements. The quality of the reconstructions is assessed by an error estimation. It turned out that an arrangement of 8 sensors in azimuthal direction for each of the 3 layers over the height (labelled as 8×3) is a good choice [7]. In this paper we will investigate if the selected sensor arrangement is able to detect the flow structure of the torsional mode and compare it with a preliminary measurement.

After a short description of CIFT and the experimental setup, an analysis of the reconstruction for a temperature difference of 2.3 K is carried out. The paper concludes with a preliminary measurement of the flow induced magnetic field for a temperature difference of 30 K.

Contactless inductive flow tomography

CIFT relies on the following integral equation system which provides the flow induced magnetic field $\vec{b}$ outside the fluid volume V for an applied magnetic field $\vec{B}_0$ under the assumption of insulating boundaries [1,2]:

$$\vec{b}(\vec{r}) = \frac{\sigma\mu_0}{4\pi} \iiint_V \frac{[\vec{v}(\vec{r}') \times \vec{B}_0(\vec{r}')] \times (\vec{r} - \vec{r}')}{|\vec{r} - \vec{r}'|^3} dV' - \frac{\sigma\mu_0}{4\pi} \oiint_S \frac{\varphi(\vec{s}')\vec{n}(\vec{s}') \times (\vec{r} - \vec{s}')}{|\vec{r} - \vec{s}'|^3} dS' \quad (1)$$

$$\varphi(\vec{s}) = \frac{1}{2\pi}\iiint\limits_V \frac{\left[\vec{v}(\vec{r}')\times\vec{B}_0(\vec{r}')\right]\cdot(\vec{s}-\vec{r}')}{|\vec{s}-\vec{r}'|^3}dV' - \frac{1}{2\pi}\oiint\limits_S \frac{\varphi(\vec{s}')\vec{n}(\vec{s}')\cdot(\vec{s}-\vec{s}')}{|\vec{s}-\vec{s}'|^3}dS' \quad (2)$$

In this equation system φ is the electric scalar potential and $\vec{v}$ denotes the velocity of the liquid metal which is assumed to have uniform conductivity σ in the volume V with the surface S. In the volume integrals, dV' is the volume element and $\vec{r}'$ the position vector in the volume. In the surface integrals, dS' denotes a surface element and $\vec{n}(\vec{s}')$ represents the normal vector of the surface at position $\vec{s}'$.
In order to reconstruct the velocity field $\vec{v}$ in the volume V this integral equation system has to be inverted. The intrinsic non-uniqueness problem of the emerging linear inverse problem, which mainly concerns the detailed depth-dependence of the velocity, is circumvented by utilizing the so-called Tikhonov regularization [1,8].

Experimental setup
The setup consists of a cylindrical column of aspect ratio one filled with GaInSn which is homogeneously heated from below through a copper plate (Fig. 1). In order to emulate the growing crystal of a Czochralski system, only partial cooling of the top is realized by means of a circular heat exchanger mounted concentrically within the upper lid. This partial cooling covers approximately 17% of the mean area modelling the growing crystal in a real puller. PID-controlled thermostats with a large reservoir supply the copper heat exchangers (heater and cooler) with coolant/heating fluid at high flow rate. Fig. 1(b) shows a photograph of the experimental cell having an aspect ratio of one (height H = diameter D = 87 mm). The transparent side wall is made of a borosilicate glass pipe because of its low heat conductivity in order to approach an adiabatic boundary condition. The whole apparatus was embedded in mineral wool during the measurements to minimize lateral heat loss.

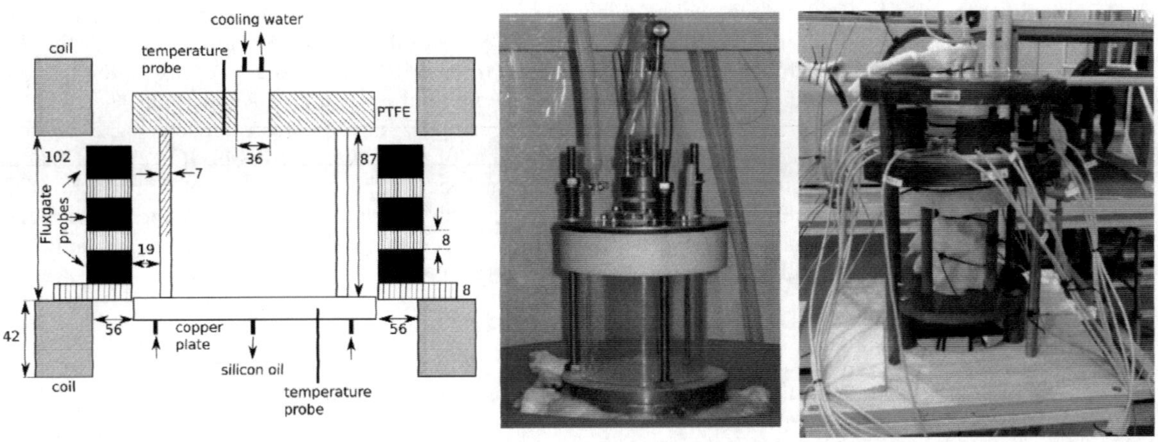

(a) Schematic sketch of the setup (b) Photo of the cell (c) Photo of the setup with instrumentation for CIFT

Fig. 1: Schematic sketch and photos of the modified Rayleigh-Bénard configuration.

The excitation magnetic field for CIFT is generated by a pair of circular coils which are fed by a current of 20 A generating a constant vertical field with approximately 4 mT in the fluid. Additionally, the coils are cooled in order to minimize temperature changes over time. The flow induced magnetic field is measured by 21 Fluxgate probes which are arranged in three layers over the height of the cylinder with 7 sensors in each layer (7 × 3). A very stiff mounting of sensors with respect to the coils is necessary to avoid any magnetic field changes resulting from geometric deformations due to thermal expansion during the experiment. The flow in the container was simulated for a temperature difference of 2.3 K using the *buoyantBoussinesqPimpleFoam* solver of the finite volume library OpenFOAM over 3000 s [3]. No turbulence models were used.

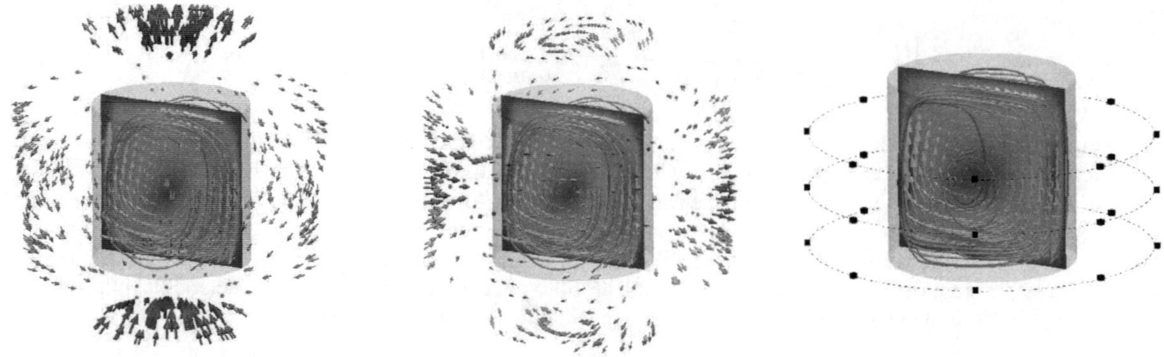

(a) $\vec{b}$ for $B_0\vec{e}_x$ (b) $\vec{b}$ for $B_0\vec{e}_z$ (c)

Fig 2. Calculated flow induced magnetic field $\vec{b}$ for horizontal (a) and vertically (b) applied field and a sketch of the sensor arrangement 8 × 3 (c)

Numerical simulations for a temperature difference of 2.3 K

In order to enable a complete three-dimensional reconstruction, two homogeneous magnetic fields in horizontal and vertical direction with $\vec{B}_{0,1}(\vec{r}) = B_0\vec{e}_z$ and $\vec{B}_{0,2}(\vec{r}) = B_0\vec{e}_x$ were applied with $B_0 = 3$ mT. The flow induced magnetic field was calculated at a distance of 36 mm from the fluid which correspond to the smallest possible distance in the experimental setup (see also [3]). As an example, Fig. 2 displays the computed fields around the vessel for a time averaged LSC, which has a flywheel structure.

In order to investigate the quality of the reconstruction of the time dependent flow, the flow induced magnetic field was calculated according to Eq. (1) and (2) for each second of a selected time period between 1600 s and 1680 s of the flow simulation. We have choosen a sensor arrangement of 8 × 3 sensors. In order to identify the dynamics of the torsional flow, the direction and amplitude of the flow close to the top and the bottom of the cylinder is important. Therefore, we averaged the velocity in a plane close to the top and the bottom of the cylinder for each reconstruction. Fig. 3(a) shows the time dependent behaviour of the direction at the top and the bottom of the cylindrical vessel. It is evident that the direction of the flow at both locations is well reconstructed. Fig. 3(b) shows the time dependent behaviours of the amplitude of the flow. It is interesting to note that there is a slight discrepancy in the amplitude between the reconstruction and the original velocity. Numerical simulations suggest that a better agreement could only be achieved, if additional sensors would be added above the top and below the bottom of the vessel.

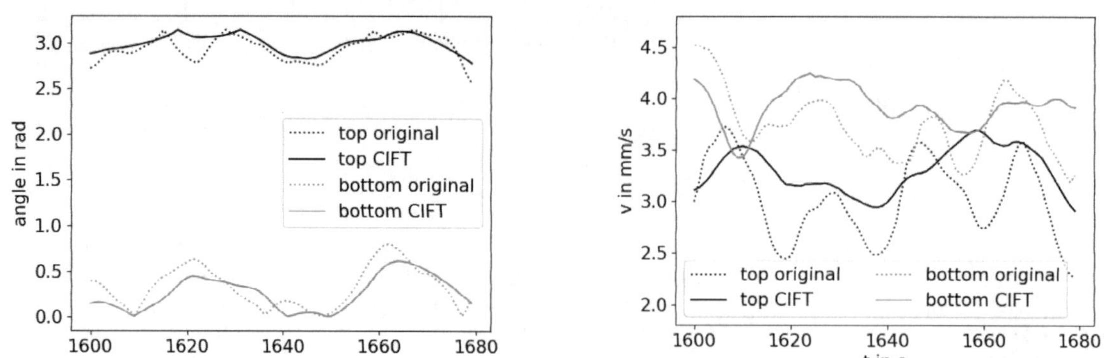

(a) (b)

Fig. 3: Comparison of the reconstructed and the original velocity regarding the direction (a) and amplitude (b) of the mean velocity at the top and at the bottom of the cylindrical vessel

Measurements for a temperature difference of 30 K

For the preliminary measurements, an arrangement of 7 sensors evenly distributed in azimuthal direction on each of the 3 layers over the height were selected. As seen in Fig. 1(a), a spacer made of a ceramic material was used to allow for a complete covering of the height of the vessel. The measurements were conducted during the night in order to avoid magnetic field changes generated by any moving ferromagnetic parts or switching currents in the experimental hall. With this precaution we were able to measure the magnetic field over 8 hours with a drift of about 20 nT. Fig. 4. shows a preliminary measurement of the flow induced magnetic field for a temperature difference of 30 K. From the 21 magnetic field sensors we selected 3 sensors at one azimuthal position and at three different heights (see also Fig. 1(a)). In Fig. 3(a) it can be clearly seen that the magnetic field rises after generating the temperature gradient at about 2 hours and is maintained until the temperature gradient approaches zero again at about 11 hours. At the end of the experiments, all 3 sensors return approximately to zero, with an offset between 20 and 40 nT. In Fig. 4(b) a time interval of 80 seconds is shown which was taken at 7.2 h after starting the measurement. It could clearly be seen that there is a phase shift between sensor 0, which is close to the top, and sensor 2, which is close to the bottom. This indicates that the flow structure of torsional modes could be indeed visualized by CIFT.

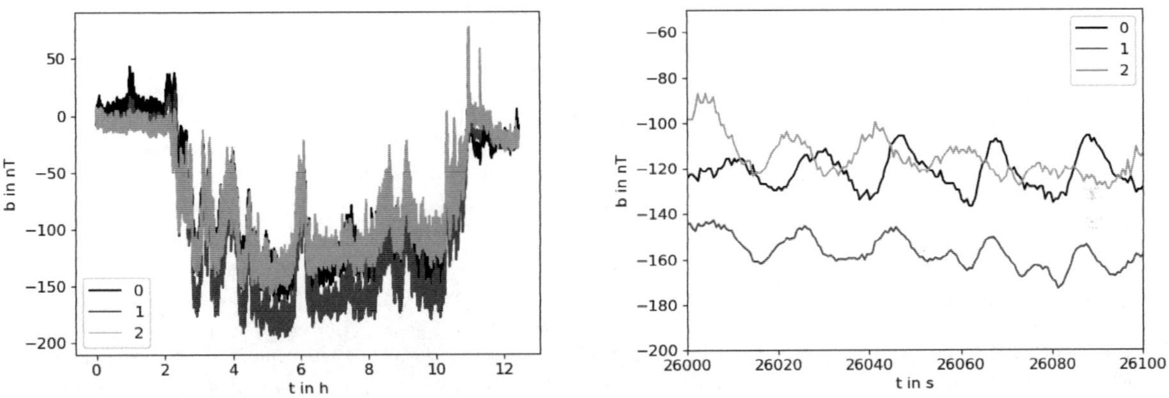

(a) (b)

Fig. 4: Measurement of the flow induced magnetic field for one experiment with a temperature difference of 30 K for 3 sensors (a) and a zoom into the measurement (b) at about 7.2 hours

Conclusions

The presented numerical and experimental investigations indicate the applicability of CIFT for the detection of the rich dynamics of the LSC of a Rayleigh-Bénard experiment using liquid metals. Beside the visualisation of azimuthal rotations and cessations of the LSC, the reconstruction of dynamics of torsional modes seems to be feasible, too. The next steps will be the reconstruction of the flow based on the measured induced magnetic fields and a comparison with ultrasound Doppler velocimetry measurements. In order to achieve a full three-dimensional reconstruction of the flow, a second pair of coils will be added to the experiment for generating a transversal magnetic field. The use of gradiometric induction coils in combination with an alternating primary magnetic field will reduce effects of fluctuations of the environmental magnetic field.

References

1. F. Stefani, T. Gundrum, G. Gerbeth, Physical Review E 70 (2004), 056306
2. M. Ratajczak, T. Gundrum, F. Stefani, T. Wondrak, Journal of Sensors 2014 (2014), 739161
3. T. Wondrak et al., Flow Measurement and Instrumentation 62 (2018), 269-280
4. D. Fünfschilling and G. Ahlers, Physical Review Letters 92 (2004), 194502
5. E. Brown and G. Ahlers, Journal of Fluid Mechanics 638 (2009), 383-400
6. H.-D. Xi et al., Physical Review Letters 102 (2009), 044503
7. T. Wondrak et al., International Journal of Applied Electromagnetics and Mechanics, submitted
8. F. Stefani, G. Gerbeth, Inverse Problems 16 (2000), 1-9

Transient Eddy Current Flow Metering: a calibration-free velocity measurement technique for liquid metals

N. Krauter[1] and F. Stefani[1]

[1] Helmholtz Zentrum Dresden Rossendorf (Department of Magnetohydrodynamics, Dresden, Germany)

Corresponding author: n.krauter@hzdr.de

Abstract

Most inductive techniques for the velocity measurement of liquid metals have the common problem that the measured signals depend on the electrical conductivity and the temperature of the liquid metal. This is particularly unfavourable for applications with strong temperature fluctuations or for measurements that have to be performed in different liquid metals with the same sensor. Usually these sensors have to be calibrated extensively to ensure accurate measurement results. We present a new measurement technique called Transient Eddy Current Flow Metering (TECFM) which overcomes the need for prior calibration of the sensors. Its principle relies on imprinting transient eddy current systems into the liquid metal and to track their movement as they are advected with the flow. Two kinds of sensors that use this new measurement principle have been developed: An external sensor [1] for the contactless velocity measurement at the boundary of liquid metal flows, and an immersed sensor [2] which allows local velocity measurements in the vicinity of the sensor. Focusing on the latter we present the results of numerical simulations as well as the results of measurements that were performed in the eutectic alloy GaInSn at room temperature and in liquid sodium at 180 °C.

Key words: flow measurement, inductive methods, calibration-free

Introduction

The measurement technique presented in this article traces back to early ideas of Zheigur and Sermons [3]. Although the external [1] and immersed [2] sensor types both track the movement of an eddy current system to determine the flow velocity, the arrangement of their sensor coils as well as the structure of the eddy currents are considerably different. The external sensor (Fig. 1), which is placed outside the boundary of the liquid metal, uses one primary coil to imprint eddy current rings into the liquid metal. This is achieved by suddenly switching a DC current on or off. In this case, the eddy currents are moving with flow velocity and this velocity can be obtained by tracking a distinctive point of the system. This is necessary because the eddy currents are decaying over time and only distinctive points, like the magnetic pole or a zero crossing point, can be identified regardless of the amplitude of the eddy currents. For the external sensor, this point is the pole of the magnetic field, which is generated by the eddy currents and can be tracked by measuring the voltage in both secondary coils [1]. Under the reasonable assumption of steady flow, the flow velocity can be calculated from two or more locations of the magnetic pole at certain points in time with: $v = \Delta x/\Delta t$.

Since the immersed sensor is the most recent development, this article will focus primarily on that type. In order to allow velocity measurements of the surrounding liquid metal, the coils have to be arranged differently than the ones of the external sensor. There are two primary and two secondary coils (Fig. 2) and the eddy current system is more complex.

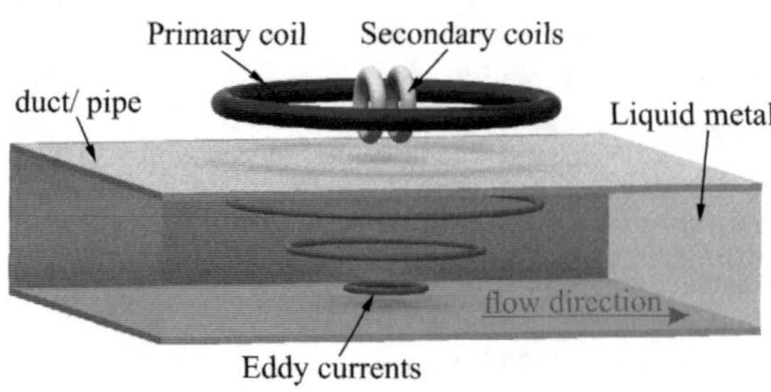

Fig. 1: Basic structure of the external sensor and sketch of the eddy current system.

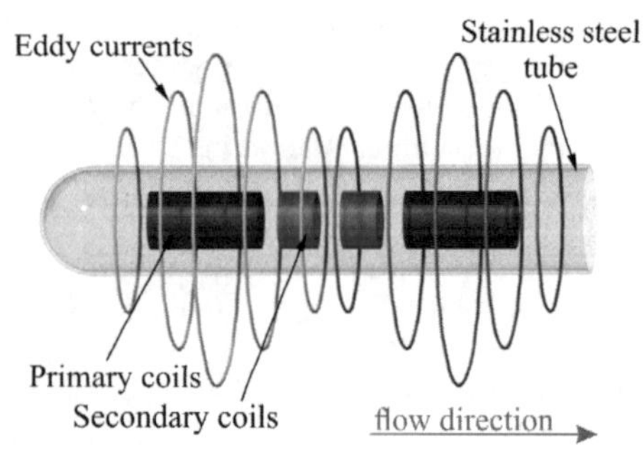

Fig. 2: Basic structure of the immersed sensor and sketch of the eddy current system.

Again, the eddy currents are imprinted into the liquid metal by switching a DC current on or off, but the DC currents in the right and left primary coil have to be directed oppositely. As a result, two eddy currents with opposite direction are induced and because they have the same amplitude, the magnetic flux density at the middle of the sensor equals zero. For a symmetric coil arrangement and eddy current distribution, the whole eddy current system and also this zero crossing point x_0 will move with the same velocity as the liquid metal. When asymmetries occur, a certain drift of x_0 must be taken into account [4]. The two secondary coils are used to track the position of x_0, which is calculated from the respective voltages, and the velocity can be obtained with the previously mentioned relation when assuming a linear motion of the eddy currents.

Numerical simulation

Numerical simulations for the immersed sensor were performed in COMSOL Multiphysics by creating an axisymmetric model of the sensor and the surrounding liquid metal. One of the most important features of an ideal immersed transient eddy current flow meter (I-TECFM) is the calibration-free character. That means that the sensor yields the same results, regardless of the electrical conductivity (or temperature) of the liquid metal, without any need for calibration. In order to evaluate this property, the parameter v_Δ is used. It represents the normalized deviation between the predefined and measured velocity and is displayed in Fig. 3 over a wide range of electrical conductivities for three different cases. Case a) shows the performance of the I-TECFM under almost ideal conditions: All sensor coils are made as small and as close together as possible, in addition there is no stainless steel cladding which separates the sensor coils from the liquid metal. It can be seen that there are nearly no variations in v_Δ for case a), the remaining deviations are below 1 % and can be attributed to the finite grid size of the simulation model. Using a smaller grid size will further decrease v_Δ but requires a much higher calculation time for a comparably small benefit. When going over to more realistic coil arrangements and sizes in case b), it can be seen that the magnitude of v_Δ is increasing with σ. By adding the stainless steel cladding and using the coil arrangement of the real sensor prototype in case c), the magnitude of v_Δ is increasing even further. Larger coil sizes for the actual sensor are necessary to accommodate a reasonable number of wire turns in order to increase the sensitivity of the sensor. For reasons of mechanical stability, the distance between adjacent coils cannot be much smaller than 1 mm.

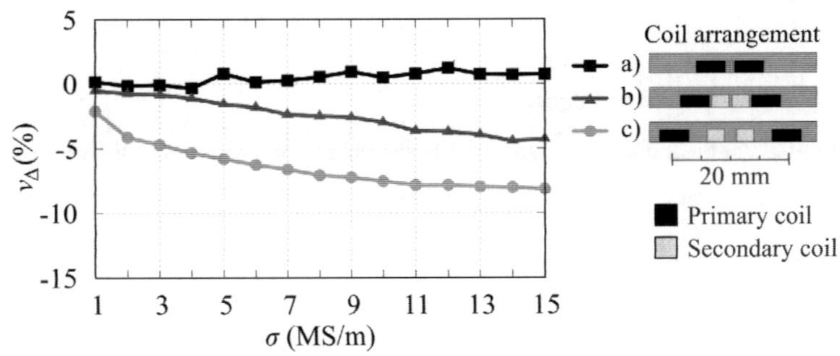

Fig. 3: Influence of the electrical conductivity of the liquid metal on the measurement error v_Δ of the velocity for different coil arrangements: a) idealized arrangement, b) very small coil distance, c) real coil arrangement of the prototype with stainless steel cladding of 2 mm thickness.

The primary reason for an increased v_Δ is the penetration depth δ (skin depth) of the magnetic fields into the liquid metal in combination with the distance between the sensor coils. For high conductivities δ is decreasing and therefore the position of the zero crossing point appears to blur when the sensor coils are too far apart because the signal strength is rapidly decreasing with the distance. This is also the reason for the differences between case b) and c) since there is a significant difference in coil size and distance for both cases. Another reason are parasitic eddy currents that are induced within the stainless steel cladding that protects the sensor from direct contact with the liquid metal. Their magnetic field is superimposed with the magnetic field of the eddy currents that are located within the liquid metal and therefore the whole eddy current system appears to move slower than it actually does. This effect is especially significant for low σ since the parasitic eddy currents have the same order of magnitude as the eddy currents within the liquid metal. This effect becomes worse with increasing wall thickness of the sensor cladding.

In order to obtain accurate measurements it is important to assure that the excitation current in the primary coils has reached its stationary value. Usually this takes between 10 μs to 100 μs after initiating the switching operation, depending on the excitation current source. Otherwise, the primary magnetic fields would interact with the eddy current system, resulting in a significant increase in v_Δ.

IOP Conf. Series: Materials Science and Engineering **424** (2018) 012008 doi:10.1088/1757-899X/424/1/012008

Measurement results

Measurements with the I-TECFM prototype have been conducted at different experimental facilities using the liquid metal alloy GaInSn at room temperature or sodium at 180 °C. The charts in Fig. 4 and Fig. 5 show the location of the calculated zero crossing x_0 in dependence on time for different pre-adjusted velocities v, and the corresponding (non-normalized) deviation v_Δ from the measured velocity, which is equivalent to the slope (obtained by linear regression) of the measurement series. Each point displayed in Fig. 4 and Fig. 5 represents the mean value of 2500 successive measurements. During production of the sensor, asymmetries resulting from manufacturing tolerances occur, and therefore the previously mentioned drift of x_0 [4] was observed during measurements. Even for very slight asymmetries, this drift cannot be ignored. Since the drift velocity is constant for each sensor, it can be determined simply by conducting measurements in a resting liquid metal. Once the drift velocity is known, it can be subtracted from each measurement series to obtain the real, compensated velocity or x_0, respectively. The following charts already contain the compensated measurement results.

The measurements that were performed in the liquid metal alloy GaInSn with an electrical conductivity of 3.3 MS/m [5] at room temperature yield promising results (Fig. 4). They were conducted at the GaInSn-Loop at HZDR. The test section has an inner diameter of 27 mm and the sensor has an outer diameter of 11.6 mm. v_Δ reaches a maximum value of 0.05 m/s. Only for the lowest velocity a comparably high relative error of around 15 % is observed; this is a general problem of the I-TECFM since the drift velocity lies within the same order of magnitude which results in higher relative measurement errors, especially at low velocities.

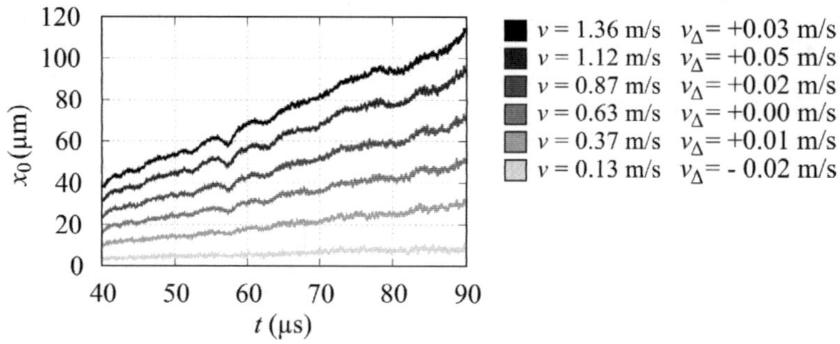

Fig. 4: Measurement results for the immersed sensor in GaInSn at room temperature for different velocities v.

The measurements in liquid Na with an electrical conductivity of 8.1 MS/m [6] at 180 °C yield similar results (Fig. 5), although v_Δ is considerably higher. There are various reasons for the higher deviation from the pre-adjusted velocity: The measurements were performed in a completely different experimental facility with different instruments, pump, flow structure inside the test section and, most importantly, a different sensor cladding with increased wall thickness to comply with security requirements. As shown in Fig. 3 a higher wall thickness of the cladding results in a loss of signal strength and quality. Since v_Δ lies around 0.1 m/s for nearly all measurements, it is also possible that some part of v_Δ does not come from measurement errors but from an unknown offset whose origin has yet to be determined.

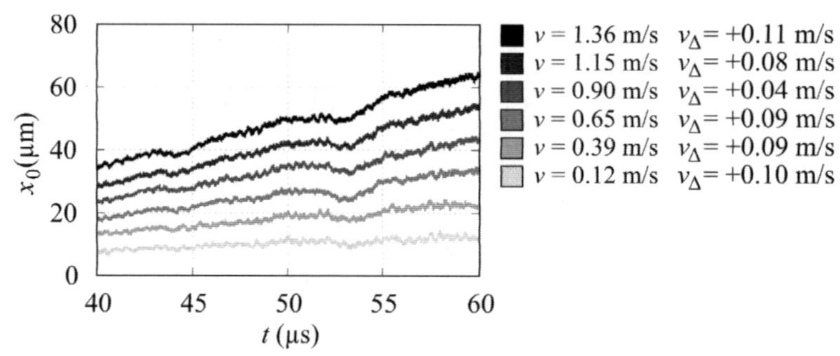

Fig. 5: Measurement results for the immersed sensor in liquid Na at 180 °C for different velocities v.

In both Fig. 4 and Fig. 5 distinct oscillations that are superimposed with the actual measurement values can be observed. They result from voltage changes within the sensor coils due to resonance oscillation of the coils which is caused by the suddenly switching off the excitation current. This effect can be mitigated by increasing the resistance of the coil wires, especially the ones of the secondary coils.

Conclusions

We have demonstrated via numerical simulations and first measurements which were conducted with sensor prototypes of the external and immersed implementations that TECFM is a viable method to determine the flow velocity of an electrically conductive fluid such as liquid metal. The distinguishing features of TECFM are the possibility for the direct measurement of the fluid velocity and the calibration-free character of this measurement technique. Although there is an additional measurement error in the range of a few percent when comparing the idealised and prototype sensor design, the flow velocity can still be determined with adequate accuracy. There are certainly more measurements needed at different electrical conductivities and temperatures to identify the reasons for increased deviations from the pre-adjusted velocity that could be observed for the measurements. However, the current implementations of the sensors already deliver promising results. Other inductive measurement techniques like the ECFM [7] can provide more accurate measurements (for the moment), as long as they have been properly calibrated for different temperatures and electrical conductivities of the liquid metal. This is the reason why TECFM is most suitable for applications where there are large fluctuations of the electrical conductivity and/ or a prior calibration of the senor is not possible or wanted.

Acknowledgment

This work was supported by CEA in the framework of the ARDECo programme.

References

1. J. Forbriger and F. Stefani, Meas. Sci. Technol. 26 (2015), 105303
2. N. Krauter and F. Stefani, Meas. Sci. Technol. 28 (2017), 105301
3. B. D. Zheigur and G. Y. Sermons, Magnetohydrodynamics 1, 101 (1965)
4. R. Looney and J. Priede, arXiv:1705.02939 (2017)
5. Y. Plevachuk et al., Journal of Chemical and Engineering Data, vol. 59, no. 3 (2014)
6. V. Sobolev, Database of thermophysical properties of liquid metal coolants for GEN-IV, SCK CEN (2011)
7. S. Sureshkumar, M. Sabih, S. Narmadha et al., Nuclear Engineering and Design 265 (2013), pp. 1223-1231

IOP Conf. Series: Materials Science and Engineering **424** (2018) 012009 doi:10.1088/1757-899X/424/1/012009

Lorentz Force Velocimetry using a bulk HTS magnet system

O. Vakaliuk, B. Halbedel

Technische Universität Ilmenau
Institute of Materials Engineering, Department of Inorganic-Nonmetallic Materials
Group of Materials research in or with Magnetic Fields
Gustav-Kirchhoff-Straße 6, D-98693 Ilmenau, Germany, Ilmenau, Germany

Corresponding author : oleksii.vakaliuk@tu-ilmenau.de.

Abstract

Design of the Lorentz Force Flow-meter (LFF-meter) for weakly-conducting and slow-flowing fluids, e.g., glass melts, place stringent requirements on the magnetic field generation and force measurements. In the scope of the Research Training Group "Lorentz Force Velocimetry (LFV) and Lorentz Force Eddy Current Testing", a LFF-meter considers: first - a magnetic field source, where NdFeB permanents magnet system is replaced with a bulk high-temperature superconductors (HTS), which promise higher flux densities (> 1 T) and thus higher output force resolution; second – a high-precision force measurements, where electromagnetic force compensation weighing balance (EMFC) setup is replaced with torsion balance based system (TFMS) , which claims an increase of the force resolution up to 1 nN. Furthermore, in order to raise the issue of the limiting total mass (which is always an issue for high-precision force measurements), the bulk HTS and TFMS are merged within the cryostat. This work discusses LFV experiments, calculations and numerical simulations, where a magnetic field is generated by the bulk HTS.

Key words : lorentz force velocimetry, bulk high-temperature superconductors, flow rate measurement, glass melts, trapped field magnets.

Introduction

The velocity or mass flux measurement in flowing substances is of great importance in industry with the regard of its ability to provide an efficient and, therefore, economic production process control. Apart, there is a growing need for non-invasive measurements, especially, in a view of a hygienic requirements in pharmacy and food; either the necessity to sustain a severe condition of opaque, hot or aggressive surroundings in a number of practical application, such as: i) liquid metals for high-quality aluminum, cooper and steel production [1]; ii) glass melts for the manufacture of high-quality glass products; iii) salt melts for solar power or storage systems [2].

The LFV have been studied extensively in the pursuit of absolutely contact-less flow measurements for electrically conducting substances [1,3–6]. It has been shown that LFV works well in liquid metals due to a high electrical conductance ($\sigma \sim 10^6$ S m^{-1}) [1]. However, for a weakly conducting and slow flowing substances LFV works only under the laboratory condition in narrow measurement range. The experimental investigation of LFV for salted water (as a model fluid) of varying electrical conductivity and flow velocity was reported in several past works [5,6]. This was realized by employing a high-precision force measurement system, i.e. electromagnetic force compensated weighing cell (Sartorius Lab Instruments), in a combination with an optimized Halbach array of permanent magnets (PMs) [7] of 1 kg. However, in the case of aggressive fluids (i.e. glass melts) the realization of close magnetic interaction with the fluid is not possible due to the large isolation walls and harsh environment.

A vital technical challenge of LFV for a weakly conducting and slow flowing substances remain in magnetic field generation and adaptation of corresponding magnet system. The conventional NdFeB permanent magnets have been considered the most practical strategy to generate a maximum possible Lorentz force with a load restriction needed to ensure force high-precision measurements [6]. However, the surface magnetic flux density of the strongest available NdFeB PMs is limited to about 0.5 T. The bulk HTSs have the ability to act as quasi-permanent trapped field magnets. Effective flux pinning permits high current densities which in turn can generate large persistent magnetic fields. The largest achieved trapped field to date is 17.6 T in silver-alloying Gd-Ba-Cu-O samples bulk using field-cooled magnetization (FC) [8], and over 5 T using compact and fast pulsed field magnetization [9]. Nonetheless, for any practical application the use of the bulk HTS magnets requires cooling and magnetization. If the bulk HTS is to be incorporated into the LFF-meter, its integration with the force measurement system is an issue to be resolved.

Experimental

The LFV measurements were conducted using a "dry calibration" set-up [10], developed at the Department of Engineering, Techniche Universität Ilmenau, but upgraded with the bulk HTSs. The idea of the "dry calibration" implies the replacement of a fluid flow with the controlled motion of a metal rod of a fixed geometry. Two types of metal rods were used: copper with an electrical conductivity of σ_{Cu} = 59 MS m^{-1} and aluminum alloy (AlMgSi) with σ_{Al} = 20 MS m^{-1}.

Fig. 1 shows a photograph of the experimental set-up used to test LFV using a bulk HTSs. The installation consists of five main components: HTS magnet system (1), force sensor (2), solid conductor (3), linear vertical drive (4) and aluminum rack (5). The experimental procedure is fully automated: an IBA-Automation environment controlls the linear drive and is used for data acquisition. The linear drive executes a repetitive motion of the metal rod with a constant velocity, that ranges from 50 to 81 mm/s. The force measurements is realized by using a commercial load cell (Model PW6D, Hottinger Baldwin Messtechnik GmbH) in a combination with a data analog measuring amplifier (SOEMER Messtechnik GmbH). The force is measured in terms of a voltage. In order to calibrate the output voltage signal from the measuring amplifier directly into a force, the masses of 5, 10, 20 and 100 g (KERN & SOHN GmbH) were used.

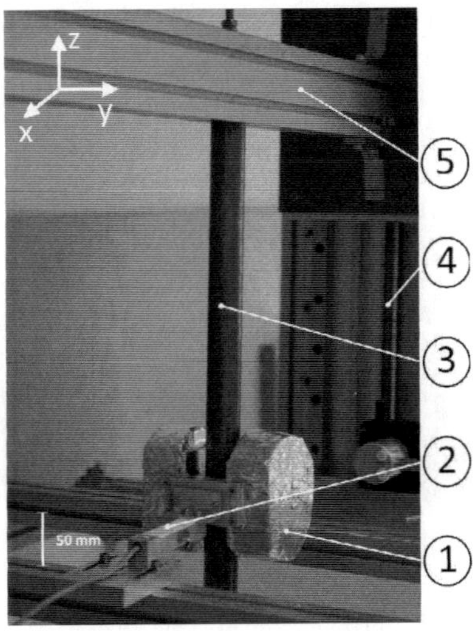

Fig. 1: Photograph of the realized LFV experimental setup, which consists of five main components: (1) HTS magnet system, (2) force sensor, (3) metal rod, (4) linear drive and (5) aluminum rack..

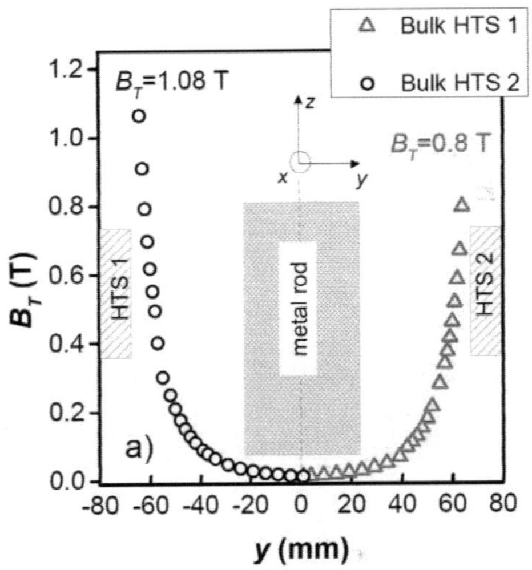

Fig. 2: The peak trapped magnetic flux density $B_T(y)$ for each bulk HTS as a function of distance (along the y-axis) with a maximum field of 1.08 T (left) and 0.8 T (right) at the bulk surfaces. The shaded rectangles indicate the measurement system: HTS bulks (left and right) and metal rod (center).

The bulk HTS magnet system consists of two Y-Ba-Cu-O samples in the form of a cylindrical disc with a diameter of 46 mm and a thickness of 16 mm, provided by Adelwitz Technologiezentrum GmbH. The Y-Ba-Cu-O bulks were encapsulated in an aluminum holder, arranged opposite each other with a distance of 64 mm between and wrapped with styrofoam. Such arrangement enables to locate the bulk HTS magnets on either side of the metal rod. After field cooling (FC) magnetization with an applied field of 1.5 T at 77 K (using liquid nitrogen), the bulk HTS magnet system provides a quasi-permanent magnetic field enabling Lorentz force measurements to be carried out with a constant velocity of the metal rod. A maximum field of about 1 T at the bulk surfaces was recorded for each bulk HTS, respectively, as indicated in Fig. 2

Numerical Model

The numerical model is implemented as a fully 3D model using COMSOL's AC/DC module, where the bulk HTS pair is assumed to be fully magnetized with an appropriate constant J_c flowing through the cross-section of each bulk to reproduce measured trapped field profiles [11, 12]. A velocity (Lorentz term, where $\boldsymbol{J} = \sigma(\boldsymbol{E} + \boldsymbol{v} \times \boldsymbol{B})$ is then prescribed for a copper rod moving between the bulk pair, reflecting the experimental setup and conditions, which then allows the Lorentz force to be calculated.

Results

The results of the measurements are plotted in Fig. 3, showing the Lorentz force linearly depends on the velocity and the electrical conductivity of the moving metal rods, which is consistent with the fundamental theory of the LFV [3, 4]. The mean velocities of the copper rod were u = [54; 64; 76; 81] mm/s. For the aluminium rod, F_L - measurements with only two velocities of u = [70; 81] mm/s were possible, due to its lower conductivity and limitations related to the resolution of the present LFV measurement system. Nevertheless, F_L values obtained with same-sized copper and aluminium rods for one prescribed velocity, e.g. u = 81 mm/s, scales as σ_{Cu}/σ_{Al}.

Besides, the obtained experimental data were used to verify a numerical model developed in COMSOL Multiphysics to predict and optimize LFV performance through numerical simulation. Fig. 4 shows the sensitivity diagram of the actual LFF-meter. It is immediately clear that the use of the bulk HTS with higher trapped fields of 3 T and 5 T enhances the resultant sensitivity approaching the values lower than 10^{-6} Nm/s. This provides evidence that the bulk HTSs are feasible and enable a further extension of the LFV application with small (σ u), thereby covering the number of industrial weakly-conducting and slow-flowing fluids, e.g. glass, molten salts melts and/ or acid. It should be noted that the F_L values are valid only for our particular design of the proof-of-principle LFV setup (see Figure 1b) and may vary for each individual system differently.

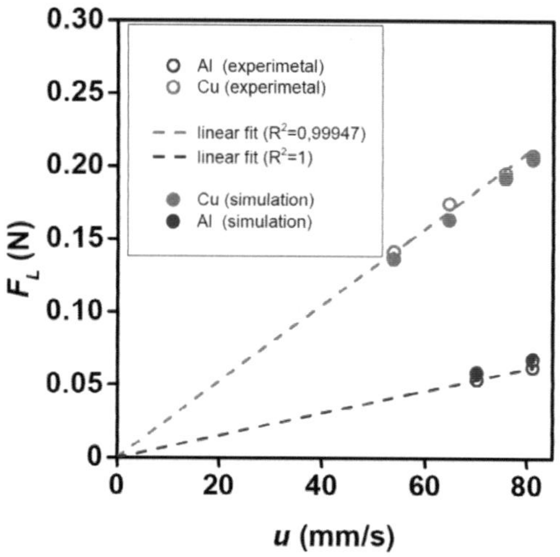

Fig. 3: Lorentz Force F_L as function of a velocity u for the copper and aluminum rods. Open circles indicate the experimental results, and the solid circles indicate the simulation results. The dashed lines indicate the linear t to the experimental results.

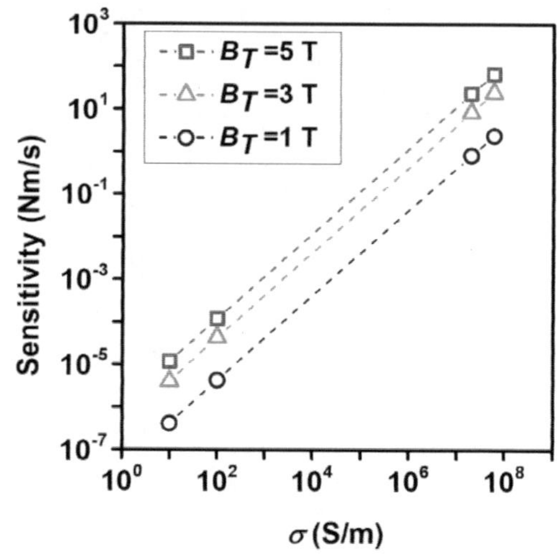

Fig. 4: Sensitivity of the LFF-meter as function of electrical conductivity for relevant fluids, e.g. solid metals, acids/base and glass melts, including the experimental results presented in this paper.

Conclusions and Outlook

Current work demonstrates the applicability of the bulk HTS in Lorentz Force Velocimetry. The bulk HTS offers higher magnetic field than previously used the NdFeB permanent magnets with the similar mass, despite it does require appropriate cooling and magnetization. Obtained experimental and numerical simulation results agree well exhibiting the linear relationship between the Lorentz force and product of the electrical conductivity and velocity, in accordance with a fundamental theory of the LFV [3, 4]. Hence, these results serve as a starting point for the future development of a new LFV prototype with improved performance.

Still, for practical application of the bulk HTS in LFV further aspects has to be resolved:

- The bulk HTS magnets require a robust and long-term operation at temperature T > 77 K. This implies that an encapsulation of bulk HTS in a vacuum cryostat is needed to increase the functional and long-term operation. Also, it allows a suitable adaptation of the cooling source (with liquid cryogen or cryogen-free technology).
- Upgrade of the force measurements from electromagnetic force compensation weighing balance setup to torsion balance based system. This promise to achieve the force resolution up to 1 nN
- In order to raise the issue of the limiting total mass of TFMS, the bulk HTS must be merged with the TFMS and be encapsulated in the integrated cryostat, where both systems are cooled together. This idea was patented by the authors [13] and the relevant portable LFV prototype is currently under construction.

Acknowledgment
This work is supported by the German Research Foundation (DFG) within the Research Training Group "Lorentz Force Velocimentry and Loretz Force Eddy Current Testing" (RTG-1567/3).

References

1. Kolesnikov, Y., Karcher, C., Thess: A.: Lorentz force flowmeter for aluminum – laboratory experiments and plant tests. Met. Trans. B., Vol. 42B, pp. 441–450, **2011**.
2. Herrmann, U.; Kelly, B.; Price, H. Two-tank molten salt storage for parabolic trough solar power plants.167 Energy **2004**, 29, 883 – 893
3. Thess A., et al.: Lorentz Force Velocimetry. Physical Rewiew Letters, Vol. 96, No. 4, pp. 164501, **2006**.
4. Thess A., et al.: Theory of the Lorentz Force flow meter. New J. Phys, Vol. 9, pp. 299, **2007**.
5. Halbedel, B.; Resagk, C.; Wegfrass, A.; Diethold, C.; Werner, M.; Hilbrunner, F.; Thess, A. A Novel Contactless Flow Rate Measurement Device for Weakly Conducting Fluids Based on Lorentz Force Velocimetry. Flow, Turbulence and Combustion **2013**, 92, 361–369.
6. Vasilyan, S.; Ebert, R.; Weidner, M.; Rivero, M.; Halbedel, B.; Resagk, C.; Fröhlich, T. Towards metering tap water by Lorentz force velocimetry. Measurement Science and Technology **2015**, 26, 115302.
7. Werner, M.; Halbedel, B. Optimization of NdFeB Magnet Arrays for Improvement of Lorentz Force Velocimetry. IEEE Transactions on Magnetics **2012**, 48, 2925–2928
8. J H Durrell, A R Dennis, J Jaroszynski, M D Ainslie, K G B Palmer, Y-H Shi, A M Campbell, J Hull, M Strasik, E E Hellstrom, and D A Cardwell. A trapped _eld of 17.6 t in melt-processed, bulk gd-ba-cu-o reinforced with shrink-_t steel. Superconductor Science and Technology, 27(8):082001, **2014**.
9. H Fujishiro, T Tateiwa, A Fujiwara, T Oka, and H Hayashi. Higher trapped field over 5t on HTSC bulk by modified pulse field magnetizing. Physica C: Superconductivity and its Applications,445-448:334 , 338, **2006**.
10. V Minchenya, C Karcher, Y Kolesnikov, and A Thess. Calibration of the Lorentz force flowmeter. Flow Measurement and Instrumentation, 22(3):242 247, **2011**.
11. Vakaliuk O , Aisnslie M D and Halbedel, Lorentz Force Velocimetry using a bulk HTS magnet system: proof-of-concept, Supercond. Sci. Technol. accepted, **2018**
12. Ainslie M D and Fujishiro H Modelling of bulk superconductor magnetization, Supercond. Sci. Technol. 28 053002, **2015**
13. B Halbedel, O Vakaliuk, T Fröhlich, N Yan. Vorrichtung zur Ermittlung von Parametern einer elektrisch leitfähigen Substanz und dazugehöriges Verfahren, DE 10 2017 005 210.7, submitted:**30.05.2017**.

A precise magnetic flux leakage method for the defects detection within the steel thin sheet

Bo WANG and Xiaodong WANG*

College of Materials Science and Opto-Electronic Technology, University of Chinese Academy of Science, Beijing 101408, China

Corresponding author: xiaodong.wang@ucas.ac.cn

Abstract

Defects detection of ferromagnetic steel thin sheet is very important in metallurgical industry. In this article, a new magnetic flux leakage (MFL) technology is used to complete this task. In our work, a tiny permanent magnet is used to replace the conventional coils to magnetize testing specimen and precise Tunnel Magneto-resistive (TMR) sensor is employed because of its superior sensitivity and linear operation range over traditional magneto-resistive sensors. Two kinds of thin sheets were selected: galvanized sheet and cold-rolled sheet. The experimental results indicate that the present method is reliable for detecting defects of steel thin sheet, and defects with diameter of 400 and 20 μm can be detected clearly for galvanized sheet and cold-rolled sheet, respectively. This accuracy fully satisfies the metallurgical production requirement. This method exhibits its potential to realize *in-situ*, online inspection process of thin sheet production.

Key words: magnetic flux leakage; permanent magnet; Tunnel Magneto-resistive; steel thin sheet

1. Introduction

In metallurgical industry, steel thin sheet is a kind of steel with large aspect ratio, its thickness usually locates at the range of 0.2 ~ 4 mm[1, 2)]. Nowadays, steel thin sheet such as galvanized sheet[3, 4)], cold-rolled sheet[5-7)], stainless steel sheet and silicon steel sheet, accounts for 50% of all rolled products. Despite the use of new construction materials, the steel thin sheet still widely used in the production of tractors and automobiles, in building, in power industry, in railroad, and elsewhere[7)]. Therefore, the defects detection of steel thin sheet is a very important process in metallurgical industry.

Ferromagnetic property is another distinct characteristic of steel thin sheet, hence the magnetic flux leakage (MFL) technology[8-13)] can be employed in its defects detection process. MFL technology is a non-destructive testing and commonly used to detect metal losses caused by corrosion in steel constructions. This technology focus on the detection of defects by detecting magnetic anomalies in magnetic flux lines using magnetic sensors.

In this article, we detect the defects of galvanized sheet and cold-rolled sheet using tunnel magneto-resistive (TMR) sensor MFL technology, and expect this technology can be applied in *in-situ*, online inspection process of steel thin sheet. The measurement principle and process are shown in the next section, and the results and discussions can be found in the section 3 and 4.

2. Measurement principle

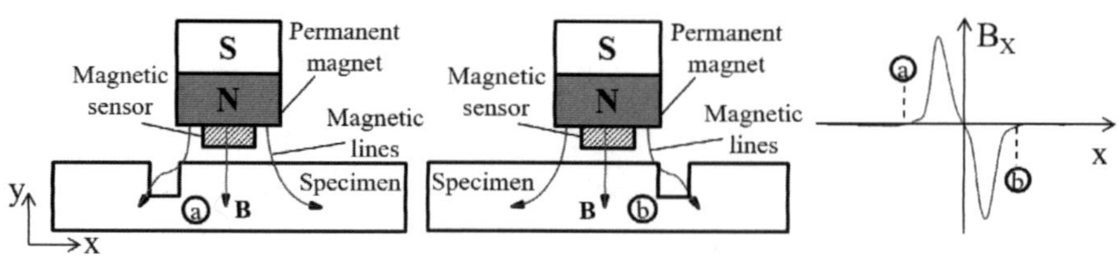

Fig. 1: The basic principle of the new MFL method used in this experiment

The basic idea of MFL technology is that the ferromagnetic material is magnetized under the applied magnetic field[8]. The principles are shown as Figure 1. When there is no defect within the specimen, the magnetic flux lines remain within the ferromagnetic material and undisturbed; when the defects are presented within specimen, the magnetic resistance increase acutely because the magnetic permeability of the defects is much smaller than that of the specimen itself, and consequently the magnetic flux in this area is distorted. The magnetic flux lines leak out of the specimen surface, and the magnetic leakage field obtained by using sensitive magnetic sensors[14, 15]. In recent years, the MFL technology has frequently used Giant Magneto Resistive (GMR)[16], Giant Magneto Impedance (GMI)[17] and Magneto Optical (MO) sensors[18] as well as Hall sensors and search coils[19]. When the defect passes underneath the magnet as shown in Figure 1, the signal will character the sinusoidal shape.

3. Experimental process and results

3.1. Experiment setup

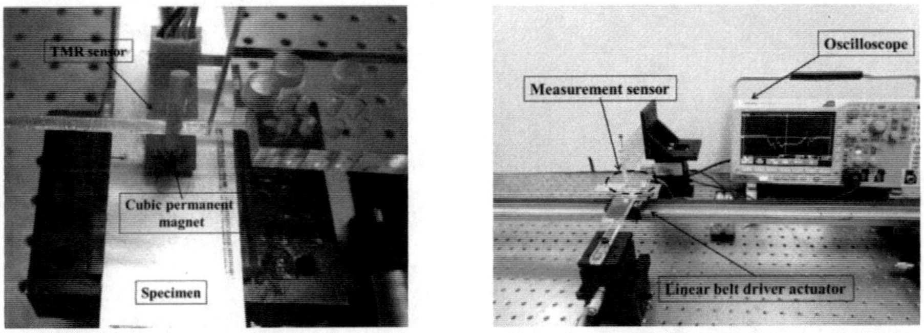

(a) (b)

Fig. 2: Experimental setup. (a) Detailed arrangement presentation. (b) General view

In order to complete the detection assignment, an experimental apparatus is built. The detail about the measuring sensor can be found in Figure 2(a). In this experiment, a cubic permanent magnet with width of 2 mm is applied to magnetize the ferromagnetic specimen, and a precise TMR sensor is employed because it has superior performance in sensitivity and linear operation range over conventional magnetic sensor.[20] It is reported that TMR possess an exceptional 70% change in resistance while GMR can only achieve 10-15%[20]. This inherent property grants TMR sensor yields very high sensitivity to the variation of magnetic field. As depicted in Figure 2(b), the steel thin sheet specimen is placed closely on a linear belt driver and move with it in a constant speed. The measuring sensor is fixed above the specimen with the interval of 0.3 mm. A sophisticated oscilloscope (MDO-3022, Tektronix Company) is used to record the signals.

3.2. Experiment results

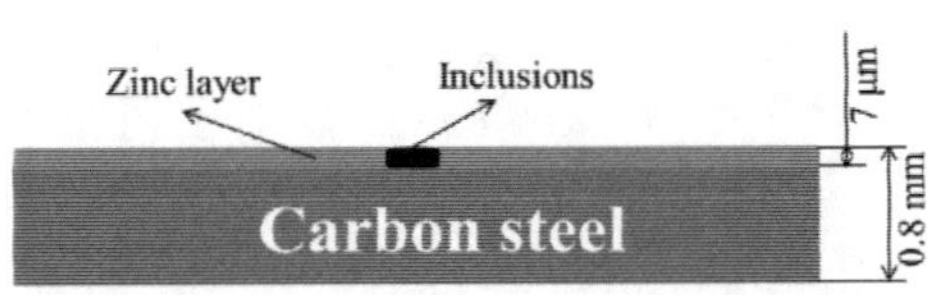

Fig. 3: The sketch of the structure of galvanized sheet

The structure of the galvanized sheet specimen, which comes from a metallurgical plant, is schematically depicted in Figure 3. Zinc layer is coated in hot-dip process in order to obtain a high corrosion resistance of the thin sheet production. During this process, the oxygen inclusions are often introduced and extruded, which will be seriously harmful to the final production quality. The thickness of galvanized sheet and

IOP Conf. Series: Materials Science and Engineering **424** (2018) 012010 doi:10.1088/1757-899X/424/1/012010

zinc layer are 800 and 7 μm, respectively. In this specimen, normally, the threshold size of inclusions is controlled under 500 μm.

In Figure 4, two galvanized sheet specimen, marked as No.1 and No.2, are employed in the experiment. For comparison, the electron microscope images of defects are shown in Figure 4(a) and (c) indicating that the defects are irregular and their equivalent diameters are 396 and 405 μm, respectively. The corresponding measurement signals obtained using MFL technology is shown in Figure 4(b) and (d). It is evident that the signals yield a high signal-to-noise ratio (SNR) (exceed 10 db) and character positive-negative peak. Fluctuations at begin and end of the signal are formed because the specimen is uneven. The measurement taken at different direction indicates the MFL technology is reliable.

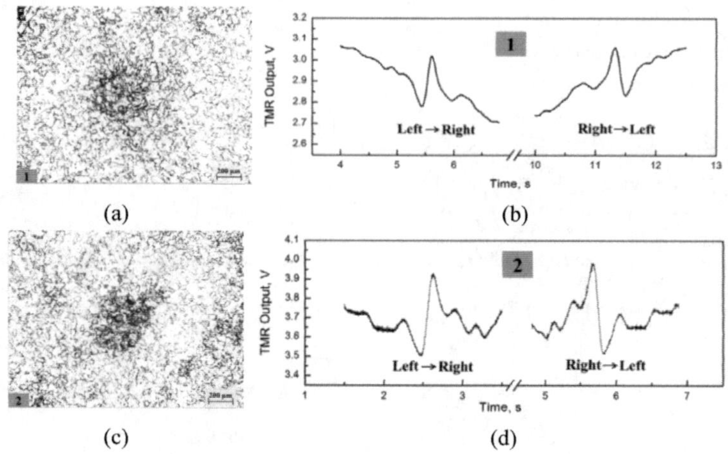

Fig. 4: The images of two galvanized sheet specimen defects (a) and (c) taken by electron microscope and the corresponding measurement signals (b) and (d)

As to cold-rolled sheet, the detection of defects with size exceed 20 μm is of important significance for production process. Similarly, another cold-rolled sheet specimen is detected using MFL technology and the result is presented in Figure 5. Upper electron microscope image show that the diameter of the defect is about 20 μm. The signals are verified via the superposition of perpendicular moving direction and they are also apparent and repeatable. This experimental result suggests that this new MFL method can satisfy the requirement of cold-rolled sheet defects detection.

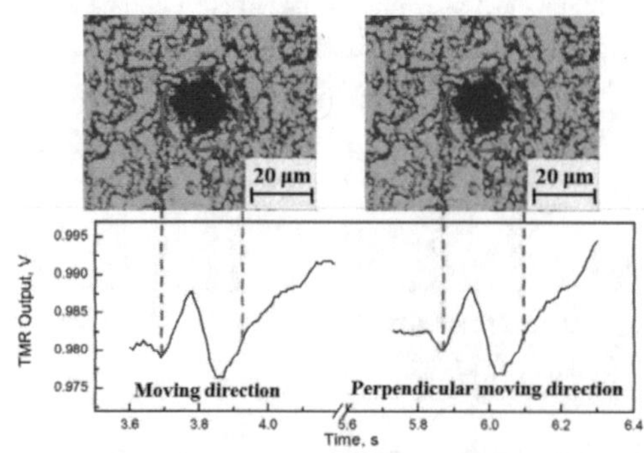

Fig. 5: The electron microscope image and signal of the 20 μm defects in cold-rolled sheet

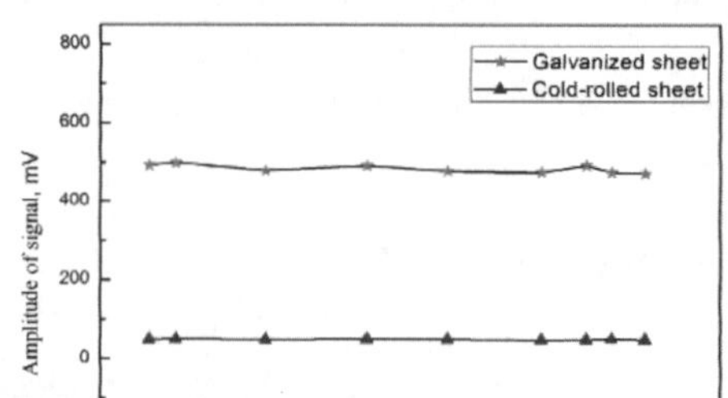

Fig. 6: The relationship between moving velocity and amplitude of signal

In present article, the relationship between amplitude of signal and specimen moving velocity is investigated as seen as Figure 6. The moving velocity is at 0~120 mm/s. For galvanized sheet and cold-rolled sheet, the moving velocity has few impacts on the amplitude of signal. This is because at low moving velocity, magnetic Reynolds number $R_m = \mu_0 \sigma u L << 1$, so that the induced magnetic field generated by eddy current can be neglected, and the leakage magnetic field mainly from the applied magnetic field.

4. Conclusions and prospects

In this study, a new MFL technology is applied to detect the defects of steel thin sheet. A tiny permanent magnet is used to substitute the conventional coils and precise TMR sensor is used to acquire the change of magnetic field. Galvanized sheet and cold-rolled sheet are selected as testing specimen, and corresponding defects with diameter of 400 and 20 μm, which can satisfy the metallurgical plant requirement, can be detected clearly. The experimental results in this article indicate that this new method is a reliable and convenient to inspect the defects of steel thin sheet. In the future, *in-situ*, online defects detection technology is an irresistible trend. Therefore, MFL-based sensor array should be applied in steel thin sheet producing process for detecting the defects. Our future research work will focus on this topic.

Acknowledgments

Financial support from the project funding (YZ201567) of the Chinese Academy of Sciences and the National Natural Science Foundation of China (51374190) is gratefully acknowledged.

References

1. M. Takahashi: *ISIJ Int.,* **55** (2015), 79-88.
2. A. B. Maksimov and I. S. Erokhina: *Steel Transl.,* **47** (2017), 752-755.
3. S. Li, Z. Yu, P. Zhang, X. Li, D. Qiu and X. Liu: *ISIJ Int.,* **56** (2016), 637-646.
4. K. Hoshino, M. Nagoshi, W. Tanimoto, Y. Yamasaki, S. Furuya, A. Matsuzaki and N. Yoshimi: *ISIJ Int.,* **57** (2017), 895-904.
5. J. Hu, W. Cao, C. Wang, H. Dong and J. Li: *ISIJ Int.,* **54** (2014), 1952-1957.
6. T. Kizu, K. Okuda, Y. Nagataki, T. Urabe and K. Seto: *ISIJ Int.,* **55** (2015), 1502-1511.
7. I. V. Doshchechkina, S. S. D'Yachenko, I. V. Ponomarenko and I. S. Tatarkina: *Steel Transl.,* **46** (2016), 364-367.
8. Y. Shi, C. Zhang, R. Li, M. Cai and G. Jia: *Sensors,* **15** (2015), 31036-31055.
9. D. A. G. Trevino, S. M. Dutta, F. H. Ghorbel and M. Karkoub: *IEEE Trans. Magn.,* **52** (2016).
10. W. Han, J. Xu, M. Zhou, G. Tian, P. Wang, X. Shen and E. Hou: *IEEE Trans. Magn.,* **52** (2016).
11. Y. Sun and Y. Kang: *Appl. Phys. Lett.,* **103** (2013).
12. Y. Sun, B. Feng, S. Liu, Z. Ye, S. Chen and Y. Kang: *J. Nondestruct. Eval.,* **34** (2015).
13. E. Yavuz and C. Mustafa: *Measurement,* 2018), 163-174.
14. Y. Zhang, Z. Ye and C. Wang: *Ndt & E International,* **42** (2009), 369-375.
15. A. N. Pechenkov, V. E. Shcherbinin and J. G. Smorodinskiy: *NDT&E Int.,* **44** (2011), 718-720.
16. J. W. Wilson and G. Y. Tian: *NDT&E Int.,* **40** (2007), 275-283.
17. M. M. Tehranchi, M. Ranjbaran and H. Eftekhari: *Sensor Actuat A: Phys.,* **170** (2011), 55-61.
18. V. Babbar and L. Clapham: *J. Nondestruct. Eval.,* **22** (2003), 117-125.
19. L. Yang, L. Jie, H. Xuezhi and Z. Duoduo, *Proceedings of the 2009 Asia Communications and Photonics Conference and Exhibition* (2009), 2.
20. B. Wu, Y. J. Wang, X. C. Liu and C. F. He: *Smart Mater Struct.,* **24** (2015).

IOP Conf. Series: Materials Science and Engineering **424** (2018) 012011 doi:10.1088/1757-899X/424/1/012011

Experimental investigation on the buoyancy-induced flow in a model of the Czochralski crystal growth process

J. Pal[1], S. Franke[2], S. Eckert[2], G. Gerbeth[2]

[1] HZDR Innovation GmbH, Bautzner Landstr. 400, 01328 Dresden, Germany
[2] Helmholtz-Zentrum Dresden-Rossendorf (HZDR), Bautzner Landstr. 400, 01328 Dresden, Germany

Corresponding author : j.pal@hzdr.de.

Abstract
Within this paper we present a model experiment focusing on investigations of the flow field in a Czochralski puller. Low melting point liquid metals as GaInSn are an important tool to investigate the flow structure for such industrial processes. The topology of the prevailing thermally-driven convection might be very complex and is mainly determined by the aspect ratio of the liquid volume and the strength of the convection described by the characteristic dimensionless Grashof number. The measurements of the fluid flow have been conducted by means of the ultrasound Doppler velocimetry (UDV) with and without the influence of external magnetic fields. Two kinds of sensor configurations were used to investigate the flow. Firstly, measurements of the radial velocity component by means of single UDV transducers were carried out shortly below the melt surface across the entire diameter of the cylindrical liquid column at various azimuthal angles. Secondly, a vertically arranged UDV array was applied at the side of the cylinder to obtain more detailed information about the radial velocities in the covered meridional plane. The results reveal the complex flow structure of natural convection in a Czochralski crucible which gains in complexity with applied external magnetic fields.

Keywords: Czochralski crystal growth, Electromagnetic stirring, Rayleigh-Bénard convection, Flow measurements

Introduction

Mono-crystalline silicon crystals are, nowadays, required in many electronic devices and for the power generation by photovoltaic systems. One of the main objectives in the manufacturing process of photovoltaic silicon wafers by the Czochralski (Cz) method is to produce crystals with low concentration of impurities such as oxygen, carbon and/or other materials. These adverse impurities might stem from the crucible and/or other inner parts of the crystal pulling equipment, transported by the melt flow and incorporated in the solidification process. Thus, the quality of the growing crystal and the yield of the wafers are crucially affected by the fluid flow. Consequently, the mass production of mono-crystalline silicon by the Cz technique has initiated a great effort of research on the convection inside the crucible. But reliable flow measurements at the real process are practically nonexistent and the validation of numerical models can only be accomplished by laboratory model experiments. Moreover, further optimizations of the process by means of applied magnetic fields are mainly based on numerical simulations of the flow, heat and mass transfer, and electromagnetic fields. Thus, the experimental modeling of the process on a laboratory scale still remains a valuable tool for the near future. For details on the growth of semiconductor silicon crystals the reader is referred to [1].

Modelling of the Czochralski technique

The simplest model of the Cz facility considers a Rayleigh-Bénard (RB) configuration, in particular a cylindrical cell characterized by a height H and diameter $2R$ with adiabatic insulated side walls. The cell is heated from the bottom and cooled at the top. The temperature gradient ΔT between the bottom and top sides generates buoyant convection.

The dynamics of the mere thermally induced convection may be described by mainly three control parameters. The dimensionless Rayleigh number Ra is the crucial parameter in modelling buoyancy and describes its strength. Instead of Ra the Grashof number Gr is also commonly used for this purpose and is coupled to Ra by the second control parameter, the Prandtl number Pr, according to $Ra = Gr \times Pr$. Generally, molten metals and semiconductor melts are low Pr number fluids in the order of 10^{-2}, which means that the thermal boundary layer thickness exceeds the viscous boundary layer by a factor in the order of 10. The third control parameter, the aspect ratio $a = H/(2R)$ concerns the geometry of the setup and affects crucially the developed convective pattern inside the melt [2,3]. The initial filling level in the real industrial Cz facility does not reach $a = 1$, it is even lower than $a = 0.5$ and decreases continuously during the process.

Other control parameter may occur if external magnetic fields are exposed to the system, such as the Taylor number Ta in case of the rotating magnetic field (RMF). The relevant parameter, Ta, describes the driving effect on the melt under investigation, which is primarily of swirling type but also causing a meridional secondary flow [4,5].

Experimental setup

The present experimental setup involves a modified RB configuration, a cylindrical melt column of variable aspect ratio homogeneously heated from the bottom. The left photo in Fig. 1 illustrates the experimental setup mounted inside the magnetic coil system MULTIMAG [6]. Besides the already mentioned RMF, it offers a variety of other magnetic field types, such as the traveling (TMF), pulsed (PMF), vertical DC or vertical DC of cusp type, and even linear superposition of them. As working fluid the ternary alloy GaInSn [7] was used because it remains liquid at room temperature and as distinguished from mercury it is non-poisonous. Moreover, its low *Pr* number is similar to that of molten silicon.

The heater was realized by an electrical heating plate embedded in a massive copper disc to achieve isothermal boundary condition. Several thermocouples were installed inside the copper disc to monitor its temperature. The upper thermal boundary condition in a Cz system is realized by a partially cooled surface. The partial cooling in our experiment covers approximately the same fraction area as the crystal does in an industrial Cz facility. It is realized with a circular heat exchanger (cold finger) mounted concentrically at the top of the experimental cell. The cold finger is optionally rotatable. The precise temperature control of the finger is guaranteed by a coolant provided from a thermostat with a large reservoir and high flow rate. Moreover, the temperature of the cold finger is monitored at various positions. During the measurements, the apparatus was embedded in mineral wool to minimize the lateral heat loss.

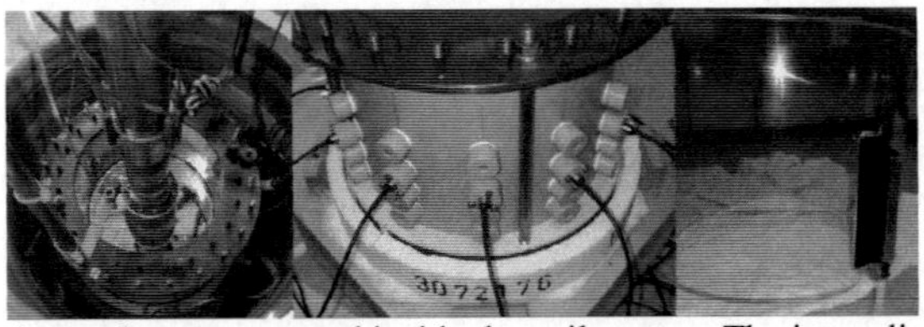

Fig. 1: Left: photo of the experimental setup mounted inside the coil system. The inner diameter of the cylindrical melt volume is 178 mm whereas that of the cold finger is 70 mm. Middle: instrumentation with single transducers which are coupled through holes directly to the melt. Right: UDV array shown at the right side measures through the wall of 2.3 mm thickness.

Flow field measurements were performed by the UDV technique. Its operation principle is described in the pioneering work by Takeda [8]. Mainly two features approve the application of UDV indispensable for the present work. Firstly, it works in opaque media including liquid metals. Secondly, it allows the quasi-simultaneous measurement of an entire profile of the local velocity component in direction of the sound propagation along the ultrasonic beam. Two different sensor arrangements were used. Most of the measurements were taken by a DOP2000 velocimeter (Signal-Processing, Lausanne, Switzerland) with single ultrasound transducers. As it can be seen on the middle part of Fig. 1 six single sensors were arranged at different azimuthal positions, each one differing by 30° to the next one. This allows gathering flow information over the whole circumference of the vessel. Furthermore, the different mounting levels allow measurements for variable aspect ratios. On the right in Fig. 1 a second setup with an attached transducer array can be seen. The UDV array consists of 25 single transducer elements arranged linearly, each of a size of 2.3×5 mm^2 with an element pitch of 2.7 mm. Thus, it allows flow measurements of the velocity component perpendicular to the transducer surface over a field width of 67 mm. Moreover, multiple array arrangements would allow measurements of further velocity components and, therefore, enable complex flow mapping in the covered plane [9,10].

Results

For the present study flow velocity measurements were performed at mainly two different aspect ratios, $a = 0.45$ ($H =$ 80 mm) and $a = 0.3$ ($H = 54$ mm), and under the influence of different types of magnetic fields (RMF, TMF, cusp). Nevertheless, only a small cutout of some selected aspects can be presented in this paper.

In the range $1.4\times10^6 \leq Gr \leq 1.6\times10^8$ two different flow regimes were identified and investigated for $a = 0.3$. The vertical position of the sensors was thereby at $h = 40$ mm. For a low temperature gradient of $\Delta T = 1$ K ($Gr = 1.6\times 10^6$) an axisymmetric flow structure occurs. The fluid heated from the bottom rises along the rim, continues towards the center axis, drops in the central region below the cold finger and closes outward to the rim. Fig. 2 on the left illustrates such a measurement done with one single transducer at the randomly selected azimuthal position $\varphi = 0°$. Already for slightly higher $\Delta T = 3.1$ K ($Gr = 5.1\times10^6$) the axisymmetric flow topology becomes instable (cf. Fig. 2-right) and for further increased Gr ($Gr = 4.2\times10^7$, $\Delta T = 22$ K) the flow develops the wind structure which is a large scale single roll circulating inside the vessel. Such a structure is well-known in general RB systems and frequently observed for $a = 1.0$.

It should be mentioned here that the preferred main flow direction of the wind develops randomly and is influenced mainly from imperfections of the experimental setup. Fig. 3 shows two time series measurements in the wind-mode atthe positions $\varphi = 0°$ and $\varphi = 90°$. The latter one corresponds thereby to the main circulation loop of the flow.

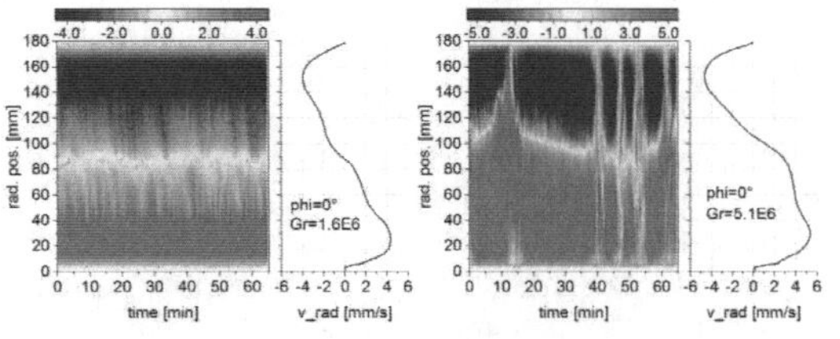

Fig. 2: Time series and the related mean profile of the radial velocity component in the axisymmetric case (left) and in the transitional range (right).

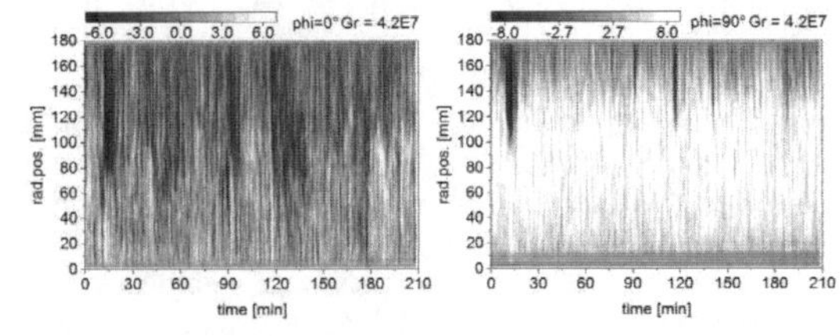

Fig. 3: Time series observed for long term measurements in the wind-mode.

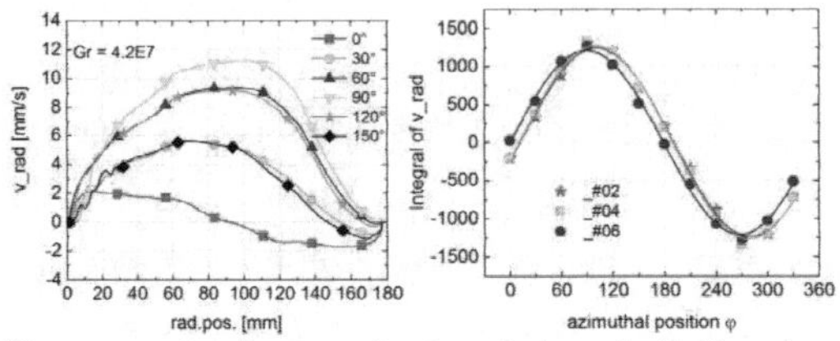

Fig. 4: Mean radial velocity profiles across the diameter for the wind-mode (left) and typical surface integrals of the mean profiles for different measurements (right).

The orientation of the wind becomes clearer from Fig. 4. The left diagram shows the mean radial velocity profiles at each of the azimuthal positions and the right side the surface integral of the respective profile. The maximum in the curve at approximately $\varphi = 90°$ indicates clearly the main circulation loop of the wind. Furthermore, the right part of Fig. 4 summarizes three different measurements performed at various days and demonstrates well the stability of the wind. It must be mentioned, however, that this was not the case for $a = 0.45$. Although not shown here, the wind was instable in the sense that the maximum of the integral curves varied in time over a large azimuthal range, even flow reversals were occasionally observed.

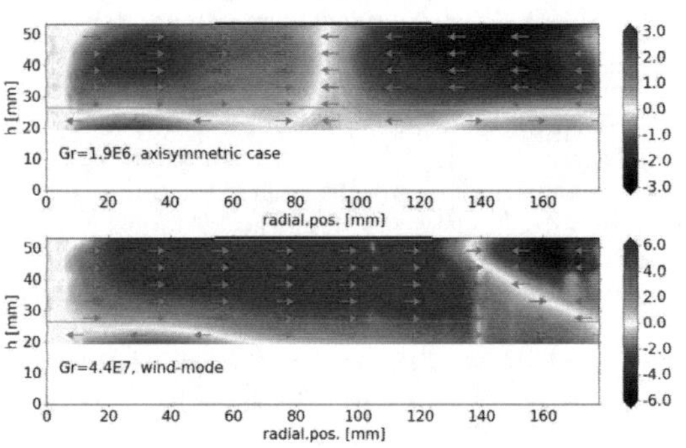

Fig. 5: Radial flow maps measured using the UDV array. Arrows indicate the radial flow direction and are not scaled. Due to design related reasons the lower 20 mm could not be measured.

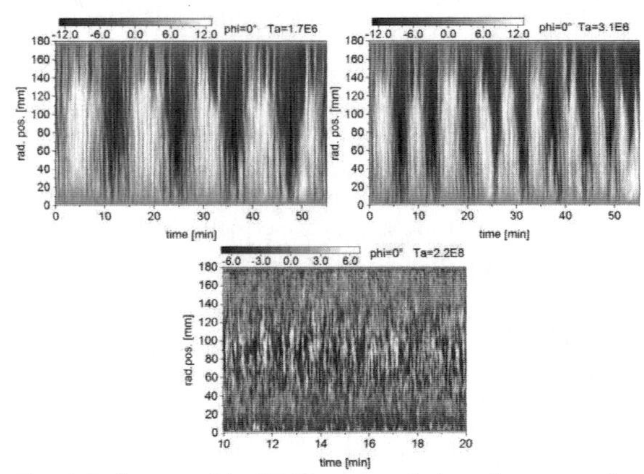

Fig. 6: Influence of the RMF on the wind-mode measured at $\varphi = 0°$ for two different, relatively low strengths of the RMF (upper part).The lower diagram shows the case for a strong RMF.

As previously mentioned, some selected measurements were done by means of the UDV array technique in a second setup. Due to the ability to measure quasi simultaneously along several sensor lines, a more informative radial flow map is achieved. Fig. 5 shows in the upper diagram such a purely buoyancy driven radial flow map in the axisymmetric low *Gr* case and for the fully developed wind-mode in the lower part. Furthermore, from the lower diagram it is apparent that the axial-radial measurement plane of the UDV array does not fully correspond with the main circulation loop of the flow. Application of an RMF to the wind-mode causes a co-rotation of the wind (cf. upper part of Fig. 6). For a sufficiently high value of *Ta* the wind completely disappears as it can be seen in the lower diagram of Fig. 6.

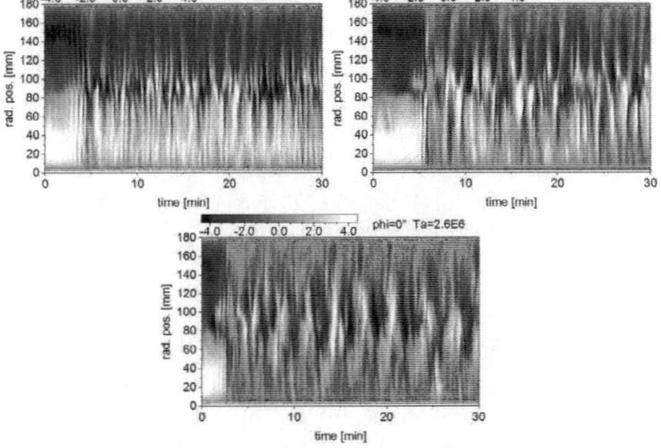

Fig. 7: Radial flow under the influence of different RMF in the axisymmetric case measured in the single transducer configuration at $h = 40$mm.

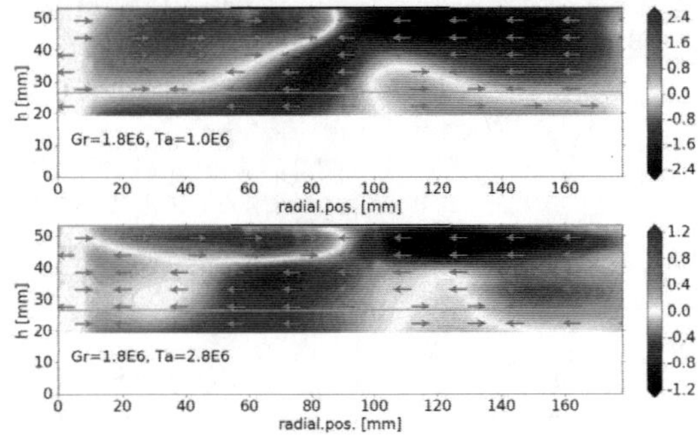

Fig. 8: Mean radial flow maps under the influence of two different RMF's.

A different influence of the RMF is observed by applying it to the axisymmetric case. The diagrams in Fig. 7 visualize the single sensor measurements for three different *Ta* numbers. At first glance, common to the diagrams are the first few minutes in which the RMF was not yet switched on. The axisymmetric structure is clearly observed here. When the RMF is switched on, some minor differences are observed between $Ta = 1.0\times10^6$ and $Ta = 1.4\times10^6$. For $Ta = 2.6\times10^6$ the axial symmetry disappears, rather a similarity to the wind structure is observed – at least along the single sensor measurement line ($h = 40$ mm). Of course it might be somewhat misleading to conclude from a single line measurement to the overall flow structure. In this regard, the major feature of the UDV array to measure simultaneously several lines becomes rather advantageous. The diagrams in Fig. 8 illustrate the mean radial flow maps for the same *Gr* and *Ta* numbers as in Fig. 7. By increasing the RMF strength rather a kind of squeezing occurs and the inward flow at the upper part of the vessel concentrates more and more to the region below the surface. Since the diagrams in Fig. 8 are the result of an averaging process over the measuring time of about 1 hour, the flow still has a time dependent and complex behaviour. Thus, care has to be taken by interpreting the overall flow topology.

Summary

A Cz-like crystal growth model exposed to different magnetic fields was the object of the present investigation. Ultrasound measurements of the fluid flow were performed by varying the strength of the RMF (*Ta*) and that of the buoyancy (*Gr*). Besides the single ultra-sound transducers, the UDV array might provide a deeper insight into the complex flow topology in a Cz setup. Furthermore, the experimental data here serve as a benchmark object for numerical codes which are still under development.

Acknowledgment

Financial support by the German Federal Ministry for Economic Affairs and Energy in the framework of the KORONA project is gratefully acknowledged.

References

1. K. Kakimoto et al., Progress in Crystal Growth and Characterization of Materials, 62 (2016), 273-285
2. F. Hébert et al., Phys. Rev. E, 81 (2001), 046318
3. J. Niemela et al., J. Fluid Mech., 449 (2001), 169–178
4. LP. Gorbatchev et al., Magnetohydrodynamics, 10 (1974), 406–414
5. P.A. Davidson, J Fluid Mech., 245 (1992), 669–699
6. J. Pal et al., Flow. Meas. Instrum., 20 (2009), 241-251
7. Y. Plevachuk et al., J. Chem. Eng. Data, 59(3) (2014), 757–763
8. Y. Takeda, Nucl. Eng. Des., 126 (1991), 277-284
9. S. Franke et al., Proceedings of ISUD7 Gothenburg/Sweden, (2010), 89-92
10. S. Franke et al., Ultrasonics, 53 (2013), 691-700

Effect of travelling magnetic field inductor characteristics on the liquid metal flow in a rectangular cell

E. Shvydkiy[1*], I. Kolesnichenko[2], R. Khalilov[2], A. Pavlinov[2] and G. Losev[2]

[1] Ural Federal University, Mira St. 19, Yekaterinburg, Russia
[2] ICMM UB RAS, 1 Akad. Korolev St., Perm, Russia

Corresponding author: e.l.shvydky@urfu.ru, kiv@icmm.ru, losev.g@icmm.ru

Abstract
Contactless liquid metal stirring by traveling magnetic field (TMF) inductor is widely applied in modern metallurgy. The quality of stirring is an actual challenge. One solution to this problem is to complicate the TMF inductor supplying current form. In present work we study the influence of the current on the hydrodynamic characteristics of the flow. The spatial distribution of the velocity field is investigated. The goal is to reduce stagnation zones in the flow and to achieve uniform stirring. Local turbulence characteristics, which has a significant effect on the stirring process, are also investigated. In this case, the purpose is to generate the flow with a sufficient intensity of the turbulence. The research is carried out by mathematical simulation and experiments. Three-dimensional calculations are performed in the Comsol Multyphysics software. The experiment is carried out on gallium eutectic. The total electromagnetic force dependencies on the supplying current characteristics are obtained taking into account the magnetic saturation. Three-dimensional flow calculations are verified on the experimental data. The flow characteristics on the TMF inductor supplying current value are determined.

Key words: electromagnetic stirring, travelling magnetic field, ultrasonic velocimeter, numerical analysis

Introduction
Electromagnetic processing of liquid metals has great potential. For example, at the stage of preparation of non-ferrous and ferrous metals, the use of magnetohydrodynamic (MHD) agitators significantly increases the energy efficiency of operations [1]. Electromagnetic stirring by rotating and travelling magnetic fields of crystallizing metals is one of the alternative ways to improve the quality of the material. In this connection, research aimed at increasing the efficiency of electromagnetic stirring by TMF can lead to a significant economic effect.
A number of both theoretical and experimental studies were devoted to electromagnetic mixing and presented the analytical dependencies of the fluid flow rate and mixing efficiency on the parameters of TMF [2 - 5]. However, many questions remain open, such as the effect of the transverse longitudinal end effect of a linear induction machine (LIM), as well as the depth of penetration of the magnetic field into the metal. Additional studies can give a more complete picture, define dependencies and give recommendations for electromagnetic stirring by a TMF. To carry out these studies, an experimental setup was built.

Experimental setup
The experimental setup consists of a vertical cell *1*, which is made from plexiglass. The cell is field with liquid alloy GaSnZn. The cell has dimensions 450×20×75 mm^3. The channel is placed on a linear inductor motor *2* of the TMF with dimensions of 480×350 mm^2. The TMF is created by the six coils (170 turns in one coil). The coils are powered by a three-phase programmable current source *3* Pasific Smart Source 360 ASX-UPC3. This power supply allows to specify the shape of the output signal and to modulate the TMF. The frequency of supplied current is 50 Hz. The measurements of the velocity of the flows in liquid metal phase are made by means of ultrasonic Doppler velocimeter *4* (UDV) DOP 2000, Signal Processing [6]. The four sensors *5* of the UDV are placed on the thin wall of the cell. Ultrasound Doppler velocimetry includes emitting ultrasound wave packages by an ultrasound Doppler sensor, their consequent adoption, and calculation of the Doppler frequency shift between the emitted and received wave packages. The ultrasonic silicon gel has to be used to provide sound contact between the probe and the wall. However, the presence of sound-reflecting particles moved by the flow in the medium is necessary. The gallium oxide particles reflect ultrasound waves in gallium eutectic. These particles are formed by oxidation of gallium in the atmosphere [7].
Measurements of the Lorenz force produced by the TMF inductor were carried out on a model installation. As a working medium, a copper plate measuring 589×15×288 mm^3 was used. The plate was suspended on inextensible threads, so that it could freely move over the inductor at a constant value of the air gap. Measurements of traction and lifting components of electromagnetic force were carried out using spring dynamometers. To measure the vertical component of the force, four dynamometers were placed between the pieces of hanging threads. A dynamometer measuring the planar component of the force was rigidly attached to the plate and stand frame. When the inductor is

turned on, the TMF generated force displaced the plate and reduced its weight. The accuracy of the measurement of the planar force component was 0.1 N, the lifting component was 1.0 N.

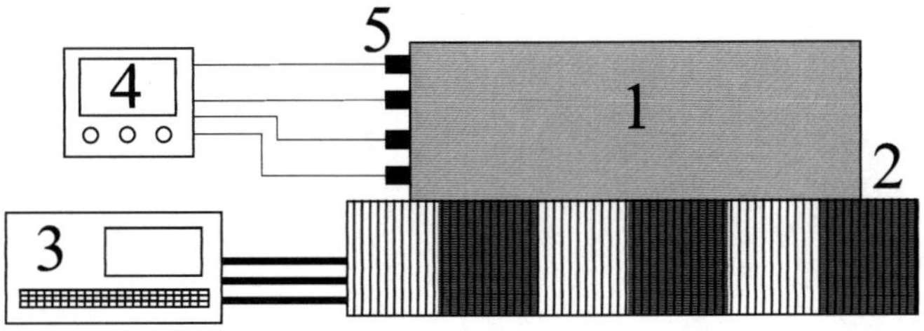

Fig. 1: Sketch of the experimental setup: 1 - channel filled with liquid metal, 2 - TMF inductor, 3 - power supply, 4 - UDV, 5 - ultrasonic sensors.

Mathematical model

3D Numerical calculations were performed by means of finite element method in Comsol Multiphysics softwere. At the first stage an electromagnetic calculation was made, based on the Maxwell equations. Considering the low magnetic Reynolds number, the induced current density was calculated Reynolds noninductive formulation [6, 8]:

$$\vec{J} = \sigma \vec{E} \tag{1}$$

where $\vec{J}$ - current density, σ - electric conductivity, $\vec{E}$ - electric field strength.

The electromagnetic force acting on a liquid metal was calculated as:

$$\vec{F}_{EM} = \vec{J} \times \vec{B} \tag{2}$$

where $\vec{B}$ - magnetic flux density.

The problem was solved in the frequency domain study for a frequency of 50 Hz. The mesh for electromagnetic part consist from 536784 tetrahedral elements. The melt lower surface was reduced due to the skin effect. Then the resulting distribution of the electromagnetic force was transferred to the hydrodynamic (HD) part as a source term of the Navier-Stokes equation:

$$\begin{aligned} &\rho\frac{\partial \vec{u}}{\partial t} + \rho(\vec{u}\cdot\nabla)\vec{u} = \nabla\cdot\left[-p\vec{I} + (\mu+\mu_T)(\nabla\vec{u} + (\nabla\vec{u})^T)\right] + \vec{F}_{EM} \\ &\rho\,\nabla\cdot\vec{u} = 0 \end{aligned} \tag{3}$$

where, ρ density; $\vec{u}$ – velocity; ∇ – nabla operator; p – pressure; μ – dynamic viscosity; $\vec{I}$ – unit matrix; μ_T – turbulent viscosity.

Calculations of the liquid metal flows were carried out using the k-ω SST RANS model. The mesh for HD calculation consists of 76554 Hexahedral uniform elements with element size 2,4 mm.

Results

The magnetic induction generated by the TMF inductor is the most important quantity. Figure 1 shows the magnetic induction curve. As can be seen, the curve is not uniform, this is due to the presence of a tooth-slot zone. The peak located in the middle of the curve can be explained by the asymmetric distribution of currents in the coils of the TMF inductor due to the longitudinal end effect. Figure 2 shows the dependence of the x and y integral Lorentz forces components on the coil current. The force is grow with increasing current as expected.

The mean velocity profile of liquid metal is shown in Fig. 3. The discrepancies between the curves obtained with the help of UDV sensors and numerical calculations lie within the permissible limits and can be explained by the assumption that the free surface is not taken into account.

The result of the numerical calculation of the velocity field under the TMF influence is shown in Fig. 5. The flow forms one main vortex, the velocity value in the lower part is greater due to the inductor proximity and the skin effect presence. It is worth noting that this flow pattern corresponds to the one described analytically in [2].

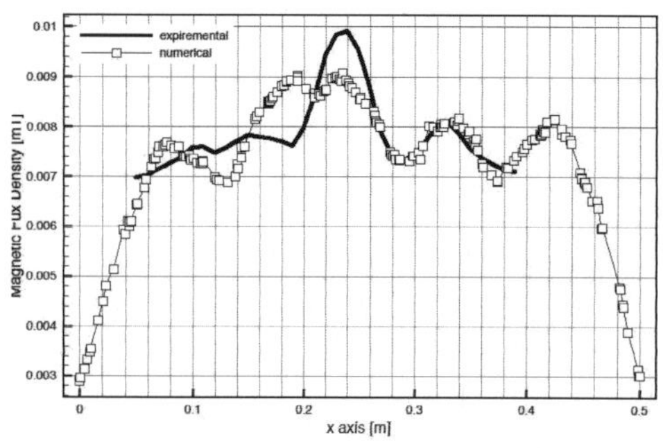

Fig. 2: Magnetic flux density on 20 mm height above TMF inductor surface

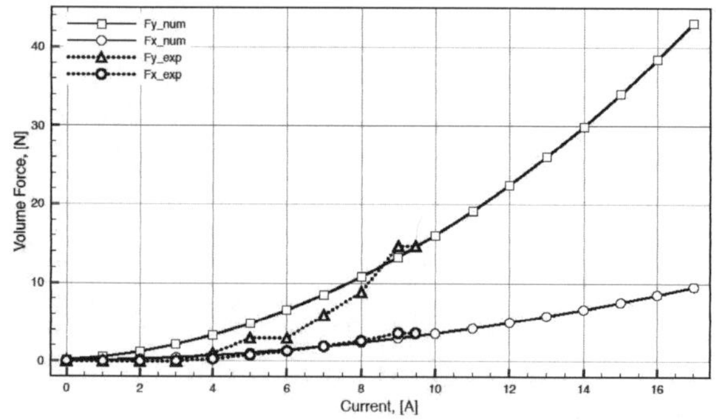

Fig. 3: Volume force verification

With an increase of magnetic flux (the supply current), the average flow velocity and hence the kinetic energy varies in different ways in different parts of the vessel (figure 6). In the first quarter, where the incoming flow creates a large-scale vortex, the average velocity is relatively weakly dependent on the magnitude of the field. In the vessel central part, the mean flow velocity dependence on the TMF is close to linear. At the vessel rear part, the average velocity reaches saturation at currents greater than 5A. Apparently, this is due to the scattering of the energy of the return flow by viscous forces.

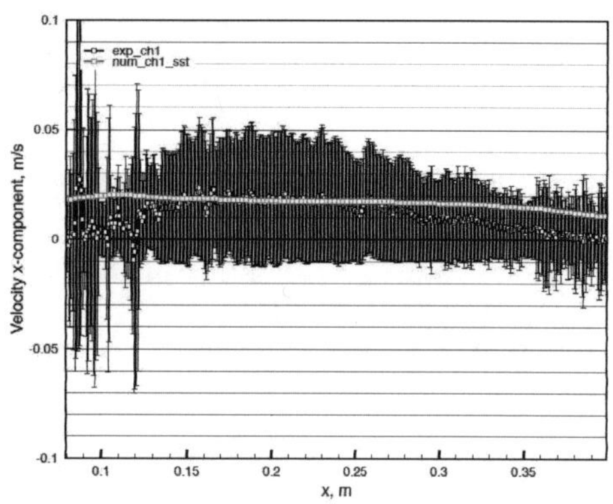

Fig. 4: Mean velocity x-component. UDV and numerical results comparison

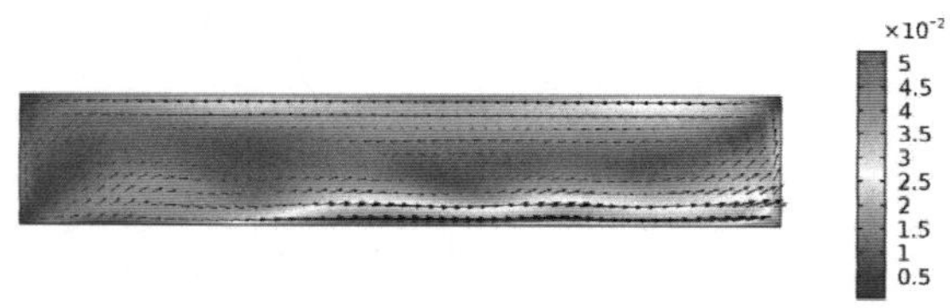

Fig. 5: Time average liquid metal flow pattern under TMF

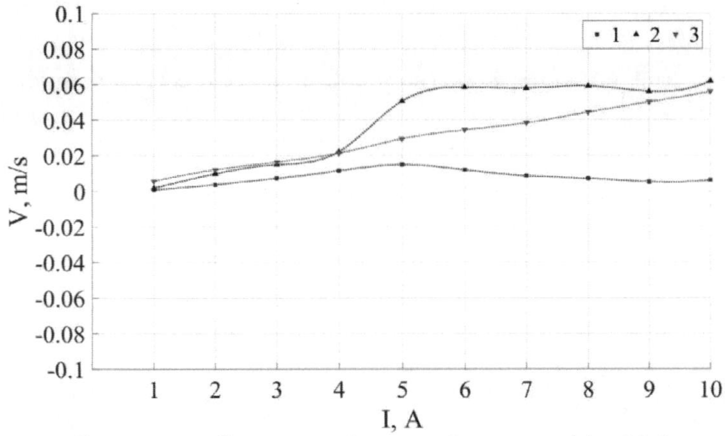

Fig. 6: Mean velocity on coil current. 1 – first quarter, 2 – second quarter and 1 – third quarter of liquid metal volume

Conclusion

An experimental setup for the study of electromagnetic mixing of TMF has been created. Experimental measurements of the magnetic field, Lorentz forces, and the flow rate of liquid metal have been carried out. A numerical model was developed and verified. Experimental and numerical experiments have shown that the flow is unsteady. Further work will be devoted to the study of flow parameters and turbulence properties under the action of a modulated magnetic field, as well as crystallization.

Acknowledgment

This work is supporting by the RFBR grant 17-48-590539_r_a.

References

1. R.B. Fdhila et al., ABB Review, 3 (2016), 45-48
2. X. Wang et al., Metall. Mater. Trans. B, 40 (2009), 104–113
3. L. Hachani et al., International Journal of Heat and Mass Transfer, 85 (2015),438–454
4. A. Kappen et al., Magnetohydrodynamics, 53 (2017), 583–593
5. M.H. Avnaim et al., International Journal of Heat and Fluid Flow, 69 (2018), 23–32
6. I. Kolesnichenko et al., Magnetohydrodynamics, 48 (2012) 221-233
7. M. Nowak, Exp. In fluids. 31 (2002), 249–255.
8. S. Denisov et al., Magnetohydrodynamics, 50 (2014) 249-265

Experimental study of liquid metal flows under volute traveling magnetic fields

K. Ueno[1], T.Kamada[2]

[1] Iwate University, 4-3-5 Ueda, Morioka 020-8551, Japan
[2] Tohoku University, 6-6-01 Aobayama Aoba-ku, Sendai 980-8579, Japan

Corresponding author: uenok@iwate-u.ac.jp

Abstract
Two kinds of volute traveling magnetic fields were simultaneously imposed on free surface of molten gallium. These fields drove an outward radial flow over 30 cm/s with small circumferential velocity component near the free surface of the molten gallium. Diverging radial flow causes upward component of the velocity near the surface. It was induced in wide area $r < 70$ mm in the vessel of 180 mm diameter. The upward component of the velocity was estimated to be 2–3 cm/s. Gallium in the vessel circulates once per 6 seconds. This surface stirrer with soft-magnetic ferrite cores performed continuous operation over 30 minutes that is much longer than the operation time of the coreless surface stirrer. This stirrer is expected to enhance interfacial reactions at slag/metal interface when it is applied to a ladle in steel making.

Key words : Electromagnetic stirrer, Traveling magnetic field, Slag/metal interface

Introduction

Reactions at slag/metal interface in ladles are remarkably influenced by stirring of melt. One of the most effective stirrers is ASEA-SKF ladle furnace [1], [2]. Its disadvantage is expensive cost of the ladle made of stainless steel that enables penetration of the traveling magnetic field to the melt. In order to avoid this difficulty, Suzuki *et al.* [3], [4] proposed an electromagnetic stirrer that drives slag/metal interface in outward direction as shown in Fig. 1. This outward flow produces new-growth interface and induces upward flow that supplies unreacted melt to the interface.

Three-phase coreless co-axial coil in Ref. [3] generates remarkable heat. Operation time of the coil is limited to 2–3 minutes under an oil cooling condition. Another disadvantage of the coreless co-axial coil is short distance between the coil and the surface of molten metal. It depends on the wavelength of the magnetic field, namely difference between the outer radius and the inner radius of the three-phase coil. In the experimental study in Ref. [3], the wavelength of the magnetic field is 28% of the bath diameter and the air gap between the coil and melt surface is 3% of the bath diameter. It is necessary to elongate the air gap when this stirrer is applied to slag/metal interface.

An electromagnetic stirrer driving molten metal surface in outward direction is developed in this study. Soft-magnetic ferrite cores are used in this system. This surface stirrer aims continuous operation by reducing heat production by coil current. This stirrer generates two kinds of volute traveling magnetic fields. The wavelength of the volute magnetic field is longer than that of the radially traveling magnetic field by coaxial coilds. Longer wavelength elongates effective distance of the magnetic field from the coils. One of the volute magnetic fields rotates in clockwise direction and the other rotates in counterclockwise direction. Superimposition of these two fields generates outward flow with small circumferential component of the velocity. This outward flow induces upward flow near the surface. This upward flow is expected to enhances reactions at slag/metal interface when the stirrer is applied to a ladle in steel making.

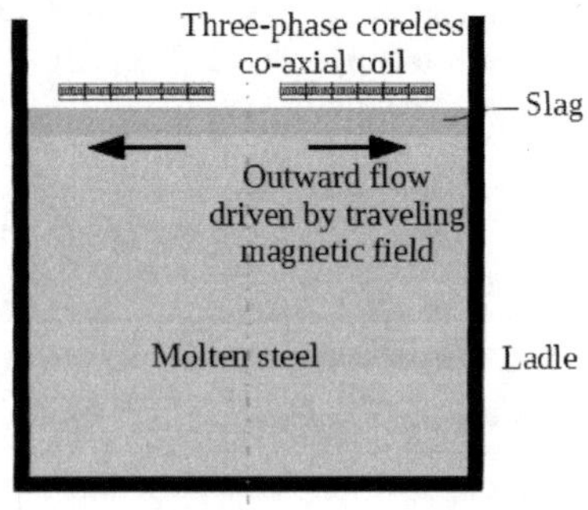

Fig. 1: Electromagnetic stirrer driving slag/metal interface in outward direction.

Experimental apparatus

Figure 2 shows coils, cores and yoke of soft-magnetic ferrite. Coils 1 consist of 6 sets of 4 coils in series connection shown by red lines in Fig. 3, while Coils 2 consist of 6 sets of 4 coils in series connection shown by blue lines in Fig. 3. The coil sets of Coils 1 are connected to a three-phase inverter system by a star connection, and the coil sets of Coils 2 are connected to another three-phase inverter system by a star connection. Magenta arrows in Fig. 3 show the traveling direction of the magnetic field by Coils 1, while cyan arrows show the traveling direction of the magnetic field by Coils 2. Two inverter systems are independent each other. Coils 1 and Coils 2 produce two-pole volute magnetic fields that rotate in opposite direction each other. Both of them are simultaneously outward traveling magnetic fields.

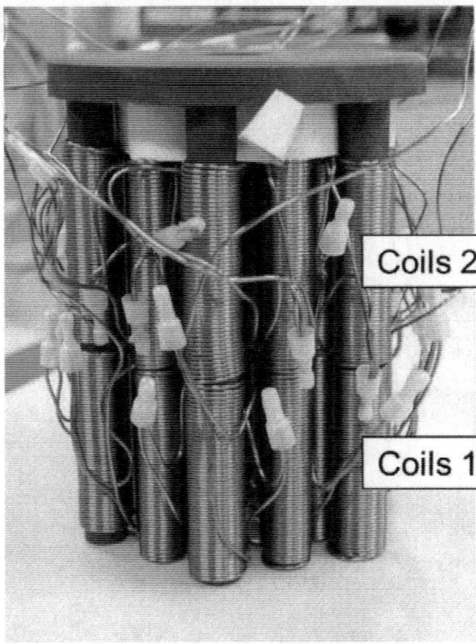

Fig. 2: Side view of the stirrer.

Fig. 3: Bottom view of the stirrer. Red and blue lines show volute connections of Coils 1 and Coils 2, respectively. Magenta and cyan arrows show traveling direction of magnetic fields by Coils 1 and Coils2, respectively.

Root-mean-square value of the vertical component B_z of magnetic flux density is measured as shown in Fig. 4 and 5. Measurable AC magnetic field by our instrument is limited to sinusoidal signal. Hence a three-phase variable transformer connected to 50 Hz power supply was used in this measurement, though two PAM inverter systems are used in stirring experiments of molten gallium as stated below.

Figure 4 shows distribution of root-mean-square value of B_z by Coils 1 below 9 mm distance from the end of core z_{core} when $I = 20$ A. Considerable nonuniformity is caused by cores. Figure 5 shows relation between the distance $z_{core} - z$ and root-mean-square value of B_z at $r = 60$ mm when $I = 20$ A. The magnetic flux density is inversely proportional to the distance in this range. Decreasing rate with z is moderate compared with the coreless co-axial coil in Ref. [3]. Value of B_z by Coils 2 is about a half of that by Coils 1, because flux leakage of Cols 2 is remarkable.

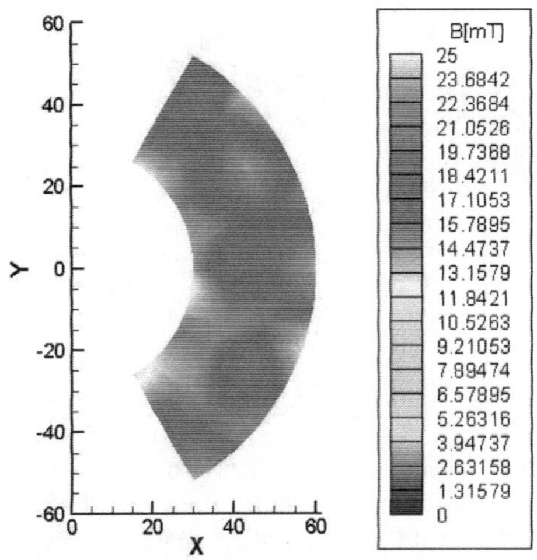

Fig. 4: Distribution of root-mean-square value of the vertical component B_z of magnetic flux density by Coil 1 below 9mm distance from the core end when $I = 20$ A.

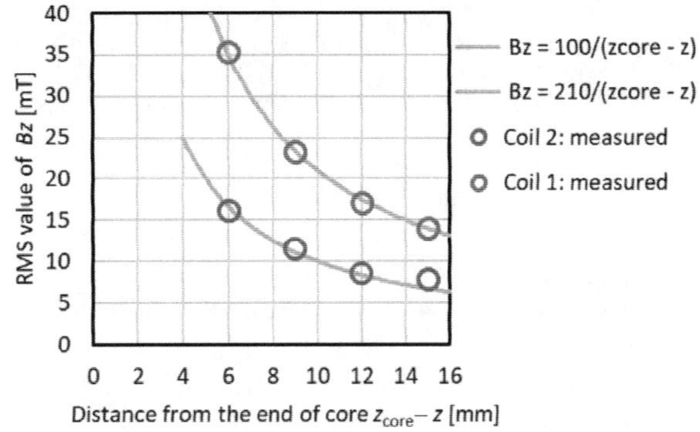

Fig. 5: Distribution of root-mean-square value of the vertical component B_z of magnetic flux density at $r = 60$ mm when $I = 20$ A,

Fig. 6: Experimental apparatus for stirring of molten gallium.

Table 1: Conditions of inverter-coil system.

Coils 1	540 Hz	43 V	10.0 A	counterclockwise
Coils 2	460 Hz	130 V	11.8 A	clockwise

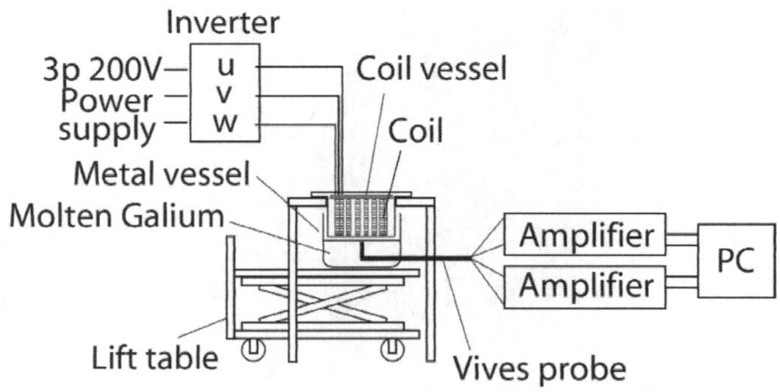

Fig. 7: Schematic of experimental system.

Figure 6 shows the experimental apparatus for stirring of molten gallium. Inner diameter of the cylindrical vessel is 180 mm and molten gallium depth is 75 mm. Distance between the end of cores and gallium surface is 6 mm. The velocity of molten gallium was measured at 5 mm below the surface. Conditions of inverter-coil system are shown in Table 1. The frequency and the voltage values in Table 1 are set points, while the currents are typical values measured in experiments. Figure 7 shows schematic of experimental system.

Radial component v_r and circumferential component v_θ of the velocity were simultaneously measured by Vives probe shown in Fig.8. This probe consists of one permanent magnet and four small electrodes as shown in Fig.9. Low pass filters in amplifiers for Vives probe were used. The cut off frequency of them is 10 Hz. Furthermore, measurement data of velocity were obtained in intervals in the following operation sequence: (35 s stirring) − (3 s interval) − (27 s stirring) − (3 s interval) − (27 s stirring) − (3 s interval). During stirring periods, velocity measurement is impossible because of noise from three-phase inverter systems.

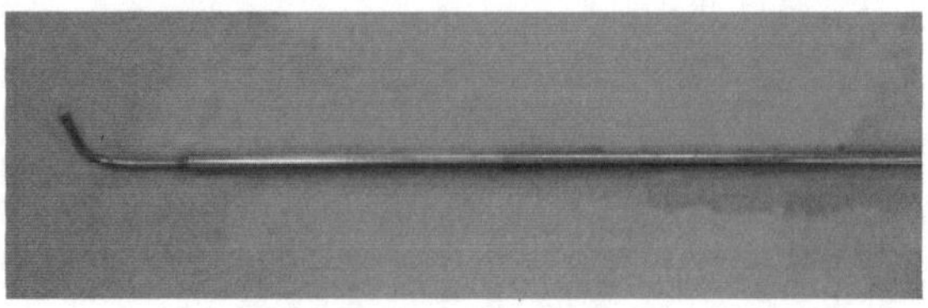

Fig. 8: Vives probe.

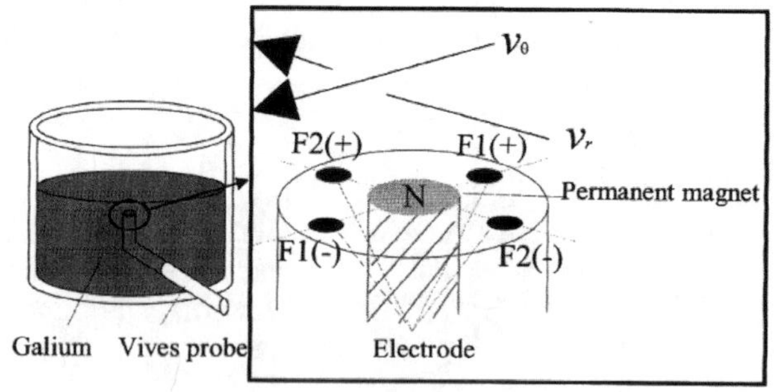

Fig. 9: Velocity components measured by Vives probe.

Experimental resutls

Figure 10 shows measurement results of the horizontal component v_x of the velocity along x axis. The radial component v_r is equal to v_x where $x > 0$, and v_r is equal to $-v_x$ where $x < 0$. The maxmum value of v_r was greater than 30 cm/s.

The averaged value of circumferential component v_θ was -0.1 cm/s and the standard deviation of v_θ was 1.5 cm/s. Rotational forces by Coils 1 and Coils 2 are almost canceled out each other in the conditions of Table 1. Consequently, hollow of free surface did not observed. This property is necessary for actual stirring of molten metal with slag layer in a ladle.

Coil current increased with temperature increasing of the coil system when voltage of inverter was fixed. Experimental data shows that there was no correlation between the coil current and the velocity of molten gallium. These facts suggest that the volute traveling magnetic field of this study is proportional not to the coil current but to the voltage of the coil. The coil current increasing is presumed to be an automatic compensation for decreasing of magnetic permeability of the ferrite cores with temperature increasing.

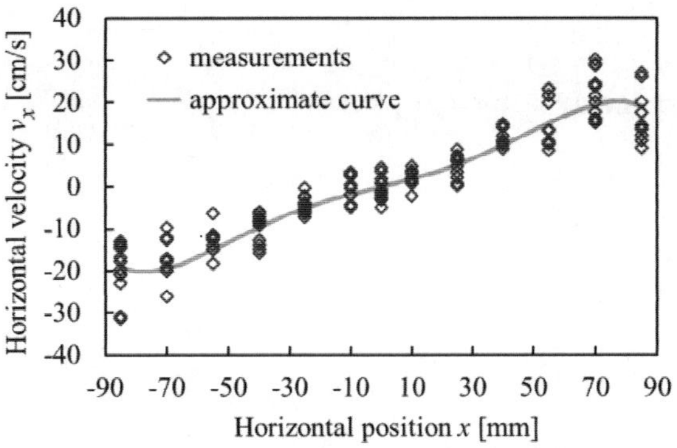

Fig. 10: Horizontal component v_x of the velocity along x axis . The radial component v_r is equal to v_x where $x > 0$, and v_r is equal to $-v_x$ where $x < 0$.

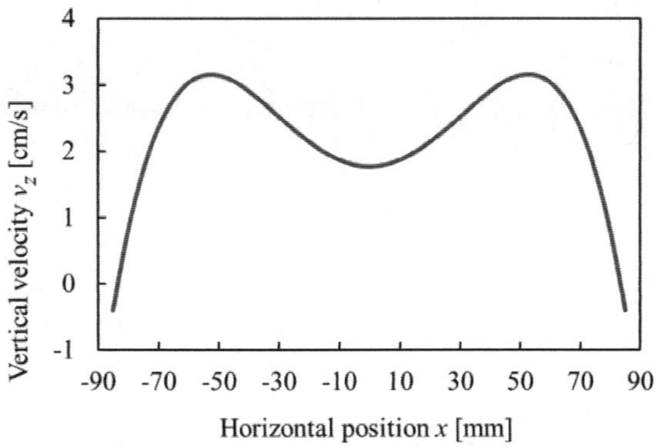

Fig. 11: Distribution of v_z that is derived from Eq.(2) and the approximated curve of v_r in Fig. 10.

The surface stirrer developed in this study performed continuous operation over 30 minutes that is much longer than the operating time of the coreless surface stirrer[3].

Let us estimate vertical component v_z. Continuity equation for axisymmetric flow is given as follows:

$$\frac{1}{r}\frac{\partial r v_r}{\partial r} + \frac{\partial v_z}{\partial z} = 0. \tag{1}$$

Integrating this equation from the position z of the Vives probe to the free surface z_{face}, we obtain

$$v_z(r,z) = v_z(r, z_{\text{face}}) + \int_z^{z_{\text{face}}} \frac{1}{r}\frac{\partial r v_r}{\partial r} \approx (z_{\text{face}} - z) \times \left(\frac{1}{r}\frac{\partial r v_r}{\partial r}\right)_{z_{\text{face}}} \tag{2}$$

where we assume that v_r does not depend on z in this range. Figure 11 shows distribution of v_z that is derived from Eq.(2) and the approximated curve of v_r in Fig. 10. The maximum value of the estimation of v_z is more than 3 m/s. The ratio of the vessel volume to the upward flowrate through the area of $r < 70$ mm gives the characteristic period of circulation. It is estimated to be 6 s.

Conclusion

An electromagnetic stirrer driving molten metal surface in outward direction is developed. Soft-magnetic ferrite cores are used in this system. This stirrer generates two kinds of volute traveling magnetic fields. One rotates in clockwise direction at 460 Hz and the other rotates in counterclockwise direction at 540 Hz. Superimposition of these two fields generated outward flow over 30 cm/s with small circumferential velocity component near the surface in molten gallium. Upward component of the velocity near the surface was induced in wide area $r < 70$ mm in the vessel of 180 mm diameter. The upward component of the velocity was estimated to be 2–3 cm/s. Gallium in the vessel circulates once per 6 seconds. This surface stirrer performed continuous operation over 30 minutes that is much longer than the operation time of the coreless surface stirrer. The electromagnetic stirrer in the present study is expected to enhance interfacial reactions at slag/metal interface when it is applied to a ladle in steel making.

Acknowledgment

The authors are grateful to Takeshi Suzuki, Tsutomu Ando, Jinya Sato, Shogo Arai, Masahide Watabe and Hiromune Morikawa for collaboration on the early stages of this work. This work was supported by JSPS KAKENHI Grant Number JP14702053.

References

1. J. Szekely and K. Nakanishi, Metallurgical Transactions B, 6 (1975), 245-256
2. S. Chung, Y. Shin and J. Yoon, ISIJ International, 32 (1992), 1287-1296
3. T. Suzuki, K. Ueno and S. Taniguchi, Proc. Int. Symp. EPM2006, (2006), 200-205
4. K. Ueno, T. Suzuki and S. Taniguchi, Proc. Int. Symp. EPM2006, (2006), 311-315

IOP Conf. Series: Materials Science and Engineering **424** (2018) 012014 doi:10.1088/1757-899X/424/1/012014

Convective end effects in annular linear induction pumps

Yves Delannoy[1], Elena Martin-Lopez[1,2], Fabrice BENOIT[2]

[1] Univ. Grenoble Alpes, CNRS, Grenoble INP, SIMaP, F-38000 Grenoble, France
[2] CEA Cadarache /DTN/STCP/LCIT, 13108 St Paul lès Durance, France

Corresponding author : yves.delannoy@simap.grenoble-inp.fr

Abstract
This paper presents a simplified theory giving the performance of an annular linear electromagnetic pump, such as the liquid sodium pumps foreseen in some GenIV nuclear reactors. End effects are taken into account with a new model enabling physical interpretation.

Introduction
An annular linear induction pump (ALIP) is an electromagnetic device producing a traveling magnetic field in an annular section, where an electrically conductive fluid is flowing against an adverse pressure gradient (produced by a dissipative hydraulic circuit). Its linear inductor, placed around the annular sodium channel, produces a radial magnetic field B_r across the sodium up to the central (passive) magnetic core. The three phase arrangement of coils produces roughly a progressive sine wave for B_r, traveling at velocity U_B along the axis, over the inductor length L.
For such large size pumps (axial length ~ meters, axial velocity ~ meters per second), the magnetic field can be significantly modified by the axial velocity U_z (high magnetic Reynolds number), especially in the end regions ($x=\pm L/2$). The performance curve of the pump is affected (stall phenomena at high slip velocity), with an unstable zone [1] where partial stall may occur. Such instabilities have been observed on a Japanese pump tested for similar applications [2].
The goal of this paper is to investigate the end effects with an analytical model simpler than [3], based on existing theories for linear motors [4]. The model results will be compared to performance curves calculated for the PEMDYN pump (Figure 1 a), installed at CEA to provide experimental results in the frame of the ASTRID project (a future fast-breeder GENIV sodium reactor). PEMDYN has a sodium flowrate up to 1500m^3/h, and a pressure up to 2.5bar.

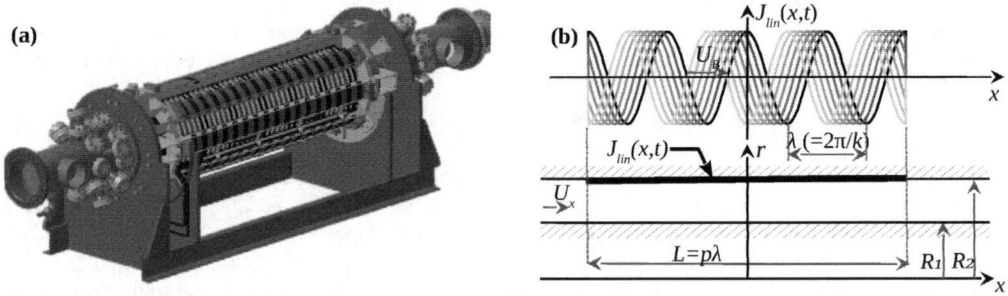

Figure 1: The PEMDYN [1] experimental pump (a), and its idealized model (b)

Electromagnetic model
The geometry considered is rotationally symmetric and axially infinite (Figure 1 b), except the inductor, simplified to a thin sheet of current (heavy line at $r=R_2$) where the linear current density follows a progressive sine wave with wavelength λ. The length of the inductor is $L=p\lambda$, where p is the "number of pole pairs", and the wavenumber is defined as $k=2\pi/\lambda$. The yokes (dashed zones in Figure 1 b) are supposed magnetically perfect ($\mu_r\sim\infty$) and in contact with the liquid sodium or current sheet at $r=R_1$ and R_2. The induction equation (1) is written using the azimuthal component A_θ of the magnetic vector potential, and the axial component U_z of the velocity. Other components of both vector fields are zero because of the rotational symmetry and the developed flow conditions along z.

$$\frac{\partial^2 A_\theta}{\partial x^2}+\frac{\partial^2 A_\theta}{\partial r^2}+\frac{1}{r}\frac{\partial A_\theta}{\partial r}-\frac{A_\theta}{r^2}=\mu_0\sigma\left(\frac{\partial A_\theta}{\partial t}+U_x\frac{\partial A_\theta}{\partial x}\right) \quad \text{with} \quad B_x=\frac{\partial A_\theta}{\partial r}+\frac{A_\theta}{r} \quad \text{and} \quad B_r=-\frac{\partial A_\theta}{\partial x} \qquad (1)$$

The boundary conditions at the magnetic yokes impose $B_z=0$ at $r=R_1$ and $B_z=\mu_0\mathrm{Re}\{J_{lin}(x)e^{-i\omega t}\}$ at $r=R_2$ (where $J_{lin}(x)=J_{peak}e^{ikx}$ for $-p\lambda/2<x<p\lambda/2$, and $J_{lin}(x)=0$ elsewhere). Looking for an harmonic solution $A_\theta(x,r,t)=\mathrm{Re}\{A_\theta(x,r)e^{-i\omega t}\}$, equation (1) and its boundary conditions can be converted to a two dimensional boundary value problem for the

complex field $A_\theta(r,t)$, written below using the dimensionless variables $\alpha=A_\theta/(\mu_0 J_{peak}/k)$, $\eta=(r-R_1)/(R_2-R_1)$ and $\xi=kx$. Introducing the velocity $U_B=\omega/k$ of the travelling field, the dimensionless flowrate $q=U_x/U_B$ to represent the fluid velocity, which is supposed independent of r in this paper, the problem can be written:

$$\begin{cases} \dfrac{\partial^2\bar{\alpha}}{\partial\eta^2}+\left(\dfrac{H_R}{1+H_R\eta}\right)\dfrac{\partial\bar{\alpha}}{\partial\eta}+\left(\dfrac{H_R}{1+H_R\eta}\right)^2\bar{\alpha}=H_\lambda^2\left[-\dfrac{\partial^2\bar{\alpha}}{\partial\xi^2}-R_{mB}\left(i\bar{\alpha}-q\dfrac{\partial\bar{\alpha}}{\partial\xi}\right)\right] \\ \dfrac{\partial\bar{\alpha}}{\partial\eta}+\left(\dfrac{H_R}{1+H_R\eta}\right)\bar{\alpha}=\begin{cases}0 & \text{for } \eta=0\\ H_\lambda\bar{\varphi}_{lin} & \text{for } \eta=1\end{cases} \quad \text{with} \quad \bar{\varphi}_{lin}(\xi)=\begin{cases}e^{i\xi} & \text{for } |\xi|<\pi p\\ 0 & \text{for } |\xi|>\pi p\end{cases} \end{cases} \tag{2}$$

where the geometrical parameters $H_R=(R_2-R_1)/R_1$, $H_\lambda=k(R_2-R_1)$ and p measure respectively the influence of axisymmetric terms ($H_R\to0$ when the domain is thin compared to its radius), the thickness of the magnetic gap compared to the reduced wavelength $\lambda/2\pi$, and the length of the inductor (in wavelengthes). The working point of the pump is represented by the magnetic Reynolds number $R_{mB}=\mu_0\sigma U_B/k$ (for a given pump and fluid, it depends on the generator pulsation ω) and by q, the fluid flowrate divided by the flowrate at synchronism.
Following [3] and [4], the problem can be solved using Fourier transforms along ξ, i.e. superposing solutions of the form $\hat{\alpha}_\kappa e^{i\kappa\xi}$ where κ is a continuously varying real wavelength and the complex amplitude of this mode. However, we will use here only three modes, one at $\kappa=1$ to represent the solution that we would have for an infinite pump, and two others with complex wavenumbers κ_u and κ_d representing the end effects generated by the inlet and outlet of the inductor and diffused upstream ($\mathrm{Im}(\kappa_u)<0$ so that this mode vanishes upstream) or convected downstream ($\mathrm{Im}(\kappa_d)>0$ so that this convected mode disapears downstream by joule dissipation). Those end effect modes are requested to verify (2) with zero source term $\bar{\varphi}_{lin}$, so they can be added to the infinite pump solution without destroying it far from the ends.
Therefore we look for solutions under the form $\bar{\alpha}=\hat{\alpha}_{iu}e^{i\kappa_u(\xi+\pi p)}$ for $\xi<-\pi p$; $\bar{\alpha}=\hat{\alpha}_{od}e^{i\kappa_d(\xi-\pi p)}$ for $\xi>\pi p$, and $\bar{\alpha}=\hat{\alpha}_{id}e^{i\kappa_d(\xi+\pi p)}+\hat{\alpha}_1 e^{i\xi}+\hat{\alpha}_{ou}e^{i\kappa_u(\xi-\pi p)}$ for $-\pi p<\xi<\pi p$ (subscripts i,o for inlet, outlet). Requesting the induction problem (2) to be fulfilled in every subdomain, we get an ordinary differential problem with complex unknown $\hat{\alpha}_\kappa(\eta)$ for each mode of wavelength κ, and we will connect the domains requesting that $\bar{\alpha}$ and $\partial\bar{\alpha}/\partial\xi$ are continuous at $\xi=\pm\pi p$.

Analytical solution
For a mode with complex wavenumber κ and amplitude $\hat{\alpha}_\kappa(\eta)$, (2) becomes:

$$\frac{d^2\hat{\alpha}_\kappa}{d\eta^2}+\left(\frac{H_R}{1+H_R\eta}\right)\frac{d\hat{\alpha}_\kappa}{d\eta}-\left(\frac{H_R}{1+H_R\eta}\right)^2\hat{\alpha}_\kappa=H_\lambda^2\left[\kappa^2-iR_{mB}(1-\kappa q)\right]\hat{\alpha}_\kappa \tag{3}$$

with

$$\frac{d\hat{\alpha}_\kappa}{d\eta}+\left(\frac{H_R}{1+H_R\eta}\right)\hat{\alpha}_\kappa=\begin{cases}0 & \text{at } \eta=0\\ 0 & \text{at } \eta=1 \text{ for } \kappa=\kappa_u,\kappa_d\\ H_\lambda & \text{at } \eta=1 \text{ for } \kappa=1\end{cases} \tag{4}$$

For $\kappa=\kappa_u$ or $\kappa=\kappa_d$, having a non zero solution (singular value problem) requires that $\kappa^2-iR_{mB}(1-\kappa q)=0$ (two solutions). Since κ_u has negative imaginary part, and writing $\sqrt{(z)}$ for the complex square root with positive real part, we get:

$$\kappa_u=-i\frac{qR_{mB}}{2}\left(1+\sqrt{1-i\frac{4}{q^2R_{mB}}}\right) \quad \text{and} \quad \kappa_d=-i\frac{qR_{mB}}{2}\left(1-\sqrt{1-i\frac{4}{q^2R_{mB}}}\right) \tag{5}$$

For large values of q^2R_{mB}, the upstream mode is damped over a short axial distance $1/(qR_{mB})$ without oscillations (counterflow diffusion), whereas the downstream mode is a weakly damped traveling wave propagating at the fluid velocity q (convection by the flow of current loops that cannot change rapidely). Such modes were yet presented in [6], but not used inside the inductor zone. A profile for the phasor of those modes can be obtained from (3),(4) without right hand side: it becomes uniform for quasi flat pumps, i.e. for a small channel height compared to its radius ($H_R\to0$).
The infinite pump solution ($\kappa=1$) of (3) with boundary conditions (4) is written analytically from the modified Bessel functions of first kind (I_0, I_1) and second kind (K_0, K_1) divided by their exponential trend at infinity: $I_{0e}(z)=I_0(z)/e^{\mathrm{Re}(z)}$, $I_{1e}(z)=I_1(z)/e^{\mathrm{Re}(z)}$, $K_{0e}(z)=K_0(z)/e^{-\mathrm{Re}(z)}$, and $K_{1e}(z)=K_1(z)/e^{-\mathrm{Re}(z)}$, using the slip magnetic Reynolds number $R_{mS}=(1-q)R_{mB}$:

$$\hat{\alpha}_1=\frac{H_\lambda}{c_1}\frac{e^{\mathrm{Re}(c_1)\eta}K_{0e}(\rho_1)I_{1e}\left[\left(\frac{1}{H_R}+\eta\right)c_1\right]+e^{-\mathrm{Re}(c_1)\eta}I_{0e}(\rho_1)K_{1e}\left[\left(\frac{1}{H_R}+\eta\right)c_1\right]}{e^{\mathrm{Re}(c_1)}I_{0e}(\rho_2)K_{0e}(\rho_1)-e^{-\mathrm{Re}(c_1)}I_{0e}(\rho_1)K_{0e}(\rho_2)} \quad \text{with} \quad \begin{cases}\rho_1=\left(\frac{1}{H_R}\right)c_1 \ ; \ \rho_2=\left(\frac{1}{H_R}+1\right)c_1\\ c_1=H_\lambda\sqrt{1-iR_{mS}}\end{cases} \tag{6}$$

Under that form, the solution can be developed for $H_R\to0$, (quasi-flat pumps) to retrieve the classical solution of (3),(4) for $H_R=0$. Furthermore, if $H_\lambda\ll1$, the complex coefficient c_1 has a small module and this quasi-flat solution may be further developed to give the thin channel solution ($H_R\ll1$, $H_\lambda\ll1$ means that the channel height is small compared to $R_1\approx R_2$ and compared to λ):

$$\hat{\alpha}_1 = \frac{H_\lambda}{c_1}\frac{\cosh(c_1\eta)}{\sinh(c_1)} \text{ for } H_R \ll 1 \text{ and } \hat{\alpha}_1 = \frac{1}{H_\lambda(1-\mathrm{i}R_{mS})} + \frac{H_\lambda\eta^2}{2} \text{ for } H_R \ll 1, H_\lambda \ll 1 \quad (7)$$

Notice that for thin channels, the distribution of $\hat{\alpha}_1$ is approximately constant along η, but its variation with η is useful to calculate B_x and then verify the boundary conditions. The continuity of $\bar{\alpha}$ and $\partial\bar{\alpha}/\partial\xi$ cannot be exactly satisfied in the general case for any value of η, because the three modes do not have the same profile along η. Writing $\bar{\alpha}(\xi,\eta)=\hat{\alpha}_1(\eta)\bar{f}(\xi)$ is thus an approximation, but it becomes precise for thin channels ($H_R \ll 1$ and $H_\lambda \ll 1$), where all profiles become uniform. We will use such a formula in all cases because real nuclear pumps have indeed $H_R \ll 1$ and their H_λ is small enough to have reasonably uniform profiles of $\hat{\alpha}$. Writing the continuity conditions without influence of the outlet on inlet ($|\exp(2\mathrm{i}\pi K_u p)| \ll 1$), we get:

$$\bar{f}(\xi) = \begin{cases} a_{iu}e^{\mathrm{i}\kappa_u(\xi+\pi p)} & \text{for } \xi < -\pi p \\ a_{id}e^{\mathrm{i}\kappa_d(\xi+\pi p)} + e^{\mathrm{i}\xi} + a_{ou}e^{\mathrm{i}\kappa_u(\xi-\pi p)} & \text{for } -\pi p < \xi < \pi p \\ a_{od}e^{\mathrm{i}\kappa_d(\xi-\pi p)} & \text{for } \xi > \pi p \end{cases}, \quad \begin{aligned} & a_{iu} = (-1)^p\frac{1-\kappa_d}{\kappa_u-\kappa_d} \;;\; a_{ou} = -a_{id} \\ & a_{id} = (-1)^p\frac{1-\kappa_u}{\kappa_u-\kappa_d} \;;\; a_{od} = -a_{id}\left(1-e^{2\mathrm{i}\pi p\kappa_d}\right) \end{aligned} \quad (8)$$

From the radial distribution (6) or (7) combined to the axial distribution (8), it is now possible to calculate the current density $\bar{j}_\theta = \sigma\left(\mathrm{i}\omega\bar{A}_\theta - U_x\partial\bar{A}_\theta/\partial x\right)$ and the radial magnetic field density $\bar{B}_r = -\partial\bar{A}_\theta/\partial x$, to get the volumetric force density $\boldsymbol{F}=\boldsymbol{j}\times\boldsymbol{B}$. The axial component of $\boldsymbol{F}$ has a mean part $F_{x\,avg}=-\mathrm{Re}(\bar{j}_\theta\bar{B}_r^*)/2$, (the star superscript stands for complex conjugate), and a part oscillating at the double of the supply frequency (DSF) $F_{x\,dsf}=-\mathrm{Re}(\bar{j}_\theta\bar{B}_r e^{-2\mathrm{i}\omega t})/2$, which has a phasor (with double bar to indicate DSF time fluctuation): $\bar{\bar{F}}_{x\,dsf} = -\bar{j}_\theta\bar{B}_r/2$. Using the dimensionless solution above, this can be written $F_{x\,avg}=F_{ref}f_{x\,avg}$ and $\bar{\bar{F}}_{x\,dsf} = F_{ref}\bar{\bar{f}}_{x\,dsf}$, with:

$$F_{ref} = \frac{k\mu_0 J_{peak}^2}{2}, \quad f_{xavg} = R_{mB}\Re\left[\left(\mathrm{i}\bar{\alpha} - u\frac{\partial\bar{\alpha}}{\partial\xi}\right)\frac{\partial\bar{\alpha}^*}{\partial\xi}\right], \quad \bar{\bar{f}}_{xdsf} = R_{mB}\left(\mathrm{i}\bar{\alpha} - u\frac{\partial\bar{\alpha}}{\partial\xi}\right)\frac{\partial\bar{\alpha}}{\partial\xi} \quad (9)$$

Those expressions can be constructed analytically from (6) or (7) and (8), to get the distribution of electromagnetic forces inside the pump. In the thin channel case, the force density will depend only on ξ and will be equilibrated by the pressure rise along the channel (and by friction forces which are typically one or two orders of magnitude lower). In other cases, the non uniform distribution of the forces across the channel is not compatible with a developed velocity profile (dU/dx=0), and it will have the velocity profile evolve towards an equilibrium profile described in [7] for infinite pumps. Since we use a uniform velocity profile, we only present results calculated with thin channel hypothesis.
Integrating the average force density $F_{x\,avg}$ over the channel length will give the pressure rise provided by the pump (neglecting friction), and integrating $F_{x\,dsf}$ will give the DSF fluctuation of this pressure rise. We use a numerical integration over the interval $-p-1<\xi/\pi<p+10$ (out of which the force density is negligible), with a step $\delta\xi=\pi/25$.

Results

The model was first applied to a pump with p=3 pairs of poles as PEMDYN, for R_{mB}=8, typical in that facility. Figure 2 shows a typical traveling wave shape at moderate slip ($1-q$=20%), with the different modes contributing to the real part of $f(\xi)$ (in black). The upstream diffusion of each end effect is hardly visible since it affects a very short length, whereas the downstream convection affects the whole inductor and a large zone after it. The inlet wave (dark red), almost compensating the infinite pump solution (light blue) at inlet, propagates downstream with damping, but since it propagates at a different speed, it can be in phase before outlet. Therefore the module of the wave does not increase monotonically from inlet downwards, as shown in Figure 3 where it is plotted for several values of q. For lower q, the velocity difference between both waves is larger and the interference effect is stronger.

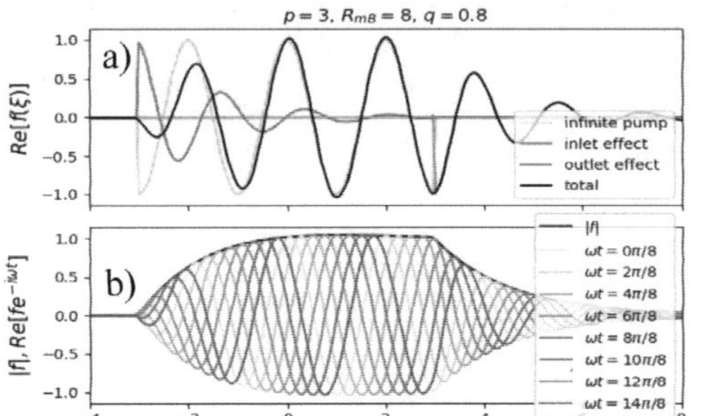

Figure 2: a) Modes added up to get the finite pump solution ; b) resulting traveling wave

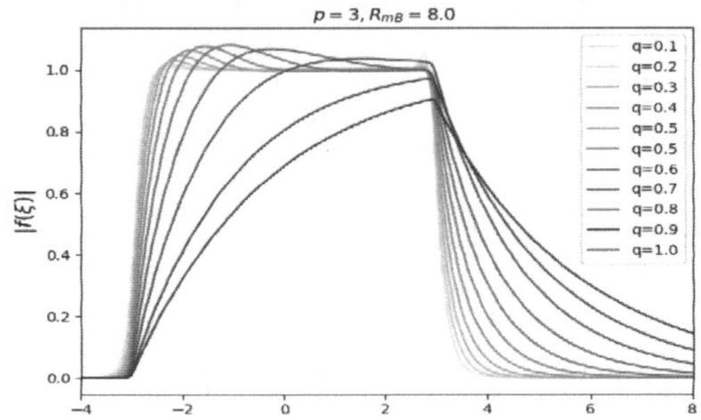

Figure 3: Module of the traveling wave for p=3, R_{mB}=8 and several values of q

Our theory was validated using a numerical axisymmetric model [8] based on the finite volume code Fluent plus a home made induction module. The configuration of Figure 1 b) was modelled with p=10, H_R=0.005, H_λ=0.45, R_{mB}=8.9 and several values of q. The finite volume code calculates the velocity profile, using a uniform profile imposed at inlet (x=−6λ), and a uniform pressure at outlet (x=15λ). Despites its uniform velocity, our theory gives a radial magnetic field close to the numerical calculation (Figure 4), and a theoretical performance curve (Figure 5) corresponding roughly to the electromagnetic term in the numerical code (the total pressure difference is slightly weaker because of friction).

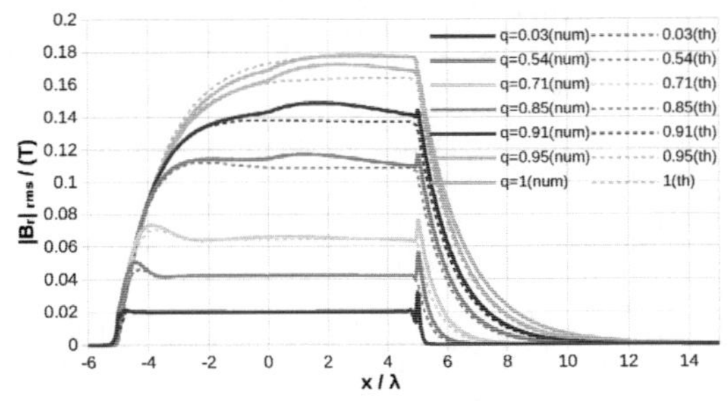

Figure 4: Radial magnetic field: theoretical and numerical

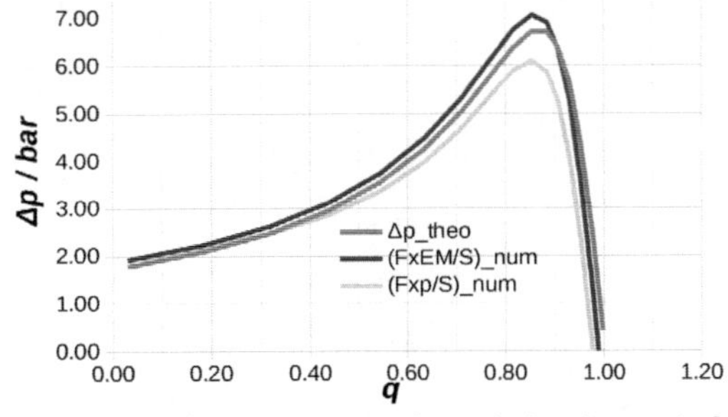

Figure 5: performance curves: theorerical and numerical.

The axial force density distribution is presented in Figure 6 for a typical PEMDYN case (p=3, R_{mB}=8), and in Figure 7 for an hypothetic pump with 10 poles (p=5) and a larger synchronism velocity, wavelength or fluid conductivity (R_{mB}=20). This last figure clearly shows the interference effect making the force density oscillate with ξ.

As previously described [3], [4], [6], the inlet end effect is responsible for a negative force (directed upstream), on a zone that appears to be longer and longer as the flowrate tends to synchronism. The electromagnetic problem seen from the fluid side is similar to transient diffusion into the channel of a suddenly applied AC field, whereas the fluid is convected downwards: the length of the inlet effect thus increases with $q.R_{mB}$ (and not R_{mS}). It appears in our model that a zone with similar length would exist after the inductor end because of the convection of the electromagnetic field after the pump outlet, if the magnetic yokes were long enough. That zone would generate few net pressure but a large amount of DSF fluctuations, and its limitation due to limited yokes will decrease pressure fluctuation but not pressure rise.

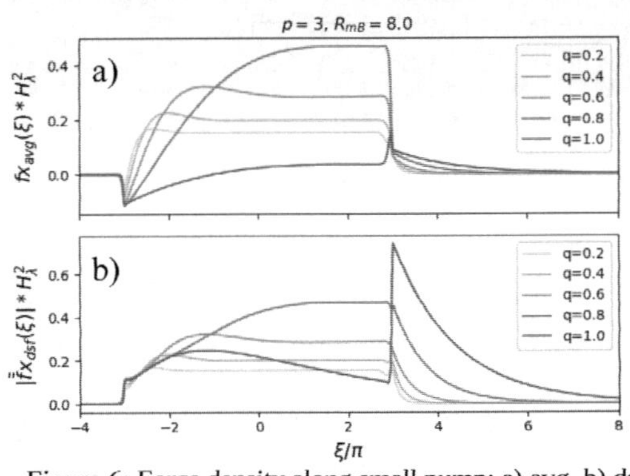

Figure 6: Force density along small pump: a) avg, b) dsf

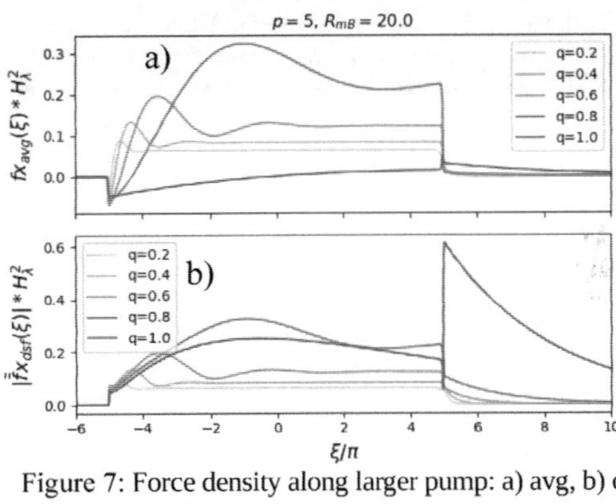

Figure 7: Force density along larger pump: a) avg, b) dsf

Integrating the force distribution, we obtain the pressure difference provided by the pump for any value of q (performance curve at constant current), but also the fluctuation of this pressure difference. Such curves are presented in Figure 8 for p=3 and several values of R_{mB}. The dimensionless values plotted in that figure should be divided by H_λ^2, and multiplied by F_{ref}/k to get pressure differences in Pascals. The DSF pressure fluctuation can be 0 even if the module of $\bar{fx}_{dsf}$ is constant along the pump, because its phase evolves along ξ. For the infinite pump solution, the phasor is proportional to $e^{2i\xi}$, and integrating from $-p\pi$ to $p\pi$ gives 0. The DSF pressure fluctuation is thus due to end effects.

An infinite pump with thin channel would give a maximum average force density at R_{mS}=1, where $H_\lambda^2 \bar{fx}_{avg}=0.5$, so that the total dimensionless force would be $H_\lambda^2 \bar{fx}^{tot}_{avg}=\pi p$ at the maximum point, if we integrate over the finite length without end effects. The maximum of the performance curves such as the ones in Figure 8 a) can be compared to those values, and this work can be repeated for other values of p. The results are presented in Figure 9, and shows that the maximum always occur at higher slip than R_{mS}=1, especially for a small number of poles where the end effects are more present. Because of the end effects, the maximum pressure provided by the pump is lower than expected from the infinite pump solution, especially if the number of pole pairs is small and expecially for large values of R_{mB}.

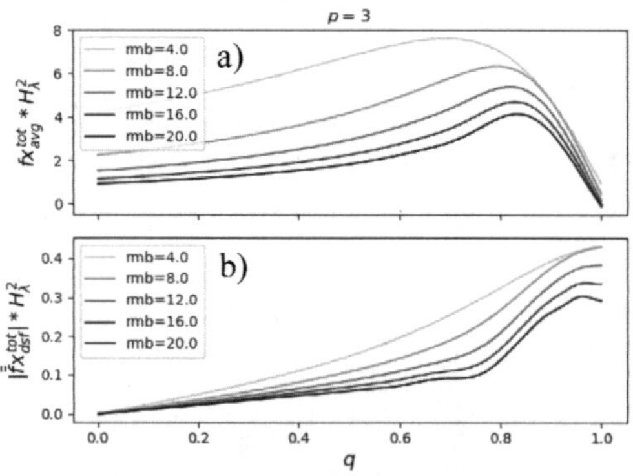

Figure 8: Pressure provided by the pump as a function of flowrate: a) mean value, b) dsf fluctuation

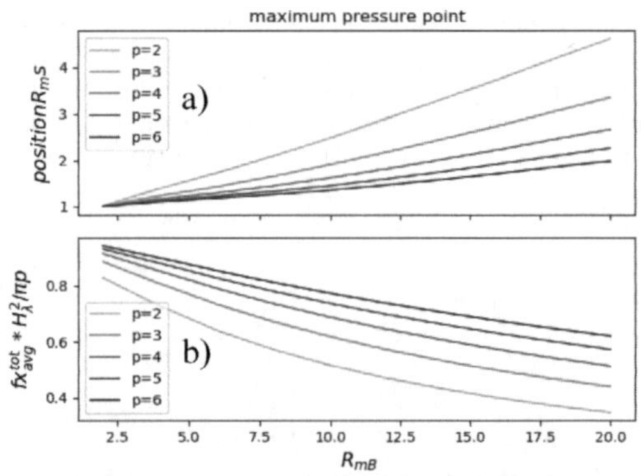

Figure 9: Maximum pressure point of an ALIP with p pairs of poles: a) position, b) value of the maximum

Conclusion

This paper presents a simple analytical model to describe the end effects in large size annular linear induction pumps. It gives the real position and value of the maximum pressure point, and thus can be used to rapidely find the limits of the stable working domain of the pump, as well as for parametric studies concerning the pump performance, or pressure fluctuations at 2ω.

References

[1] A. Gailitis and O. Lielausis, Magnetohydrodynamics 11 (1975) 69-79.
[2] H.Araseki, I.R.Kirillov, G.V.Preslitsky, A.P.Ogorodnikov, Nuclear Engineering and Design 227 (2004) 29-50.
[3] H.R. Kim, Annals of Nuclear Energy 62 (2013), 103-108
[4] M.Pouloujadoff, B.Morel, A.Bolopion, IEEE Transactions on Power Apparatus and Systems 99 (1980), 1172-1180
[5] L.Goldsteins, PhD, Univ. Grenoble Alpes, https://tel.archives-ouvertes.fr/tel-01232570 (2015)
[6] F.Werkoff, Experimental thermal and fluid science 4 (1991) pp 166-170
[7] Y.Delannoy, E.Martin-Lopez, F.Benoit, 18th Int. Symp. on Appl. Electromagn. and Mechanics, Chamonix (2017).
[8] E.Martin-Lopez, Y.Delannoy, F.Benoit, A.Muñoz Medina, R.Martinie, Int.Cong.Adv.Nucl.Pow.Plants, Charlotte, USA (2018)

IOP Conf. Series: Materials Science and Engineering **424** (2018) 012015 doi:10.1088/1757-899X/424/1/012015

Online Flow Control with Mold Flow Measurements and Simultaneous EM Braking and Stirring

M. Sedén[1], N. Jacobson[1]

[1] ABB AB, Industrial Automation, Process Industries, Metallurgy, Vasteras, Sweden

Corresponding author: martin.seden@se.abb.com

Abstract
To obtain flow control of the molten metal in the mold where the initial solidification steps of continuous casting occur, a versatile electromagnetic actuator is required to enable both flow acceleration and stabilization to counteract different undesired flow phenomena. Various mold monitoring systems instantly give insights into the dynamics of the meniscus level, solidification status and molten steel flow characteristics during casting. An online mold flow sensor in combination with an electromagnetic actuator capable of applying time varying braking and stirring forces simultaneously, enable closed loop mold flow control with automatic handling of regular as well as sudden variations in the casting conditions in real time.
A control package for the FC Mold G3, a combined two-level electromagnetic actuator with a brake in the bottom of the mold and a combined stirrer and brake in the top of the mold, reacting on varying macroscopic casting conditions has been used in steel plant production to minimize quality and operational issues. The control relies on predictions of the amount of surface defects based on experience and numerical simulations of the fluid flow. Plant quality results, collected during a year of full production, show a surface defect reduction of more than 50% with FC Mold G3 on critical ULC grades for exposed automotive body panels [1]. In the next step, real time information taken directly from process monitoring will dynamically control the FC Mold with further improved slab cleanliness and surface quality as a result.

Key words: Flow control, electromagnetic stirring (EMS), electromagnetic braking (EMBR), FC Mold, slab, surface quality, cleanliness, FBG, temperature measurement, level measurement

Introduction

Building on the success of the FC Mold to stabilize and optimize the mold flow in the continuous slab casting process with resulting enhanced slab quality, the subsequent step is to incorporate a more sophisticated mold flow detection device in combination with an increasingly flexible EM actuator to be able to catch and react on localized casting deviations. Local meniscus fluctuations, undesired fluid flow velocities or patterns and crack indications are examples of events that may call for an adjustment in the electromagnetic field distribution imposed on the molten metal in the mold during casting. Furthermore, the mold fluid flow may display asymmetric features due to deviations from ideal conditions in the entry nozzle or mold. Examples of this are inhomogeneous SEN clogging, asymmetric stopper or slide-gate positioning, or asymmetric argon injection. Even with a perfectly aligned and symmetric geometry, the turbulence of the fluid flow in the SEN and mold induces flow variations that cause asymmetric flow patterns. These asymmetric flows may lead to local variations of the metal end product quality, e.g. one side of a slab may contain large clusters of non-metallic inclusions close to the surface due to violent meniscus behavior and mold powder entrainment on this side.

By applying asymmetric mold flow control by means of a flexible EM device, a more stable and symmetric casting process can be maintained. E.g., excessive meniscus fluctuations and flow speeds on one side of the mold can be mitigated by extra stabilization and braking in this area, or an uneven speed relationship between the SEN jets due to SEN clogging can be homogenized by applying more braking or stirring on one side of the mold. A homogeneous and stable casting process yields a homogeneous solidified end product, with both quality and financial benefits for the steelmaker and the end-product consumer.

FC Mold

The FC Mold is a two-level electromagnetic device producing an upper field above the SEN outlet ports, close to the meniscus, and a lower field below the SEN ports in the lower portion of the mold. In the lower level, a static DC magnetic field is created across the mold thickness to brake the SEN jets, reduce downward penetration, and stabilize the flow pattern in the mold. For modern and flexible FC Mold units, another DC field is applied in the upper level either by itself, or simultaneously with AC stirring. This allows the near meniscus flow to be braked if flow speeds are too high, or accelerated with stirring should the flow speeds in this region be too low. In the Combi Mode, both DC and AC fields are applied and the meniscus fluctuations can be stabilized at the same time as a steady rotative stirring keeps the fluid moving in a robust way [7]. In modern FC Mold versions, the magnetic core and electrical drives allow full flexibility in controlling the magnetic fields in both the upper and lower levels, as well as on left and right sides independently.

Mold Flow Measurements

A high resolution thermal mold plate monitoring method based on optical Fiber Bragg Grating (FBG) sensors has been developed, and a pilot installed in one broad face copper plate in a continuous slab caster at the TATA Steel, IJmuiden works [2]. This technology allows a multitude of simultaneous temperature sensors along a single optical fiber, and by mounting a set of fibers inside a mold plate, more than 2500 FBG temperature sensors are distributed over the mold plate. The sensors measure the local temperatures and the system provides a sharp thermal image of the mold, updated during each scanning period of 0.5 s. Out of this pilot sensor setup, the OptiMold Monitor® system for mold thermal and fluid flow monitoring has evolved.

The operation principle of the FBG sensor technology is broadband pulses of light incident on the fiber gratings which produces a reflection wavelength spectrum based on the configuration of the gratings along the fiber. When exposed to the heat of the molten steel in the mold, the well-defined thermal expansion along each optical fiber produces a shift in the wavelengths of each grating. By the material parameters of the fiber, the wavelength shift can be directly translated to a specific temperature increase.

Utilizing a vertical sensor point spacing of 5 mm, where each sensor point has a measurement error less than 0.7 °C, local thermal phenomena such as the meniscus shape can be clearly resolved as visualized in the thermal distribution image in Fig. 1. The meniscus shape is a finger print of the flow pattern in the mold and the flow velocities in the molten steel near the meniscus level. E.g., in a standard double roll flow pattern commonly seen in slab casters for medium to high casting speeds, the flow momentum from the SEN jets impinges on the mold narrow faces and is then partially split upwards and downwards along the narrow face as illustrated in Fig. 2. The upward flow creates a wave crest and is then redirected and follows the meniscus toward the SEN in a standing wave shape where the flow speed is related to the height of the wave. This means that by measuring the meniscus shape, the flow speed of the meniscus can be indirectly estimated. A set of nailboards have been submerged into the meniscus flow to tune and quantify this correlation [3].

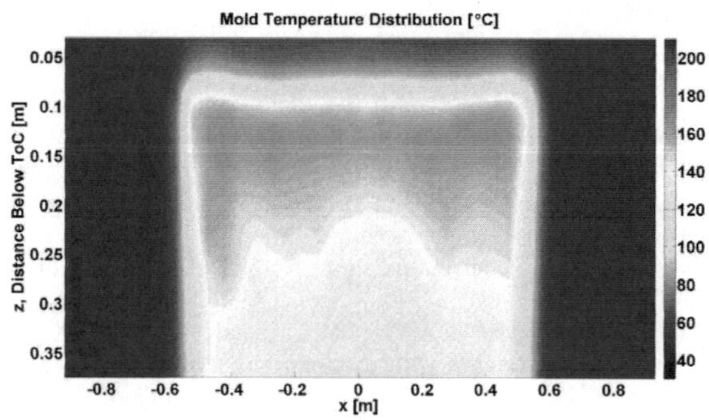

Fig. 1: Measured high resolution temperature map of broad face mold plate, 237x1100 mm, 1.0 m/min.

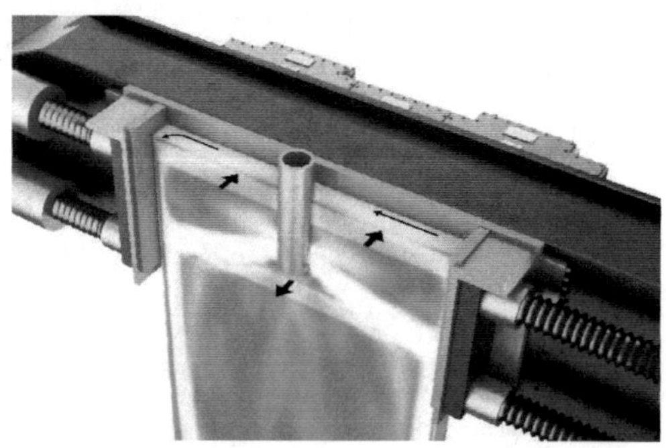

Fig. 2: Double roll fluid flow pattern in caster strand and magnetic field locations of the FC Mold.

Over a measurement campaign during production at TATA Steel IJmuiden, mold temperature distributions, and indirectly meniscus wave and speed profiles, were collected for more than 1700 cast slabs of selected steel grades. In the same caster, an FC Mold, a double level DC electromagnetic brake is installed [4]. In roughly half of the cast slabs of the test period, the static magnetic fields of the FC Mold were engaged, and in the rest of the slabs the FC Mold was turned off. Real time meniscus wave profile characteristics such as wave height and flow speed on both left and right sides of the mold were logged continuously and then averaged over the casting time of each slab. In this trend analysis, each slab provides a single point in the set of data from all cast slabs of the measurement campaign.

An asymmetry index, AI, of the meniscus wave shape of the fluid steel is defined as

$$AI = 100\frac{h_L - h_R}{h_{ref}} \tag{1}$$

where h_L and h_R are the standing wave heights of the meniscus on the left and right sides respectively, related to a reference height h_{ref} of 20 mm. If the absolute value of *AI* is constantly high, then there is a stable meniscus shape asymmetry caused by biased mold flow indicating e.g. asymmetric clogging in the SEN. If *AI* is varying with a large σ over time, then this indicates a swaying biased flow. Both of these asymmetric flow features may be detrimental to the end quality of the steel as an inhomogeneous solidification is taking place, and excessive flow speeds or oscillations may pull mold powder into the bulk of fluid steel.

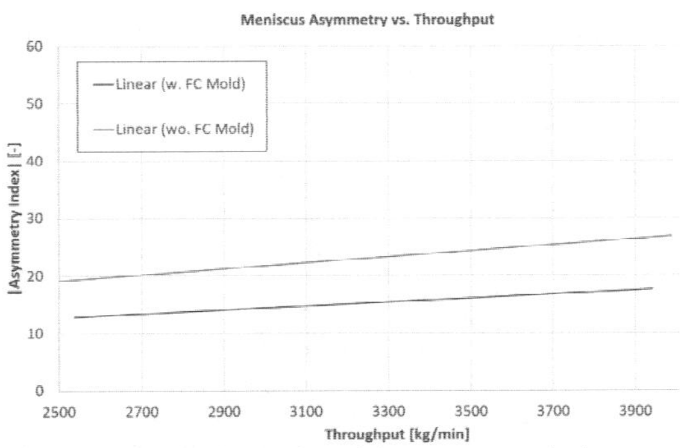

Fig. 3: Mean absolute value of the asymmetry index vs. casting throughput. With and without FC Mold.

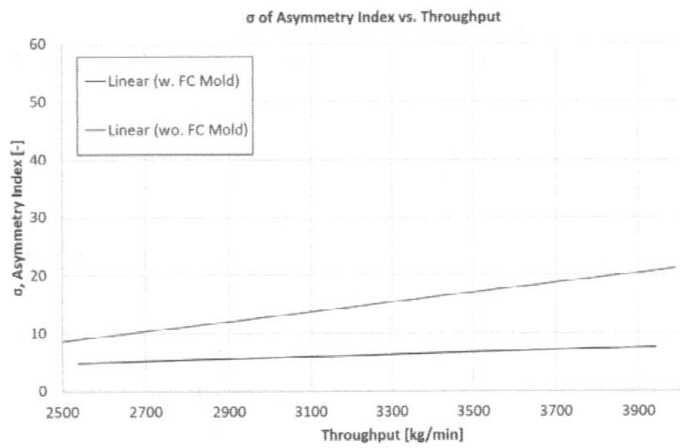

Fig. 4: Standard deviation of the asymmetry index over time within a slab. With and without FC Mold.

Based on the full data set from the measurements, a linear regression of the average absolute value of the asymmetry index per slab, |*AI*|, is in Fig. 3 plotted against throughput. The slabs composing the measurement campaign are cast with a wide variety of conditions, but all slabs used in the analysis are cast with constant casting conditions within each slab. With upper and lower level DC magnetic fields from the FC Mold acting across the mold thickness to brake the SEN jets and stabilize the flow pattern in the lower mold region and to brake meniscus speeds and stabilize the flow in the upper region, the results in Fig. 3 show that the mean meniscus asymmetry is reduced for all throughputs. In Fig. 4, a linear regression of the measured σ of the asymmetry index *AI*, i.e. the time standard deviation of *AI* per slab, is shown versus throughput. The stabilizing functionality of the DC magnetic fields of the FC Mold reduces the swaying over time in the biased meniscus flow generally, but also makes variations in asymmetry nearly independent of casting throughput.

Numerical Simulations of Flow Asymmetry when Exposed to FC Mold EM Fields

A numerical CFD simulation has been conducted to illustrate the effect of asymmetric magnetic fields on an asymmetric flow field. Using a fictive skew inlet flow condition over a cross section in the SEN for a 237x1100 mm slab format cast at 1.9 m/min, the effects of a partially clogged nozzle have been modelled. A snap-shot of the simulated velocity distribution in the center slice of the mold during these conditions is shown on the left of Fig. 5.

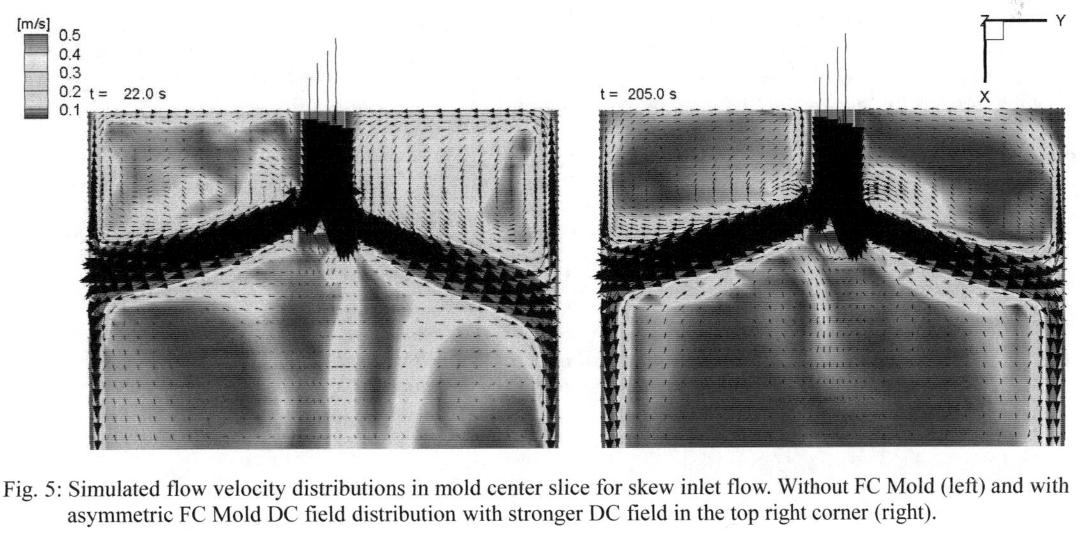

Fig. 5: Simulated flow velocity distributions in mold center slice for skew inlet flow. Without FC Mold (left) and with asymmetric FC Mold DC field distribution with stronger DC field in the top right corner (right).

In Fig. 6 the transiently simulated meniscus flow speeds at the center thickness of the mold, $y = \pm 0.385$ m for the left and right sides, are shown for a time period of 210 s. In the first 60 s of the simulation, the FC Mold is turned off. From here until t=75 s, horizontally homogeneous and symmetric DC magnetic field distributions in both levels of the FC Mold are ramped up, with the result that between 75 and 135 s, the meniscus flow speeds are reduced and fluctuations stabilized. However, the gap between the left and right flow speed curves of Fig. 6 is only closed in the last 60 s of the simulation when the biased flow is symmetrized. This is the result of a horizontally asymmetric field distribution in the upper FC Mold level where a 25% boost is applied to the DC magnetic flux density on the top right. The stronger flow speed on the right side of the meniscus is in this way subject to extra braking to obtain the same flow speed as on the left side. The flow speed pattern in the mold during asymmetric braking is illustrated on the right in Fig. 5.

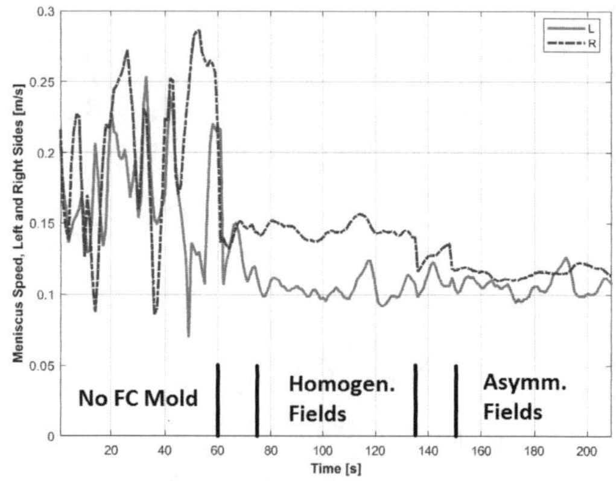

Fig. 6: CFD simulation results of meniscus flow speeds on left (L) and right (R) sides in modelled SEN clogging case for varying electromagnetic field configurations;
0-60 s: No field,
75-135 s: Homogeneous FC Mold DC fields,
150-210 s: Asymmetric FC Mold DC fields counteracting L/R flow asymmetry.

FC Mold Control

By accurately measuring the flow speeds of the meniscus in the mold with the high resolution method described above, and at the same time monitoring asymmetric flow pattern deviations, the FC Mold is via a closed-loop control system in the FC Mold Autonomous Control Mode set up to apply varying braking and stirring electromagnetic fields to counter-act meniscus speeds that are too low or excessive. In the same way, the control works together with the flexible FC Mold to mitigate flow pattern asymmetries. E.g. a stronger flow speed in one side of the mold is suppressed by a locally strengthened DC field, or an insufficient flow in one side of the mold is accelerated by locally controlled AC stirring to create a homogeneous flow speed distribution over the full width of the slab. Robust PI-methods are used in the controller implementation. Control can also be carried out against an electromagnetic level sensor where high frequency feedback can be utilized to obtain detailed level and fluctuation information in the probing locations. This enables stability control of the meniscus level and the upper part of the mold.

Summary

Flow asymmetry induced by e.g. SEN clogging can according to simulations be counter-acted by an asymmetric FC Mold magnetic field distribution to achieve a homogeneous flow speed distribution over the entire meniscus level. With modern FC Mold equipment, the DC and AC magnetic fields are configurable independently not only in the upper and lower levels, but also on the left and right hand side regions. This, in combination with the high resolution flow monitoring capabilities of the OptiMold Monitor® makes it possible to control the magnetic fields of the FC Mold autonomously against different online flow measurement parameters such as the meniscus speed on the left and right sides of the mold. The dynamic control of the FC Mold guided by the process monitoring of OptiMold Monitor® unlocks potential for further improved slab cleanliness and surface quality of the cast steel.

References

1. N. Jacobson et. al., Flow Control by FC Mold G3, Operation and Results from Slab Casting Process, Proceedings, 9th European Continuous Casting Conference, Vienna, Austria (2017), 199-208
2. A.L. Spierings et al., "Development and application of Fiber Bragg Gratings for slab casting", Proceedings, AISTech2017, Nashville (2017)
3. R. Liu et al., Measurement of molten steel surface velocity with SVC and nail dipping during continuous casting, Proceedings, TMS (2011), 51-58
4. M. Sedén et. al., Securing Dynamic Mold Flow Control with FC Mold and OptiMold Monitor, Proceedings, 9th European Continuous Casting Conference, Vienna, Austria (2017), 67-75
5. S. Kenichiro et al., Flow field control of liquid steel in the mold of a continuous slab caster by an electromagnetic brake, Tetsu-to-Hagane (1983), Vol. 69, No.12, 912
6. Y. Miki, S. Takeuchi, Internal defects of continuous casting slabs caused by asymmetric unbalanced steel flow in mold, ISIJ International (2003), Vol. 43, No. 10, 1548-1555
7. M. Sedén et. al., Control of Flow Behavior by FC Mold G3 in Slab Casting Process, 8th ECCC, Graz, Austria (2014), 558-569

Linear stability of parallel flow of liquid metal in a rectangular duct driven by a constant pressure gradient under the influence of a uniform magnetic field

T. Tagawa[1]

[1] Department of Aeronautics and Astronautics, Tokyo Metropolitan University, Asahigaoka 6-6, Hino, Japan

Corresponding author: tagawa-toshio@tmu.ac.jp

Abstract
Linear stability of the plane Poiseuille flow of liquid metal under a uniform magnetic field is numerically studied. The liquid metal is driven by a constant pressure gradient in a rectangular duct. The induced magnetic field and the Joule heating are neglected in this analysis. The governing equations are the continuity of mass, momentum equation, Ohm's law and conservation of electric charge. The solution of basic state is well known as the Hartmann flow in the limit of the large aspect ratio of the cross-section of the duct. It is supposed that disturbance has periodicity along the direction of the basic flow. The linear stability of the basic flow that depends only on the Hartmann number is analyzed by a finite difference method discretized by a fourth order central difference scheme together with the use of HSMAC algorithm. We obtained the phase velocity of Tollmien-Schlichting wave and the Reynolds number as the neutral stability state when input parameters such as the aspect ratio, the Hartmann number and the wavenumber were given. It is predicted that the critical Reynolds number at the onset of instability is about 4.0 x 10^5 and the wavenumber is about 1.15 when the Hartmann number is 5.

Key words: Linear stability analysis, Hartmann number, liquid metal flow

Introduction

Use of magnetic field is one of the most promising method for controlling liquid metal flows, which are encountered in steel making process, semiconductor crystal growth and so on. The static magnetic field can be used to stabilize liquid metal flows, whereas the alternating magnetic field can be used to drive and stir liquid metal flows as well as to generate Joule heating. Our interest here is the stability of liquid metal flows when uniform static magnetic field is imposed. Among various magneto-hydrodynamic flows, the plane Poiseuille flow in a duct having rectangular cross section under the uniform magnetic field has been selected and focused. This kind of magnetohydrodynamic flow is call as Hartmann flow. The linear stability of such a parallel flow under the influence of a magnetic field was firstly studied by Lock [1]. It was concluded that the critical Reynolds number is found to rise very rapidly with increasing the Hartmann umber. Kakutani [2] investigated the stability of the modified plane Couette flow in the presence of a transverse magnetic field. It was found that the critical Reynold number takes its minimum value of 3.8 x 10^5 when Ha = 5.4. The stability of duct flow with rectangular cross-section depends on the aspect ratio of the cross-section of the duct. Tatsumi and Yoshimura [3] investigated the linear stability of the duct flow for the range of aspect ratio from 1 to 10 and the Reynold number up to 50000. They reported that the duct flow could be unstable in the range of aspect ratio more than 3.2. Priede *et al.* [4] analyzed linear stability of Hunt's flow which is a liquid metal flow in a square duct subjected to a transverse magnetic field. In this study, linear stability analyses were carried out for a duct flow of liquid metal under the influence of uniform magnetic field.

Configuration of problem and governing equations

Figure 1 shows the schematic model considered in this study as one of the simplest model among various magnetohydrodynamic duct flows. The duct is rectangular and the aspect ratio of the cross-section is assumed to be large since the duct flow for small aspect ratio is linearly stable [3]. The primary flow direction is z as shown in Fig. 1 and a uniform magnetic field is applied in x-direction. We consider the case that all the walls are electrically insulated.

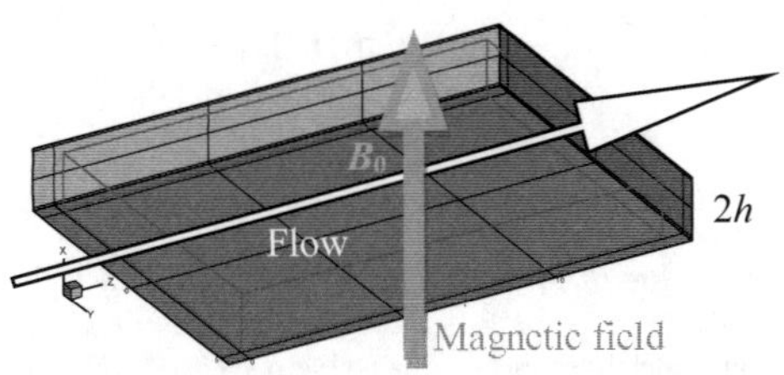

Fig. 1: Sketch of the duct flow problem with magnetic field

It is assumed that the fluid is an incompressible Newtonian fluid and the heat source such as viscous dissipation and Joule heating are neglected. In addition to this, since the magnetic Reynolds number is usually small in this kind of problems, the induced magnetic field is neglected in this analysis. The governing equations are written as follows.

$$\partial_j u_j = 0, \tag{1}$$

$$\rho\left(\partial_t u_i + u_j \partial_j u_i\right) = -\partial_i p + \mu \partial_j^{\,2} u_i + \varepsilon_{ijk} j_j b_k, \tag{2}$$

$$\partial_j j_j = 0, \tag{3}$$

$$j_i = \sigma\left(-\partial_i \phi + \varepsilon_{ijk} u_j b_k\right). \tag{4}$$

The boundary conditions are as follows. Here, $2h$ is the height of the duct and $2a$ is spanwise length of duct.

$$x = \pm h: \quad u = v = w = j_x = 0, \quad \text{and} \quad y = \pm a: \quad u = v = w = j_y = 0 \tag{5}$$

Basic state and disturbance equations

In the present analyses, we consider the case that the duct length along the primary flow is sufficiently long. The basic states without having any disturbance are determined by the force balance between the pressure gradient, viscous force and electromagnetic force by numerically solving the following three equations expressed in a dimensionless form.

$$\left(\partial_X^2 + \partial_Y^2\right)\overline{W} + 2 - Ha^2 \overline{J}_Y = 0, \tag{6}$$

$$\partial_X \overline{J}_X + \partial_Y \overline{J}_Y = 0, \tag{7}$$

$$\overline{J}_X = -\partial_X \overline{\Phi}, \quad \overline{J}_Y = -\partial_Y \overline{\Phi} + \overline{W}, \quad \overline{J}_Z = 0. \tag{8}$$

It is noted that this basic state is not related to the Reynolds number but to the Hartmann number, whose square value is the ratio between the electromagnetic force to the viscous force.

In order to examine the linear stability of the basic flow, it is assumed that disturbance has periodicity along the direction of basic flow and therefore the variables such as velocity, pressure, electric current density and potential are presumed as the summation of the basic state and the disturbance as follows:

$$\varphi(x,y,z,t) = \overline{\varphi}(x,y) + \tilde{\varphi}(x,y)\exp(i\alpha z + st), \quad \text{where} \quad \varphi = \left(u_i, p, j_i, \phi\right) \tag{9}$$

Here, α represents the wavenumber, and s represents the complex number, whose real part indicates the linear growth rate and imaginary part is the angular frequency. If $s_r > 0$ the disturbance grows and the basic flow is unstable while if $s_r < 0$ it attenuates and the basic flow is stable. After substituting eq. (9) into eqs. (1) to (5), and making non-dimensionalization, we have the following dimensionless disturbance equations linearized.

$$\partial_X \tilde{U} + \partial_Y \tilde{V} + ik\tilde{W} = 0, \tag{10}$$

$$S\tilde{U} + i\overline{W}k\tilde{U} = -\partial_X \tilde{P} + Re^{-1}\left(\partial_X^2 + \partial_Y^2 - k^2\right)\tilde{U}, \tag{11}$$

$$S\tilde{V} + i\overline{W}k\tilde{V} = -\partial_Y \tilde{P} + Re^{-1}\left(\partial_X^2 + \partial_Y^2 - k^2\right)\tilde{V} + Ha^2 Re^{-1}\tilde{J}_Z, \tag{12}$$

$$S\tilde{W} + \tilde{U}\partial_X \overline{W} + \tilde{V}\partial_Y \overline{W} + i\overline{W}k\tilde{W} = -ik\tilde{P} + Re^{-1}\left(\partial_X^2 + \partial_Y^2 - k^2\right)\tilde{W} - Ha^2 Re^{-1}\tilde{J}_Y, \tag{13}$$

$$\partial_X \tilde{J}_X + \partial_Y \tilde{J}_Y + ik\tilde{J}_Z = 0, \tag{14}$$

$$\tilde{J}_X = -\partial_X \tilde{\Phi}, \quad \tilde{J}_Y = -\partial_Y \tilde{\Phi} + \tilde{W}, \quad \tilde{J}_Z = -ik\tilde{\Phi} - \tilde{V}. \tag{15}$$

The definition of the characteristic velocity, dimensionless variables and non-dimensional number are shows as follows:

$$u_{ref} = -\frac{(d\overline{p}/dz)h^2}{2\rho\nu}, \quad (X,Y,Z) = \frac{(x,y,z)}{h}, \quad \left(\overline{W},\tilde{U},\tilde{V},\tilde{W}\right) = \frac{\left(\overline{w},\tilde{u},\tilde{v},\tilde{w}\right)}{u_{ref}}, \quad \left(\tilde{J}_X,\tilde{J}_Y,\tilde{J}_Z\right) = \frac{\left(\tilde{j}_x,\tilde{j}_y,\tilde{j}_z\right)}{\sigma u_{ref} B_0},$$
$$P = \frac{p}{\rho u_{ref}^2}, \quad \left(\overline{\Phi},\tilde{\Phi}\right) = \frac{\left(\overline{\phi},\tilde{\phi}\right)}{u_{ref} B_0 h}, \quad Re = \frac{u_{ref} h}{\nu}, \quad Ha = B_0 h\sqrt{\frac{\sigma}{\rho\nu}}, \quad A = \frac{a}{h}, \quad k = \alpha h, \quad S = \frac{s}{u_{ref}/h} \tag{16}$$

The characteristic length is the half value of the duct height, and the characteristic velocity is taken as the maximum velocity of the basic flow. Here, *Re* represents the Reynolds number, *Ha* the Hartmann number, *A* the aspect ratio, *k* the wavenumber and *S* the complex angular frequency. The system of the above equations and the boundary condition constitutes a generalized eigenvalue problem.

Numerical strategy and validation of the present code

The simultaneous ordinary differential equations were discretized on a two-dimensional equidistant staggered mesh system within a cross-section of the duct and solved with a finite difference method using the fourth order central difference method. First, the basic state such as the velocity, potential and electric current density are obtained for a given Hartmann number. Second, after having given the wavenumber, the disturbance equations are solved by using the HSMAC method, meanwhile both growth rate (real part of *S*) and angular frequency (imaginary part of *S*) of the most unstable mode should be determined. For determining the complex eigenvalue, we have used a Newton-Raphson method. Before moving to the MHD flow instability, it would be necessary to examine the validity of the present numerical code. One of the most famous critical values for the onset of instability in fluid mechanics is that for the plane Poiseuille flow, of which critical Reynold number is known as 5772 with the critical wavenumber 1.02. Hence, we have implemented one dimensional computation neglecting *y*-directional meshing and the effect of electromagnetic field. Table 1 shows the neutral values of the Reynolds number, angular frequency and phase velocity when the critical wavenumber 1.0205474, which is believed to be exact with high accuracy, was given in the present analyses. It is judged that the present code can obtain the correct values when the number of meshes is sufficient.

Table 1: Dependency of number of meshes for the obtained neutral values (k = 1.0205474 and S_R = 0)

Number of meshes	Re	S_I	C_{ph}
100	5683.448	0.2701748	0.2647532
200	5756.642	0.2695493	0.2641222
400	5769.967	0.2694423	0.2640174
800	5771.919	0.2694271	0.2640025
Exact value	5772.222	0.2694248	0.2640003

Case in the finite aspect ratio

Figure 2 shows the example of visualization of the flow mode for two wavelengths at a neutral stability when the input parameters such as aspect ratio of cross-section, Hartmann number, wavenumber are given as shown in the figure caption. The uniform magnetic field is imposed in the *X*-direction, but its strength herein is limited to a very small case (Ha = 0). Figure 2 (a) indicates the Tollmien-Schlighting (T-S) wave in the core region far from the sidewalls. In the limit of large aspect ratio for the cross-section, the critical Reynolds number takes the value of 5772. For a finite aspect ratio, the critical Reynolds increases as decrease in the aspect ratio until its value is about 3.2 as indicated in the reference of Tatsumi & Yoshimura [3]. Figure 2 (b) indicates that high values of potential appear near the sidewall and almost zero in the core region. It is imagined that when the aspect ratio is very large, the effect of disturbed electric potential is negligible. Even in that situation, we take the insulating sidewall into account since the basic state of electric potential or electric field are crucial in this problem. In the next section, we will see the influence of the Hartmann number on the liquid metal flow driven by a constant pressure gradient when the aspect ratio of cross section is infinity.

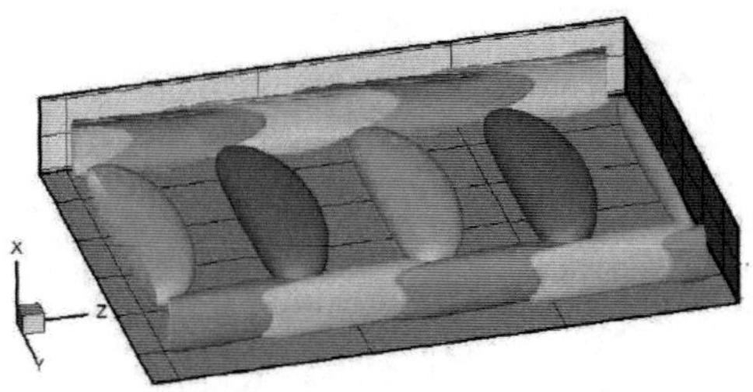

(a) *X*-directional component of velocity

(b) Electric potential

Fig. 2: Iso-surfaces of each eigenfunction for the neutral stability condition at A = 5, Ha = 0, k = 0.91, Re_n = 9844 and S_I = 0.2135. The number of meshes employed for the computation is 70 x 350 for the cross section.

Influence of the strength of magnetic field

Figure 3 shows the velocity profile of the basic flows for several Hartmann numbers when the sidewalls are perfectly insulated. As increase in the Hartmann number, the core velocity tends to be flattened. This profile can be theoretically given by the equation shown below.

$$\overline{W} = 2\left\{Ha \cdot \tanh\left(Ha\right)\right\}^{-1}\left\{1-\cosh\left(Ha \cdot X\right)/\cosh\left(Ha\right)\right\} \quad (17)$$

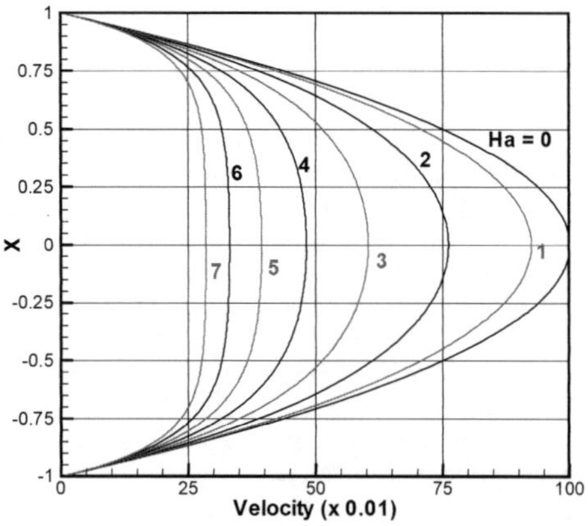

Fig. 3: The velocity profile of basic flows for several Hartmann numbers.

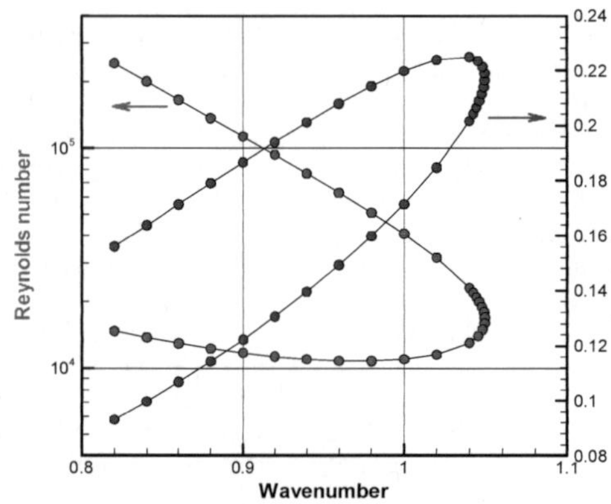

Fig. 4: Neutral stability curves in the vicinity of the critical point for $Ha = 1$.

Table 2 shows the summary of the values of critical point for several Hartmann numbers when the one-dimensional linear stability analysis is conducted with the number of meshes of 200. As increase in the Hartmann number, once the wavenumber decreases, but the Reynolds number increases monotonously. We confirmed that grid-size dependency occurs especially in the case of large Hartmann number. Up to $Ha = 5$, however, the discrepancy of the results between 200 meshes and 400 meshes is rather small. Figure 4 shows the neutral stability curves in the vicinity of the critical point for $Ha = 1$ together with the angular frequency. When the wavenumber exceeds the value of 1.05, the basic flow is always stable. Figure 5 shows the streamlines of the disturbed flow for two wavelengths at the critical state. The thickness of the boundary layer becomes thinner with increasing the Hartmann number. Therefore, finer mesh system is required to maintain the accuracy for the large Hartmann number.

Table 2: Summary of the computation results for the critical wavenumber, Reynold number, angular frequency and phase speed for several Hartmann numbers.

Ha	k	Re	S_I	C_{ph}
0	1.021059	5756.612	0.2697457	0.2641823
1	0.972707	10785.30	0.2119994	0.2179479
2	0.930397	37036.39	0.1366636	0.1468874
3	0.965232	104653.2	0.0993330	0.1029111
4	1.049365	222651.6	0.0819787	0.0781222
5	1.153168	400723.1	0.0721515	0.0625681

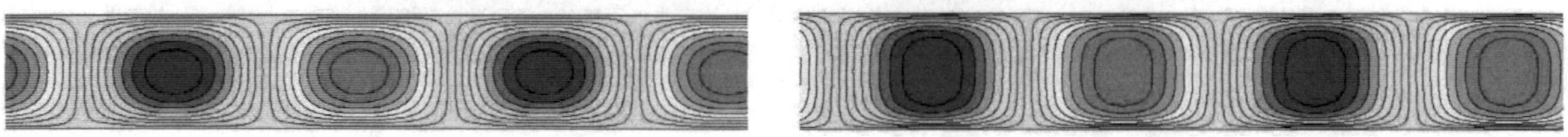

(a) $Ha = 1$ (b) $Ha = 3$

Fig. 5: The streamlines of flow for $Ha = 1$ and 3 at each critical point shown in Table 2.

References

1. R. C. Lock, The stability of the flow of an electrically conducting fluid between parallel planes under a transverse magnetic field, *Proc. R. Soc. Lond. A* 233.1192 (1955): 105-125.
2. T. Kakutani, The hydromagnetic stability of the modified plane Couette flow in the presence of a transverse magnetic field, *J. Phys. Soc. of Japan* 19.6 (1964): 1041-1057.
3. T. Tatsumi and T. Yoshimura, Stability of the laminar flow in a rectangular duct, *J. Fluid Mech.* 212 (1990): 437-449.
4. J. Priede, S. Aleksandrova and S. Molokov, Linear stability of Hunt's flow, *J. Fluid Mech.* 649 (2010): 115-134.

IOP Conf. Series: Materials Science and Engineering **424** (2018) 012017 doi:10.1088/1757-899X/424/1/012017

Performance maps of a high flowrate EM pump
Experimental and numerical analysis

S. Vitry[1*], E. Martin Lopez[1,] F. Benoit[1], L. Cachon[1] and L. Goldsteins[2]

[1] CEA Cadarache /DTN/STCP/LCIT, 13108 St Paul lès Durance, France
[2] Institute of Physics University of Latvia, LV-2169, Salaspils, Latvia

* *Corresponding author: Sylvain.Vitry@CEA.fr*

Abstract

In the framework of the Fast Reactor R&D programme, an experimental electromagnetic (EM) induction pump has been designed with similar magnetohydrodynamic (MHD) characteristics to those expected for the secondary cooling system. This pump will provide us with experimental results that are needed to validate the numerical models used for the preliminary design 2, with the main objective being to develop and qualify a finite-element model of MHD instabilities of Annular Linear Induction Pump (ALIP). This paper describes the current status of the experimental and numerical analysis carried out to detect and investigate liquid sodium flow instabilities in the innovative ALIP, installed in the PEMDYN loop with the objective of better managing the EM pump and preventing MHD instability. In this paper we present the experimental results collected during the operation of the test campaign aiming establishing at performance maps of the high flowrate ALIP. It also describes the numerical finite-element models that have been built to gain a more in-depth understanding of the interdependent MHD phenomena 1.

Key words: Magnetohydrodynamic, induction pump, sodium-cooled fast reactor, performance curve.

I. INTRODUCTION

In the framework of the French Alternative Energies and Atomic Energy Commission's (CEA) Sodium Fast Reactor R&D programme, it has been proposed to use an electromagnetic linear induction pump (EMIP) in the secondary cooling system. EMIP are an interesting alternative to mechanical pumps for the transportation of liquid metals such as sodium in secondary cooling systems of fast reactors, due to its superior safety features, e.g. no moving parts in liquid metal and a completely leaktight construction. However, detailed studies will be required to improve control EMIP operation and prevent the undesirable phenomenon called MHD instability. The CEA has been in charge of the design, engineering and construction of a new sodium loop named PEMDYN intended for studies on experimental high flowrate EMIP. The ALIP (Fig 1) will provide us with experimental results that are needed to validate the numerical models used for the preliminary design, with the main objective being to develop and qualify a finite-element model of electromagnetic instabilities inside an annular linear induction pump for a maximum flowrate of 1500m^3/h and a pressure of 3bar 4.

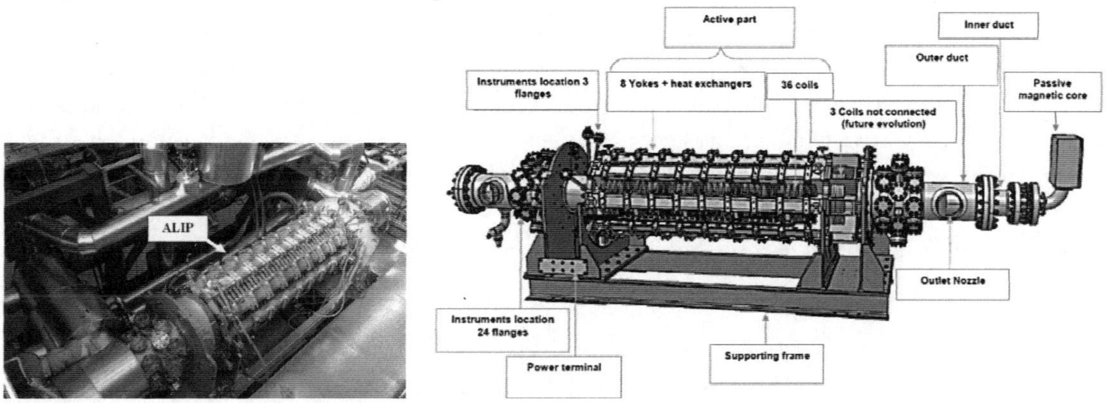

Fig. 1: PEMDYN experimental pump

That ALIP can be described as a three-phase alternating current machine with a cylindrical inductor generating a magnetic field, variable over time and in the pump channel space, that induces currents directly in the volume of the conducting fluid (ig. 2). The above described ALIP is allows for studies involving different magnetic system configurations and hydraulic configurations. The inner core can be removed to implement sensors or it can be replaced with a second inductor where coils are embedded in a laminated ferromagnetic core. These configurations should help us gain a better overall understanding, as well as quantify any additional MHD effects and instabilities of double-supply-frequency ALIPs 1 and 4.

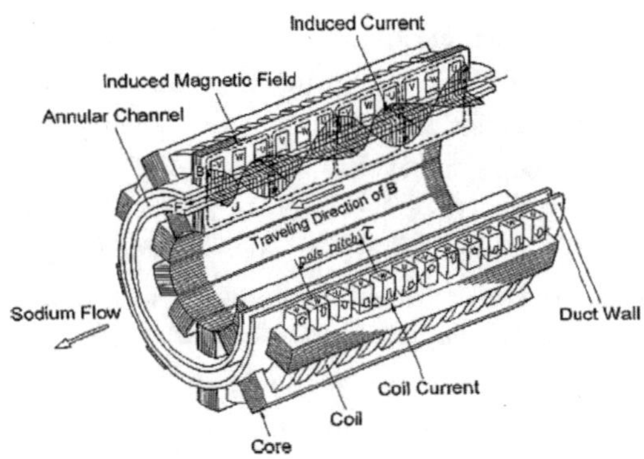

Fig. 2: Working principle of an ALIP

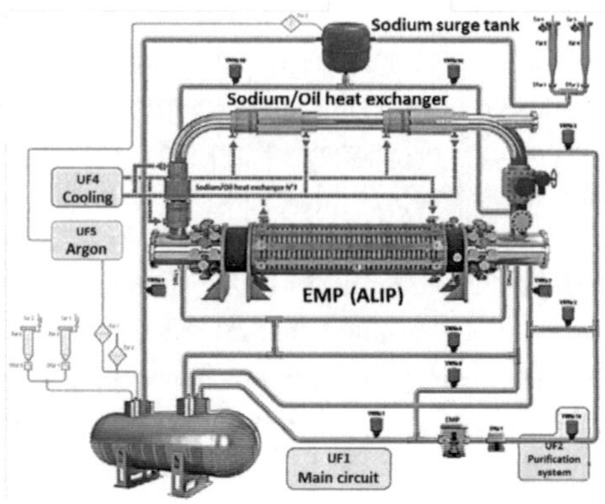

Fig. 3: Diagram of PEMDYN facility

PEMDYN design comprises one external inductor where three phase coils are embedded in laminated ferromagnetic core, an annular channel where liquid metal flows in the direction of the traveling magnetic field and an inner ferromagnetic core, which can be removed and replaced by an inner inductor equipped with sensors. The aim of this construction is to generate a magnetic field perpendicular to the direction of the flow (radial component). From cross product of azimuthal induced currents in sodium and radial component of magnetic field, axial electromagnetic force in the direction of travelling field is induced in the liquid metal. This force is responsible for pumping. The PEMDYN diagram presented in Fig. 3 is composed of the ALIP and a ball valve implemented to regulate and adjust the pressure drop of the circuit, to determine the performance curves of the ALIP. Three oil/sodium heat exchangers are implemented on the circuit to maintain the sodium temperature constant. The flowrate is measured with an ultrasound flowmeter.

II. ANALYTICAL

The MHD system of equations governing an ALIP is the same as presented in 1 and 4. It is composed by coupled Navier-Stokes (1) equation and induction equation in terms of vector potential A taking into account Coulomb gauge. It can be written in the following form (2). The terms responsible for the coupling of the system are $j \times B$ in equation (1), which consists of adding the volumetric force creating the pumping, and $-u \times \nabla \times A$ in equation (2), which is the convective term in transport equation for A. As system of (1) and (2) is non-linear it can be precisely solved using only iterative methods. However, in praxis so called solid body or electrodynamic approximation is often used, neglecting (2) and using only mean velocity of the flow. Basic model of electromagnetic induction pump is so called ideal machine Ideal model of EMIP (Fig. 4).

$$\rho\frac{\partial u}{\partial t} + \rho(u\cdot\nabla)\cdot u = -\nabla P + \rho\nu\nabla^2 u + j \times B \quad (1)$$

$$\Delta A = \mu_0\sigma\left[\frac{\partial A}{\partial t} - (v \times \nabla \times A)\right] - \mu_0 J_e \quad (2)$$

$$\nabla \times A = B \quad (3) \qquad \nabla A = 0 \quad (4)$$

$$\Delta p = \frac{\sigma B_0^2 v_B s l}{2}\cdot k_w k_l - \lambda\frac{l}{D_h}\frac{\rho v_m^2}{2} \quad (5)$$

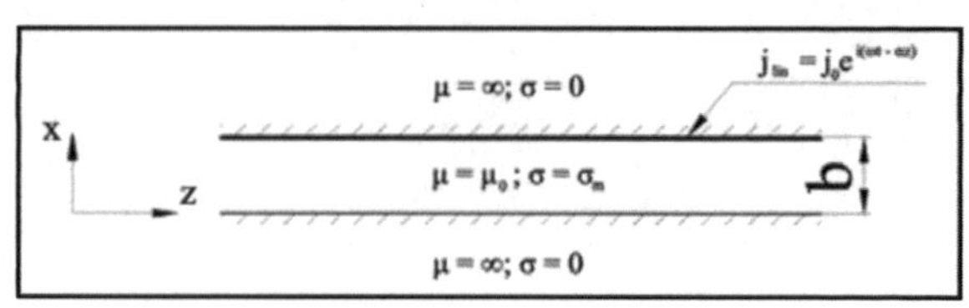

Fig. 4 Ideal model of EMIP

It consists of two infinite length perfect ferromagnetic plates and between them conductive media, sodium, moves with constant velocity. On surface of upper ferromagnetic plate linear current is applied in harmonic form. For more details on MHD refer to 2. The analytical formula for estimation of quasi stationary developed pressure using solid body approximation is given in (5), where the last part represents pressure losses due to the walls friction. The solid body approximation quantify in integral characteristic estimations, and if $R/b \geq 2$ and $\tau/b \geq 2$, the influence of cylindrical system curvature on integral parameters is negligible (order of 1 – 2 %). Nevertheless, it was found that due to the fluid secondary circuit, in the case of high magnetic Reynolds number, MHD machine can experience instability [4]. While in the case of ALIP this effect is strongly connected with a break of axial symmetry, it should not be forgotten that hydrodynamic part can play significant role in such machine and should be taken into consideration.

III. EXPERIMENTAL

The objective of this test programme was to measure operating points under steady-state conditions so as to plot the flow and pressure performance curves of the EM pump. This involved measuring the flowrate and pressure during a steady-state flow at given power supply parameters. Fig. 5 compares trace the variation of the EM pump impedance for the simulation results and the experimental measurements 3. The simulation results and experimental measurements for each operating point of the EM pump are consistent in terms of impedance.

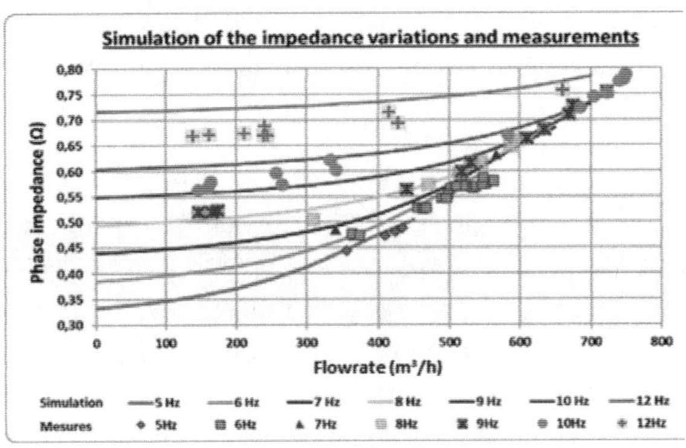

Fig. 5: Measurements and simulation of the impedance variations

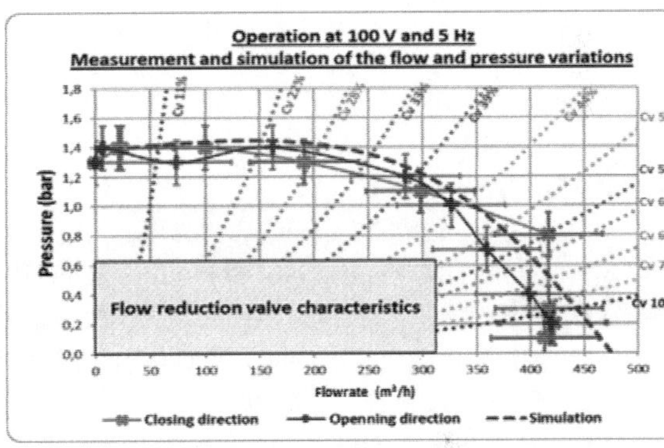

Fig. 6: Measurement and simulation of the flow and pressure variations

In order to obtain the experimental data required to plot the EM pump's characteristic at low power, the campaign involved performing tests by operating the flow reduction valve and by imposing the EM pump voltage and frequency. The measurements were post-processed to produce the following curves in Fig. 6: The dotted curve shows the pressure drop in the circuit according to the CFD assumptions indexed to the opening of the flow reduction valve expressed as a percentage. The dashed curve shows the simulation of the flow and pressure variations in the EM pump. The measurement points are represented by markers and interconnected to illustrate the variations in the series of measurements. The experimental follow the simulation results consistently, the action to close the flow reduction valve reduces the flow and increases the pressure drops, therefore the EM pump adapts well to variations without any regulating action on the power supply.

The campaign for the high flow rate involved performing tests by operating the voltage and frequency of the EM pump by imposing the flow reduction valve characteristics (Fig. 7). The measurement points are represented by marker, the dashed curve shows the linear regression and the standard deviation 10%. Fig. 8 present the first experimental data collected during the operative condition of the test programme aiming to establish performance maps for high flowrate. The dashed curve shows the simulation of the flow and pressure variations in the EM pump as function of the increase in the effective voltage at 15Hz. Markers represent the measurement points obtained during the operative condition. The experimental load curve at high power follows the simulation results consistently.

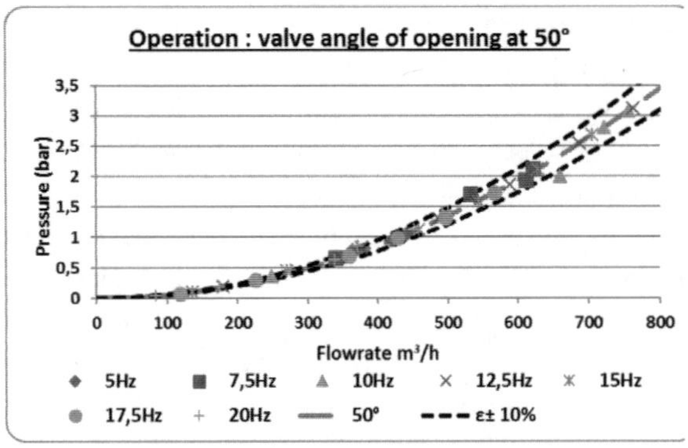

Fig. 7: Load curve for a valve angle at 50°

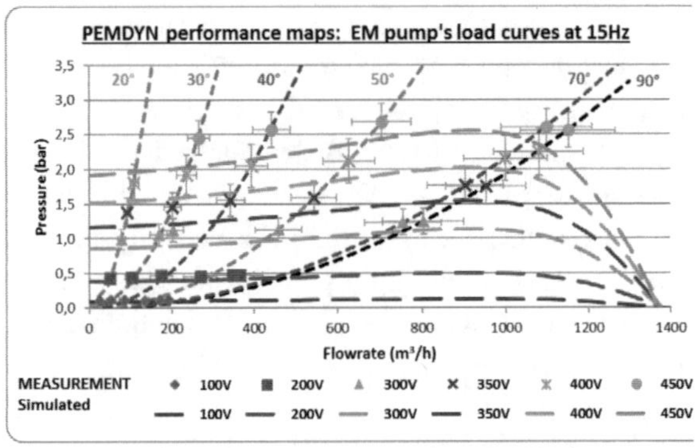

Fig. 8: PEMDYN load curve at 15Hz

IV. CONCLUSION

The results of the test campaigns performed validate the fact that the flow and the pressure vary consistently with the results of simulations performed beforehand. However, the actual sensor are not accurate enough to permit the analyses foreseen by the experimental programme, particularly with respect to post-processing the electrotechnical quantities and analysing the magnetohydrodynamic behaviour of the sodium flow. A dedicated sodium instrumentation was manufactured to be able to measure inlet and outlet static pressures and local flowrate of the EMP.

REFERENCES

1. E. Martin Lopez, Y. Delannoy, F. Benoit, A. Muñoz Medina, R. Martini, S. Vitry. *Analytical and numerical study of the stalling phenomenon in an electromagnetic pump for a sodium fast reactor* ICAPP2018 April 8-11, 2018
2. E. Martin Lopez, Y. Delannoy, F. Benoit, R. Martinie, S. Vitry, "*of electromagnetic annular linear induction pump*," Proceedings of ICAPP 2017, Kyoto, Japan (2017).
3. S. Vitry, L. Goldsteins, C. Biscarrat, F. Benoit, S. Madeleine, F. Dechelette, O. gastaldi. *Experimental and numerical analysis of magnetic field spatial measurements inside an electromagnetic pump channel duct. PAMIR 10, June 20-24, 2016 Cagliari, Italy*
4. L. GOLDSTEINS, *"Experimental and numerical Analysis of behavior of electromagnetic annular linear induction pump, "* PHD, (2015).

IOP Conf. Series: Materials Science and Engineering **424** (2018) 012018 doi:10.1088/1757-899X/424/1/012018

Electron beam melting toward inclusion-free titanium alloys

H. Funagane[1]

[1] Advanced Technology Research Laboratories, Nippon Steel and Sumitomo Metal Corporation, 20-1 Shintomi, Futtu Chiba, 293-8511 Japan

Corresponding author : funagane.zq3.hitoshi@jp.nssmc.com

Abstract
Titanium ingots for airplanes must satisfy high standards of quality, removal of inclusions is essential. For the purpose of reducing the risk of contamination by LDI, a new way of removal of LDI has been studied. In the present study, marangoni flow induced by Intensive Heating of electron beam (IH) has been considered to control a state of a flow of molten metal in a cold crucible. Numerical simulations and an experiment have verified that LDI bounding for a lip was not able to go through the IH and stayed in a hearth.

Key words : Titanium, Electron beam melting, Removal of inclusions

1. Introduction
Titanium ingots or titanium alloys ingots for airplanes must satisfy high standards of quality, removal of inclusions is essential because they cause cracks in aero parts [1]. Since inclusions stay in raw material, removing them in the melt process is necessary. As one of the ways of melting raw material, electron beam melting with a cold crucible is used (see Fig.1) [2]. On the one hand, inclusions heavier than molten metal (called High Density Inclusions (HDI)) such as tungsten carbide sink in molten metal and stay in a hearth, so that the chance of contamination is almost zero. On the other hand, since inclusions like titanium nitrides and titanium oxides have porous structure, though their true densities are greater than molten metal, some of them move on the surface of molten metal and flow into a mold. This kind of inclusions is called Low Density Inclusions (LDI). Therefore, demands on reducing the risk of contamination by LDI are still high even these days [3, 4]. For the purpose of manufacturing inclusion-free ingots, a new way of removal of LDI has been studied.

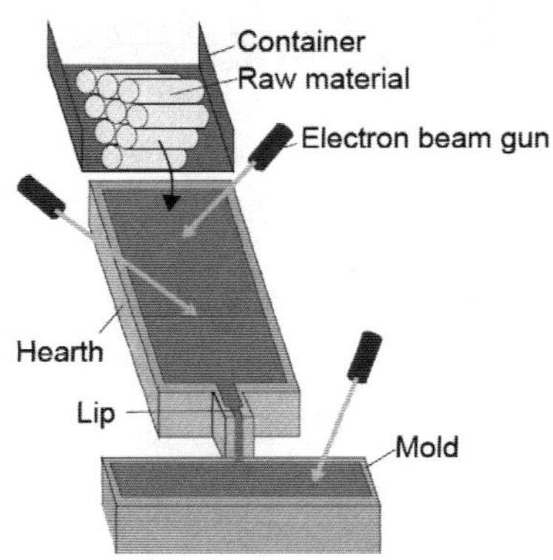

Fig. 1: An example of an electron beam furnace.

2. Methodology
It is well known that if there is a temperature gradient on the surface of fluid, Marangoni flow is induced. In the case of titanium, it flows from a high temperature place to a low temperature one. Thus, if some place of molten metal is intensively heated by electron beam (hereafter, we call Intensive Heating (IH)), marangoni flow originating there is induced because temperature of the heating place is higher than its surroundings. We thought that IH might be useful to control a state of a flow of molten metal in a cold crucible. More precisely, if molten metal in the crucible is heated without any reason except for keeping the temperature, since a uniform flow from the upstream side to the downstream side is formed, LDI moving on the surface of molten metal flow into a mold. But, if IH shown in Fig.2 is carried out, marangoni flow bounding for the upstream side from the IH zone is induced. As a result, LDI bounding for the mold stay in front of the IH zone (see Fig.3) and may be trapped or dissolved in the course of time.

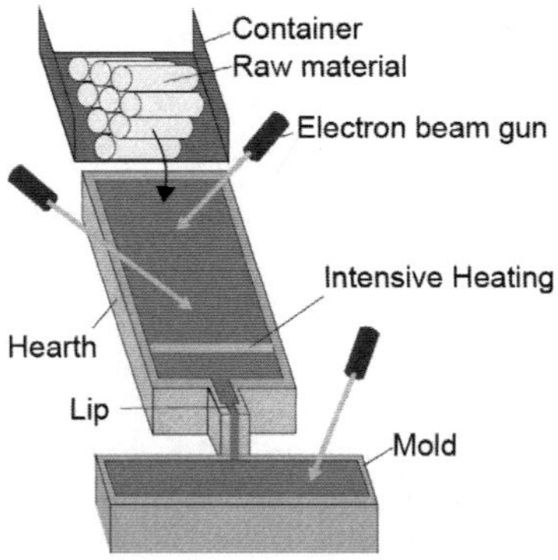

Fig. 2: An example of a beam path of Intensive Heating.

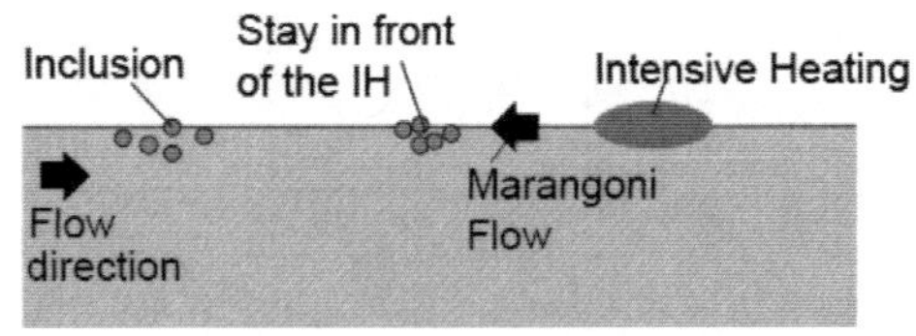

Fig. 3: Mechanism of removal of inclusions by Intensive Heating.

3. Numerical simulation

In order to verify whether Marangoni flow induced by IH can stop LDI moving on the surface of molten metal, a computational model describing a molten metal flow and an inclusion behavior in a cold crucible has developed [5-7] and numerical simulations have been carried out.

3.1. Computational model

Molten metal flow is described by the following equations:

$$\frac{\partial \rho}{\partial t} + \nabla \cdot (\rho \vec{v}) = 0 \tag{1}$$

$$\frac{\partial}{\partial t}(\rho \vec{v}) + \nabla \cdot (\rho \vec{v}\vec{v}) = -\nabla p + \nabla \cdot (\bar{\bar{\tau}}) + \rho \vec{g} - \rho \beta'(T - T_0)\vec{g} + \frac{(1-\beta)^2}{(\beta^3 + \varepsilon)} A_{mesh} \vec{v} \tag{2}$$

$$\bar{\bar{\tau}} = \mu[(\nabla \vec{v} + \nabla \vec{v}^T) - \frac{2}{3} \nabla \cdot \vec{v} I] \tag{3}$$

$$\frac{\partial}{\partial t}(\rho H) + \nabla \cdot (\rho \vec{v} H) = \nabla \cdot (k \nabla T) \tag{4}$$

$$H = c_p T + \beta L \tag{5}$$

$$\begin{cases} \beta = 0 & \text{if } T < T_S \\ \beta = 1 & \text{if } T > T_L \\ \beta = \dfrac{T - T_S}{T_L - T_S} & \text{if } T_S < T < T_L \end{cases} \tag{6}$$

Here, t is time, ρ is density, $\vec{v}$ is velocity, p is pressure, $\bar{\bar{\tau}}$ is stress tensor, μ is viscosity, I is unit matrix, $\vec{g}$ is gravitational acceleration, β' is thermal expansion coefficient, T_0 is reference temperature, β is liquid fraction, $\varepsilon = 10^{-3}$, $A_{mesh} = 10^5$, k is thermal conductivity, T is temperature, c_p is specific heat, L is latent heat, T_S is solidus temperature, T_L is liquidus temperature.

Behavior of inclusion is written by the below equation:

$$\frac{d\vec{u}_p}{dt} = \frac{\vec{v} - \vec{u}_p}{\tau_r} + \frac{\vec{g}(\rho_p - \rho)}{\rho_p} \tag{7}$$

$$\tau_r = \frac{\rho_p d_p^2}{18\mu} \frac{24}{C_d Re} \tag{8}$$

$$Re = \frac{\rho d_p |\vec{u}_p - \vec{u}|}{\mu} \tag{9}$$

Here, the first term of the right hand side of Eq. (7) represents drag force. And, τ_r is relaxation time, $\vec{u}_p$ is particle velocity, ρ_p is particle density, d_p is particle diameter, C_d is drag force coefficient [8], Re is relative Reynolds number.

3.2. Numerical conditions and results

Behavior of LDI moving on the surface of molten metal has been investigated by the computational model under the following assumptions:

- Electron beam
  - ① Amount of heat: 0.25 MW
  - ② Scan velocity of an electron beam gun: 1.6 m/s
- LDI
  - ① Diameter: 5.0 mm
  - ② Density: -10 % lower than that of molten metal

A numerical result of behavior of LDI without IH is shown in Fig.4. In it, molten metal flows from the left hand side to the right hand side, so that the trajectory of LDI shown by a solid line is reaching to the right edge. This means that LDI can flow into a mold.

Numerical results of the case with IH are shown in Fig.5. In the figures of temperature, some place where temperature becomes highest is corresponding to a beam spot. In other words, in state 1, the beam spot is located at one of the edge of a beam path. In state 2, the spot is located at the center of the path. In state 3, the spot is located at other edge of the path. In the figures of behavior of LDI, the trajectories of LDI shown by a solid line stop at near the center of the hearth. That is, IH can stop LDI. In other words, IH can reduce the risk of contamination.

State	Surface temperature	Behavior of LDI
1		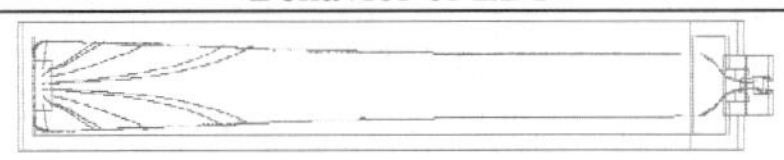
	300 K – 2200 K	

Fig. 4: Behavior of LDI without Intensive Heating.

State	Surface temperature	Behavior of LDI
1	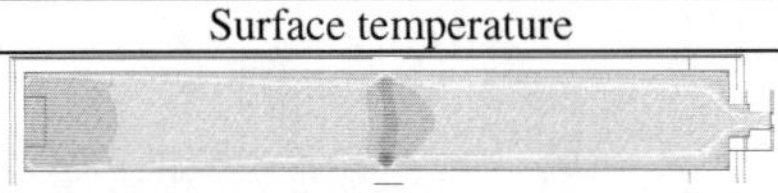	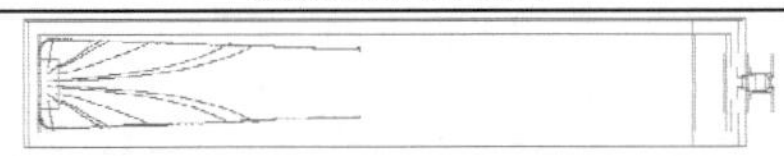
2	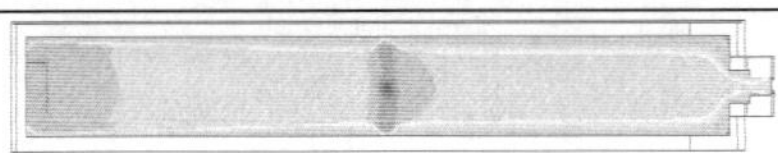	
3	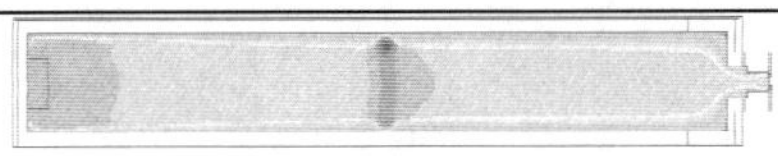	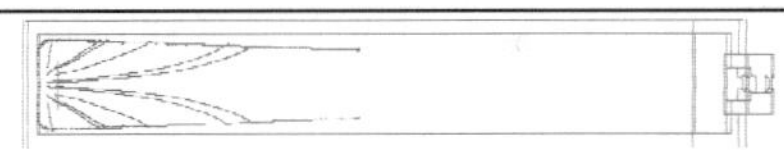
	300 K – 2700 K	

Fig. 5: Behavior of LDI with Intensive Heating.

4. Experiment

In order to verify whether Marangoni flow induced by IH can stop LDI moving on the surface of molten metal even in an actual furnace, an experiment carried out. In it, a piece of graphite mixing into raw material was used as a tracer because the density was lower than that of titanium and the boiling point was more than 3000 ℃. Time transition of a piece of graphite without IH is shown in Fig. 6. The graphite moving on the surface of molten metal from the upstream side to the downstream side finally flowed into a mold. Meanwhile, that with IH is shown in Fig. 7. The graphite was stopped at IH zone and was not able to go through. That is, the effect of IH has been verified.

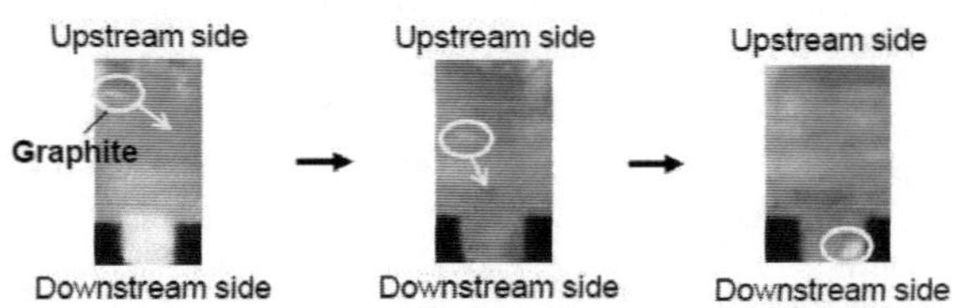

Fig. 6: Time transition of a piece of graphite without IH.

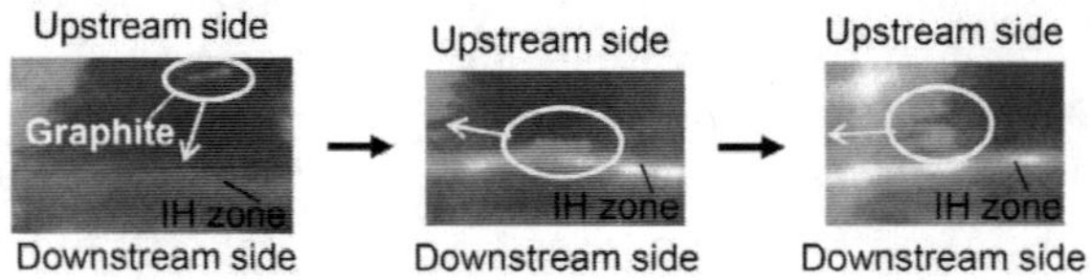

Fig. 7: Time transition of a piece of graphite with IH.

Acknowledgment
The authors appreciate all coworkers of this study for their support, encouragement and advices. Above all, help of Mr. Takeda and engineers at Naoetsu Works for the experiment was essential, so that we thank them enough.

References

1. A. Mitchell, ACTA METALLURGICA SINICA (ENGLISH LETTERS), 12 (1999), 283-296.
2. A. Mitchell, Materials Science and Engineering A, 263 (1999), 217-223.
3. J.P. Bellot, B. Foster, S. Hans, E. Hess, D. Ablitzer and A. Mitchell, Metallurgical and Materials Transactions B, 28B (1997), 1001-1010.
4. J.P. Bellot, E. Hess and D. Ablitzer, Metallurgical and Materials Transactions B, 31B (2000), 845-854.
5. J.P. Bellot, V. Descotes and A. Jardy, Numerical Modeling, 65 (2013), 1164-1172.
6. X. Zhao, Master's thesis of the University of British Columbia, (2006).
7. Z. Zhang, Master's thesis of the University of British Columbia, (2013).
8. S.A. Morsi and A.J. Alexandar, J. Fluid Mech., 55 (1972), 193-208.

Electroslag process for better titanium deposition morphology

E. Platacis[1], I. Kaldre[1], E. Blumbergs[1], V. Serga[2]

[1]Institute of Physics, University of Latvia, Miera 32, Salaspils, LV-2169, Latvia
[2]Institute of Inorganic Chemistry, Riga Technical University, 3/7 P. Valdena, Riga, LV-1048, Latvia

Corresponding author: Imants.kaldre@lu.lv

Abstract
Most common process for titanium production nowadays is Kroll process. This process has several drawbacks. Process is cyclical, and titanium comes out in sponge like form. This means that process is long, reaction is uncontrollable and requires post-processing of titanium sponge, which is energy consuming. In this work we present the idea to use Kroll process combined with electroslag process to produce titanium in continuous, more efficient process. Main idea is to use heated reactor where the initial reaction products (titanium tetrachloride and magnesium) are injected in liquid and gaseous phases, and reaction takes place above liquid slag in gaseous phase. Slag layer ensures titanium separation from unreacted magnesium and magnesium chloride, and other reaction waste products. Titanium droplets filtrate through the slag and is separated from the other fractions. In this paper we test the feasibility of this idea by studying the electroslag remelting of a titanium electrode in a specially prepared laboratory reactor. Chemical analysis is done to measure the composition of reaction products.

Key words : Electroslag melting, Kroll process, Titanium

Introduction

An essential part of the Kroll process titanium production is post processing of the titanium sponge, which forms after the reaction [1]. Basic reaction is magnesium thermal reduction of titanium chloride.

$$TiCl_4 \text{ (gas)} + 2\,Mg \text{ (liquid/gas)} = Ti \text{ (solid)} + 2\,MgCl_2 \text{ (liquid)} \quad (1)$$

This reaction is exothermal, the thermal dissipation for reaction (1) at 1100 K is equal to 686 kJ/mol [2]. This heat is enough to conduct the process without heat supply from the outside in industrial scale reactor. The lowest temperature limit of the reduction is restricted to the temperature of $MgCl_2$ melting (714°C) and the highest temperature limit is 975°C, when the temperature is higher, titanium actively interacts with iron, thus steel walls cannot be used. Practically, the process is conducted at 800-900°C in 1-3 m^3 large cylindrical reactors made of heat resistant steel [1-3]. At the same time, the optimal temperature for the reaction is 1200-1400°C [4]. At such a high temperature reaction takes place in a fully gaseous phase, and reaction products condenses on cooler surfaces. To ensure such temperature regime extra energy should be injected. One of the ways is insert additional heat via electric arch, which allows to increase temperature in the core of the reactor. Electric arch also improves the mixing of the reaction products and induces significant motion into the liquid slag. At temperature above 1200 °C reaction at gas state takes place and titanium and reaction products forms as a powder at the surface of the liquid slag. Due to the density difference titanium droplets sinks through the slag, while $MgCl_2$ remains at the surface and can be periodically drained from the reactor. Thus, in this case slag layer acts as a smart membrane separating the reaction products.

In this work preliminary experimental results acquired by small laboratory scale experimental setup is presented and numerical and analytical results showing the potential of this technology is given. Basic idea is to use high temperature reactor where titanium tetrachloride is reduced with magnesium, while liquid slag layer is placed bellow reaction zone as shown in principal scheme in Fig.1. Reactor walls are water cooled stainless steel and solid slag layer is forming on the walls thus protecting them from contact with titanium at high temperature. As solid slag is a poor thermal conductor, this layer also serves as thermal insulation limiting heat flux through the walls. For the elaboration of the technical parameters of the process in the first experiments NaF was used as a slag, but it is planned to investigate the performance and compare several different slags (CaF_2, NaCl and others).

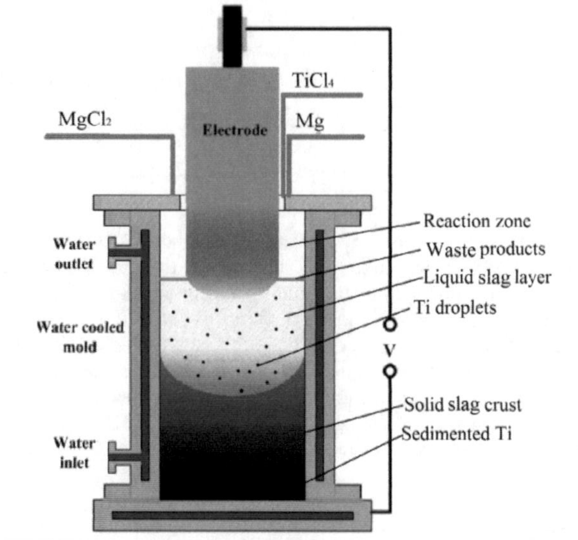

Fig. 1. Scheme of the combined Kroll and electroslag processes

Results and discussion

Small scale experimental facility has been designed and built to test the reaction kinetics at elevated temperature and to test the validity of the idea to use slag mixture as semi-transparent membrane for reaction product separation. First experiments are done using titanium electrode, which partly melts. The aim of these experiments is to test the hypothesis about titanium droplet sinking through the slag layer and to measure the morphology and chemical composition of sedimented products at the bottom. Inner diameter of the reactor is 60 mm. Reactor is filled with liquid slag at height of 120 mm. Titanium electrode of 14 mm in diameter is immersed into the slag. Argon was fed to the chamber at a rate of 10 l/min, maintaining small overpressure. Overall scheme of the experimental setup is given in Fig. 2(a).

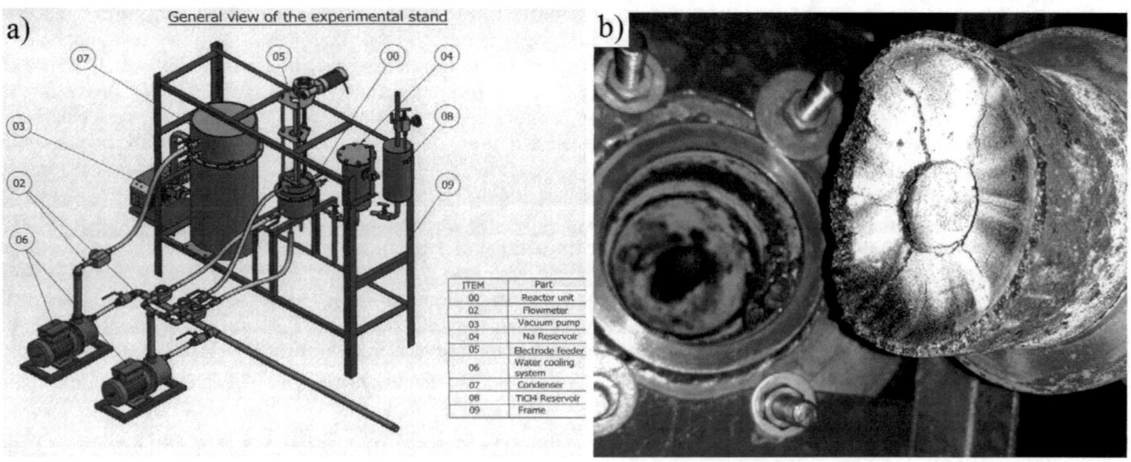

ITEM	Part
00	Reactor unit
02	Flowmeter
03	Vacuum pump
04	Na Reservoir
05	Electrode feeder
06	Water cooling system
07	Condenser
08	TiCl4 Reservoir
09	Frame

Fig. 2. a) Scheme of the experimental setup; b) Titanium reactor and solidified content after reaction

Solidified NaF slag after the experiment is shown in Fig.2(b). On the right side is solid NaF, and on the left side Ti electrode and inner surface of the reactor with a slag crust is visible. Approximately 10 kW (20 V, 500 A) of electric power is applied during experiment. Titanium electrode starts to melt in this process and titanium sinks through the slag layer. Titanium droplets at the bottom were analyzed by X-ray diffractometry, and results show that droplets consist of 80 % titanium and 20 % of titanium oxide TiO. According to the XRD analysis (Fig. 3), material contains a titanium as the main phase of the titanium and TiO as an impurity [5]. Elemental analysis of slag showed the presence of small amounts of titanium in it (0.06-0.50 wt.%). This may be explained by the fact that small droplets sink slowly and was trapped in slag during solidification after the heating was turned off.

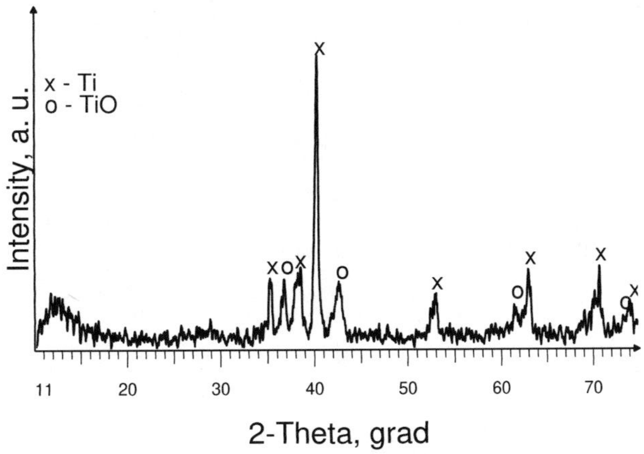

Fig. 3. XRD pattern of the titanium droplets

Slag layer may significantly change the morphology of the titanium deposition as titanium droplets are filtering through it rather than solidified directly in the reactor. It is shown that this process can be done in small semi-continuous reactor by electric arc heating. Numerical simulation demonstrates that reaction products can be heated up to 1500 °C by electric arch (Fig.4(b,c)), while reactor walls are protected by thin solid crust of the slag material. Numerical models to describe thermal regime, fluid flow and electromagnetic processes in reactor are developed. Axisymmetric Comsol 5.0 numerical model with 8000 cells was developed. Density of the slag is described by following linear function: $\rho[kg/m^3]=3070-0.39\cdot T[^oC]$. Thermal conductivity of slag is 1 W/m·K, electric conductivity 150 sim/m, viscosity 0.006 Pa·s [6].

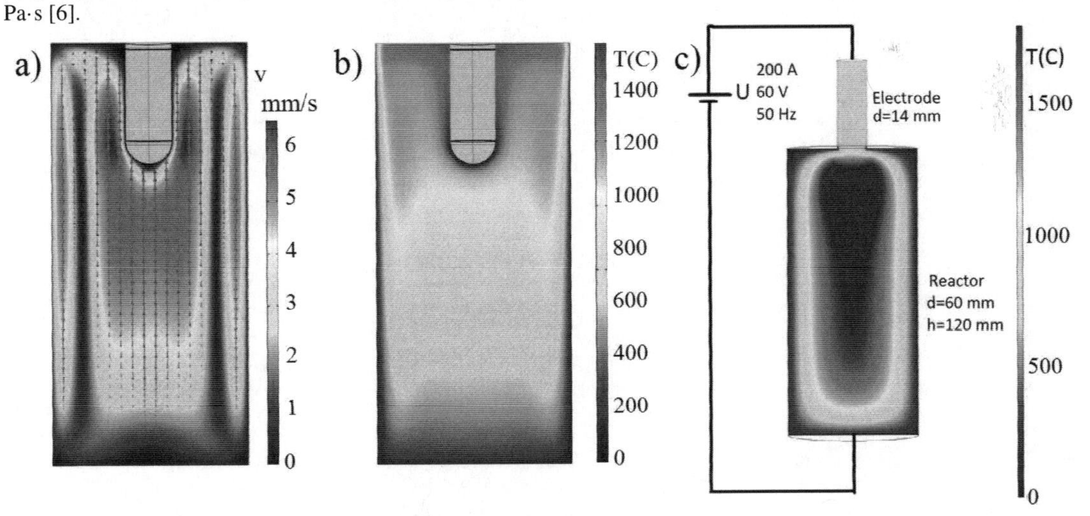

Fig. 4. a) Numerically calculated velocity distribution of the liquid slag; b) Temperature distribution in reactor chamber; c) temperature distribution if the slag is solid

Numerical models show that significant natural convection appears in the crucible if slag is fully liquid. Convective flow of 5 mm/s completely change the temperature distribution in the reactor as shown in Fig. 4(a).

Numerical simulations show that in case of stationary volume high temperature can be reached near the electrode as shown in Fig. 4(c). Still the precise location of solidification interface is difficult to locate, but solid crust layer is estimated to be 3-5 mm thick. Natural convection of the slag significantly affects the temperature distribution in reactor. Nevertheless, numerical model confirms that even with convection expected temperature is reached near the electrode. One of the problems is that data on the properties of the slag is fragmentary and they may vary significantly depending on temperature, impurities and other factors.

Conclusions

It has been shown that Kroll process can be replicated in small scale reactor with additional electric arch heating. Series of experiments has been done, showing that reactions take place. First experiments with liquid slag layer shows that titanium droplets sink to the bottom of the reactor through the liquid slag layer. Effective removal of the reaction waste products remains one of the experimental obstacles, because slag layer at the surface significantly decreases the process efficiency. Numerical simulations of temperature distribution and liquid slag motion has been done. It is shown that temperature distribution is largely affected by liquid slag natural convection which arises due to localized heat distribution near the electrode. Numerical models confirm that we can expect that reactor walls are protected by solid slag layer. The obtained data demonstrate the possibility of combining the Kroll process and the electroslag process with the installation described in this paper. Within current research project it is planned to developed larger experimental setup to test the whole process in laboratory scale.

Acknowledgment

This work was supported by the ERDF project No. 1.1.1.1/16/A/085. "Electroslag process for better titanium sediment morphology"

References

1. R.B. Subramanyam, Bulletin Mater. Sci, 16 (1993), 433-451
2. C.R.V.S. Nagesh, C.S. Rao, N.B. Ballal et al. Metall and Materi Trans B, 35 (2004), 65-74
3. A.V. Tarasov, Titanium Metallurgy. Moscow, Academkniga, 2003, 328 p.
4. Yu.N. Olkhov, S.V. Ogurtsov et al, Investigations in Titanium Chloric Metallurgy (Moscow, Metallurgiya, 1969, 302.
5. S. B. Ptel, Journal of Materials Chemistry B, 23 (2014), 3559-3706
6. Q. Wang, Z. He, B. Li et al. Metall and Materi Trans B, 45 (2014), 2425-2441

IOP Conf. Series: Materials Science and Engineering **424** (2018) 012020 doi:10.1088/1757-899X/424/1/012020

Pulse electromagnetic force microstructural control at continuous billet casting 7xxx Al-alloys

M.A. Slazhniev[1], K.H. Kim[1], H.S. Sim[1], S.W. Kim[1], W.J. Kim[1]

[1]Dong San Tech. Co., ltd., 169 Hamansandanro, Hamangun, Gyeongnam, Korea

Corresponding author: slazhnevmykola@gmail.com.

Abstract

In this study, the original equipment and methods for the macro/microstructure managed formation in the crystallizers of continuous casting processes by the applying of the amplitude modulated (AM) pulsing magnetic fields (PMF) it the electromagnetic stirrer (EMS) has been developed and tested.

Method has provide the complex electrodynamic, hydrodynamic and acoustic pulse-force processing of unsolidified casting product portion for providing equalizing of solidification front in the billet mushy zone. Resulted non-dendritic type of microstructure has been provide by the means applying: Pulse magnetic field with periodically changing in time amplitude (AM-PMF); Pulse amplitude-modulated rotating melt stirring (AM-PRMF); Pulse amplitude-modulated 3D-helicoidally rotating melt stirring.

Study efficiency applying of unidirectional axial pulse magnetic field (AM-PMF) with pulse frequency in range 5~15Hz, carried frequency in range 22~55Hz, at intensity of MF 0.05~0.1T, for semi-continuous 7xxx alloys (analogue 7075 alloy) 7'billet casting on micro- and macrostructure billet formation has been show reaching the stabile effect for non-dendrite state microstructure formation in all billet cross-section. As well-achieved uniform and homogeneity microstructure state with average grain size not more 100μm, and for middle and central billet part in range 60~80μm, in 3~5 times lowest instead grain sizes 200~650μm at DC casting without EMS, or instead 150μm grain sizes as for conventional rotating EMS. As well, fully solved problems of axial and structural segregation, hot-crack ability or micro-cracks on grains edges and micro-porosity occurring.

Key words: High-strength aluminum alloys, continuous billet casting, electromagnetic stirring, pulse magnetic field, microstructure, grain size

Introduction

As known at the continuous casting of the billets from the 6xxx or 7xxx series high-strength aluminum alloys are very important to provide the special hydrodynamic, temperature and heat-force impact conditions, that there are playing leading role at the macro- and microstructure formation at the billet solidification [1]. It has comprised in the creation of the required conditions of the solid melt skin formation at the water-cooled crystallizer walls, at that to provide high intensity cooling of the solidifying billet, included removing heat from the melt volume in the full relationships with the billet pulling speed at the continuous casting process.

Depend from the billet size 2~10 inches the formation equiaxial by billet cross-section of microstructure are so very important problem [1, 2]. Solution of the dispersed fine microstructure in the all billet cross section are comprised in the creation into liquid state alloy in crystallizer the special 2D~3D contactless melt stirring for creation small different by the temperature and chemical compositions of liquid phase behind front of solidification [1, 2]. This effect can achieved through the conversion of the solidification front surface of billet mushy zone from crater/parabolic type into a smoothed linear, flat front. As well, no less important role of the electromagnetic melt stirring play the force hydrodynamic interaction in the mush zone [1, 3], where are melt take the semi-solid state.

As known at the continuous casting from 7xxx-series of high-strength wrought aluminum alloys on today widely apply electromagnetic stirrer (EMS) by the AC electromagnetic fields for the creation into melt of electromagnetic forces for melt stirring. Most applied for that it is rotating, traveling or combined electromagnetic fields, that using in the 3-phase's electromagnetic stirrers of radial, running (linear) or combined constructive executing.

However, at the applying contactless melt stirring so much important the electric current shape and mode, that in result are predict mode and direction intermixing, as well melt speed stirring, which also have some limitation. The last interesting trend in the EMS scientific researches take placing into creation of the pulse mode impact to the melt into continuous casting crystallizers by the changing in time magnetic field intensity, with different frequency, amplitude and shapes of pulses, using poly-harmonic AC, modulated by amplitude, frequency, phases and et.

Developing method and device for pulse EM-force microstructural control at continuous billet casting 7xxx alloy

Conceptual approach for the creation improved special regimes (modes) of 3D-microturbulent's submerged electromagnetic melt stirring for EMS-crystallizer conclude in creation of pulse magnetic fields/forces, at using amplitude modulation 3-phase's AC-magnetic fields with modulation ratio (M), frequency (f), magnetic intensity (B).

For practical researching in this study has selected for three main modes: - Pulse magnetic field with periodically changing in time amplitude (AM-PMF); - Pulse amplitude-modulated rotating melt stirring (AM-PRMF); - Pulse amplitude-modulated 3D-helicoidally rotating melt stirring.

In contrast to existed 3-phase's usually applying radial 3~9 poles EMS devices, there are original design of 3-phases's EMS-mold (fig. 1, 3), comprised: vertical water-cooling round short crystallizer (150~225 mm length); 6-poles EMS electromagnetic system, contained by three individual, mechanically and electrically divided each by one, C-shaped 2-poles electromagnets. Each one of it are located in perpendicular to longitude axes of crystallizer under angle 120° (180° or 90° accordingly for double or fourth EM-system design) has been developed and tested (fig. 1~3).

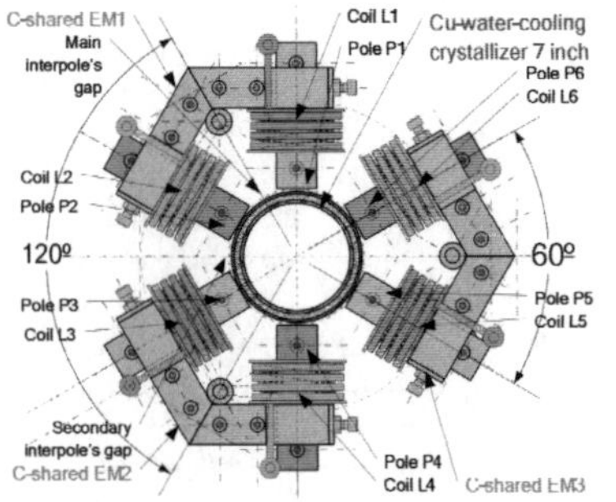

Fig. 1. Scheme of 6-pole's EMS device for electromagnetic stirring.

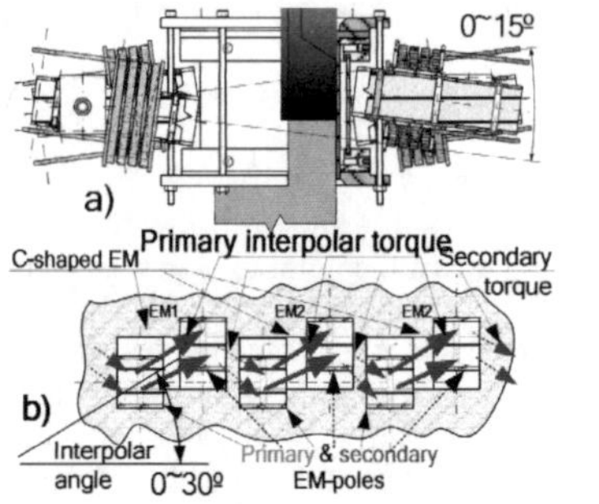

Fig. 2. Tilting EMS-poles (a) and its spatial distribution (b) at helicoidally torque.

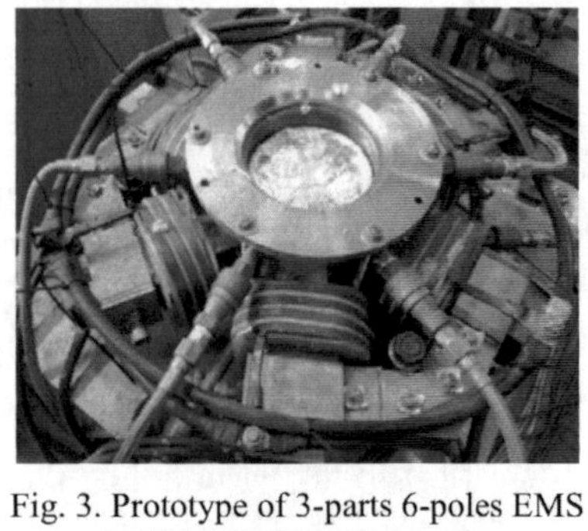

Fig. 3. Prototype of 3-parts 6-poles EMS in AM-PRMF testing mode.

Each one C-shaped electromagnet (fig. 1) consisted from main C-type magnetic core, placed around crystallizer and have by two pole's, adjustable by lengths and vertical tilting angles between each other (±0~15°), relative to the horizontal plane (fig. 2 a, b). Each nearby located poles has directed perpendicularly to longitude axes of crystallizer (billet) with angle between each for 60°, optionally.

Unlike to existed on today methods at creation poly-harmonic AC-magnetic fields for EMS, comprising amplitude modulated mode, in present development has applied method of algebraic superposition of two harmonic magnetic fields $B_{1n}\sin(\omega_{1n}t+\varphi_1)$, $B_{2n}\sin(\omega_{2n}t+\varphi_2)$, with different intensity ($B_{1n}\neq B_{2n}$) and frequency ($f_{1n}\neq f_{2n}$). Each one magnetic fields $B_{1n}(t)$ and $B_{2n}(t)$ generating by the two nearest located electromagnetic poles (P1n and P2n) in its inter-poles gaps for each C_n-shaped electromagnet, at the supplying it's coils (L_{1n}, L_{2n}) by the AC currents (I_{1n}, I_{2n}) with different frequencies and voltages (U_{1n}, U_{2nn}). For that the special EMS-power supply unit, consisting as minimum two electrically divided 3-phases AC power supplying channels (Ch1, Ch2), with adjustable and controlled voltages (U_1, U_2), frequency (f_1, f_2) and current (I_1, I_2), based on the standard frequency inverters (0.1~400Hz) has been developed.

The pulse magnetic field with periodically changing in time amplitude (AM-PMF) mode has been create at the supplying EMS-coils (L1, L3, L5), connected by star and located on left's poles each C-shaped electromagnets, and coils (L2, L4, L5), are placed on it right poles, by the harmonic AC-electrical currents $I_1(t)=I_1\sin(2\pi f_1 t)$ and $I_2(t)=I_2\sin(2\pi f_2 t)$. Each AC has different frequency ($f_1\neq f_2$, but $f_1<f_2$) but equal value for magnetic flux intensity $B_{1n}=B_{2n}=(U_n/f_n)$. At the its interaction in the C-shaped electromagnet inter-poles gap, and also in the gap between nearby located second and third C-shaped electromagnets (fig. 2b), has been created amplitude-modulated magnetic field $B_{AM}(t)$ (fig. 4a), as a result superposition of two magnetic fields $B_{1n}\sin(2\pi f_{1n}t+\varphi_1)$, $B_{2n}\sin(2\pi f_{2n}t+\varphi_2)$ with different frequencies (1).

$$B_{AM}(t) = B_{1n}\sin(2\pi f_{1n}t)+B_{2n}\sin(2\pi f_{2n}t) = 2B[\sin(2\pi\{f_{1n}+f_{2n}\}t)\times\cos(2\pi\{f_{1n}-f_{2n}\}t)] \quad (1)$$

Wherein shape, frequency and amplitude dedicated by the result of the algebraic product of two main harmonic component, with summed frequency $\sin(2\pi\{f_{1n}+f_{2n}\}t)$, - carrier frequency, and difference $\cos(2\pi\{f_{1n}-f_{2n}\}t)$ - envelope.

At the interaction resulted pulse magnetic field $B_{AM}(t)$ (fig. 4a) with the unsolidified casting product portion (liquid molten metal) in crystallizer, there are create the pulse amplitude modulated electromagnetic force (fig. 4b) by the Lenz's law, that are directed from around side to the ingot center in horizontal plane (fig. 5a, b), presented as (2).

$$F_{AM}(t) \approx kB_{AM}\,(1 - 0.5\cos(4\pi f_{1n}t) - 0.5\cos(4\pi f_{2n}t) - \cos(2\pi\{f_{1n} - f_{2n}\}t) + \cos(2\pi\{f_{1n} + f_{2n}\}t) \quad (2)$$

Pulse magnetic force $F_{AM}(t)$ (2) have an 5 components: 1) "1" constant component, that are create EM-pressure, providing reducing friction effect of billet with wall; 2) and 3) $0.5\cos(4\pi f_{1n}t)$, $0.5\cos(4\pi f_{2n}t)$ - as two half-amplitude double frequency vibratory components, accordingly for the summing harmonic magnetic fields B_{1n}, B_{2n};

4) $\cos(2\pi\{f_{1n} + f_{2n}\}t)$ - vibratory component with summed frequency and 5) $\cos(2\pi\{f_{1n} - f_{2n}\}t)$ - vibratory component with difference frequency, predicted shape, waveform and frequency of the electromagnetic force pulsation, - named as an modulation frequency. On fig. 4b an illustrated time diagram of influencing modulated by amplitude resulted electromagnetic force divided on normal and tangential components, for conditions f_1=50Hz, f_2=60Hz, M=1, B_{AM}<0.1T.

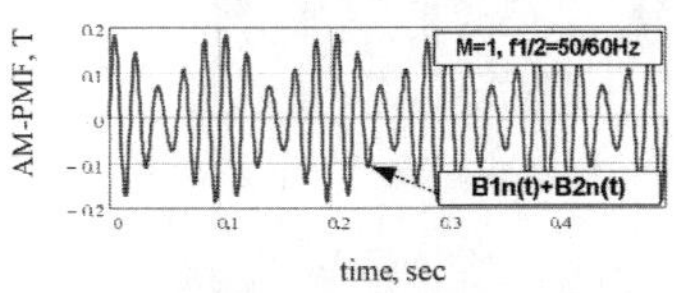

a) AMPM field, Δf=10Hz, M=1

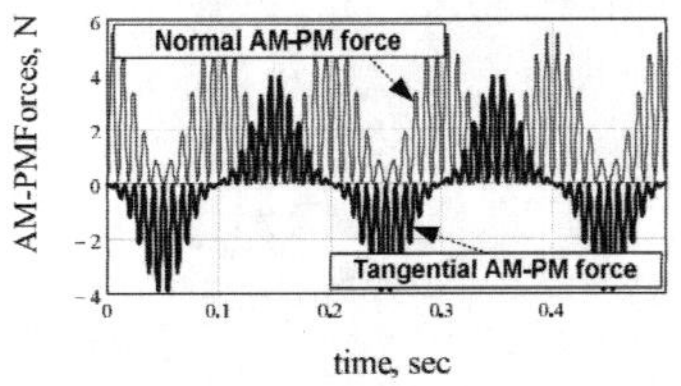

b) AM-PM force

Fig. 4: Time diagrams AM-PM field (a) and normal and tangential components of resulted AM-PM forces (b).

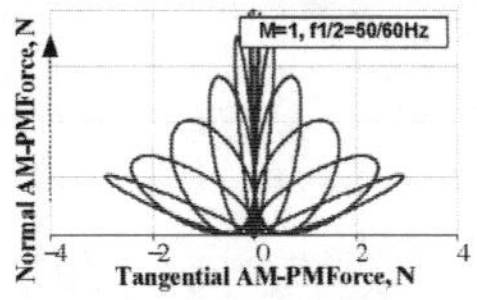

a) 2D-interpole PEMF

b)

Fig. 5: 2D-diagram impact direction for AM-PM force (a) and axial/radial melt stirring in crystallizer (b).

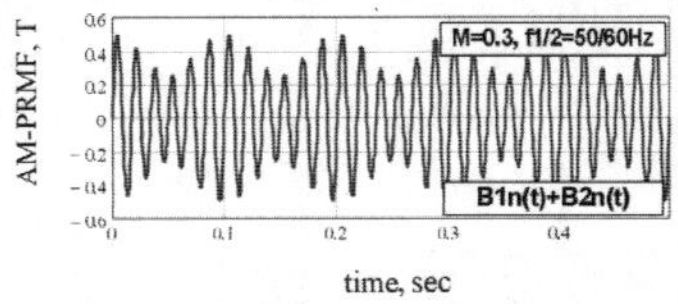

a) AMPRM field, Δf=10Hz, M=0.3

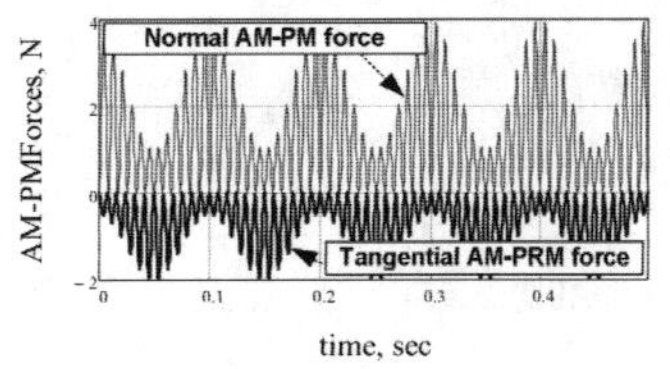

b) AMPM force

Fig. 6: Time diagrams AM-RPM field (a) and normal and tangential components of resulted AM-PM forces (b).

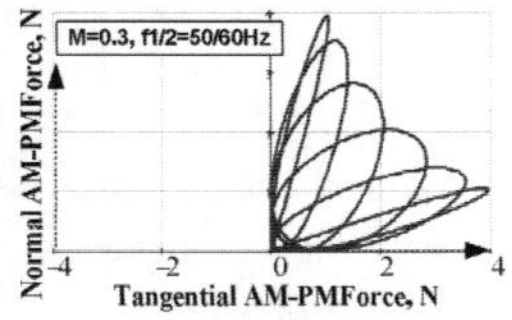

a) 2D-interpole PREMF

b)

Fig. 7: 2D-diagram impact direction for AM-RPM force (a) and axial/radial melt stirring in crystallizer (b).

5th of low frequency vibratory component (2) predict the main pulsation force hydrodynamic impact on the melt into crystallizer, and have perpendicularly direction to it center (fig. 5). As well, at the using two magnetic fields with different frequencies but equal intensity, summed interaction describing as periodical in time mode, when at the coincidence its phases – electromagnet working in the opposite mode, and after (πn) time transferring to the agreed mode (fig. 4b, 5a,b), that to explaining of tangential shaking stirring mode for case AM-PM forces with M=1.

For creation pulse amplitude-modulated rotating melt stirring (AM-PRMF) fig. 6~7, at combination axial pulse and radial pulse rotation stirring mode realizing at not equal value of the magnetic flux intensity $B_{1n} < B_{2n}$ and $M = B_{1n}/B_{2n}$= 0.7~0.3. In this case creating component of the amplitude-modulated pulse magnetic forces (AM-PM force), directed from around to ingot center and component of running (pulse rotating magnetic field) (fig. 7a). Resulted magnetic field (fig. 6. a) having additional tangential component's - $(B_{1n} - B_{2n})\sin(2\pi f_{1n}t)$ (see fig. 6b – blue colored line), activated of pulse rotation magnetic field (PRMF), (fig. 7) in the direction dedicated by the shift angles of the 3-phase electric current feeding coils L1, L3, L5 of EMS. As well, one from both summing harmonic fields, having an higher intensity (0.15T and 0.05T accordingly), also including that using 3-phase electric current, the higher components are creating the rotary torque, that are presented on fig. 7 as tangential component.

In additions, at adjusting of vertical tilting angles (±0~15°) for both poles of the C-shaped electromagnets (fig. 2), there are providing creation of tilted up rotation torque at pulse stirring mode. Additional AM-PRM forces having an vertical components have creating as minimum 3 (three), arranged around crystallizer pulse rotary torque with vertical (up stronger and down smaller), at performing helicoidally pulse rotary melt stirring mode into crystallizer.

Experimental evaluation microstructural control by pulse magnetic field for 7xxx alloys billet casting

Experimental researches efficiency applying of pulse magnetic field with periodically changing in time amplitude for providing the similar by grain size of micro- and macrostructure in the central, middle and surface billet parts as tested alloy was selected analog of the 7075 wrought aluminum alloy at it semi-continuous billet casting with 7inch diameter.

As a main vessel was used water-cooling coper cylinder crystallizer, having lengths 225 mm and internal diameter 185 mm. In the bottom side for the starting billet pulling process at first time was attached the starting head, further move down with managed speed. For the temperature range 7075 alloy 690~710°C accordingly to the cooling rate facilities of crystallizer 7inch billet pulling speed has been applied in range 60~80 mm/min.

For study of AM-EMS processing efficiency has been produced 3-types billets: 1st - DC casting process (fig. 8 a, b); 2nd - with applying pulse amplitude modulated electromagnetic unidirectional processing 10 Hz oscillation and carried frequency 55Hz (fig. 8 c, d), at the supplying electromagnetic coils of EMS by 60Hz and 50Hz electric current with equal magnetic field intensity. As 3rd mode, has been applied the two-components unidirectional pulse and pulse rotation stirring mode at modulation ratio M=0.3. For deeply studying of pulse hydrodynamic and acoustic impact for microstructural processing in researches the 50~60Hz magnetic fields range was selected, wherein on billet depth more 22 mm from crystallizer wall its intensity and electrodynamic impact have reduced in 2 times value.

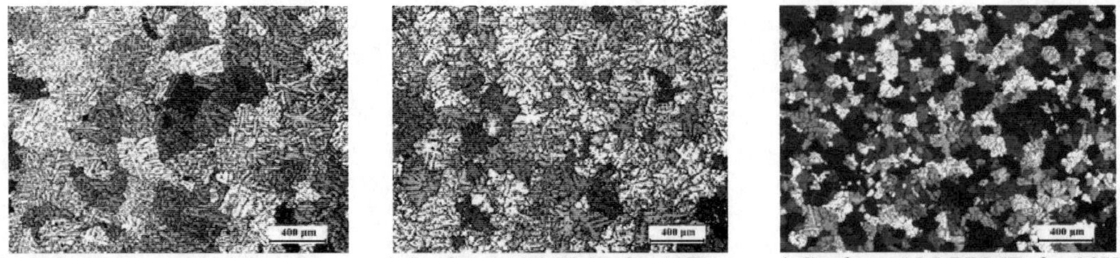

a) DC, surface c) Surface, AM-PMF, f_m=10Hz e) Surface, AM-PRMF, f_m=10Hz

b) DC, midle/center d) Midle/center, AM-PMF, f_m=10Hz f) Midle/center, AM-PRMF, f_m=10Hz

Fig. 8: Microstructure of 7inch billet from 7075 in casting state: - (a, b) DC casting (no EMS); - (c, d) pulse AM electromagnetic unidirectional processing (10 Hz); - (e, f) two-components pulse EM-rotation torque mode (10Hz).

For case DC casting fig. 8a on surface observing large dendrite microstructure with no uniform grain shapes and sizes in ranges 200~650μm. On depths from surface of 40~60mm appear more compact microstructure, similar to the non-dendrite type with characterized sizes of grains 80~250μm (fig. 8b). However microstructure characterized by micro-cracks, that further predicting occurs central macro-cracks, explained by large temperature gradient between surface and central part of billet at it solidification using continuous casting method.
Another picture have case applying pulse amplitude modulated electromagnetic unidirectional processing 10Hz oscillation and carried frequency 55Hz (fig 8 c, d). Surface microstructure characterized as mainly dendrite type (fig. 8c), however more compact with grain 50~300μm, that explained by the pulse force hydrodynamic impact on the front solidification in mushy zone in semi-solid state. But started from 20 mm from billet surface to center (fig. 8d) are presented homogeneity microstructure in non-dendrite state with grain sizes 60~130μm without micro-cracks or micro-porosity.
The more pure and fine structure was obtained for case applying of two components unidirectional pulse amplitude modulated and pulse electromagnetic rotation torque mode processing at 10Hz oscillation and carried frequency 55Hz (fig. 8 e, f). Additional pulse rotation torque in combination with unidirectional pulse impact has provide like more uniform and in 2 time more fine on billet surface microstructure (fig. 8e) grain size less than 150μm, and have view transformation dendrite type to non-dendrite state. Middle and central billet part (fig. 8 e) characterized as non-dendrite uniform and homogeneity state with grain size lowest 100μm, with no evident any cracks or porosity.

Conclusion

The original equipment and methods for the macro/microstructure managed formation in the crystallizers of continuous casting processes by the applying of the amplitude modulated (AM) pulsing magnetic fields (PMF) it the radial EMS has been developed and tested. Method of complex electrodynamic, hydrodynamic and acoustic pulse-force processing provide equalizing solidification front in the billet mushy zone. Study efficiency of unidirectional axial pulse magnetic field (AM-PMF) with pulse frequency 5~15Hz, carried frequency 50~57Hz, at intensity of MF 0.05~0.1T, for semi-continuous 7xxx alloys (analogue 7075 alloy) 7inches billet casting has been show reaching the stabile effect for non-dendrite microstructure formation in all billet cross-section. Two components EMS-processing by unidirectional AM-pulse and pulse electromagnetic rotation torque mode at 10Hz oscillation to provide uniform and homogeneity billet microstructure with average grain size not more 100μm, and for middle/central billet part 60~80μm, in 3~5 times lowest instead 200~650μm at DC casting, or instead 150μm for conventional rotating EMS. Fully solved problems of axial and structural segregation, hot-crack ability or micro-cracks and porosity occurring.

References

1. Y. Zuo, H. Nagaumi, J. Cui. Journal of materials processing technology, Vol. 197 (2008), 109-115.
2. L. Puzhaylo,V. Gavrilyuk, S. Seryi & et. (2012), Casting processes 5, Kyiv (Ukraine), 54~60.
3. V. Dubodelov, V. Fixsen, M. Slazhnev & et. (2005) Joint 15th Riga and 6th PAMIR International Conference on Fundamental and Applied MHD. – Riga, Jurmala, Latvia, – V. 2. – P. 77-82

MHD-physical modification effect on microstructure control of the 7xxx high-strength Al-alloys

S.W. Kim, M.A. Slazhniev, K.H. Kim, H.S. Sim, W.J. Kim

Dong San Tech. Co., ltd., 169 Hamansandanro, Hamangun, Gyeongnam, Korea

Corresponding author: slazhnevmykola@gmail.com

Abstract
In this study, the possibilities of the microstructure refinements effect of processing by electromagnetic and magneto-hydrodynamic means on high-strength casting aluminium alloys Al-Zn-Mg-Cu/Ni (0.2%Zr) system has been studied.
In the case of direct manufacturing of high-strength casting Al-alloy by the MHD-physical affecting, the unidirectional 120Hz pulse electromagnetic force(70~143.5N, 2.5~3.5m/s) created in working area(WA) of the magneto-dynamic installation electromagnetic pump has provide a fine emulsion distribution of intermetallic particles in liquid melt during intensive MHD-processing for melt capacity of 150~500kg. Microstructure of permanent chill-mold specimens (1`` rod and 3`` billet) has shown the unique non-dendrite microstructure with compact grains in range of 50~100μm, and distributed homogeneity by all cross-section, instead an over 250μm dendrites for the non-treated state at sampling from conventional melting furnace.
As well, applying MHD-processed alloy for low pressure casting method at manufacturing 3`` billet for 550mm lengths into graphite tube have an mainly compact globular microstructure, characterized as non-dendrite state with 80~120μm grain size and 50~80μm in the central part of billet. Achieved effect can be explained by an artificial heat-force effect of the pulse unidirected electromagnetic force, created in the MDI working area which affecting the solidification structure. In addition just 10 minutes for 150kg alloy by artificial pumping mode MHD melt processing was found to increase cast-ability of the Al-Zn-Mg-Cu casting high-strength alloy at pouring temperature of 680~720°C, wherein hot-cracks disappear, and hot-cracks effect is not occurs for the different conditions of the pouring speed.
Tensile stress test of the developed high-strength casting Al-Zn-Mg-Cu alloys has been shown 547~566MPa, 516~547 yield stress and 2.9~3.4% elongation for the case of chill mold gravity casting.

Key words : High-strength aluminum alloy, electromagnetic stirring, pulse magnetic field, microstructure, grain size

Introduction

In high strength aluminum casting, maintaining high level of the strength characteristics (450MPa~600MPa, >3% elongation) while reducing the main alloying elements (Zn, Mg, Mn, Cu in range 5~15% wt.) are very important technical challenge for creation of special casting method of super-high-strength aluminum alloys for wide applying for Production industry of the car, plane and others details by the casting methods.
For the case of 7xxx-series alloys casting application, the special contactless pre-casting processing was devised[1] for providing fine dispersed non-dendrite microstructure in state, and also for prevention of the casting defects, such as a hot-cracks, porosity in low pressure casting to the chill mold.
These effects can be achieved by basic two stages contactless multifunctional MHD/EM processing in liquid (preparation) and semisolid (at solidification) states. Secondly, by applying Ni and Fe elements, instead of Cu, for improving cast-ability and structure formation at the chill-mold casting especially for Al-Zn-Mg system.
For solving complex problem of manufacturing, refining by means of contactless EM/MHD impacts which using pulse electromagnetic forces and AC magnetic fields using on process in industrial scale (50~500kg alloy) are known. and such processing is implemented by existing magneto dynamic installations (MDI) for Al-alloys [2~4] (fig. 1). MDI is an induction channel furnace with electromagnetic pump that have been widely applied in industrial casting, melting and complex processing of non-ferrous metals and alloys [1, 2]. The main working principle of MHD-equipment based on the interaction of AC-current (5~20kA) with external AC-magnetic field (B) in the working area (WA) (fig. 1).
As result of the Fleming's Left Hand Rule in WA, the unidirectional pulsating (100~120Hz) and electromagnetic force (70~285N) are created [4]. It produce pumping of the molten metal from WA through central channel, follow or into crucible by circulation mode or charging into casting molds by EM pump [1] with managed pressure of 7~35kPa and flow rate (0.3~10kg/sec) by the low electromagnetic pressure principle. As well, the wide contactless electrodynamic (density of electric current 10~60A/mm^2), magneto-dynamic (pressure of 10~35kPa) and hydrodynamic (melt moving velocity of 0.1~5 m/sec) MHD-means can be applied for the preparation, processing and casting real product. This process is also used for low pressure casting [2, 3], dosing of melt by managed time, weight and flow rate and also for full cycle preparation and pouring of different Al-based functional material [3, 4].

Researching of MHD-physical modification effect on microstructure refinement on the Al-Zn-Mg-Cu (0.2%Zr) high-strength casting aluminum alloy 7xxx-series

The alloy used in first stage of the researches is high-strength aluminum alloys Al-Zn-Mg-Cu system (analog 7075) with different contents of Zn(5.5~8.5), Mg(2~3%) and Cu(1.5~2.3) and no more 0.2%Zr. For manufacturing alloy, unique methods are used. When MDI furnace started by pure aluminum in quantity 50kg and all necessary component was added step by step into MDI crucible in the master alloy state. The final capacity of melted alloy reached for 300kg as for test mode.

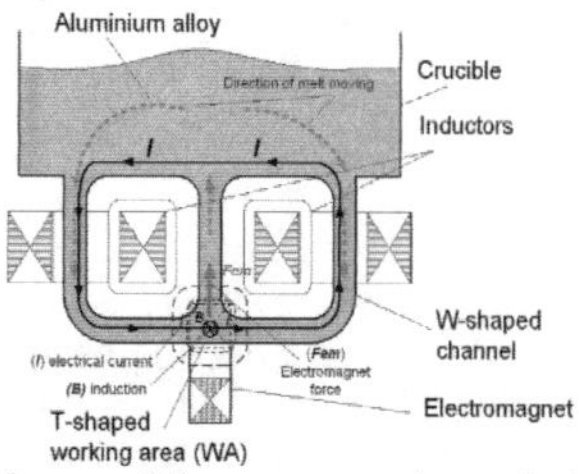

Fig. 1: Scheme of the magneto-dynamic installation (MDI) and operating principle

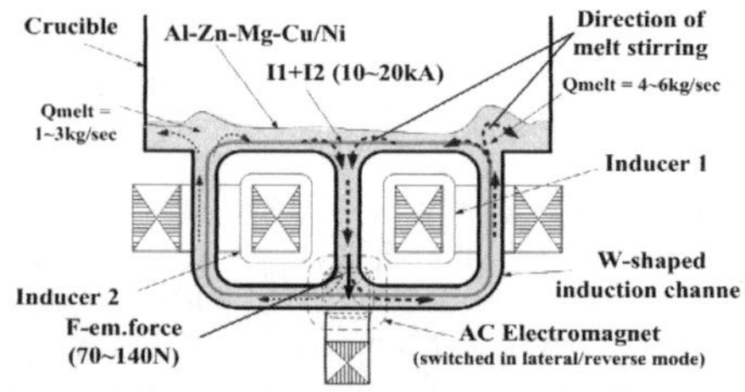

Fig. 2: Scheme of Al-Zn-Mg-Cu/Ni alloy MHD-processing in MDI at circulation mode (100~500kg)

For homogenization analysis of the alloy, 2 main working modes were used. initially permanent holding mode electro-magnetic stirring(slow melt flow of 2~3 kg/sec) and intake reverse mode intensive stirring was applied. In this mode, pulsing of 100~120Hz, electromagnetic force (70~143.5N) with pressure of 20kPa are applied during 10minutes (fig. 2). Temperature of alloy was maintained at set level 710~720°C. After each stages of alloy processing, a specimens were produced by the chill mold and 3`` graphite round mold for the microstructure analysis, tensile strength test, as for simulation of billet solidification conditions.

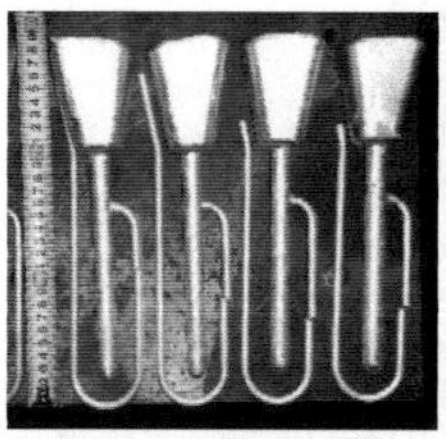
a) non-treated, U-type, ∅6

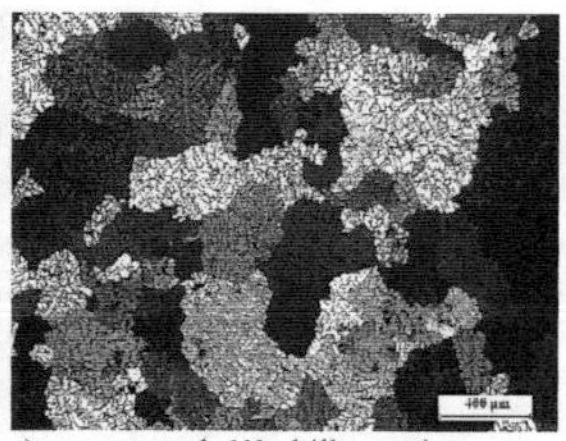
a) non-treated, 1`` chill, gravity

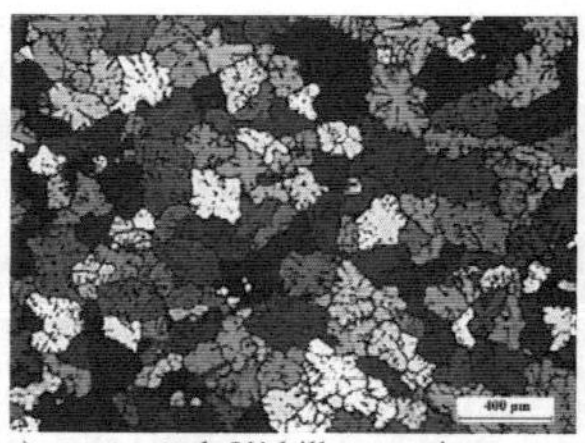
a) non-treated, 3`` billet, gravity

a) 3`` billet, 550mm, LEMP

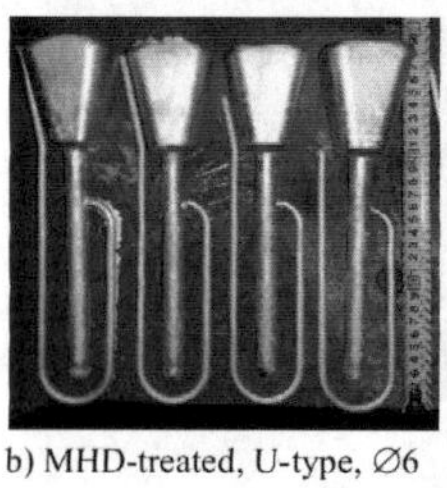
b) MHD-treated, U-type, ∅6

b) MHD-treated, 1`` chill, gravity

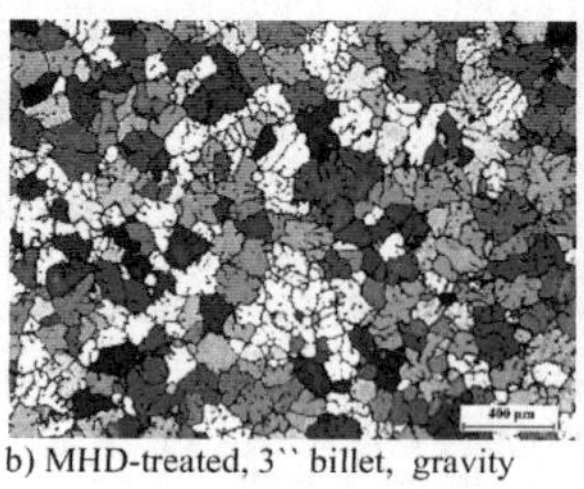
b) MHD-treated, 3`` billet, gravity

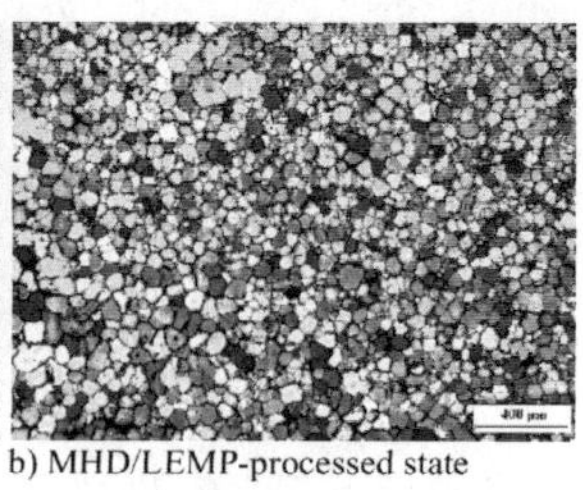
b) MHD/LEMP-processed state

Fig. 3. Cast/crack-ability U-type chill mold test: (a) - after holding and (b) - MHD-treatment, 100kg

Fig. 4. Microstructure for 1`` chill mold specimens of Al-Zn-Mg-Cu alloy: (a) - non-treated, (b) - and after 10 min MHD-processing

Fig. 5. Microstructure for 3`` graphite mold from Al-Zn-Mg-Cu alloy: (a) - non-treated, (b) - after 10 min MHD-processing

Fig. 6. Photo of 3`` 550 mm billet by LEMP casting (a) its billet microstructure, classified as 2 - stages MHD-processed state

Especially for the analysis and studying the effect of the physical modification by the MHD-means, the sampling of 4 main types of specimens was performed, that classified by the 4 different cooling rates and casting methods most applied in the foundry practices. 1) cast/crack-ability ∅6mm U-type mold test (fig. 3a, b) for test casting and hot-crack ability properties of processed alloy 2) 1`` permanent mold (horizontal chill mold with developed vertical riser) speci-mens (fig. 4a, b) 3) 3`` billet shape sample, with 70 mm diameter and height of 150~200 mm, gravity casted to graphite tube mold (fig. 5 a, b). Additionally, for comparison artificial impact of MHD low electromagnetic pressure (LEMP) to microstructure formation, 500~600 mm height, 3`` billet has produced by graphite mold (fig. 6a, b).

According to the analysis of U-type permanent chill mold tests (fig. 3), after holding (fig. 3a), and additional artificial electromagnetic MHD-processing (fig. 3b) of 100kg alloys during 10 minutes, the hot cracks that has been occurred for non-treating mode disappeared. As well, after MHD-processing, hot-cracks effect is not occurs for the different conditions of the pouring speed is ranges of 0.2~0.8kg/sec at a temperature of 680~720°C. Al-Zn-Mg-Cu alloy.
The unique effect of hot-crack disappearing has been researched by microstructure of horizontal 1`` specimens from MHD-processed Al-Zn-Mg-Cu (0.2%Zr) high-strength casting specimens by chill permanent mold at gravity casting (fig. 4). Its microstructure has not a big different characteristics of overall dendrite with grain size 250~400μm (fig. 4a, b). However, the factors that make a big difference in hot-crack-ability, as shown in fig. 4b are very small micro grains with grain size of 5~25μm, located between big dendrites and fully filed space between each for MHD-processing alloy. As well, the more compact microstructural density fill the space between primary dendrites in this case that explain solving problem of hot-cracks occurring. Analysis microstructure of Al-7%Zn-2.5%Mg-1.6%Cu-0.2Zr alloy (fig. 5), sampled from the MDI crucible after 10 minutes reverse mode stirring, has shown the tendency of non-dendrite microstructure forming (fig. 5b) with compact grains range of 50~100μm. As well all grains has distributed homogeneity by all cross-section of 3`` billet, in comparison with non-treated billet from conventional furnace (150~350μm) (fig. 3a).
The most interesting effects (fig. 6) were found in the low electromagnetic casting process for 3`` billet. Most grains appeared compact, homogeneous and small grain size in the range of 80~120 μm presented in the middle and 50~80μm in the central part of 3`` billet in the range from 5 to the 35 mm (fig. 6b). This effect can be explained by an artificial heat-force MHD-impact of the pulse unidirected electromagnetic force [3, 4] that created below into MDI working area and influenced directly on the solidification of casting structure. As well, vibratory, hydrodynamic and acoustic impacts also affect the growing crystals into mushy zone during the solidification of the billet. From the practical point of view, this effect is useful for the casting of real product from high-strength aluminum alloys with improved properties. The properties of developed Al-Zn-Mg-Cu (Zr) high-strength aluminum alloy after heat-treatment was 547MPa tensile strength, 516MPa yield strength at 3.4% elongation in gravity manual casting by the permanent chill mold. In the low pressure casting condition that used in the industrial conditions at the manufacturing detail part for machinery from high-strength aluminum alloys, tested properties showed more high value such as 566MPa tensile strength, 547MPa yield strength at 2.9% elongation after applying MHD-processing.

Researching of MHD-physical modification effect on microstructure refinement of the Al-Zn-Mg-Ni (<1%Ni) high-strength casting aluminum alloy
The previous MHD processing for Al-Zn-Mg system alloys was applied for the case of second type of the high-strength aluminum alloys without Cu and with Ni in quantity no more than 1%. In this alloy, the main strength phase is participate intermetallic composition as Al_3Ni, and the temperature range of pouring is 720~680°C. In this case, modification effect for the microstructure building also similar. 4-main types specimens were tested (fig. 7~10).

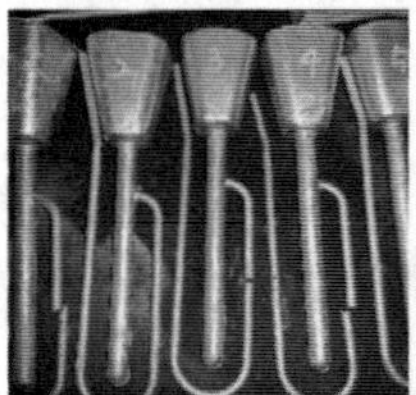
a) non-treated, U-type, ∅6

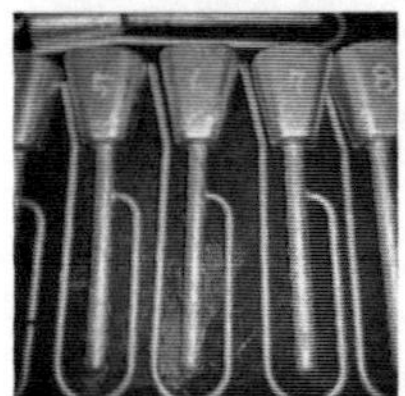
b) MHD-treated, U-type, ∅6

Fig. 7. Cast/crack-ability U-type chill mold test: (a) - after holding and (b) - MHD-treatment, 100kg

a) non-treated, 1`` chill, gravity

b) MHD-treated, 1`` chill, gravity

Fig. 8. Microstructure for 1" chill mold specimens of Al-Zn-Mg-Ni alloy: (a) - non-treated, (b) - and after 10 min MHD-processing

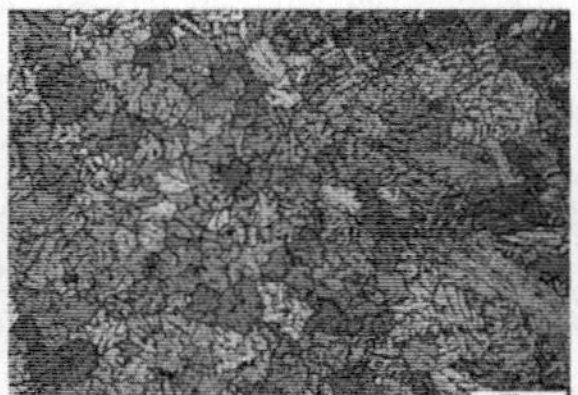
a) non-treated, 3`` billet, gravity

b) MHD-treated, 3`` billet, gravity

Fig. 9. Microstructure for 3" graphite mold from Al-Zn-Mg-Ni alloy: (a) - non-treated, (b) - and after 10 min MHD-processing

a) 3`` billet, 550mm, LEMP

b) MHD/LEMP-processed state

Fig. 10. Photo of 3`` 550mm billet by LEMP casting (a) its billet microstructure, classified as 2-stages MHD-processed state

Artificial MHD-processing in the reverse pumping mode for the Al-Zn-Mg-Ni alloy confirm again the effect of hot-cracks dis-occurring (fig. 7 a, b), and showed reduced grain size and presence of small dendrites filled space between main cells (fig. 8 a, b).

Analysis of Al-Zn-Mg-Ni alloy after melting, refining and holding into magneto dynamic installation for case 3`` billet (fig. 9), poured from MDI crucible by the ladle by gravity method showed very uniform microstructure on surface and also in middle and center part of billet (fig. 9b.). Non-treated (Al-Zn-Mg-Ni) alloy showed same typical dendrite structure at the middle part and center (fig. 9a), however some compact shapes located in each cells.

At talking about MHD-physical modification effect, relations of the mechanical properties with each stage of processing need to be considered. As well, for the case of non-treated Nickalin (fig. 9a), tensile strength after heat treatment by the 2-stages solid solution (450°C (3hr) + 530°C (3hr)) and one stage aging (130°C (6hr)) showed the maximal tensile strength of 456~460MPa and elongation of 1.28%~1.38%.

However, same alloy melted and processed in MDI showed 5~7% increased UTS (474~479MPa), but elongation was reduced to the 1%(fig. 9b).

When the Al-Zn-Mg-Ni alloy processed by the pumping mode (LEMP) and casted under low electromagnetic pressure (fig. 10b), it showed 8%~10% increased UTS (490~496MPa) and elongation was reached to the 1.2~1.4%.

However, in terms of solving the hot-crack ability problem of this alloy is more great advantage of MHD-processing, especially when we talking about foundry practical application indicated alloy for producing real parts, for example for automobile or machinery building.

Conclusions

Developed and tested MHD-technologies for manufacture and high intensive processing of the high-strength aluminum alloy series (Al-Zn-Mg-Cu-Zr) comprised of melting, refining and processing in quantity from 100 to 650kg.

Complex magneto-hydro-dynamical processing of Al-Zn-Mg-Cu/Ni (Zr) alloys at 700~720°C allowed traditional dendrites shape transformed to the one body no uniform grains without distributed branches.

In the direct manufacturing of high-strength casting Al-alloy, the means of MHD-physical affecting, by Unidirectional 120Hz pulse electromagnetic force (70~143.5N, 2.5~3.5m/s) created in working area (WA) of the magneto-dynamic installation electromagnetic pump has provided an fine emulsion distribution of intermetallic particles in liquid melt during intensive MHD-processing for melt capacity 150~500kg. Microstructure of permanent chill-mold specimens (1`` rod and 3`` billet) for Al-Zn-Mg-Cu alloy has shown the unique non-dendrite microstructure with compact grains in range 50~100μm, distributed homogeneity by all cross-section, instead over 250μm grains of the non-treated specimen which sampling from conventional melting furnace.

As well, applying MHD-processed alloy for low pressure casting method at manufacturing 3`` billet for 550mm lengths into graphite tube have an mainly compact globular microstructure, characterized as non-dendrite state with 80~120μm grain size and 50~80μm in the central part of billet. Achieved effect can be explained by an artificial heat-force effect of the pulse unidirected electromagnetic force, created below into MDI working area which affecting the solidification structure. In addition just 10 minutes artificial pumping mode MHD melt processing for 150kg alloy was found to increase cast-ability of the Al-Zn-Mg-Cu casting high-strength alloy at pouring temperature of 680~720°C, wherein hot-cracks disappear, and hot-cracks effect is not occurs for the different conditions of the pouring speed.

Tensile strength test of the developed high-strength casting Al-Zn-Mg-Cu alloys has been shown 544~582MPa, 516~555 yield strength and 3.4~5% elongation for the case of chill-mold gravity casting. Especially for improving castability of high-strength aluminum casting alloys Al-Zn-Mg system, Ni and Fe applied (without Cu) chill-mold casting has performed. Real MHD-processing effect of Al-Zn-Mg-Ni alloy provide 8~10% improved tensile strength over the conventional melting mode, and no less than tensile strength of gravity casting producing mode(490~496MPa).

References

1. N. Barman, P. Kumar and P. Dutta, (2009), Journal of Materials Processing Technology, 5912–5923
2. V. Polishchuk, M. Tsin, R. Horn & et., (1989), Kyiv (Ukraine), Naykova dumka, 256
3. V. Dubodelov, N. Slazhnev, V. Fixsen and et., (2006), The 5th International Symposium on Electromagnetic Processing of Materials “EPM 2006”, Sendai (Japan), P. 265-270
4. V. Dubodelov, V. Fikssen, M. Slazhniev and et., (2012), Magnetohydrodynamics, V48, No. 2. – P. 379-386

IOP Conf. Series: Materials Science and Engineering **424** (2018) 012022 doi:10.1088/1757-899X/424/1/012022

Nucleation and crystal growth in magnetoelectrodeposition

A.L. Daltin, M. Benaissa, J.P. Chopart

LISM, URCA, B.P. 1039, 51687 Reims Cedex 2, France

Corresponding author: al.daltin@univ-reims.fr

Abstract
The aim of this work is to overview effects of a homogeneous magnetic field on the nucleation and growth crystal in electrodeposition processes, according to its amplitude and orientation with respect to the substrate. In this presentation, examination of such interest is given after a brief resume of a model that is generally used to describe the current transients corresponding to the early stages of nucleation and growth. Magnetic field superimposition during electrodeposition generates magnetohydrodynamic (MHD) convection and micro-MHD effect that interact on nucleation and growth phenomena from the very first moment of deposit.

Keywords : electrodeposition, magnetic field, magnetoelectrodeposition, nucleation, growth, magnetohydrodynamic convection.

Introduction

The use of magnetic fields on deposition obtained by electrochemical process makes it possible to modify many parameters among which the deposition rate, the homogeneity, the structure and the physicochemical properties of the deposits. Although the action of magnetic fields on the characteristics of electrochemical deposits has been studied by many authors, the details of the mechanisms involved in their synthesis have concerned a smaller number of papers. The interest of magnetoelectrodeposition lies in developing new materials with specific properties such as catalytic or magnetic and/or the use of electrolytic bath without organic molecules that are introduced as levelling agents. The nucleation and growth during electrodeposition are fundamental in obtaining controlled microstructure deposition with specific properties. The thermodynamics and kinetics aspects are therefore very important in electrodeposition technique. The nucleation depends on the characteristics of the substrate surface such as the crystal structure, the lattice orientation, the crystallographic lattice mismatch between the nuclei and the substrate surface, and also the specific free surface energy and the adhesion energy. In order to achieve information concerning the kinetics of the deposit formation by electrochemical processes, it is usual to analyze the current density transients. This has been done for several materials electrodeposited under magnetic field in order to highlight its influence on the nucleation kinetics and growth control and subsequently on characteristics of deposits.
Here early stages of electrodepositions of some metals, alloys, oxides and inorganic materials are examined and compared with other examples from the literature [1-9] to highlight the impact of magnetic field, that is usually more prominent when it is applied perpendicularly to the electric field. Micro-MHD effect, due to a non-uniform distribution of the current lines, caused by the edge effects of the electrode surface and the presence of gas bubbles, is also important in other configurations [1].

Magnetoelectrodeposition

When magnetic field is applied parallel to the cathode surface during electrodeposition, the mass transport of electrolyte ions to this surface is increased and electrochemical reaction can be modified. This MHD effect is due to forces, which increase the convection, among which the most important Lorentz force due to the interaction between the magnetic and electric field (maximal when the magnetic field is imposed perpendicular to the direction of the cathodic current), and the paramagnetic force coming from the potential energy possessed by the magnetic dipoles placed in a magnetic field, which acts inside the diffusion layer during a strong concentration gradient in electroactive species due to the electrochemical reactions at the working electrode [10]. The macroscopic stirring is accompanied by microscopic minute vortexes which emerge inside the boundary layer, called the micro-MHD effect. The forces involved in magnetoelectrodeposition may be of electromagnetic origin (Lorentz force) and / or magnetic (paramagnetic) depending on the nature of the involved ions. The intensity of these convections varies depending on the relative orientation of the electric fields and imposed magnetic fields, and concentrations of ionic species c in the solution.

The forces involved when the magnetic field B is parallel to a plane electrode are:

- the Lorentz force : $\vec{F}_L = \vec{j} \times \vec{B}$ (1)

where j is the current density and B the magnetic field intensity

- the coulombian force : $\vec{F}_C = \rho \times \vec{E}_H$ (2)

where ρ is the local space charge and E_H the tangential electrical field.

The forces involved when the magnetic field B is perpendicular to a plane electrode are:

- the magnetic gradient force : $F_B = \chi_m cB\nabla B / \mu_0$ (3)
- the paramagnetic force : $F_P = \chi_m B^2 \nabla c / 2\mu_0$ (4)

where χ_m and c are respectively the molar magnetic susceptibility and the concentration of the involved species, ∇c the concentration gradient, ∇B the magnetic field gradient, and μ_0 the magnetic permeability of free space.

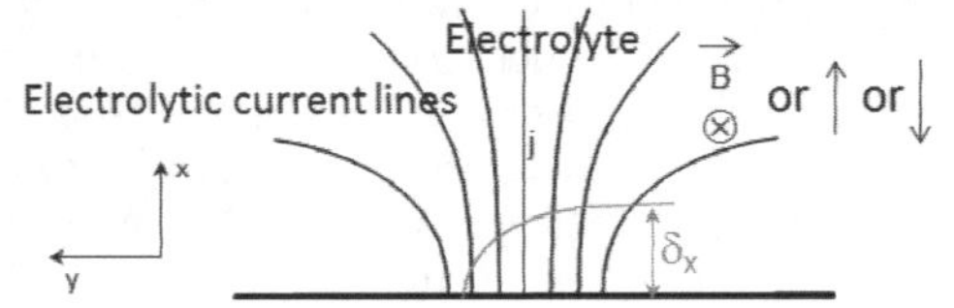

Fig. 1: Electrolytic current lines in the case of a horizontal upward substrate surface

Chronoamperometric transients

Depending on the nucleation process during potentiostatic depositions, some typical relationships between the current and the time can be obtained. The growth type can be one, two, or three dimensional. The origin control can be a kinetic one or a diffusion one and the nucleation type can be instantaneous or gradual that means progressive. The instantaneous nucleation corresponds to a slow growth of nuclei with a quite small number of active sites appearing at the same moment. In this case, nuclei are born before being affected by the spreading of nucleation exclusion zones. The progressive nucleation, corresponds to a fast growth of nuclei on many active sites which are progressively born; in that case the break in nucleation process is dominated by spreading of nucleation exclusion zones.

Potentiostatic chronoamperometric transient modelling is usually used to study nucleation process. A classical treatment of the experimental data leads to numerous models [11-14], reviewed by Hyde and Compton [15], that considers that the nuclei are formed according to the equation:

$$\frac{dN}{dt} = AN_0 e^{(-At)} \quad \text{or} \quad N = N_0(1 - e^{-At}) \qquad (5)$$

where t is the time since the overpotential is applied to the electrode, N is the number of nuclei, N_0 is the saturation nucleus density corresponding to the number of active sites and A is the nucleation rate constant. In order to consider the overlap of diffusion zones, the Avrami's theorem can be applied. It considers collisions and overlaps of the growing centres in a process of multiple nucleations and growth. According to the Scharifker and Hills' model [16] considering that the nuclei have a hemispherical form, and with diffusion process during the nucleation, two limiting nucleation mechanisms instantaneous and progressive are described. Normalized transients (corresponding to the ratio of the total current I per the maximum current I_m versus the ratio of the time t per the time t_m corresponding to I_m) are used to discriminate nucleation kinetics.

These curves are representation of some expressions summarized in Table 1 which are typical for the different nucleation processes.

Tab. 1: Expressions resulting from the analysis of the current maxima from [14].

Instantaneous nucleation	Progressive nucleation
$t_m = \frac{1.2564}{N\pi kD}$	$t_m = \left(\frac{4.6733}{AN_\infty \pi k' D}\right)^{1/2}$
$I_m = 0.63832 zFDc(kN)^{1/2}$	$I_m = 0.4615 zFD^{3/4}c(k' AN_\infty)^{1/4}$
$I_m^{\,2} t_m = 0.1629(zFc)^2 D$	$I_m^{\,2} t_m = 0.2598(zFc)^2 D$
$\frac{I^2}{I_m^2} = \frac{1.9542}{t/t_m} \left\{1 - \exp\left[-1.2564\left(t/t_m\right)\right]\right\}^2$	$\frac{I^2}{I_m^2} = \frac{1.2254}{t/t_m} \left\{1 - \exp\left[-2.3367\left(t/t_m\right)^2\right]\right\}^2$

This model was used in different studies concerning potentiostatic deposition of materials under the superimposition of magnetic field. Usually, the nucleation rate of crystals increases with magnetic flux density as for Nickel or Hydroxyapatite [2-3] or sometimes decreases as for Zinc deposition in overpotential regions [4]. For Ag, Fe and CoFe or CoNiMn alloys electrocrystallisation, no effect of a magnetic field on the nucleation processes has been noticed while the growth of nuclei can be affected [5-6]. Cu_2O crystal nucleation seems to become instantaneous in the presence of the magnetic field whereas it appears progressive with no magnetic field [7] as it can be seen in figure 2 compared with normalized transients of the two limiting nucleation mechanisms.

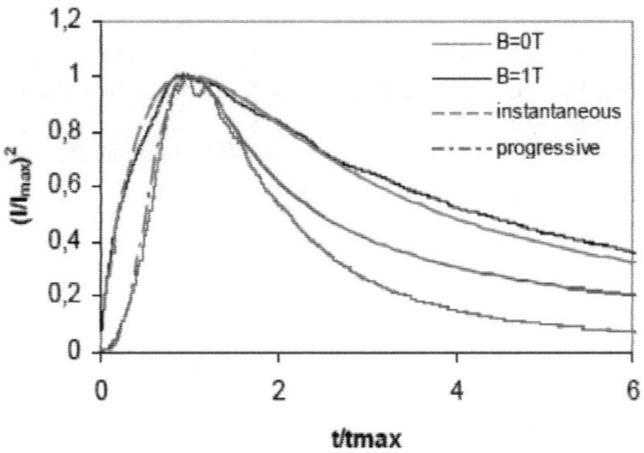

Fig. 2: Normalized potentiostatic transients for cuprous oxide electrodeposition on horizontal upward stainless steel electrode // B (B ⊥ j).

In a potentiostatic regime and for discrete germs without coalescence, a set of current-time relations for a nucleation independent from the substrate surface state and a growth of germs controlled either by the mass transport, or by the electronic transfer kinetics, has been proposed by Thirsk and Harrisson [17]. The general expression of the current that depends only on the reaction mechanism takes then the following shape:

$$I = kt^n, \quad (6)$$

where k is a constant that depends on the deposited material characteristics and on the reduction mechanism, n depends on the nucleation/growth type, and on the electrochemical regime control as presented in Table 2. In the case of Mn doped Cu_2O deposition, two opposite effects on nucleation are observed considering doping or magnetic field superimposition: Mn^{2+} ion promotes progressive mode, and magnetic field promotes instantaneous mode.

Tab. 2: Value of the exponent *n* corresponding to the chronoamperometric curves according to the mechanism of the electrodeposition from [17].

1	instantaneous	2D	kinetic
2	gradual	2D	kinetic
2	instantaneous	3D	kinetic
3	gradual	3D	kinetic
1	gradual	2D	diffusion
1/2	instantaneous	3D	diffusion
3/2	gradual	3D	diffusion

netic fields we obtain the same (100) preferential orientation.

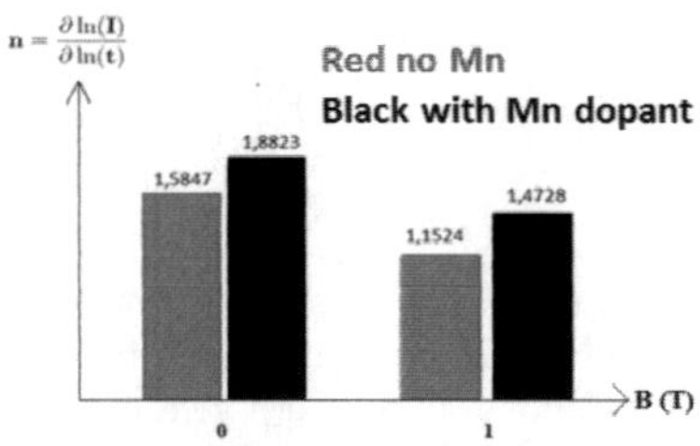

Fig. 3: n value for Mn doped Cu_2O deposition with and without doping under 0 or 1 T magnetic field.

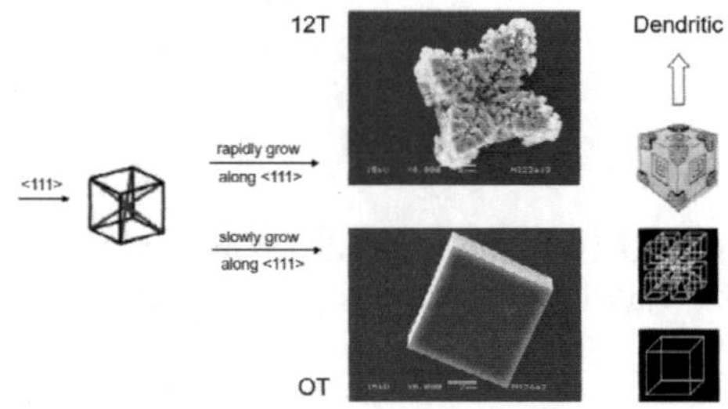

Fig. 4: Growth mechanism of Cu_2O crystal under no or a 12 T magnetic field superimposition

Concerning the growth process, the MHD convection could principally act on the branching growth without modifying the Cu_2O crystal habit (Figure 4) [8]. Due to forces and induced convection, magnetic field can act on the mass transport process at the very beginning of nucleation process and modify the mechanism of crystal or grain growth. Also magnetic field could influence the kinetics of chemical reactions in the electrical double layer [9].

Conclusion

Both magnetohydrodynamic convection and micro-MHD effect induce modifications from the very beginning of the growth process.

References

1. R. Aogaki, R. Morimoto, M. Asanuma, J. Magn. Magn. Mater., 322, (2010) 1664–1668
2. A. Ispas et al., J. Electroanal. Chem, 626 (2009), 174-182
3. C. Liu, A. Tian, H. Yang, Q. Xu, X. Xue, Applied Surface Science 287 (2013) 218–222
4. H. Matsushima, A Ispas, A Bund, B Bozzini, J. Electroanal.Chem. 615 (2008) 191–196
5. J.A. Koza, M. Mogi, , K. Tschulik, M. Uhlemann, C. Mickel, A. Gebert, L. Schultz, Electrochim. Acta, 55, 22, (2010) 6533-6541
6. L. Gang et al., Surface Review and Letters, 25, 4 (2018) 1850037
7. A.L. Daltin, F. Bohr, J.P. Chopart, Electrochim. Acta 54 (2009) 5813
8. A.L. Daltin, A. Addad, P. Baudart, J.-P. Chopart, CrystEngComm 13, 10 (2011) 3373-3377
9. L.M.A. Monzon, L. Klodt, J.M.D, Coey, J. Phys. Chem. C, 116, 34 (2012) 18308-18317
10. N. Leventis, A. Dass, J. Am. Chem. Soc., 127 (2005) 4988-4989
11. B.R. Scharifker, J. Mostany, J. Electroanal. Chem. 177 (1984) 13
12. M. Sluyters-Rehbach, J.H.O.J. Wijenberg, E. Bosco, J.H. Sluyters, J. Electroanal.Chem. 236 (1987) 1
13. M.V. Mirkin, A.P. Nilov, J. Electroanal. Chem. 283 (1990) 35
14. L. Heerman, A. Tarallo, J. Electroanal. Chem. 470 (1999) 70
15. M.E. Hyde, R.G. Compton, J. Electroanal. Chem. 581 (2005) 224
16. B.R. Scharifker, G.J. Hills, Electrochim. Acta 28 (1983) 879
17. H.R. Thirsk, J.A. Harisson. (Academic Press, New York, 1972) 118

Determination of the properties of ionic vacancy by magnetic field

R. Aogaki [1,10]*, A. Sugiyama [2,3,10], M. Miura [4], Y. Oshikiri [5], M. Miura [6], R. Morimoto [7], I. Mogi [8], S. Takagi [9], Y. Yamauchi [10]

[1]Polytechnic Univ., [2]Yoshino Denka Kogyo, Inc., [3]Waseda Univ. Res. Org. Nano & Life Innovation, [4]Hokkaido Polytechnic College, [5]Yamagata College of Industry and Technology, [6]Polytechnic Center Kimitsu, [7]Saitama Industrial Technology Center, [8]Tohoku Univ., [9]Koriyama Technical Academy, [10]National Institute for Materials Science

Corresponding author: AOGAKI.Ryoichi@nims.go.jp

Abstract
Ionic vacancy is, being produced as a byproduct in electrode reaction, a polarized free space of the order of 0.1 nm surrounded by oppositely charged ionic cloud. As a perfect gas in electrolyte solution, it does not interplay with other solution particles. On the other hand, magnetic field provides useful means to examine the natures of it. Using a special electrode operated in a magnetic field called cyclotron MHD electrode (CMHDE), we first determined the lifetime of vacancy in copper deposition as 1.2 sec, which is extraordinarily long in comparison with the collision interval of solution particles. Then, using kinetic theory of gas, we measured the collision radius of vacancy gas to be 0.70 nm, which was consistent with theoretical calculation 0.65 nm.

Keywords: ionic vacancy, perfect gas, nanobubble, magneto-dendrite effect, cyclotron MHD electrode.

Introduction
Magnetoelectrochemistry is the electrochemistry in magnetic field. In order to study electrode reactions in magnetic fields, some special electrodes are operated under magnetohydrodynamic (MHD) flow induced by Lorentz force, e.g., under vertical magnetic field, a macroscopic tornado-like vortex called vertical MHD flow arises over a disk-electrode (VMHDE) (Fig.1a), and a circulating solution emerges between a pair of concentric cylinders with electrodes (CMHDE) (Fig. 1b).

In various electrode reactions, unrelated with hydrogen and oxygen evolution, a large number of microbubbles containing dissolved N_2 gas emerge on a VMHDE (Fig.2a,b), which experimentally validates the fact that ionic vacancy is a byproduct of electrode reaction [1-3]. Ionic vacancy is a polarized free space of the order of 0.1 nm surrounded by oppositely charged ionic cloud (Fig. 2a), which is created by electron transfer in electrochemical reaction, emitted to solution side from the conservation of momentum and charge during (Fig. 3A,B) [4]. As shown in Fig.1a, the macroscopic rotation of vertical MHD flow is characterized by two areas, inner and outer regions. In the outer region, the fluid particles rotating at a large distance from the electrode are in hydrodynamic equilibrium under the centrifugal force balanced with the radial pressure gradient. In the inner region, due to friction, the peripheral velocity of the particles near the electrode at rest is reduced, decreasing the centrifugal force materially, whereas the radial pressure gradient directed towards the axis remains the same.

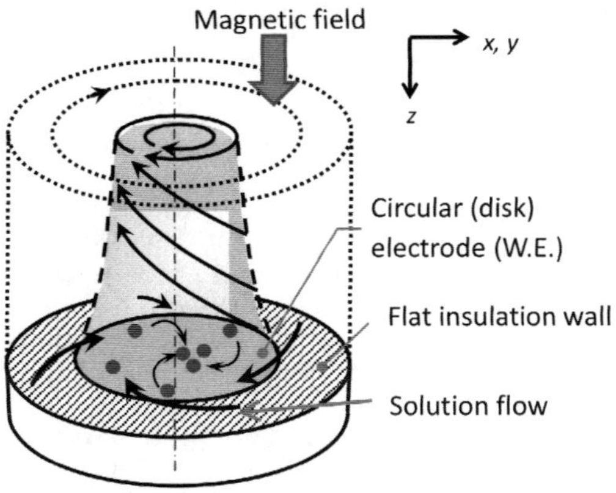

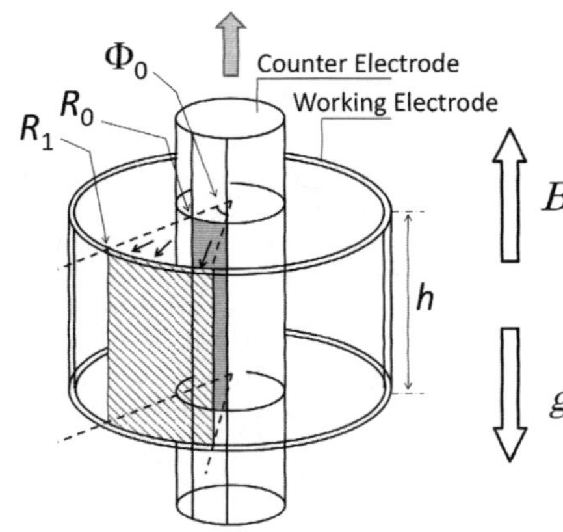

Fig. 1: MHD electrodes. (a), Vertical MHD electrode (VMHDE). ○, Ionic vacancy. (b), Cyclotron MHD electrode (CMHDE). R_0; inner radius, R_1; outer radius, h; the electrode height, Φ_0; the angle of the arc electrode surfaces, B_z; the magnetic flux density, g; gravitational acceleration.

This set of circumstances causes the particles near the electrode to flow radially inwards, and for reasons of continuity that motion must be compensated by an axial flow upwards. Ionic vacancies produced on the electrode are carried with the secondary flow to the center, so that coalescence by collision transforms them into microbubbles. Furthermore, in copper electrodeposition under vertical MHD flow, numerous microscopic vortexes called micro-MHD flows are induced on the electrode surface, which create screw dislocations with chirality acting a chiral catalyst for enantiomeric reaction [5].

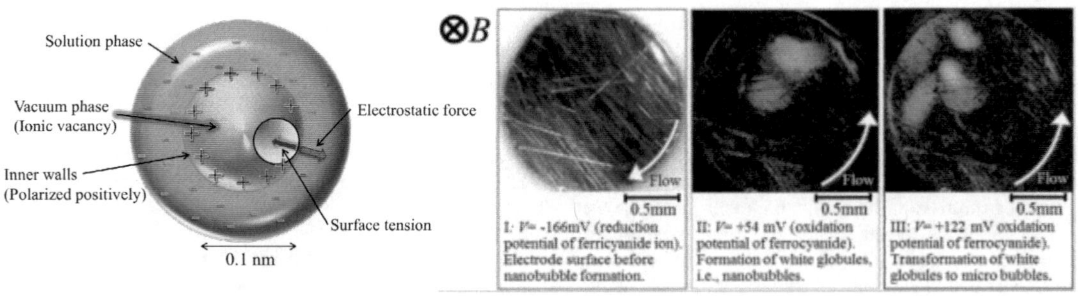

(a) (b)

Fig. 2: Ionic vacancy and microbubble evolution [1]. (a) Schematic of a positive vacancy. (b) Microbubbles on VMHDE. I, Electrode surface during the reduction at an overpotential V = -166 mV (+264 mV vs. NHE); II, Nanobubble-layer formation with refractive variation at V = +37 mV (+467 mV vs. NHE); III, Microbubble formation at V = +122 mV (+552 mV vs. NHE). The images are subtracted and painted by yellow.

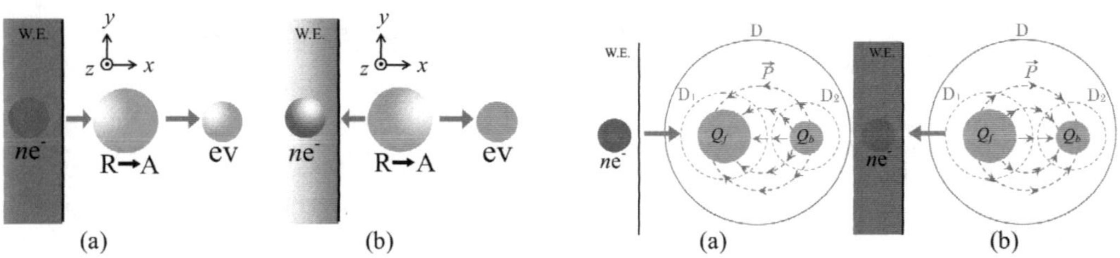

(A) (B)

Fig. 3: Emissions of embryo vacancy in electrode reactions [4]. (A) Emission of embryo vacancy in electrode reaction. (a), Cathodic reaction; (b), Anodic reaction. (B) Charge transfer process by electric polarization.(a), Cathodic reaction; (b), Anodic reaction. ne^-, electrons transferring in the reaction; R, reactant; A, activated complex; ev, embryo vacancy; W.E., working electrode.

Lifetime of ionic vacancy

CMHDE was first developed for measuring the lifetime of ionic vacancy [6]. As shown in Fig. 1b, electrolyte solution is placed between concentric cylinders under a vertical magnetic field. Both cylinders have electrodes, and except for the electrodes, the walls are insulated. As electrolytic current flows, Lorentz force makes the solution rotate between the cylinders. In a weak magnetic field, due to low rotational speed, ionic vacancies disappear on the way returning to the electrode of birth, whereas in a strong magnetic field, due to high rotational speed, they can survive the travel, covering the cylinder walls to form vacancy-rich layers. As a result, the solution flow changes from viscid to inviscid with magnetic field. Figure 4a shows the plots of the limiting diffusion current against magnetic field strength in CMHDE for copper electrodeposition. As the magnetic field increases, the plot changes from a straight line with a slope of 1/2 to a straight line with a slope of 1, corresponding to viscid and inviscid flows, respectively. From the crossing point (critical point), the lifetime of ionic vacancy τ was determined, which is, in Fig. 4b, plotted against the collision efficiency γ. An intrinsic lifetime of ca. 1 sec is obtained in the case without collision for $\gamma = 0$, whereas in the case of one by one collision for $\gamma = 1$, the collision interval with another vacancy τ_{colli} = ca. 10^{-2} sec is obtained. A 1 sec lifetime is extraordinarily long in comparison with a 10^{-10} sec corrosion interval between other solution particles. Due to free space without solid matter, ionic vacancy migrates by repeating creation and extinction together with ionic cloud. In the creation, polarized charge of a vacancy core first induces surrounding ionic cloud. Then, using the electrostatic energy emitted from the cloud, the core is dynamically expanded up to a given size. In the extinction, disassembling ionic cloud withdraws the energy from dwindling core. In this process, ionic vacancy transfers without entropy production. This implies that ionic vacancy does not interplay with other solution particles but with other ionic vacancies; in other word, ionic vacancy behaves as a perfect gas in electrolyte solution.

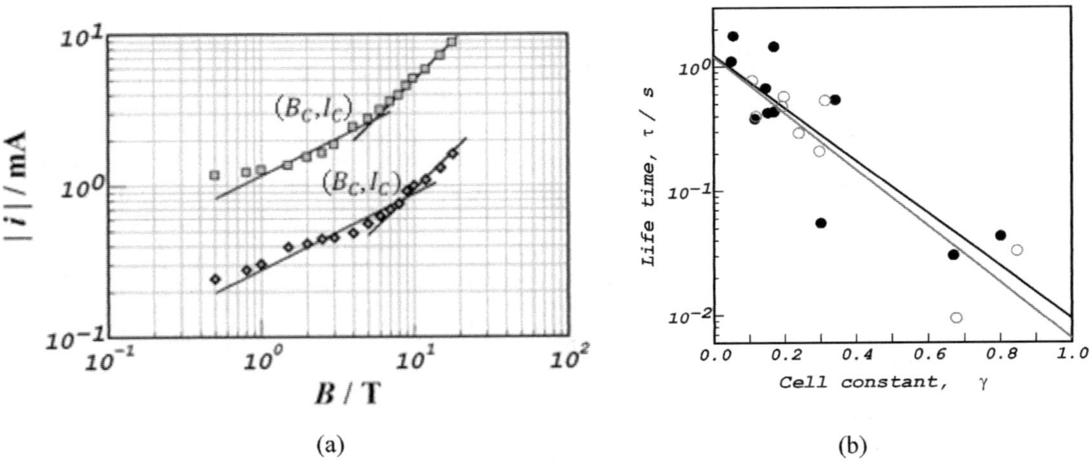

Fig. 4: Measurement of the lifetime of vacancies by CMHDE [6]. (a) Plot of current vs. magnetic flux density for copper deposition in a 100 mol m^{-3} H_2SO_4 solution. I_c and B_c are the critical current and critical magnetic flux density, respectively. (b) Plots of the lifetime of ionic vacancy vs. cell constant (collision efficiency). ○, Ferrocyanide-Ferricyanide redox reaction, and ●, Copper deposition.

Magneto-dendrite effect

Owing to continuous supply of newly created vacancies, using CMHDE as well as VMHDE, we cannot always control the vacancy concentration in the vacancy layer on the electrode, but also examine the chemical nature of ionic vacancy in electrode reactions. Figure 5A,B shows a 'magneto-dendrite effect' observed in copper deposition using CMHDE, where instead of hydrogen molecules, nanobubbles arising from coalescent vacancies are adsorbed on copper 3D nuclei, blocking nucleation [7]. As a result, unrelated with hydrogen gas evolution, characteristic copper dendritic growth occurs with a drastic anodic potential shift of rising current from hydrogen evolution potential (Fig. 5b).

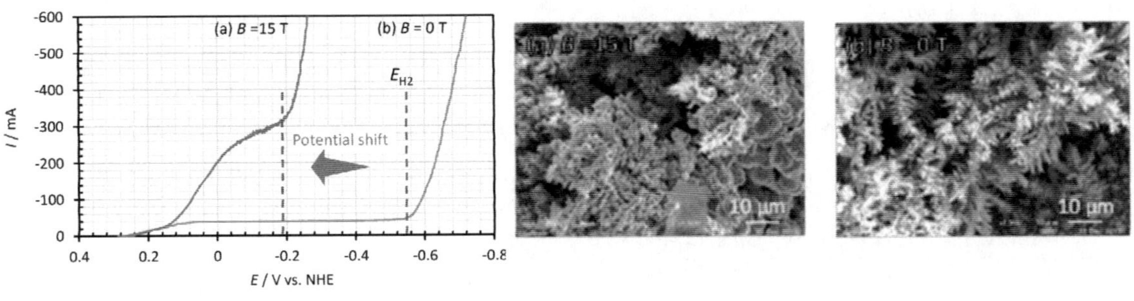

Fig. 5: Potential shift in anodic direction observed in copper deposition by cathodic potential sweep at a 15T magnetic field and SEM images of the corresponding dendritic deposition [7]. (A) Current *vs.* potential curves; (a) Red solid line, B = 15 T; (b) Green solid line, B = 0 T. E_{H2}, Hydrogen evolution potential; [$CuSO_4$], 300 mol m^{-3}; [H_2SO_4], 500 mol m^{-3}; Sweep rate, 5 mVs^{-1}. Electrode, Copper-CMHDE. (B) SEM images of copper dendrites obtained after the potential sweeps. (a) B = 15 T without hydrogen gas evolution. (b) B = 0 T with hydrogen gas evolution.

Ionic vacancy as a perfect gas

The extraordinarily long lifetime measured in Fig. 4b strongly suggests that ionic vacancy behaves as a perfect gas. As a result, using kinetic theory of gas, we can measure the collision radius of ionic vacancy from the experimental data of CMHDE. When an ionic vacancy that is circulating around the cylinder walls collides with other vacancies newly created on the electrode, its kinetic energy as well as linear momentum changes after the collision. By means of the kinetic model, we calculate the frequency with which vacancy collisions occur and the distance a vacancy travels on average between collisions $V_V\tau$, where V_V and τ are the mean speed and lifetime of vacancy, respectively. We count a 'hit' whenever the center of two vacancies come within a distance *d* of each other, where *d*, the collision diameter, is of the order of the actual diameter of vacancy. In Fig. 6a, b, a collision tube swept out by a returning vacancy and its collision process in CMHDE are exhibited. In Fig. 6b, the vacancy layer takes a very thin thickness of several μm depending on the diffusion coefficient and the collision lifetime. Due to the gas behaviour without interaction with other solution particles, the vacancy layer can circulate without viscosity, so that the neighbouring part of solution can take a fluid motion without friction. This is the reason why as shown in Fig. 4a, inviscid flow corresponding the plot of a slope 1 emerges in CMHDE. Figure 6c exhibits the plot of collision radius vs. collision efficiency obtained in copper deposition in copper sulfate + sulfuric acid solution. For comparison, theoretical value is also plotted by a dotted line. The experimental data 0.70 ±0.07 nm are in good agreement with the theoretical estimation 0.65 nm, which assures the validity of the theoretical analysis of vacancy gas.

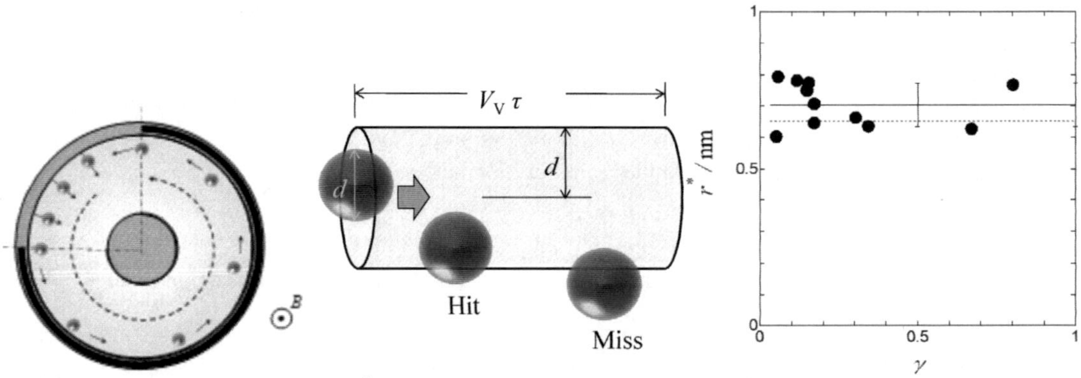

(a) (b) (c)

Fig. 6: Collision process of vacancy gas. (a) Collision tube swept out by a returning vacancy. V_V, mean speed; τ, lifetime; *d*, collision diameter. (b) Collision of ionic vacancies at the critical state in CMHDE. *B*, applied magnetic field. (c) Plot of the collision radius r^* vs. collision efficiency γ for the ionic vacancy created in copper deposition. Solid line is the average value, and dotted line is the

Acknowledgement

The authors thank the Tsukuba Magnet Laboratory, National Institute for Materials Science (NIMS), Tsukuba Japan and the High Field Laboratory for Superconducting Materials, Institute for Materials Research (IMR), Tohoku University for financial support and access to superconducting magnets.

References

1. A. Sugiyama, *et al., Electrochemistry* **81** (2013), 890-892.
2. M. Miura, *et al., Electrochemistry* **82** (2014), 654-657.
3. Y. Oshikiri, *et al., Electrochemistry* **83** (2015), 549-553.
4. R. Aogaki *et al, Sci. Rep.*, **6** (2016), 28927.
5. I. Mogi *et al, Sci. Rep.*, **3**, (2013), 2574.
6. A. Sugiyama *et al, Sci. Rep.*, **6** (2016), 19795.
7. M. Miura *et al, Sci. Rep.*, **7** (2017), 45511.

IOP Conf. Series: Materials Science and Engineering **424** (2018) 012024 doi:10.1088/1757-899X/424/1/012024

Effects of Vertical MHD Flows and Cell Rotation on Surface Chirality in Magnetoelectrodeposition

I. Mogi[1], R. Morimoto[2], R. Aogaki[3], K. Takahashi[1]

[1]IMR, Tohoku University (Katahira, Aoba-ku, Sendai 980-8577, Japan)
[2] SAITEC (Kawaguchi, Saitama 333-0844, Japan)
[3]Polytechnical University (Sumida-ku Tokyo 130-0026, Japan)

Corresponding author: mogi@imr.tohoku.ac.jp

Abstract
The combination of vertical MHD flows and micro-MHD vortices can produce chiral surfaces in magnetoelectrodeposition (MED) of metals. Here we show two effects on the chiral surface formation in MED of copper. The first one is the effect of perturbation of vertical MHD flow using a tube wall around the electrode. The MHD flow penetrated into the tube and affected the chiral surface formation, as a result, the surface chirality depended on the height of wall. The other is the effect of rotation of electrochemical cell in magnetic fields. The rotational direction allowed the control of surface chirality.

Key words : magnetoelectrodeposition, micro-MHD vortex, vertical MHD flow, chirality, rotation, amino acid

Introduction

Chiral surfaces of materials can serve as enantioselective catalysts, which are of great significance in pharmaceutical industries. Magnetoelectrodeposition (MED) produces chiral surfaces on metal films, and such films exhibit chiral recognition of amino acids [1-5].

When a magnetic field is imposed perpendicularly to a working electrode surface, the Lorentz force generates two types of MHD convections around the electrode [6,7]. One is a macroscopic vertical MHD flow around the electrode edge, the other is a micro-MHD vortex around a local bump on the deposit surface. The latter affects the numbers of right- and left-handed screw dislocations, leading to chiral surfaces. We have examined the chiral surface formation in the MED of copper with various conditions. The results showed that the surface chirality depends on several parameters, such as magnetic fields, deposition currents, electrode diameters and the specific adsorption of chloride ions [2-4]. In this study, we have investigated the effects of vertical MHD flow and cell rotation on the chiral surface formation of MED copper films.

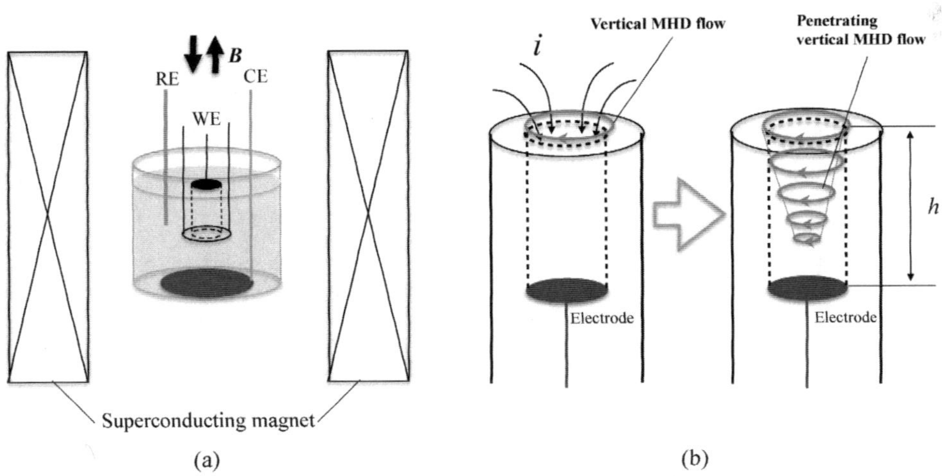

Fig. 1. (a) Schematic of magnetoelectrodeposition in a superconducting magnet. WE; a Pt working electrode, CE; a Cu counter electrode, RE; a Ag/AgCl reference electrode. The WE is embedded in a Teflon tube with the wall height h of up to 12 mm. (b) Schematic of the vertical MHD flow in the tube. The MHD flow is generated around the entrance (left) and then it penetrates into the tube (right).

IOP Conf. Series: Materials Science and Engineering **424** (2018) 012024 doi:10.1088/1757-899X/424/1/012024

Effects of vertical MHD flow

The vertical MHD flows have aymmetric influence on the micro-MHD vortices [7]: The cyclonic micro-MHD vortices are stable, whereas anticyclonic ones become unstable. In order to examine the effects of vertical MHD flows, the MED of Cu films was conducted on a working electrode embedded in a Teflon tube, where the tube wall could be expected to suppress the vertical MHD flow.

The electrochemical cell was placed at the bore center in a cryocooled superconducting magnet (Sumitomo Heavy Industries Ltd.), which generated a magnetic field of 5 T, as shown in Fig.1a. The applied magnetic field was parallel (+5 T) or antiparallel (–5 T) to the faradaic current ***i***. The cell consists of conventional three electrodes: a polycrystalline Pt disc working electrode with a diameter of 3 mm (ALS Co. Ltd.), a Cu plate counter electrode, and a Ag | AgCl | 3 M (M = mol dm^{-3}) NaCl reference electrode. The working electrode was embedded in a Teflon tube, whose length was varied from 2 to 12 mm. The Cu films were electrodeposited up to the passing charge of 0.4 C cm^{-2} (the thickness of ~150 nm) on the working electrode via the potentiostatic electrodeposition at constant potentials of –0.05 - –0.45 V in a 50 mM $CuSO_4$ aqueous solution containing 0.5 M H_2SO_4. The films prepared in a magnetic field of +5 T or –5 T are termed the +5T-film or the –5T-film, respectively. In this configuration, the vertical MHD flow is generated around the tube entrance (Fig. 1b left), and then it could penetrate into the tube with damping (Fig. 1b right). Thereby, the influence of vertical MHD flows on the MED film surfaces depends on the height of tube wall *h*.

The surface chirality of MED films was examined by the measurements of voltammograms of alanine enantiomers (20 mM) on the MED film electrodes in a 0.1 M NaOH aqueous solution. To evaluate the chirality in voltammograms, an enantiomeric excess (*ee*) ratio can be defined as $ee = (i_p^L - i_p^D) / (i_p^L + i_p^D)$, where i_p^L and i_p^D represent the peak currents of L- and D-alanines, respectively. The positive sign of the *ee* ratio represents L-activity, and the negative one represents D-activity.

Figs. 2a and 2b show the *ee* ratio profiles of the +5T-film and –5T-film electrodes prepared without tube wall at various deposition potentials. The +5T-film exhibits D-activity in the small overpotential region around –0.1 V, and then L-activity in the larger overpotentials around –0.4 V (Fig. 2a). On the contrary, the –5T-film exhibits L-activity around –0.3 V and D-activity around –0.4 V. These results demonstrate that the chiral sign depends on both the

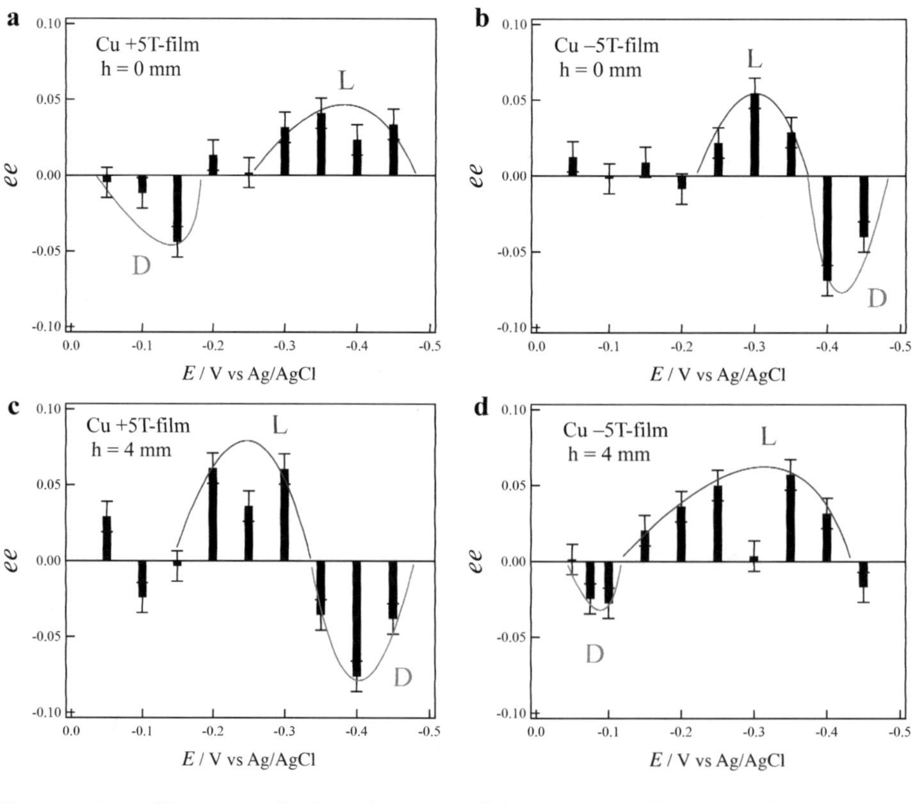

Fig. 2. The *ee* ratio profiles versus the deposition potentials. **a**: the +5T-film electrode without wall (h = 0 mm), **b**: the –5T-film electrode without wall, **c**: the +5T-film electrode prepared with wall at h = 4 mm and d: the –5T-film electrode with wall at h = 4 mm.

deposition potential and the polarity of magnetic field.

Our previous paper [4] showed that the surface chirality of MED Cu film depends on the deposition current in the galvanostatic MED. This current dependence of the chirality is due to the difference in the rate-determining step of the MED process. The potential dependence of the chirality in Figs. 2a and 2b can be similarly understood. The change of rate-determining step in Cu MED at 5 T occurs around –0.3 - –0.4 V [4], the electrode process is a kinetic-controlled at more positive potentials and mass-transfer-controlled at more negative potentials. Such a change in the rate-determining step corresponds to that in the chiral signs of MED films.

On the other hand, the magnetic-field-polarity dependence of chiral sign is easily understood by considering the direction of vertical MHD flow. As mentioned before, under the vertical MHD flow, the cyclonic micro-MHD vortices becomes stable, and the anticyclonic ones becomes unstable [8]. The reversal of magnetic field polarity reverses the vertical MHD flow and thus changes the chiral signs of the MED films.

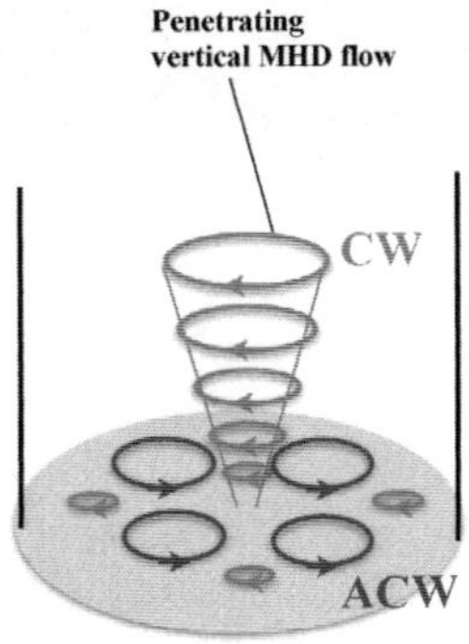

Fig. 3. Schematic of the reversal of the vertical MHD flow near the electrode surface in the tube wall.

Figs. 2c and 2d show the *ee* profiles of the +5T-film and –5T-film electrodes prepared with the tube wall at h = 4 mm. It is surprising that the *ee* profile of the +5T-film (Fig. 2c) is almost the same as that of the –5T-film without wall (Fig. 2b), and the profile of the –5T-film (Fig. 2d) is the same as that of the +5T-film without wall (Fig. 2a). These facts imply that the vertical MHD flow around the electrode surface at h = 4 mm is opposite to that without wall. Fig. 3 shows a possible scheme for such reversal of vertical MHD flow. When the vertical MHF flow penetrates into the tube with damping and its front would reach near the electrode surface, the reverse flows could be caused around the penetrating MHD flow such that the adjoining flows never conflict with each other. The effects of tube wall were examined with taller walls. The surface chirality almost disappeared at h = 12 mm, indicating that the penetrating MHD flow could not reach the electrode surface. These results demonstrate that the vertical MHD flow plays significant roles in the chiral surface formation and that the tube wall around the electrode would be useful for the control of surface chirality.

Effects of cell rotation

The rotation of electrochemical cell in magnetic fields could break the symmetry of micro-MHD vortices through the Coriolis force, producing chiral surfaces [7,9]. We have conducted such rotational MED experiments and examined the surface chirality of the MED copper films.

The electrochemical cell was rotated in the bore of a cryocooled superconducting magnet, and the magnetic field is imposed perpendicularly to the working electrode embedded in a tube (h = 20 mm). Clockwise (CW) and anticlockwise (ACW) rotations were generated by a geared motor with a frequency of 4 Hz.

Figs. 4a and 4b show the voltammograms of alanine enantiomers on the +5T-film electrodes prepared in the CW and ACW rotations, respectively. The CW-film exhibits D-activity, and the ACW-film exhibits L-activity. This result indicates that the cell rotation allows control of the surface chirality of MED films.

On the other hands, the comparison between the voltammograms of the same enantiomer on CW- and ACW-film electrodes shows the influence of the rotational direction on the chiral site formation. Fig. 5a shows the voltammograms L-alanine on the –5T-film electrodes prepared with CW and ACW rotations, and Fig. 5b shows the voltammograms D-

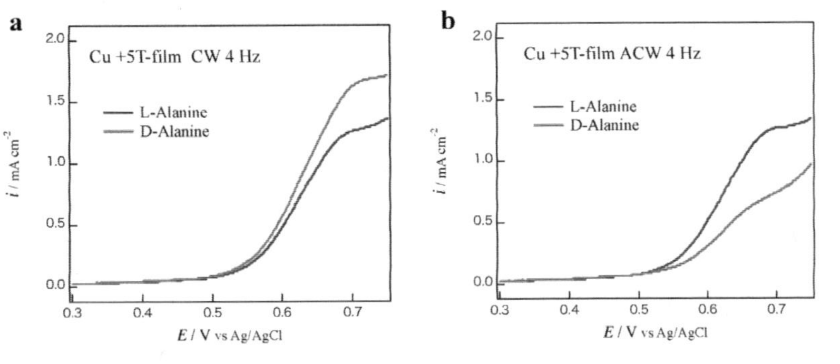

Fig. 4. Voltammograms of L- and D-alanines on the Cu +5T-film electrodes prepared in rotational MED at 4 Hz. **a**: the CW-film electrode and **b**: ACW-film electrode.

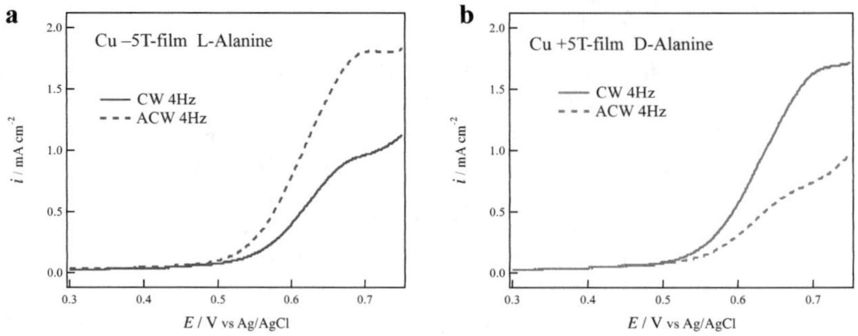

Fig. 5. Voltammograms of alanine on the CW-film and ACW-film electrodes prepared in the rotationsl MED at 4 Hz. **a**: L-alanine on the –5T-film electrode, b: D-alanine on the +5T-film electrode.

alanine on the +5T-film electrodes with CW and ACW rotations. In Fig. 5a the ACW-film exhibits greater currents of L-alanine than the CW-film. This means that in –5 T the ACW rotation increases the number of L-chiral site and the CW one deceases it. On the contrary, Fig. 5b exhibits that in +5 T the CW rotation increases the number of D-chiral site and the ACW one decreases it. These results demonstrate that there exists beautiful symmetry among the magnetic field polarity, rotational direction and surface chirality. Hence, the rotational MED could be a useful tool for the control of surface chirality of metal films.

Acknowledgment
This work was partially supported by JSPS KAKENHI Grant-in-Aid for Scientific Research (C) No. 16K04913. The authors thank the staff members of the High Field Laboratory for Superconducting Materials of IMR Tohoku University for the use of cryocooled superconducting magnets.

References

1. I. Mogi, K. Watanabe, ISIJ Int., 47 (2007), 585-587.
2. I. Mogi, K. Watanabe, Int. J. Electrochem., (2011), 239637.
3. I. Mogi, K. Watanabe, Magnetohydrodynamics, 48 (2012), 251-259.
4. I. Mogi, R. Aogaki, K. Watanabe, Bull. Chem. Soc. Jpn., 88 (2015) 1479-1485.
5. I. Mogi, R. Morimoto, R. Aogaki, Curr. Opin. Electrochem., 7 (2018) 1-6.
6. R. Aogaki, Magnetohydrodynamics, 39 (2003), 453-460.
7. R. Aogaki, R. Morimoto, Heat and Mass Transfer: Modeling and Simulation, ed. by M. Hossain, InTech, Croatia, (2011), pp. 189-216.
8. J. Sommeria, S. D. Meyers. H. L. Swinney, Nature, 331 (1988), 689-693.
9. I. Mogi, R. Morimoto, R. Aogaki, K. Watanabe, Sci. Rep., 3 (2013) 2574.

Electrodeposition of doped ZnO under a constant magnetic field

M. Stübner[1], J.P. Chopart[1] and A.L. Daltin[1*]

[1] LISM, URCA, B.P. 1039, 51687 Reims Cedex 02, France

Corresponding author: al.daltin@univ-reims.fr

Abstract

Magnetoelectrodeposition is a synthesis technique which allows to work with a superimposed magnetic field while the sample is usually electrodeposited. In this technique the magnetic field can modify the convection in the solution during the deposition via the Lorentz force convection also known as magnetohydrodynamic (MHD).

In this paper, a diluted magnetic semiconductor (DMS) material, a ferromagnetic transition metal doped with ZnO, is desired to be obtain; to determine if the superimposition of a constant magnetic field change some properties, such as its ferromagnetic properties. This type of material could be use in spintronics as well as in optoelectronics because of ZnO interesting optical intrinsic properties.

Key words : Electrodeposition, constant magnetic field, magnetohydrodynamic, diluted magnetic semiconductor, ZnO

Introduction

ZnO is one of the most used semi-conductors in a broad range of applications such as dye-sensitized solar cells or light emitting diode [1]. This hexagonal wurtzite structure material presents typically n-type electrical conductivity and a wide band gap (Eg ~ 3.3 eV at 300 K). It also possesses a large exciton binding energy (60 meV) which is much higher than traditional GaN (21 meV) and ZnS (20 meV) semi-conductor [1]. Since Dietl et al. [2] have predicted room temperature ferromagnetism (RT-FM) for Mn doped ZnO (T_c>300 K) transition metals (TM)-doped ZnO are more focused. In fact, ZnO finds application in optoelectronics as well as in spintronics. Theoretically, its properties relie on charge carriers properties, but also controllable spins (contribution of the dopant), revealing ferromagnetic properties [2]. There are numerous methods, both physical and chemical to obtain zinc oxide and it can be synthesized under various forms. Magnetic properties are strongly reliant on the preparation method employed. And several reports exist about electrodeposition of ZnO with ferromagnetic properties. So, we have chosen to electrodeposit ZnO because this technique has the benefits of low cost and low temperature preparation. Furthermore, doping is direct in this approach. Here, the retained ions are cobalt II (Co^{2+}) or nickel II (Ni^{2+}), which are ferromagnetic TM. The feature of the preparation of our thin films Co-doped ZnO is with the superimposed constant magnetic field during the deposition.

The chosen substrate is titanium. Indeed, a huge number of study are performed on Indium Tin Oxide (ITO) or Fluorine Tin Oxide (FTO) substrates [3]. But some studies report these substrate materials can show ferromagnetic behavior. To avoid risks of signal contamination, we culled a diamagnetic material, titanium, which possess a $\chi_m = 2.3 x 10^{-4}$ magnetic susceptibility. Effects of a magnetic field on an electrochemical reaction have been studied taking in account magnetohydrodynamics (MHD) convection induced by a Lorentz force. Several authors recognize that magnetohydrodynamics (MHD) induces effects on morphology and/or crystallographic orientation of the deposits [4].

The mechanism allowing diluted magnetic semiconductors to be ferromagnetic is a bit blurring. Plenty of theories, among which the most widespread concern oxygen vacancies or spin jump in interaction exchange (direct, indirect, superexchange...) are studied. Generally, if the techniques of material synthesis affect the properties of the latter, it is especially true for DMS. Several studies concerning the synthesis of doped zinc oxide (nickel or cobalt) by electrodeposition [5] have already been carried out and they have given satisfactory results regarding their magnetic properties. So here a Co or Ni-doped ZnO thin film is deposited, in order to obtain a DMS. But, at the same time, experiments are carried out under a superimposed constant magnetic field to know if the convective Lorentz force modifies some nucleation/growth process or some orbital interaction and finally on a more direct scale enhances the temperature interval of the ferromagnetism or decreases it.

Experimental details

Pure and cobalt or nickel-doped ZnO films were synthesized by electrochemical deposition on titanium substrates. First, these substrates have been clean during 5 minutes in acetone, then 5 minutes in distilled water using an ultrasonic-bath. The electrolyte consists of 0.1 mol/L concentration of zinc nitrate hexahydrate ($Zn(NO_3)_2.6H_2O$, purity 99%, Fluka) dissolved into distilled water. The average pH of the solution is around 5. The Co doping was carried out by adding cobalt nitrate hexahydrate ($Co(NO_3)_2.6H_2O$, purity 99%, Fischer Scientific). A cobalt concentration of 0.1 mmol/L has been retained. For nickel-doped ZnO, samples have been doped up to 5 mmol/L and 0.5 mmol/L with nickel nitrate hexahydrate ($Ni(NO_3)_2.6H_2O$, purity 96.5%). The electrodeposition was performed in galvanostatic mode using a

Versastat 4, Princeton Applied Physics potentiostat. A constant cathodic current of I = -5 mA has been applied on the 1.1 cm² surface area of the substrate (i.e. the current density is $j = -4.55x10^{-3}$ A/cm²). The setup is a classical 3-electrodes configuration. While working electrode is a titanium one, counter electrode is a platinum wire. The reference electrode chosen is a saturated Ag/AgCl one. Experiments were done at 80 °C during 100 min. Some experiments have been carried out under a vertically downward magnetic field parallel to the substrate (from 1 to 4 T) using a Cryogenic Superconducting Magnet (SMS120C). Samples were put into the center of the magnetic field to get a constant magnetic field on the substrate.

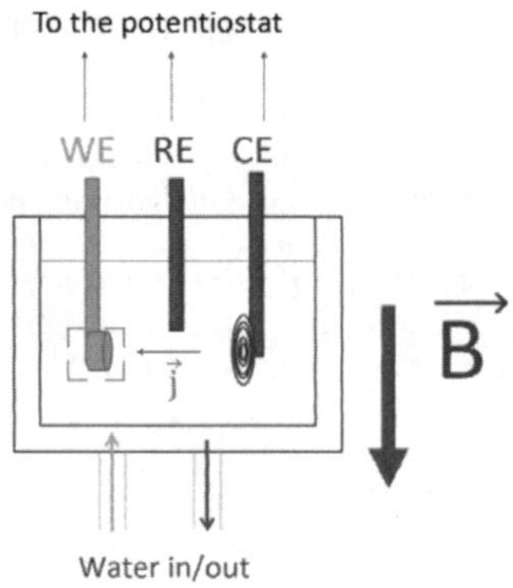

Fig. 1: Sketch of experimental setting.

After each deposition, samples have been washed with distilled water to remove Zn^{2+} and Co^{2+} left from the solution. The structural characterization of the material was investigated using X-ray diffraction (XRD) D8 Advance, Bruker. A glancing angle of 3°, a step size of 6 sec /0.06° with CuK_α (1.542 Å) radiation and a 2 θ ranging from 20 to 70° have been used. Morphologic characterization was mainly performed with scanning electron microscope (SEM) JEOL JSM 6460LA or with a JEOL7900F and chemical compositions were determined by an EDS JEL 1300 Microprobe coupled with the SEM. A micro-Raman spectrometer was used for the characterization of the Co-doped ZnO samples. The magnetic characteristics have been studied with SQUID and VSM measurements. The diamagnetic signal of the capsule was directly subtracted for M-H loop on ZnO samples. In order to determine chemical bounding and elemental constituents FTIR PerkinElmer Uatr Two has been used.

Results and discussion

The structure of Co-doped ZnO and pure ZnO thin films has been determined using X-ray diffraction. Fig. 1 (a) shows XRD spectra of ZnO:Co thin films prepared on titanium substrate from 30 to 70°. Some peaks from the substrate are noticeable and marked with a star. Other peaks result directly from the ZnO phase (JCPDS 36-1451, *c/a*: 1,5871) and no other phase is present. Films are (002) oriented.

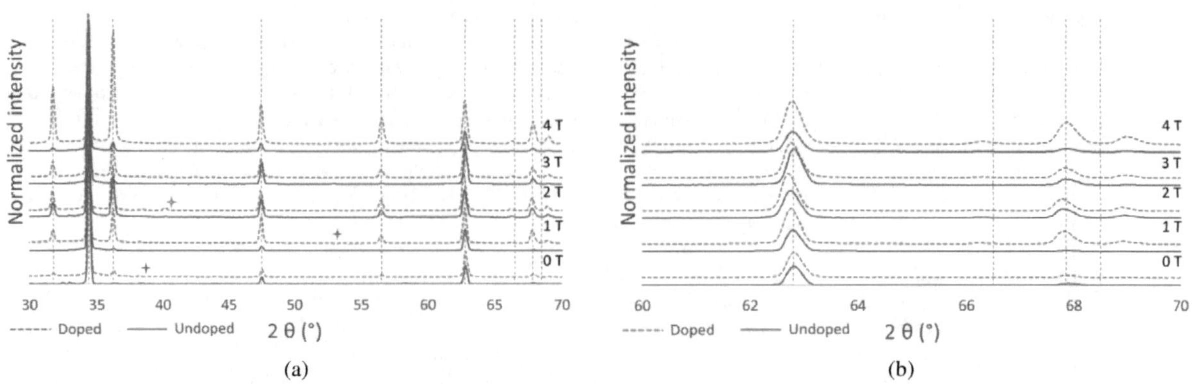

Fig. 2: (a) XRD pattern of pure and Co-doped ZnO under different magnetic field intensities (b) a zoom between 60-70°.

A focus between 60 and 70° on Co-doped ZnO samples (Fig. 1 (b)) shows a slight wide angles shift depending on magnetic field intensity. The affected angles are (103), (200), (112) and (201). This offset seems to be inexistent for undoped samples. The shift is not steady with the increasing of the field. It first goes toward the small angles and from 2 T until 4 T, it returns to its initial place. Magnetic field seems to slightly influence the structural properties of the samples (doped and undoped).

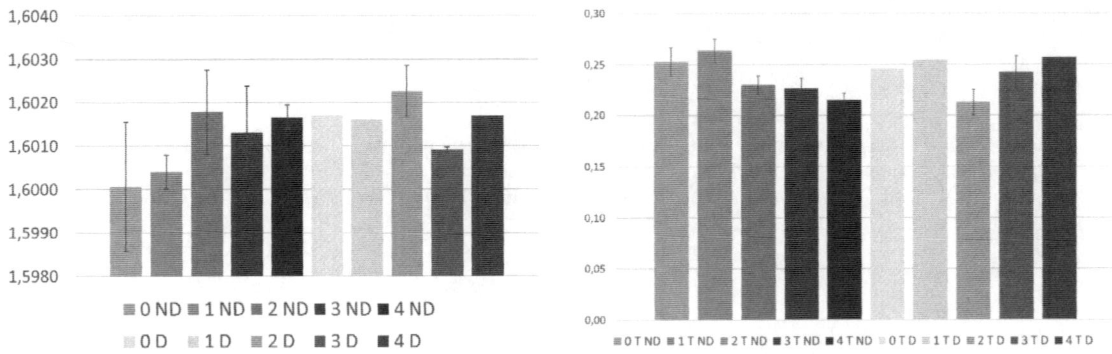

(a) (b)

Fig 3: (a) ZnO hexagonal c/a ratio and (b) full width at half maximum (FWHM) of Co-doped ZnO (D) and undoped ZnO (ND) samples depending on magnetic field intensity.

Looking at c/a ratio where c and a (a=b, because of hexagonal structure) are the lattice parameters which define the unit cell geometry, doping does not seem to affect the ratio but the magnetic field seems to do it. The same inference can be made with the full width at half maximum. Co-doped and undoped samples in terms of FWHM behave the same way. Sample surfaces are not completely homogeneous and hence there is some uncertainty on measurements which are visible on XRD error bars.

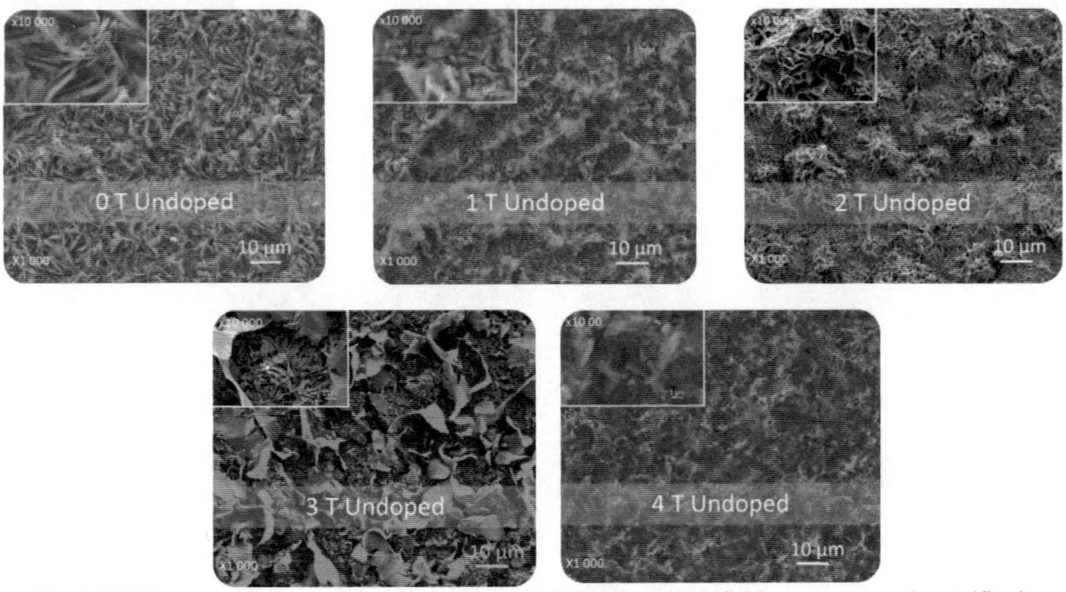

Fig. 4: SEM images of ZnO samples under a constant 1, 2, 3, 4 T magnetic field or not, at several magnifications (x1000 and inset (x10000).

Even if their surfaces are not homogeneous, here, SEM shots (Fig. 3) depict the behavior of most of samples. Those morphologies of pure ZnO under different magnetic field amplitudes are different. As before, the field seems to act on deposit, but in a not linear mode. The morphology goes to slips to big leaves.

Conclusion

Based on this these measurement, magnetic field seems to play a part in the growth process of the deposit since morphological and structural modifications are noticeable. This is a first step in understanding the influence, if any, of the magnetic field on doped ZnO deposits.

Acknowledgment

Financial support from Grand Est Region under SYMEDOX grant is gratefully acknowledged.

References

1. Ü. Özgür et al., J. Appl. Phys., 4 (2005) 041301
2. T. Dietl, H. Ohno, F. Matsukura, J. Cibert, D. Ferrand, Science, 287 (2000) 1019
3. H. Majumdar, S. Majumdar, D. Tobjörk, R. Österbacka, Synthetic Metals, 160 (2010), 303-306
4. A.L. Daltin, A. Addad, P. Baudart, J.P. Chopart, CrystEngComm, 10 (2011) 3373-3377
5. A. Simimol, et al., J. Appl. Phys., 21 (2015) 214310

***** Formatting Issue - Best Available Paper/Graphic *****

IOP Conf. Series: Materials Science and Engineering **424** (2018) 012026 doi:10.1088/1757-899X/424/1/012026

The depositing cell was kept at 90°C. The depositing time duration was controlled until the total charge for whole area of the cathode plate was integrated up to 1C, 3C, 5C, 8C and 15C, respectively.

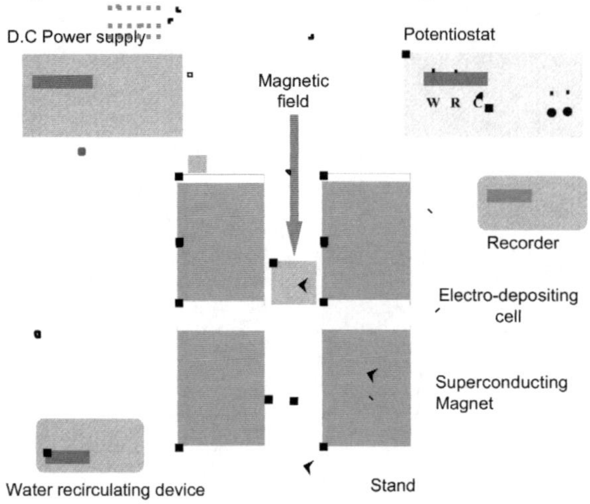

Fig. 1 Total system of experimental apparatus.

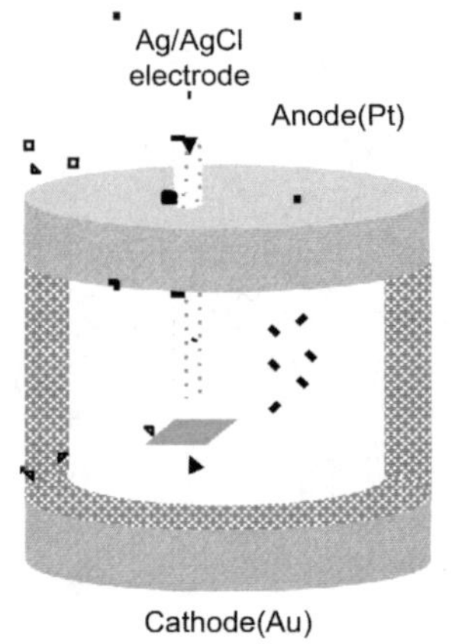

Fig. 2 Schematic of depositing cell.

Fig.3 shows the typical XRD pattern of deposited ZnTe film of the charge of 8C without magnetic field. ZnTe has sphalerite type cubic structure, and originally (111) plane is preferable depositing plane. However, crystal orientation by magnetic field imposition cannot be expected because of magnetic isotropy of cubic structure. In this experiment, actually, magnetic field has only micro MHD effect to promote the deposition behavior which deteriorate the performance of deposited film due to metal Te deposition.

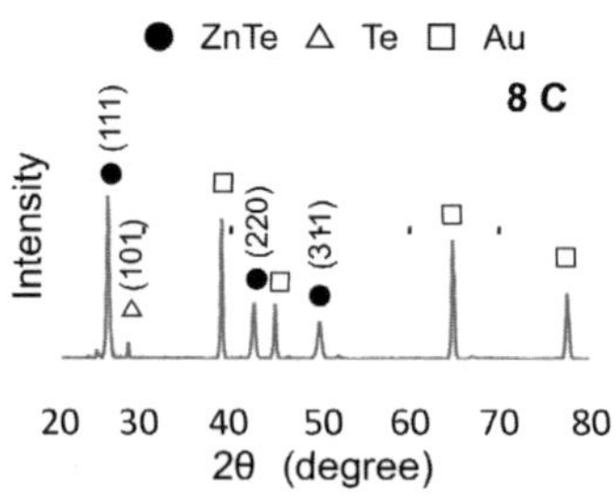

Fig. 3 Typical XRD pattern of electro-deposited ZnTe film.

From the results of these experiments, this research proposes a buffer layer deposition before ZnTe deposition and adopts Te film as a buffer layer because the structure has magnetic anisotropy. It is expected that ZnTe deposited on oriented crystal layer should have somehow crystal orientation.

Results and Discussion

Fig.4 shows the XRD patterns of the buffer layer under the magnetic field condition of (a) no magnetic field, (b) 5T of perpendicular field and (c) 5T of parallel field. Perpendicular magnetic field can increase (003) peak because the c-axis is preferable direction of magnetization. And it is found that the optimum condition is 5T of perpendicular magnetic field and 8C of the total charge amount of depositing electric current. After the buffer layer deposition, ZnTe is deposited on the buffer layer without magnetic field. The XRD pattern of ZnTe film is shown in Fig.5. As increase of total charge amount, ZnTe (111) peak becomes strong. However, in the cases of 8C and 15C, other peaks are also strong because of a disturbance due to long time deposition. So in this research, it can be concluded that optimum condition for obtaining high performance ZnTe film is the combination of Te buffer layer with 8C under 5T perpendicular magnetic field and the ZnTe layer with 5C under no magnetic field. In these conditions crystal orientation is evaluated by orientation index given by Eq. (1).

IOP Conf. Series: Materials Science and Engineering **424** (2018) 012026 doi:10.1088/1757-899X/424/1/012026

***** Formatting Issue - Best Available Paper/Graphic *****

$$\text{Orientation Index, N} = \frac{I_{(111)}}{\sum I} \bigg/ \frac{PDF \cdot I_{(111)}}{\sum PDF \cdot I} \tag{1}$$

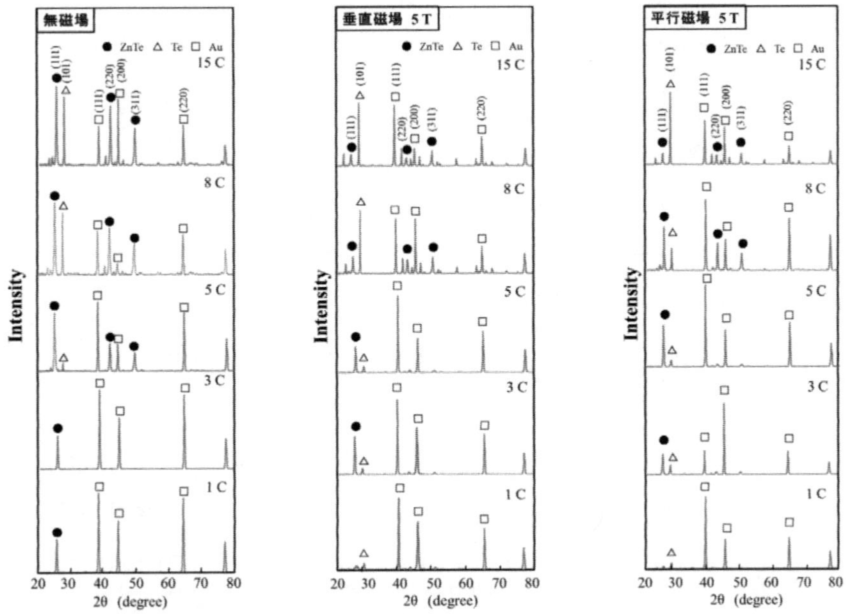

(a) no magnetic field (b) 5T of perpendicular field (c) 5T of parallel field

Fig. 4 XRD patterns of electro-deposited Te film under various magnetic field conditions.

Fig. 6 shows the orientation index of three types of samples with (a) no buffer layer and 5C of total charge, (b) Te buffer layer of 8C without field and 5C for ZnTe film and (c) Te buffer layer oriented by magnetic field imposition of 8C and 5C for ZnTe film. The index of oriented buffer layer increases to 1.48 compered with 1.30 of no buffer layer.

IOP Conf. Series: Materials Science and Engineering **424** (2018) 012026 doi:10.1088/1757-899X/424/1/012026
***** Formatting Issue - Best Available Paper/Graphic *****

Finally, performance of ZnTe semi-conducting film can be evaluated by band gap value. The ideal value for the optimum energy conversion ratio is 2.26 eV. The obtained energy band gap value and composition of the film in this research are shown in Table 2. The energy band gap was measured using Tauc plot based on the reflection factor.

Table 2 Obtained results relating the performances of electro-deposited ZnTe film.

	no buffer layer	Te buffer layer
Composition (Zn:Te)	42.5 : 57.5	44.1 : 55.9
Energy band gap	1.14 eV	1.53 eV

Conclusions

In the electro-deposition technique of ZnTe semi-conducting film, new deposition method applying strong magnetic field is proposed and obtained results are followings.

1. Imposition of strong magnetic field gives bad effect for the composition of deposited film because of micro MHD effect which promote metal Te deposition.
2. Crystal orientation can be realized by inserting buffer layer which is well oriented by strong magnetic field.
3. Using Te buffer layer, 1.53eV ZnTe film can be obtained.

References

1. Dong-Hun Han, Shin-Jung Choi and Su-Moon Park, Journal of The Electrochemical Society, 150(2003), C342-C346
2. T. Mahalingam, V. Dhanasekaran, K. Sundram, A. Kathalingam, Jin-Koo Rhee, International Journal of Ionics 18(2012), 299-306.
3. C. Winnewisser, P. Uhd Jepsen, M. Schall, V. Schyja and H. Helm, Appl. Phys.Lett. 70(1997) 3069.
4. D.Lee, J.Zucker, M.D.Divino, R.F. Austin, R.D. Feldman, K.L. Jpnes and A.M. Johnson, Appl. Phys. Lett. 59(1997).

Overview of Electromagnetic Forces to Control Flow During Continuous Casting of Steel

B.G. Thomas[1] and S.M. Cho[1]

[1] Colorado School of Mines, Mechanical Engineering, 1610 Illinois St., Golden, CO 80401, USA

Corresponding author: bgthomas@mines.edu

Abstract
During continuous casting of steel slabs, electromagnetic forces can be applied to help control the flow pattern in the mold, augmenting the effects of nozzle and strand geometry, casting speed, and argon injection. Specific configurations include: local, single-ruler, and double-ruler electromagnetic braking, traveling magnetic fields which actively drive flow in the upper mold, (speeding up, slowing, or stirring), and recently combinations of both systems together. Electromagnetic forces have the advantage of being changeable with casting conditions, and naturally adjust with flow variations. This gives this important flow-control tool the potential to stabilize flow in the mold, and to control molten steel velocities to lessen defects. To achieve these benefits in practice, however, requires good understanding of how the defects form, and the details of how the electromagnetic field(s) affect the transient flow. The best way to gain this understanding is by combining validated computational models with plant measurements. This paper shows recent advances using these tools to understand how different configurations affect both steady and transient flow behaviour, the formation of defects from level fluctuations, meniscus freezing, inclusion entrapment, and other mechanisms. These findings suggest promising ways to implement electromagnetic forces to lower defects to improve the process.

Key words: slab casting, electromagnetic braking, turbulence, transient flow, surface defects, inclusions

Introduction

The quality of continuous-cast steel depends greatly on fluid flow in the mold region of the process. Thus, the mold flow pattern must be controlled within acceptable process windows to avoid excessive surface velocities, large level fluctuations, meniscus stagnation, shell thinning, inclusion entrapment, and many other problems [1,2]. Together with the flow control system (stopper rod or slide gate), nozzle geometry, (port angle, shape, etc.), strand cross-section, submergence depth, casting speed, and argon injection, optimized electromagnetic forces offer an important way to maintain the flow pattern and velocity field within the optimal range for a given caster.

Schematics of the different electromagnetic system configurations that have been used in slab casting, and the magnetic fields which they produce, are shown in Fig. 1. ElectroMagnetic Braking (EMBr) applies a static magnetic field (using direct current) to restrict the flow through certain regions of the mold. The magnetic field strength through the strand thickness increases with the applied current according to the coil geometry. Movement of the conducting molten steel through this field induces a current in proportion to the field strength, but perpendicular to the field direction. The induced current in turn interacts with the flowing steel to generate a Lorentz force that increases with the steel velocity, and is rotated a further 90 degrees, so that it directly opposes the flow direction. Thus, the force naturally adjusts according to flow variations, giving it the potential to stabilize the flow.

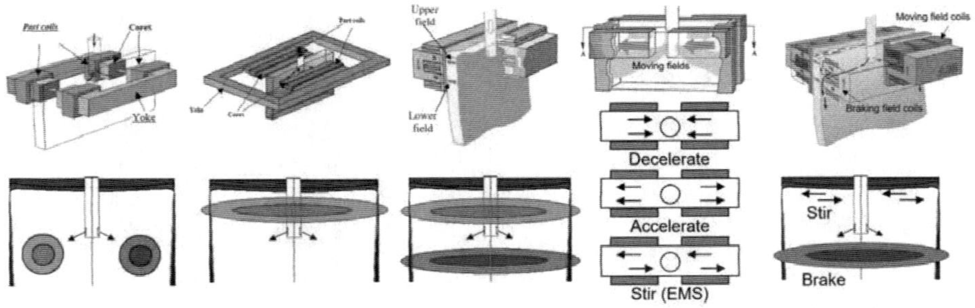

(a) Local EMBr (b) Ruler EMBr (c) Double-ruler EMBr (d) Moving fields (e) Combined

Fig. 1. Types of electro-magnetic mold flow control systems showing hardware (top) and field shape (below).

IOP Conf. Series: Materials Science and Engineering **424** (2018) 012027 doi:10.1088/1757-899X/424/1/012027

Tools to Study Electromagnetic Effects

The best way to investigate electromagnetic effects is with validated computational models and plant measurements [3]. In addition to solving the Navier Stokes equations for the fluid flow, adding electromagnetics requires finding the magnetic field, by calculation or measurement, and solving Ohm's law and Maxwell's equations for the current and Lorentz forces [4]. Plant measurements include surface velocities, found from the solidified lumps on inserted nails [5], or deflection of inserted rods [6], and the surface level profile, found by tracing oscillation marks around the mold perimeter [7]. Lab experiments with flowing molten metal are helpful too [8, 9]. Finally, long-term plant studies to correlate defect incidence with casting conditions are essential.

Effect on Fluid Flow

The best steel quality is reported with a stable double-roll flow pattern in the mold [1,2], with jets impinging and flowing up the narrow face, with surface velocities back towards the SEN falling within an optimal range. That range has been reported as 0.2 – 0.5 m/s [10], but depends on the specific geometry and casting conditions. Single-roll flow patterns, where the jet first impinges on the top surface and flows to the narrow faces, generate higher level fluctuations and defects, while complex flow patterns which are always unstable give the worst quality. In addition to maintaining an optimal time-averaged flow pattern, it is important to achieve stable flow that minimizes detrimental transient fluctuations, which are inherent in turbulent flow. Electromagnetic forces should be designed to avoid surface velocities that are either too fast or too slow, with an optimal turbulence level to promote mixing and melting of the slag layers, while improving flow stability.

Local EMBr fields create roughly circular regions of high flow resistance, positioned near the nozzle ports as shown in Fig. 2. They are designed to slow down jet flow, as it passes through these regions [11]. In practice, however, these regions tend to deflect the jets either up or down, according to the current submergence depth, which changes the port location relative to orientation of the circular magnetic field regions [7]. If the jet is below the region, it retains its momentum to impinge strongly against the narrow face, leading to high-speed surface flows and deep penetration of large particles. Worse is if the jet is above this region, so it will be deflected upwards to cause a detrimental complex flow pattern. Optimizing the field strength requires consideration of the submergence depth and mold width.

Single-ruler EMBr systems generally apply a static field across the entire top region of the mold, (Fig. 1b)) in order to slow down the flow, which is often useful to avoid excessive surface velocities in high-speed thin slab casters. However, the danger of a non-optimal single-ruler system is that surface velocities may become too slow if the field current is too high for the flow conditions. This can result in excessive cooling of the meniscus region, insufficient mixing of the mold slag layers, leading to meniscus freezing, where the accompanying subsurface hooks can capture inclusion-laden bubbles and mold slag droplets into the solidification front, leading to an increase in surface defects. Locating the single ruler below the nozzle ports can increase surface velocity and stabilize the flow [8,12,13]. In addition, vortex formation at the mold top surface, caused by biased surface flow due to nozzle misalignment, can be suppressed with single-rule EMBr together with optimized argon injection, field strength, and ruler location beneath the nozzle ports [14]. Care should be taken to avoid centering the magnetic field over the nozzle ports, as this creates flow instability [8,12,13].

Double-ruler EMBr sytems, also called "FC-Mold" [9], applies two rectangular-shaped magnetic fields across the mold width, with one above and the other below the nozzle ports, as shown in Fig. 1c). As pictured in Fig. 3, the upper ruler tends to slow surface velocities, decrease level profile variations, decrease level fluctuations, and dissipate high-frequency turbulent fluctuations if the field strength is increased [6] or if the ruler is located just above the nozzle ports, so it deflects the jet downwards [6,13,15-17].

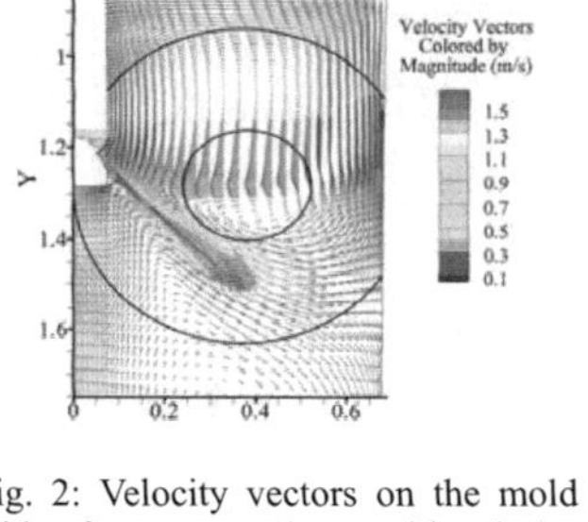

Fig. 2: Velocity vectors on the mold wide face centerplane with circles showing local EMBr field strength, and downward jet deflection [7].

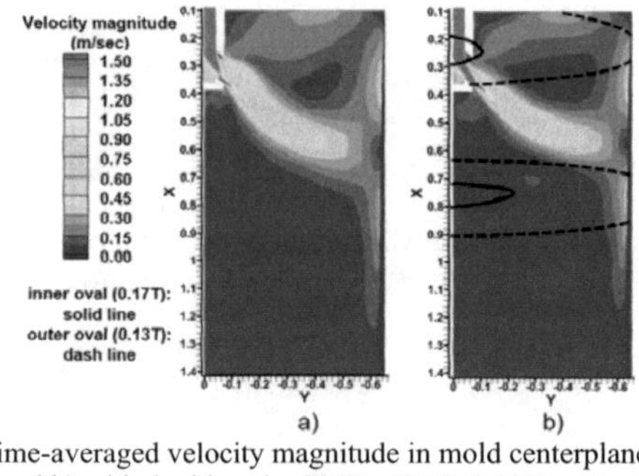

Fig. 3 Time-averaged velocity magnitude in mold centerplane a) without and b) with double-ruler EMBr, (0.17, 0.13T contours) [17].

The upper EMBr ruler also has an important strong effect on stabilizing flow through the nozzle, making it less subject to variations in swirl direction [17] or even suppressing swirl completely [6]. This helps to avoid surface velocity variations, which have been shown to originate with changes in swirl direction exiting the nozzle ports [17]. Increasing the strength of the lower ruler field tends to deflect the jet upwards, causing surface velocities and turbulence to increase. Thus, the two ruler strengths should be adjusted to achieve optimal surface velocity.
Moving magnetic fields are created from alternating current through a series of magnets, each having a different phase shift in order to create an electric motor effect that generates a force to actively drive the flow tangentially across the wide face surfaces. The strength of the force depends on both the applied current and the effective frequency of the phase shift. By adjusting the direction of the phase shift (and force), moving fields can cause the flow to accelerate (EMLA), to slow down, (EMLS), or to create a stirring motion around the mold perimeter, (EMS), as shown in Fig. 1d) [1,18]. Stirring should increase the chances of large particles being washed along the dendritic solidification front without being captured, until they can safely be removed into the top slag layers. This hypothesized mechanism needs to be quantified and validated by modeling and plant testing.
Combining EMBr and moving field systems together is recently creating new systems, such as shown in Fig. 1e), which offer even more ways to adjust the electromagnetic forces that change the mold flow pattern [19]. One combination with good potential is to combine a lower ruler brake, with stirring around the upper regions of the mold. The lower ruler should stabilize the jet flow impinging on the narrow faces, by deflecting the jets upwards in proportion to their downward deviations, while stirring should lessen inclusion entrapment.

Effect on Surface Defects
In addition to average surface velocities, transient flow fluctuations greatly affect surface defects. Level fluctuations lead to slag entrapment at the meniscus, and large level profile variations cause slag infiltration problems [17]. The effect of double-ruler EMBr on surface velocities and level profiles are shown in Figs. 4 and 5. Both the Large Eddy Simulations (LES) and measurements show that EMBr slightly decreases the surface level fluctuations, and consequently makes the level flatter. The decrease in variations with EMBr is indicated by comparing the error bars on the measurements, and the range of the calculated instantaneous profiles.

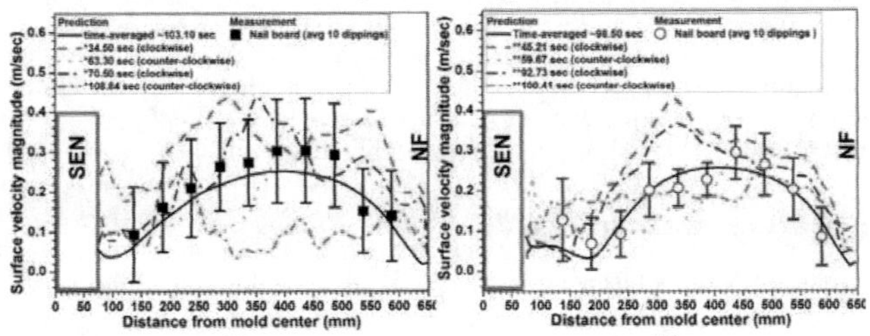

(a) (b)

Fig. 4. Surface velocity profiles comparing LES modeling and measurements a) without and b) with EMBr [17].

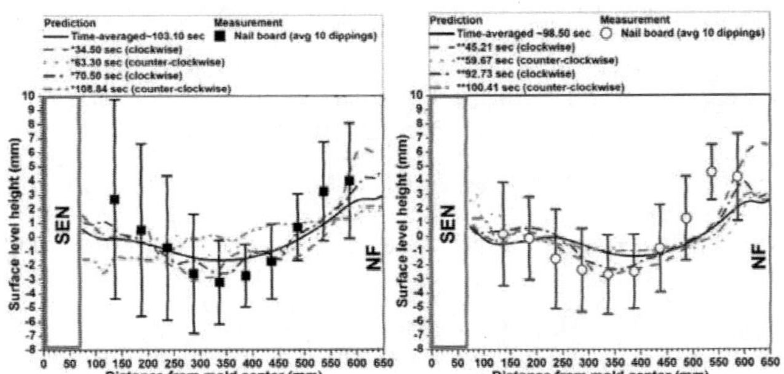

(a) (b)

Fig. 5. Surface level variations comparing LES modeling and measurements a) without and b) with EMBr [17].

Effect on Internal Inclusion Defects
Inclusions transported into the lower recirculation regions are generally entrapped, leading to internal defects. Optimal application of lower EMBr ruler can slow the velocity of jets down the narrow faces, lessening the penetration of large inclusion-laden bubbles and slag droplets into the lower regions of the caster, thereby improving internal quality. LES results show that EMBr can also decrease the amount and rate of entrapment of smaller inclusions slightly, as shown in Fig. 6 [20].

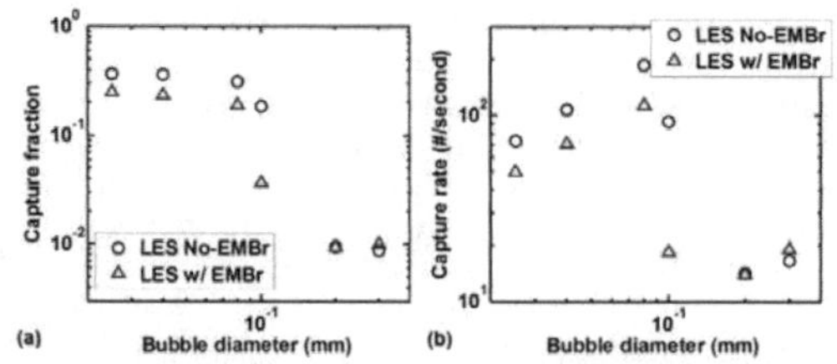

Fig. 6. Effect of EMBr on (a) capture fractions and (b) capture rates of bubbles [20].

Future Developments

Unlike most parameters that control mold flow, such as nozzle geometry, electromagnetic forces are capable of active real-time adjustment in the steel plant. Based on differences in the flow conditions, the magnetic field current can be adjusted to respond to changes in the flow pattern. For example, the expected field requirements can be set up according to the current casting conditions, (speed, submergence depth, mold width, etc.) based on offline parametric studies with validated computational models. Another pioneering method, to adjust the field to respond to unpredictable local variations associated with turbulence and argon gas, is to infer the flow conditions from real time measurements of the surface level profile using two eddy-current sensors on the same side of the mold [1]. Alternatively, the surface level profile and other phenomena may be inferred from thermocouples or high-resolution mold temperature maps using fiber-Bragg-grating sensor systems [21,22].

Conclusions

1)Computational models, validated with plant measurements of surface velocities and surface level profiles, can accurately predict transient fluid flow in continuous casting systems, including the effects of electromagnetic forces. This is providing better understanding of surface defects and particle capture, which can enable optimization of caster operation conditions, including electromagnetic flow-control systems.

2)Electromagnetic forces can significantly affect flow in the mold, with benefits to lowering surface and/or internal defects if implemented properly. Recent developments in electromagnetic systems which combine braking and stirring together, high-resolution mold temperature sensors, and advanced computational models, offer great potential for electromagnetic systems to further improve steel quality.

Acknowledgements

The sponsors of the Continuous Casting Center at the Colorado School of Mines are gratefully acknowledged. We also thank the National Science Foundation Grant CMMI 18-08731 for support that enabled this research.

References

1. Dauby, P.H., Revue de Métallurgie, 109, (2012), 113-136
2. B. G. Thomas, Chapter 14. in Making, Shaping and Treating of Steel, AISE, (2003)
3. B. G. Thomas, Steel Res. Int., 89 (2018), 1700312: 1-21
4. R. Chaudhary, S. P. Vanka, and B. G. Thomas, Physics of Fluids, 22:7, (2010), 075102
5. R. Liu, B. G. Thomas, J. Sengupta, S. D. Chung, and M. Trinh, ISIJ Int., 54, No. 10, (2014), 2314-2323
6. Jin, K., S.P. Vanka, and B.G. Thomas, Metall. Mater. Trans. B, 48B:1, (2017), 162-178
7. K. Cukierski and B. G. Thomas: Metall. Mater. Trans B, 39B:1, (2008), 94-107
8. B.G. Thomas, R. Singh, S.P. Vanka, K. Timmel, et al, J. Manufac. Sci. Prod., 15:1, (2015), 93-104
9. A. Idogawa, M. Sugizawa, S. Takeuchi, K. Sorimachi, and T. Fujii, Mat. Sci. & Eng., A173, (1993), 293-297
10. J. Kubota, K. Okimoto, A. Shirayama, and H. Murakami, in Mold Operation for Quality and Productivity, ISS, Warrendalte, PA, (1991), 197-205
11. Y. Wang and L. Zhang: Metall. Mater. Trans B, 42B: 6, (2011), 1319-1351
12. R. Singh, B. G. Thomas and S. P. Vanka, Metall. Mater. Trans. B, 44B:5 (2013), 1201-1221
13. R. Chaudhary, B. G. Thomas and S. P. Vanka, Metall Mater Trans B, 43:3, (2012), 532-553
14. B. Li, T. Okane, and T. Umeda, Metall Mater Trans B, 32:6, (2001), 1053-1066.
15. S.-M. Cho, S.-H. Kim and B. G. Thomas, ISIJ Int, 54:4, (2014), 855-864
16. R. Singh, B. G. Thomas and S. P. Vanka, Metall. Mater. Trans. B, 45B:3 (2014), 1098-1115
17. S.-M. Cho, B. G. Thomas, and S.-H. Kim, Metall. Mater. Trans. B, 47B:5 (2016), 3080-3098
18. S. Kunstreich and P.H. Dauby, Ironmaking & Steelmaking, 32:1, (2005), 80-86
19. H. Yang, M. Sedén, N. Jacobson, J-E. Eriksson, and H. Hackl, AISTech Conf. Proc., AIST, (2012)
20. Jin, K., S.P. Vanka, and B.G. Thomas, Metall. Mater. Trans. B, 49B:3, (2018), 1360-1377
21. M. Sedén, ABB, Sweden, Personal Communication, (2017)
22. G. Hedin, A. Kamperman, M. Sedén, K. Fröjdh, and J. Pejnefors, SCANMET V Conf. Proc., Swerea, (2016)

Application of static magnetic field to casting of steel

Y. Miki[1*], K. Furumai[2] and T. Odagaki[2]

[1] JFE Steel Corp., 1 Kawasaki-cho, Chuo-ku, Chiba-shi, Japan
[2] JFE Steel Corp., 1 Kokan-cho, Fukuyama-shi, Japan

*Corresponding author: y-miki@jfe-steel.co.jp

Abstract
A magnetic field was applied in the continuous caster in steelmaking. A static magnetic field has advantages in the steelmaking processes, in that the apparatus for generating a static magnetic field is simpler than alternating magnetic coils, and stabilization of the flow is enhanced by a static magnetic field because a larger force acts on a faster steel flow. Therefore, application to industrial processes has been investigated.
In this presentation, two applications are discussed, as follows:
1) Steel flow control in mold
The steel flow is controlled in the mold to produce high quality steel products with high productivity. An unbalanced steel flow causes surface defects. A static magnetic field is effective for stabilizing the steel flow in the mold. Instability becomes more severe with larger throughput or a thinner mold. A static magnetic field is effective for reducing this instability. Furthermore, because an insufficient steel velocity causes entrapment of bubbles and inclusions on the solidified shell, optimization of the steel flow velocity by controlling the strength of the two domain magnetic fields is important.
2) Microstructure control of cast metal
A static magnetic field suppresses the liquid flow, and the microstructure can be controlled by this flow control. In a Sn-Zn alloy casting test, the columnar grain ratio increased when a magnetic field stronger than 0.1 T was applied.
These phenomena occurred because thermal convection was reduced by the magnetic brake, followed by less supercooling at the dendrite end. The equiaxed crystal size increases with a magnetic field. Since the magnetic field in the early stage has a larger effect, nuclear generation of equiaxed crystals is considered to be restrained by reducing the steel flow and the temperature gradient.

Keywords
Static magnetic field, casting, steel, surface defects, microstructure, control

1. Steel flow control in mold

It is important to optimize the flow velocity in a mold in order to reduce surface defects of cast steel. One cause of surface defects is the existence of an asymmetric unbalanced time-dependent flow. This means that the steel flow rate from one spout is not always the same as that from the other[1)].

To prevent penetration of molten steel and bubbles into the deep portion of the strands, the downward flow velocity along the narrow face is reduced by two types of electromagnetic flow control systems. The FC mold (Flow Control Mold) utilizes a static magnetic field, as shown in Fig. 1.

The FC mold, which consists of two magnetic coils, generates a homogeneous magnetic field over a large area. The upper field stabilizes the top free surface, and the lower field seems to be effective for suppressing the downward steel flow. Therefore, the effect of a static magnetic field on the time-dependent flow was also estimated by numerical simulation and measurement of cast slabs.

Because the dendrite angle changes depending on the steel flow at the solid/liquid interface, the flow velocity can be estimated by measuring the dendrite angle.

Fig. 2 shows the calculated results of the flow velocity with magnetic densities of 0.1 T or 0.2 T. The momentum of the steel jet from the nozzle spout is estimated by the flow velocity. The momentum is predicted by using the Stuart number, as shown at the right. The Stuart number, N, is the ratio of magnetic force to momentum force and is defined as follows.

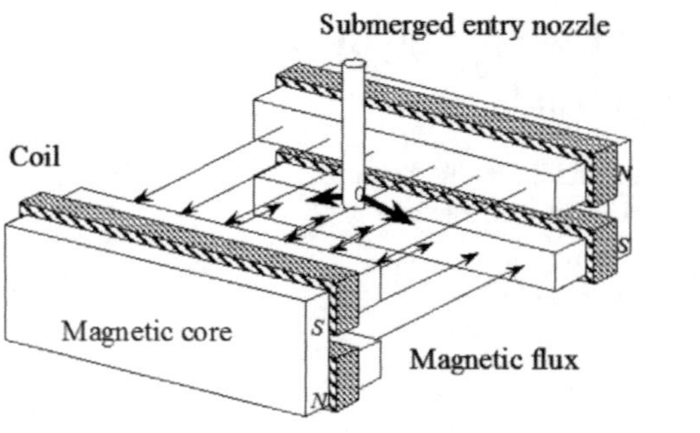

Fig. 1 Schematic illustration of the FC mold.

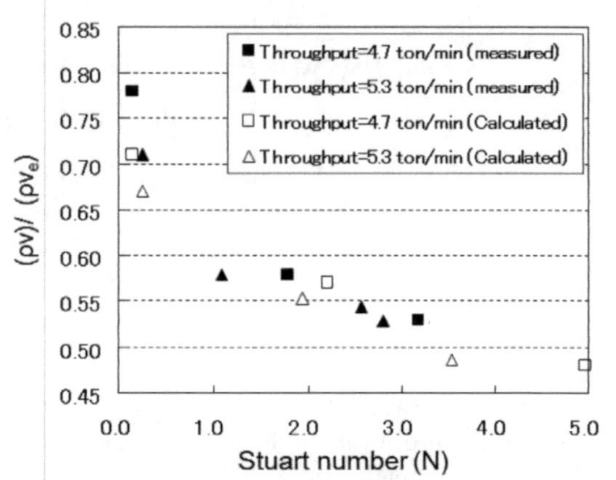

Fig. 2 Reduction of momentum to narrow face by static magnetic field [2)]

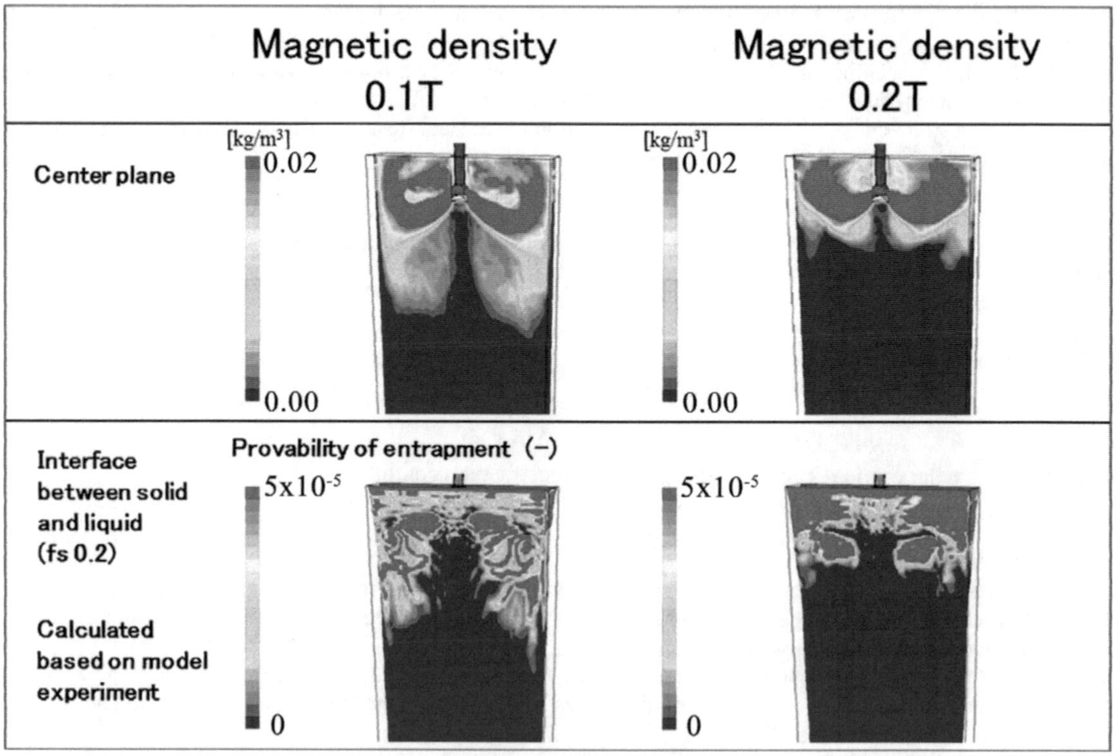

Fig. 3 Calculated probability of entrapment of 0.5 mm argon bubble in molten steel with static magnetic fields of 0.1 T and 0.2 T. (1500 mm width, 260 mm thickness, and throughput 4.5 t/min) [3)]

$$N = \frac{B^2 \sigma L}{\rho v}$$

Where, B : magnetic density, σ : , L : length, ρ : density of steel, v : flow velocity

The measured value of the effect of the electromagnetic brake in decreasing the molten steel momentum was consistent with the calculated value. A correlation existed between this effect and the Stuart number under different throughput and magnetic flux density conditions. The molten steel momentum could be reduced by more than 50% when the Stuart number was more than 3.5.

Furthermore, since an insufficient steel velocity causes entrapment of bubbles and inclusions on the solidified shell, optimization of the steel flow velocity by controlling the strength of the two domain magnetic fields is important.

Entrapment of bubbles and inclusions can be suppressed by adjusting the magnetic flux density. Fig. 3 shows the calculated probability of entrapment of 0.5 mm argon bubbles in molten steel with static magnetic fields of 0.1 T and 0.2 T. In this case, entrapment is reduced with the higher magnetic density of 0.2 T.

2. Microstructure Control

The microstructure of cast steel affects the properties of final steel products. The microstructure also changes the efficiency of soft reduction to reduce center segregation.

Therefore, the effect of a static magnetic field on the solidification microstructure of a Sn-Zn alloy is examined in this study. A Sn-Zn alloy casting test was carried out using the apparatus shown in Fig. 4. Fig. 5 shows the microstructure on the vertical plane. The columnar grain ratio increased with a magnetic field stronger than 0.1 T. These phenomena occurred because thermal convection was reduced by the magnetic brake, followed by less supercooling at the dendrite end. The effect of the magnetic brake for steel was estimated by the Hartmann number. The result showed that a magnetic field stronger than 0.17 T is effective for increasing the temperature gradient and promoting the columnar grain formation. The experimental results of the columnar ratio shown in Fig. 6 can be explained by the above discussion.

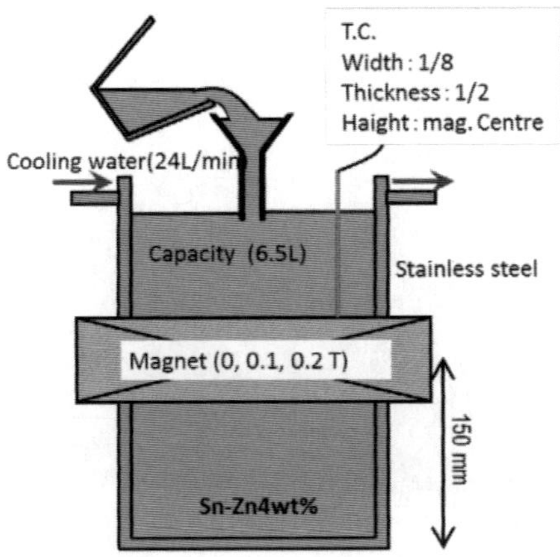

Fig. 4 Experimental apparatus

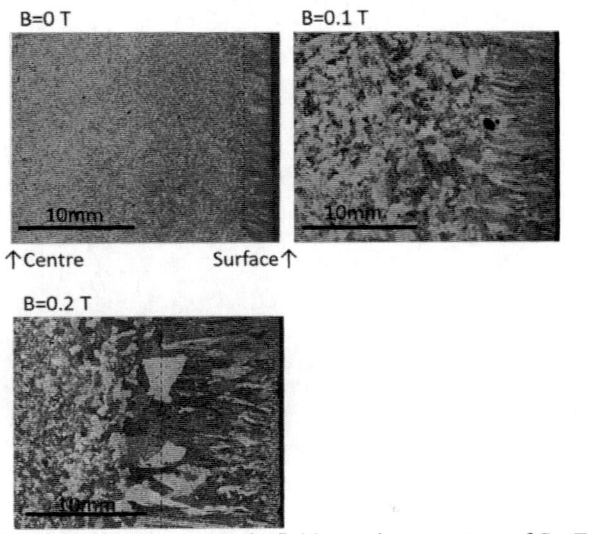

Fig. 5 Effect of static magnetic field on microstructure of Sn-Zn alloy ingot (vertical section) [4]

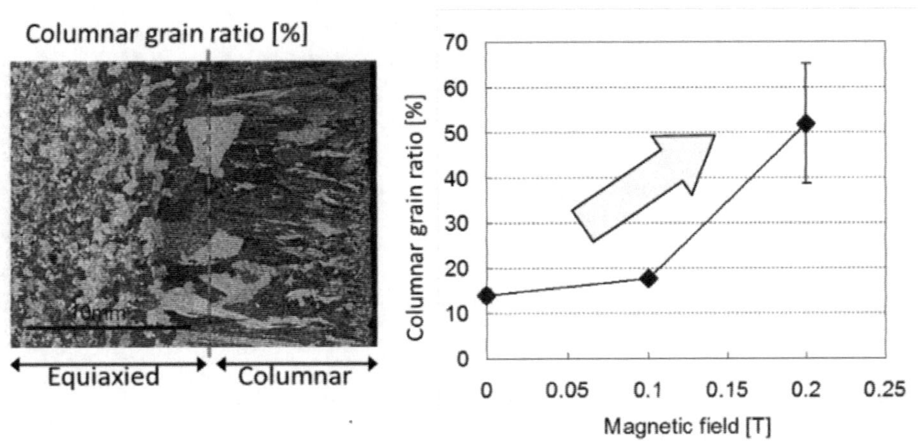

Fig. 6 Relationship between magnetic density and columnar grain ratio [4]

The equiaxed crystal size increases with a magnetic field. The magnetic field in the early stage has a larger effect, so nuclear generation of the equiaxed crystal is considered to be restrained by reducing the steel flow and the temperature gradient.

IOP Conf. Series: Materials Science and Engineering **424** (2018) 012028 doi:10.1088/1757-899X/424/1/012028

References

1. Y. Miki and S. Takeuchi: ISIJ Int., 43 (2003)10, 1548-1555.
2. K. Furumai, Y. Matsui, T. Murai, Y. Miki: Tetsu-to-Hagane, 100(2014)4, 563-570.
3. Y. Miki, H. Ohno, Y. Kishimoto, S. Tanaka: Tetsu-to-Hagane, 97(2011)8, 423-432.
4. T. Odagaki, N. Aramaki, N. Kikuchi and Y. Miki: CAMP-ISIJ, 31(2017), 202.

IOP Conf. Series: Materials Science and Engineering **424** (2018) 012029 doi:10.1088/1757-899X/424/1/012029

Electromagnetic particle separation in the cold crucible melting with novel type bottom pouring nozzle

V. Bojarevics and K. Pericleous

University of Greenwich, London SE10 9LS, UK

Corresponding author: v.bojarevics@gre.ac.uk

Abstract

The cold crucible melting technique is proven to be most effective to obtain reactive metal castings and produce high quality metal powders for aerospace, automotive and medical applications. A weak point of this technology is the graphite/ceramic nozzle used to pour the molten alloy through the bottom opening, which has very limited life time (several minutes) and adds contamination to the high quality final material. The paper uses the mathematical modelling technique, previously validated on multiple similar cases, to investigate a new type of non-consumable nozzle made of copper segments. During the dynamic melting the nozzle entrance is protected by a thin solidifying layer of the same alloy as the main melt. The AC electromagnetic (EM) field effects are investigated for the possibility to extract particles (impurities, precipitates, oxides, bubbles, etc.) from the melt. The dynamics of particles are affected by the hydrodynamic drag, buoyancy, turbulent fluctuations and the EM force via the local pressure distribution. The EM force leads to high mixing rates, transitional flow structures and turbulence of the melt, contributing to the particle dispersion, transport and separation to a desired location. The paper demonstrates effects of flow on the time dependent particle 'swarm' transport and separation to the side solidified 'skull'.

Key words: titanium alloys, cold crucible, electromagnetic mixing, melting, particle tracking

Introduction

Titanium alloys are the light structural metals attractive for aerospace and automotive applications. Research efforts are aimed at developing new melting techniques and practices to achieve desirable mechanical properties for critical components [1]. One major barrier to casting these components is melting and subsequently casting homogeneous alloys free of contamination and severe segregation, or alternatively adding uniformly dispersed micro to nano particles to create a composite alloy of enhanced mechanical properties [2]. The additional stirring created by electromagnetic forces enhances the particle dispersion and eases the restrictions on the cooling rate during the casting process. It is well known that forced stirring of the melt by electromagnetic forces can significantly promote the detachment of dendrite arms and transporting the crystal fragments in a disperse fashion due to the presence of high turbulence. The cold crucible technique, alternatively known as the induction skull melting (ISM), is proven to be most effective to obtain reactive metal castings and produce high quality metal powders. There is a variety of mathematical modelling attempts of the ISM to mention just the few representative publications [3-5] and the references therein. The present paper continues the attempts of the fully dynamic simulation for the melting including the dynamic free surface resolution during the full operational cycle of the cold crucible operation by using the previously validated spectral code SPHINX [4,6]. The code is supplemented with the Lagrangian particle tracking module in attempts to give insight to the various inclusion behavior and possible separation in the ISM process.

The large scale electromagnetically driven flow circulation exerts a drag force which contributes to the particle transport [7]. The particles travel due to the hydrodynamic drag, buoyancy, turbulent fluctuations and the electromagnetic force action via the local pressure distribution around the particle. A significant contribution to the force balance arises from the so called electro-magnetophoretic forces [8], which are well known for a variety of MHD applications, such as removal of inclusions from steel melt [9], concentrating electrically insulating bubbles or particles towards the surface [10], and for many other purposes of separation or mixing [7]. The electromagnetic force acts directly onto electrically conducting inclusions, however the electromagnetic force in the surrounding fluid creates the pressure distribution gradient leading to the integral force on the non-conducting inclusions as well. Experimental observations of the particles of various shapes and electrical conductivities in the liquid metal carrying electrical current are described in [11], where the expression similar to Leenov, Kolin's [8] was derived, but for the conditions of gradient magnetic field, which results in the increase of the effective force on the particles. In the AC magnetic field the time average force effects are often treated using the same formulae as for the constant uniform fields [7-10]. The latter contradicts the skin-layer distribution of the AC field magnitude exponentially rising near the metal surface. Moreover, the total force in AC field contains also the pulsating part contributing to the particle drag by the 'history' and 'added mass' contributions [12,13]. The high mixing rates ensure the intense turbulence in the bulk of molten metal, which additionally affects the particle dispersion due to the turbulent velocity pulsations. This paper demonstrates the importance of the AC magnetic interaction effects to the particle transport and the eventual separation during the cold crucible operation accounting for the solidifying skull formation and the melt outflow to the nozzle at the bottom. A novel type of the non-consumable nozzle made of copper segments is proposed which could help to solve the problem of the short life cycle of traditional graphite/ceramics type of nozzles used in the existing ISM devices [1].

Mathematical modelling

The mathematical modelling of the complex time dependent problem starts initially with a cylindrical solid body heating, leading to the gradual melting process and subsequent liquid metal magnetic confinement dynamics, internal fluid flow and the temperature evolution. This is accomplished using the spectral code SPHINX developed at the University of Greenwich [4,6,14]. Within the temperature limit below melting the solid body heating is solved for the continuously adjusted electromagnetic field from the power source coils. In the present ISM setup two different coils are used: one at the side (lower frequency ~ 5 kHz) and the second smaller at the nozzle (~50 kHz). The time dependent Joule heating eventually leads to melting of the solid ingot. When the solidus temperature is reached, the model accounts for the phase transition and the following liquid material shape change due to the action of the EM force distribution adjusted to the moving solid/liquid and the free interface. The fluid flow is solved using the Reynolds averaged turbulent flow model for an incompressible fluid with the effective viscosity ν_e and the turbulent thermal diffusivity α_e modified heat transfer equations given in detail elsewhere [4,6]. The boundary conditions are prescribed dynamically on the moving liquid interface. The free surface boundary conditions are determined by the tangential and normal stress components accounting for the temperature dependent surface tension $\gamma(T)$. At the contact to the solid cold crucible walls the no-slip condition for velocity is imposed. The thermal boundary conditions account for the radiation losses, the high vapor pressure alloy component evaporation, and the heat loss to the water-cooled solid surface of the crucible and the nozzle [4]. The numerical solution method is based on the continuous co-ordinate transformation adapting to the free surface and the containing vessel shape at the consecutive time steps.

The present code implementation includes the Lagrangian passive particle tracking algorithm. The position **R**(x,y,z,t) of a large number (N~1000) of individual particles can be determined following their paths and the variable total force **F**(x,y,z,t) acting on each of the particles by solving the set of two equations [12]:

$$\partial_t \mathbf{R} = \mathbf{u}_p, \quad m_p \partial_t \mathbf{u}_p = \mathbf{F}, \qquad (1), (2)$$

where $\mathbf{u}_p$ is the particle velocity and m_p its mass. The force **F** acting locally on the spherical particle can be decomposed into the instantaneous fluid drag force $\mathbf{F}_d$ (depending on the local Reynolds number Re_p), the memory Basset force, the added mass force (see [15]), the buoyancy force $\mathbf{F}_g$ and the effective electromagnetic force $\mathbf{F}_e$. The electromagnetic force acts on the induced electric current in the particle [11] and due to the fluid pressure redistribution on the surface of the particle. For the case of electrically non-conducting particles (oxides, carbides, dendrite fragments, bubbles etc.) the effective electromagnetic force action in the AC skin-layer can be derived as $\mathbf{F}_e = -V_p(\frac{3}{2}\bar{\mathbf{f}}_e + \mathbf{f}'(t))$, the sum of the time average and the oscillating components in the fluid at the location of the particle [15]. Note, the Leenov, Kolin [8] expression derived for DC constant fields gives two times smaller force acting on a non-conducting particle.

The typical mixing flow in the cold crucible is turbulent (Re ~ 10^3 - 10^4), which requires a modification for the $\mathbf{F}_d$ calculation due to the stochastic part of velocity in accordance to the resolved turbulent kinetic energy k and the local eddy life time (or the particle transit time in that eddy, whichever is the shortest). The use of k- ω turbulence model in the SPHINX code facilitates to obtain these quantities, which are locally interpolated to the particle position at each time step. The numerical integration of equations (1) and (2) is done for each individual particle of various properties depending on the initial seeding locations. The exponential numerical scheme, proposed in [14], permits stable time integration of the particle tracks using adjustable time steps of the order 0.1 - 1ms used for the unsteady fluid flow solution in the examples considered below.

Melting and particle tracking results

In the considered cold crucible the two coils coaxially surround the segmented, water cooled copper walls and the liquid zone, as shown in the Figs. 1 and 2 for a symmetric half of the crucible section. The computed EM force distribution shown in the figures correspond to a medium frequency (5 kHz) fed side coil which provides most of the Joule heating and the melt confinement force via the induced current in the melt. The bottom nozzle region is surrounded by the second coil to impose the localized high intensity heating produced by the high frequency AC (50 kHz). After the initial melting stage practically all volume of the titanium aluminide alloy is in liquid phase shaped in a typical dome (Figure 3a). At this stage the action of the bottom coil helps to maintain the nozzle free from solid phase, except from the extremely thin solidified skull acting as a protecting film between the out flowing liquid metal and the copper segmented wall of the nozzle (Fig. 3b). The action of the EM force in the nozzle region affects the outflow rate which is monitored during the full simulation cycle. The Fig. 4 shows the predicted outflow rate and the moderate superheat of about 20 deg C

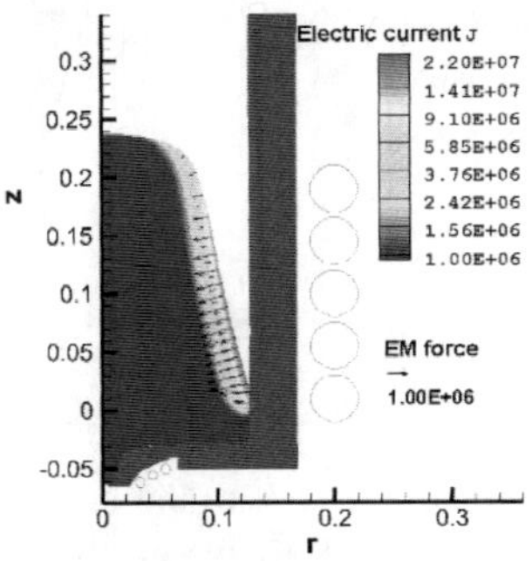

Fig. 1: The time average force and induced electric current from the main coil of cold crucible.

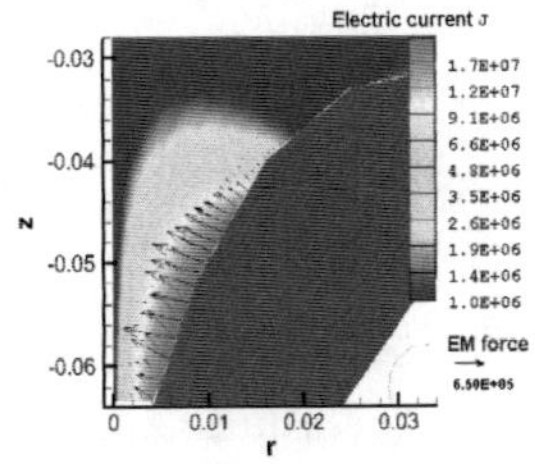

Fig. 2: The zoom-in view to the nozzle with the force and induced current from the bottom coil.

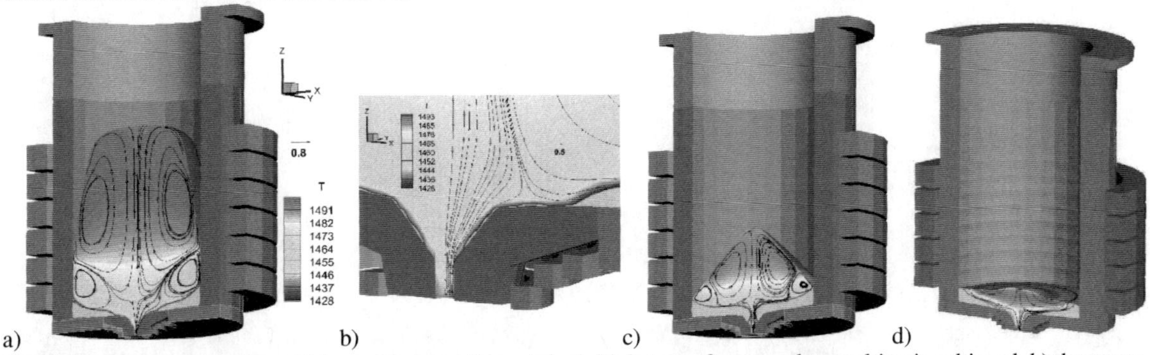

Fig. 3: Dynamic stages of the cold crucible operation: a) the initial stage after complete melting is achieved, b) the zoom-in view to the nozzle region at initial stage when pouring starts, c) the late stage of pouring with reduced power coupling, d) the final stage just before the end of pouring.

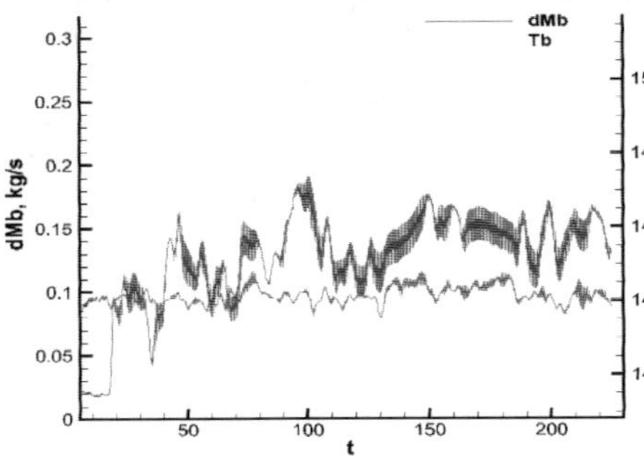

Fig. 4: The liquid metal pouring rate in kg/s and the temperature in C° at the exit nozzle.

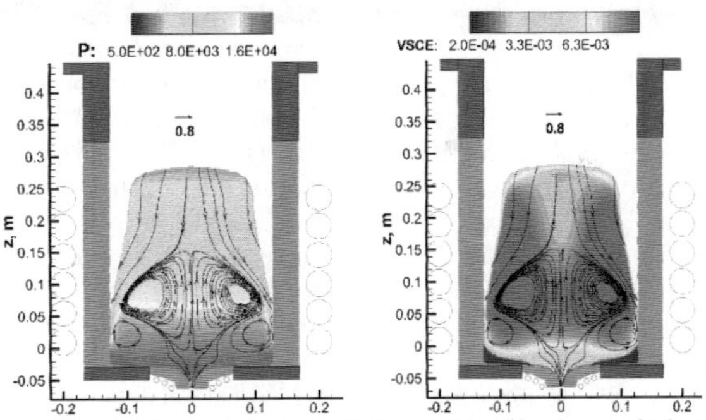

Fig. 5: The pressure P (Pa) and the effective turbulent viscosity (m^2/s) distribution in the melt.

above the solidification temperature for this alloy. The flow is highly time dependent and turbulent, as demonstrated by the curves at the Fig. 4. The distribution of the computed effective turbulent viscosity is shown in the Fig. 5. The high intensity of the turbulence corresponds to the regions where the EM force drives the fluid flow, while the turbulence effects are damped in the vicinity of the solidified skull and towards the narrow exit nozzle. Fig. 5 shows also the pressure distribution at a particular time moment demonstrating the nearly hydrostatic distribution modified by the EM force confinement action in the external skin-layer. The Figures 3c and 3d show final stages of the emptying crucible when the EM force action gradually reduces due to the decreased coupling with the EM field producing side coil. The solid skull layer becomes thicker at the bottom, leaving only the area at the nozzle free from the solid due to the action of the bottom coil EM field. This particular crucible design permits to empty about 38 kg of liquid titanium alloy in 4 minutes of the pouring action. The impurity particle distribution were simulated by adding the aluminium oxide particles on the surface of the liquid metal at the fully molten stage. 3*300 particles of three sizes (1, 20 and 50 μm) were dynamically followed during the full pouring cycle in order to follow their tracks, extraction or separation routes. There is a remarkable difference in the particle behavior depending on EM force effect on particles.

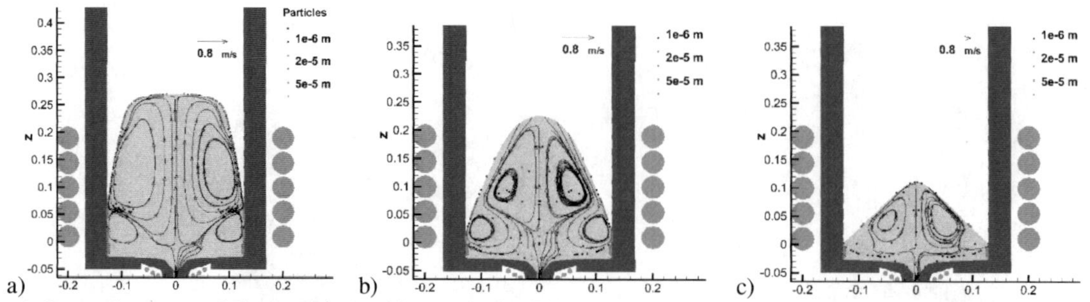

Fig. 6: Dynamic stages of the particle tracking *with the EM pressure effect* on the particles: a) the initial stage after complete melting, particles introduced on the surface are retained from spreading into the melt, b) only the small size 1-20 μm particles are partially dispersed, c) at the late stage of pouring large particles are extracted to the 'skull' and a limited number of small particles travel to the nozzle.

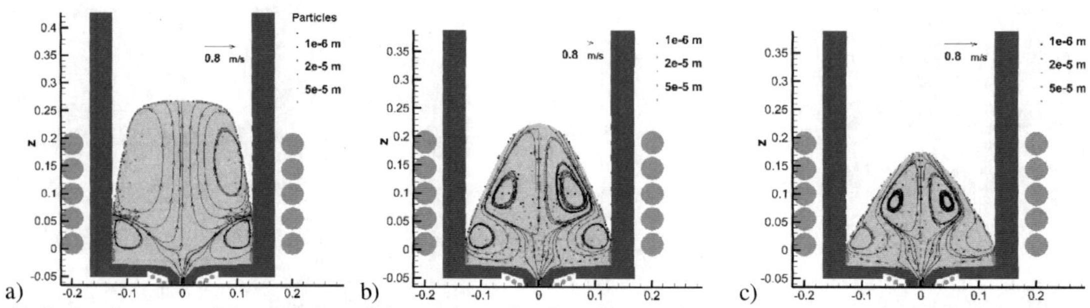

Fig. 7: Dynamic stages of the particle tracking *without the EM effect* on the particles: a) the initial stage after complete melting, particles introduced on the surface start spreading into the melt, b) the particles are dispersed at mid-stage of the pouring, c) at the late stage of pouring all sizes of particles travel through the nozzle.

The sequence of the particle distribution can be compared from the Figures 6 and 7. The largest particles of 50 μm size are completely separated to the solid side skull region and remain there in spite of the vigorous mixing in the melt when the EM effects are accounted computing the time dependent forces during the particle propagation in the melt. A limited amount of the smaller particles (1 and 20 μm) eventually pass to the nozzle exit, while the majority are captured in the bottom and side solidified skull (Figure 6). This behavior is in sharp contrast to the particle distribution computed in exactly the same way, except excluding the effective electromagnetic force $\mathbf{F}_e$ action on the particles in the equation (2).

Conclusion. The beneficial features of the cold crucible melting to purify the melt are explained using the numerical modelling with the dynamic particle tracking. The non-consumable nozzle design is presented in principle, however leaving the specific parameters open to optimization and further discussion.

References

1. R.E. Haun, JOM, **69**, N 12 (2017) 2615-2620.
2. W.H. Sillekens, et al., Metall.Materials Trans., 45A, (2014) 3349–3361.
3. T. Okumura, K. Yamamoto, M. Shibata, Proc. 6th Int. Symp. Electromagn. Proc. Materials, Dresden, Germany (2009) 521-524.
4. V. Bojarevics, R.A. Harding, K. Pericleous and M. Wickins, Metall. Materials Trans., 35B (2004) 785-803.
5. Y. Yang, R. Chen, J. Guo, Y. Su, H. Ding, H. Fu, Metall.Materials Trans., 49B (2018) 28 – 33.
6. V. Bojarevics, K. Pericleous, ISIJ International, vol. 43 (2003), N 6, pp. 890-898.
7. L.Zhang, Sh. Wang, A. Dong, J. Gao, L.N.W. Damoah, Metall. Materials Trans., 45B (2014) 2153-2185.
8. D. Leenov and A. Kolin, Journ. Chem. Phys., 22, N 4 (1954) 683-688.
9. T. Toh, et al., ISIJ International, 47, No 11 (2007) 1625-1632.
10. S. Taniguchi, A. Kikuchi, Proc. 3rd Int. Symp. Electromagn. Proc. Materials, ISIJ, Nagoya, (2000) 315-320.
11. V. Bojarevics, J. Freibergs, E. Shilova, E. Shcherbinin, Electrically Induced Vortical Flows, Kluwer Academic Publishers, Dordrecht, Boston, London, (1989).
12. R. Clift, J.R. Grace,M.E. Weber, Bubbles, Drops, and Particles, Dover Publications, Mineola, New York, (2005).
13. L.D. Landau and E.M. Lifshitz, Fluid Mechanics, Pergamon Press, 1987.
14. V. Bojarevics, K. Pericleous and R. Brooks, Metall.Materials Trans., 40B, N 3 (2009) 328 – 336.
15. V. Bojarevics, K. Pericleous, Proc. 8th Int. Conf. Electromagn. Proc. Materials, Cannes, France (2015) 283-286.

IOP Conf. Series: Materials Science and Engineering **424** (2018) 012030 doi:10.1088/1757-899X/424/1/012030

Numerical simulation of surface deformations in a three-fluid process stirred by low frequency magnetic field

R. Bourrou[1, 2], A. Gagnoud[1], O. Budenkova[1], P. Charvin[2], C. Lafon[2]

[1] SIMaP, EPM, 1340 rue de la piscine, 38402 St. Martin d'Hères, France
[2] CEA, DEN, DE2D, SEVT, F-30207 Bagnols sur Cèze, France

Corresponding author: remi.bourrou@simap.grenoble-inp.fr

Abstract
A process allowing the treatment of solid low-activity nuclear wastes (metals and organic matter) is currently being developed by the CEA (French Alternative Energies and Atomic Energy Commission). During elaboration part of the process, some metal covered by conditioning glass is melted and stirred by low-frequency magnetic field. The pressure originating from Lorentz forces creates a steep dome on the surface of the metal, which can emerge from the glass. It is important to know the position of the interfaces between the fluids to determine heat and momentum transfers through them in order to get better insight for further development of the process. The goal of the presented numerical study is to obtain the positions of the three interfaces (glass-air, glass-metal and metal-air). It consists of a coupling between two commercial software: COMSOL Multiphysics to calculate magnetic field, and ANSYS Fluent for the three-phase flow (VOF). The results obtained have been validated by some measurements on the prototype.

Key words: numerical simulation, three-phase, electromagnetic stirring, low frequency, nuclear waste

Introduction

Each year, nuclear electricity production generates a large quantity of radioactive wastes. The larger volumes come from technological wastes (pumps, filters…) contaminated during the transformations and the handling of nuclear combustibles. Generally, these wastes are sorted in order to be treated in different ways: organic matter is incinerated then incorporated in melted glass, and metallic parts are often compressed and cast in concrete. The packages thus produced are then stored in underground sites in order to prevent the contamination of the environment [1]. The CEA is currently developing a process allowing the treatment of these two types of waste simultaneously. In the fusion part of this process, two immiscible materials are melted and stirred by low frequency electromagnetic induction: some glass in which are incorporated ashes of organic wastes, and the metal contained in the wastes.

The number of physical phenomena involved and the complexity of their coupling makes the operating of the prototype difficult. Hence, a numerical study is carried out to improve the understanding of these phenomena.

The major difficulty encountered when modelling this type of process is the representation of the deformation of the liquid surfaces by the electromagnetic forces. This phenomenon impacts directly the others physical phenomena like the fluid flow and heat transfers, and also the performance of the process by modifying the power repartition between the conductive elements. In this work, after a brief presentation of the process configuration and the physical phenomena considered, the numerical model developed to calculate the shape of the interfaces is presented and the results are compared with measurements on the actual prototype.

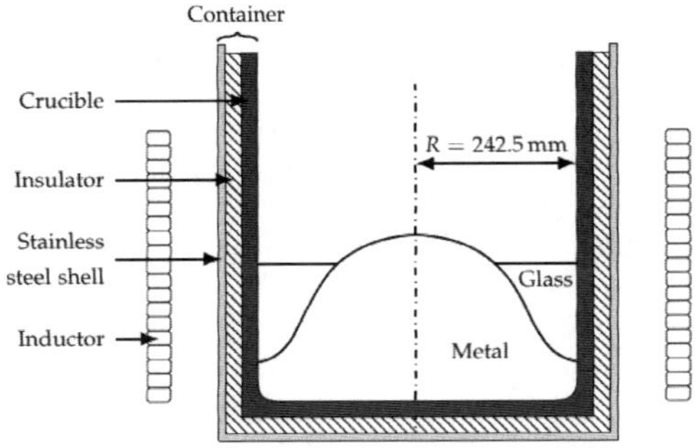

Fig. 1: Schematic of the process.

	Viscosity	Density	Electrical conductivity
	η (Pa.s)	ρ (kg/m^3)	σ (S/m)
Glass	1.3	2500	20
Metal	6×10^{-3}	7000	10^6
Air	5×10^{-5}	1	0

Tab. 1: Physical properties of the fluids used in the numerical model.

Process configuration

The studied part of the process consists in a container composed of a ceramic crucible of internal radius R, some insulator and a stainless steel shell (Fig. 1). This container holds the liquids during the treatment of the wastes and is also the first barrier between radionuclides and the environment for the storage. The whole package is placed in a helical inductor powered by an AC generator producing a current intensity I in the range 2800–4200A and in the frequency range 30–50Hz. Tab. 1 regroups the physical properties of the liquids taken for this numerical study, estimated from the materials used on the prototype. The properties of the metal, composed of stainless steel and copper, are estimated using the mass ratio of these two components. Concerning the glass, it consists of a conditioning glass developed specifically for the process. The values are taken at 1350°C, corresponding to the average temperature measured during the operating of the prototype.

Hypothesis

As in many processes of this type, four principal physical phenomena are coupled during the operating of the process: electromagnetism, fluid flow, heat transfer, and the deformation of the interfaces between the fluids. Some of these phenomena can be neglected in the numerical model when focusing exclusively on the shape of the surfaces of the liquids. First of all, the geometry is considered to be 2D axisymmetrical (r, θ, z) because the helical aspect of the inductor is slight, and it can be assimilated to a stack of concentric rings. Second, due to the high thermal conductivity of the metal, the temperature in this phase is homogenous. Furthermore, measurements in the melted glass have shown that its temperature is also uniform at the exception of the area near the free surface, due to strong radiative transfers. It has been chosen to neglect this effect for this study, therefore, thermal gradients and their influence on the physical properties of the liquids are not modeled. Further approximations are obtained using dimension theory. The deformation of the surface of the liquids is defined by the pressure equilibrium at the interfaces between the fluids as:

$$P_{magnetic} + P_{dynamic} + P_{static} + P_{surface\ tension} = constant$$

$$\frac{B_t^2}{2\mu_0} + \frac{\rho U_t^2}{2} + \rho g z + \gamma K_c = C$$

Where B_t and U_t are components of the magnetic flux and fluid velocity tangential to the surface of the metal, respectively, μ_0 the vacuum permeability, g the standard gravity acceleration, γ the surface tension between the fluids, and K_C the mean curvature of the surface (sum of principal curvatures).

An order of magnitude for the magnetic flux (B_0) can be estimated using Biot-Savart law applied to a point at the center of a coil. In the functioning range of the generator $B_0 \sim 10^{-1}$ T, giving a magnetic pressure of approximately 10^3 Pa.

A characteristic velocity of the liquid metal U_0 can be determined using Alfvén velocity along with experimental results from Fautrelle et al. [2]:

$$U_0 = \frac{0.2\, B_0}{\sqrt{\mu_0 \rho}}$$

This equation gives estimates the order of magnitude of the velocity to 10^{-1} m/s, leading to $P_{dynamic} \sim 10^1$ Pa which is negligible in comparison with the magnetic pressure. This characteristic velocity also gives a magnetic Reynolds number $Rm = R\mu_0 U_0 \sigma$ smaller than unity, so the advection of the magnetic field by the metal flow is negligible.

For this study, surface tension effects are also neglected because the surfaces are wide (few decimeters) compared to the curvature radius necessary to obtain a significant pressure (few millimeters).

In this frequency range, the electrical conductivity of the glass makes it transparent to magnetic field. With the assumptions made above, only static pressure determines the shape of the free surface of this material and it is therefore expected to be flat $(\rho g z = constant)$.

Numerical model

The numerical model developed to calculate the shape of the interfaces between the fluids consists in a coupling between two commercial software. Electromagnetic induction is modeled in COMSOL Multiphysics (finite element) using harmonic magnetic potential equation. The glass being unaffected by the magnetic field, only the metallic phase is represented in this model. Incompressible transient Navier-Stokes equations are modeled with ANSYS Fluent based on finite volume. Volume of fluid (VOF) method is used to determine the deformations of the interfaces of the liquids [3-5]. The principle of this method is to represent materials by their volume fractions α in each mesh cell, allowing to solve a single set of Navier-Stokes equations for all the phases. The physical properties of the mixture are proportional to α, e.g. $\rho = \sum \alpha_i \rho_i$, and a continuity equation is solved for the volume fractions in order to track the interfaces.

Since the shape of the surfaces is supposed to be independent from the fluid flow, the viscosity of the fluids has been increased artificially in the model to reduce the time necessary to reach the stabilization of the surface and avoid using a turbulence model. The viscosity has been taken to 60 Pa.s for the metal and 10^{-2} Pa.s for the air.
For each time step, the two programs calculate their part simultaneously, and exchange metal volume fraction and time averaged Lorentz force density at the end of the step. This allows to optimize transfer time between the programs by realizing data transfer and the interpolations between the meshes on COMSOL's side because the computing time is shorter than in Fluent. With this parallel operating, there is always a lag between the two programs: the calculation of Lorentz forces uses the position of the metal surface from previous time step and not current time step, and it is the same for the calculation of the position with the forces. Therefore, the time step has been chosen to 5 ms to have only little variations between two consecutive steps and thus ensure stability.
At the beginning of the calculation the current intensity in the inductor is increased from zero to the aimed value I_s to avoid oscillations of the surfaces caused by the quasi instantaneous deformation in response to the magnetic field. The height of the metal surface being function of the square of the magnetic field, which is itself proportional to the current in the inductor, the intensity is increased as the square root of the time to produce a smooth rising of the surface:

$$I = I_s \cdot \begin{cases} \sqrt{t/t_r} & \text{if } t < t_r \\ 1 & \text{otherwise} \end{cases} \tag{1}$$

The duration of the ramp t_r was chosen to 4s to completely avoid oscillations of the metal surface during the calculation in most cases (see Fig. 2). In addition to the convergence of hydrodynamic equations estimated with the residuals calculated by Fluent at each iteration, another ,convergence C criterion has been implemented to monitor the stabilization of the metal surface and to stop the calculations. It represents the average rate of variation of the metal volume fraction repartition between iteration $N-1$ and N:

$$C_N = \frac{\sum|\alpha_N - \alpha_{N-1}|}{Nb\ cells \cdot \Delta t} \tag{2}$$

Large oscillations on the convergence plot (Fig. 3) corresponds to global movements of the metal surface, and the little ones are due to the small fluctuations created by the glass moving off of (slipping down) it. The square root case seems to converge as fast as the case with no ramp but allows to prevent the metallic phase to possibly reach the top of the computational domain.

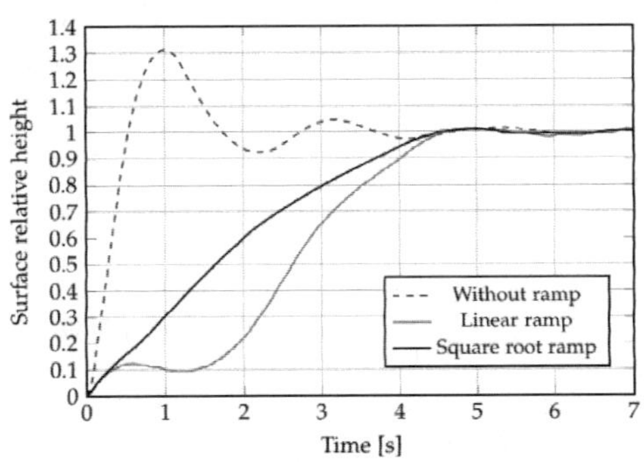

Fig. 2: Total height of the metal surface as a function of the ramp on the intensity at the beginning of the calculation.

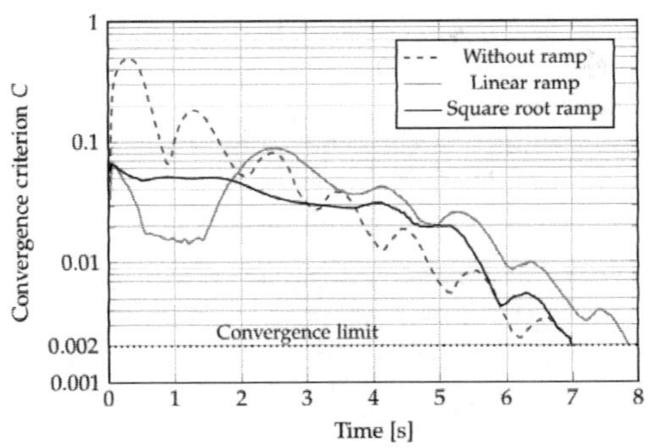

Fig. 3: Evolution of the convergence criterion during the calculation (Equation 2).

Comparison with experimental results
The shape of the metal surface was measured on the prototype by manually immersing a metallic rake in the liquids during three different experiments. The cooling provoked by the rake solidifies a layer of liquid on its fingers and it is then possible to measure the height of the marks left. It is impossible to know if the rake reached the bottom of the crucible, therefore, the measurement is corrected to obtain the same maximal height as the model, and the comparison is made on the relative height only.
The results of the second measurement are shown in Fig. 4. The average error on the height is approximately 6.5 mm for this case, and the average relative error (absolute error divided by the height of the deformation) are about 5.9 %,

4 %, and 3.8 % for the three measurements respectively. It has been observed that the maximal errors are located at the interface glass-metal, and probably comes from the large dispersion of the measurements in this area.
The numerical model has also been validated using experimental results from M. Kirpo Ph.D [6]. This case contains one liquid only (Wood's metal) but the measurements are much more precise than those with the rake on the prototype. The results of the numerical model are very close to the experimental values, the error close to the crucible may be explained by the approximations made on some parameters like the thickness of the crucible (Fig. 5). For this case, the average absolute error is about 0.65 mm leading to a relative error of 3.6 %.

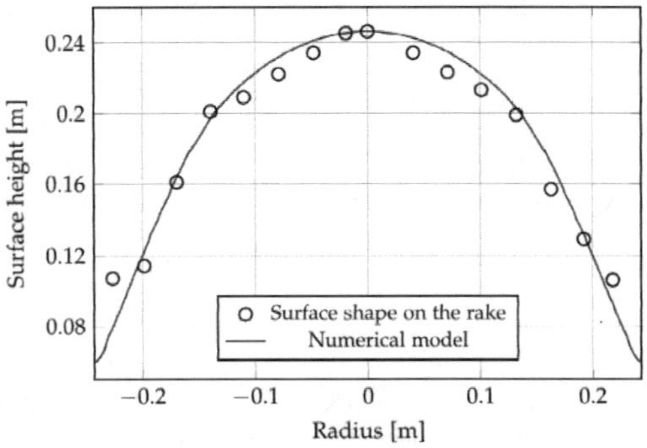

Fig. 4: Comparison between the numerical model and a measurement on the prototype.

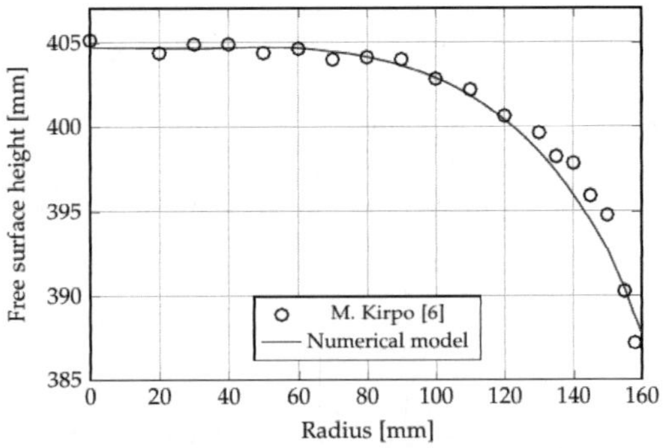

Fig. 5: Comparison between the numerical model and a case from the literature [6]

Conclusion
A model allowing the calculation of the position of the interfaces between the fluids in a process containing melted metal and glass under alternating magnetic field has been developed. It consists of a coupling between COMSOL Multiphysics – solving harmonic magnetic potential – and ANSYS Fluent – for the three phase flow using VOF method. The shape of the surfaces has been assumed to depend only on the magnetic and the static pressure, therefore some assumptions were made, including the artificial increasing of the fluids viscosity. A convergence criterion has been set up to end the calculation, based on the movement of the metal phase. The model has been validated with some measurements on the prototype of the process, and the results were found precise concerning the shape of the metal surface (relative error of approximately 4 %). Concerning the glass, comparisons are difficult due to the formation of air bubbles in this material. To characterize this phenomenon, a model taking turbulence into account, real viscosities of the fluid, and with a 3D geometry will be developed.

Acknowledgments
This work was financially supported by a program "Investing for the Future" from the SGPI granted to the ANDRA, by ORANO (former AREVA) and by the CEA.

References
1. CEA/DEN, Le conditionnement des déchets nucléaires, Éditions du Moniteur, (2008)
2. Y. Fautrelle, F. Debray, and J. Etay, Transfer Phenomena in Magnetohydrodynamic and Electroconducting Flows: Selected papers of the PAMIR Conference held in Aussois, (1997)
3. S. Spitans, A. Jakovičs, E. Baake, and B. Nacke, Magnetohydrodynamics, 46.4 (2010), 425-436
4. A. Bansal, P. Chapelle, Y. Delannoy, E. Waz, P. Le Brun, and JP. Bellot, Metall. Mater. Trans. B, 46.5 (2015), 2096-2109
5. P. Bulińnski, J. Smołka, S. Golak, and R. Przyłucki, Przegląd Elektrotechniczny, 92.3 (2016), 49-52
6. M. Kirpo, Ph.D. thesis, University of Latvia, (2008)

Development of EMBr/EMS Multifunction Mold

Nobuhiro OKADA[1], Masayuki KAWAMOTO[2] and Shintaro OHGA[3]

[1] Nippon Steel & Sumitomo Metal Corporation, R & D Laboratories
20-1 Shintomi, Futtsu, Chiba Prefecture, 293-8511 Japan
[2] Nippon Steel & Sumitomo Metal Corporation, Osaka Steel Works,
1-109 Shimaya 5-chome, Konohana, Osaka Prefecture, 554-0024 Japan
[3] Nippon Steel & Sumitomo Metal Corporation, Kashima Works,
3 Hikari, Kashima, Ibaraki Prefecture, 314-0014 Japan

Corresponding author: okada.7c8.nobuhiro@jp.nssmc.com

Abstract
EMBr (ElectroMagnetic Brake) and in-mold EMS (ElectroMagnetic Stirring) have been developed to control molten steel flow in a continuous casting mold. EMBr is able to stabilize the molten flow in the mold. EMBr is usually used to prevent breakout during high speed casting of medium carbon steel. In-mold EMS is effectively to improve surface quality of the cast such as interstitial free steel by forced flow of the molten steel in front of initially solidifying shell. Suitable molten flow pattern in the mold is depends on its casting condition and components. However, ordinary electromagnetic device has only one function of EMBr or EMS, so a multifunction device which can be selected EMBr or EMS in the mold is desirable. It is difficult to obtain sufficient performance both EMBr and EMS in only one electromagnetic device because the necessary structure of them is different. EMBr needs to generate large magnetic flux density by large teeth of iron core. A lot of small teeth are necessary for EMS to generate travelling magnetic field along the mold side uniformly. The structure and current applied method of the EMBr/EMS multifunction mold has been developed by numerical analysis of electromagnetic field and fluid flow dynamics. The EMBr/EMS multifunction mold has been applied to Kashima No.3 CC.

Key words : EMBr, EMS, continuous casting, multifunction mold

1. Introduction
EMBr and EMS are used widely to control molten steel flow in a continuous casting mold. EMBr is used to prevent breakout during high speed casting of medium carbon steel, and EMS is suitable for casting of interstitial free steel by forced flow of the molten steel in front of initially solidifying shell. Suitable molten flow pattern in the mold is depends on its casting condition and components. However, ordinary electromagnetic device has only one function of EMBr or EMS, so a multifunction device which can be selected EMBr or EMS in the mold is desirable. We have been developed the EMBr/EMS multifunction mold by numerical analysis of electromagnetic field and fluid flow dynamics.

2. Numerical analysis
2.1 Electromagnetic field analysis
The equation of electromagnetic analysis applied Galerkin's method is given by the following equations:

$$G_i = \int_V (\nabla \times \mathbf{N_i}) \cdot (\nu \nabla \times \mathbf{A}) dV - \int_V \mathbf{N_i} \cdot \mathbf{J_0} dV + \int_V \mathbf{N_i} \cdot \sigma \left(\frac{\partial \mathbf{A}}{\partial t} + \nabla \Phi \right) dV \quad (1)$$

$$G_{di} = -\int_V \nabla N_i \cdot \left\{ \sigma \left(\frac{\partial \mathbf{A}}{\partial t} + \nabla \Phi \right) \right\} dV \quad (2)$$

G_i and G_{di} are the residual at the i th edge and i th node, $\mathbf{N}_i$ is the vector shape function, N_i is the scalar shape function and V is volume of an element. ν is magnetic reluctivity, $\mathbf{A}$ is magnetic vector potential, $\mathbf{J}_0$ is external current density, σ is electric conductivity and Φ is magnetic electric scalar potential. Magnetic flux density vector $\mathbf{B}$ is given by $\mathbf{B}=\nabla\times\mathbf{A}$.

For the analysis of the induction heating using alternating current, it is useful to represent $\mathbf{A}$ and Φ as a complex number. Therefore, the derivative of $\mathbf{A}$ with respect time is replaced by the $j\omega\hat{\mathbf{A}}$, where j is the imaginary number and ω is the angular frequency. Then, eddy current density $\mathbf{J_e}$, Joule heat Q and Lorentz force $\mathbf{F}$ are given as following equations:

$$\hat{\mathbf{J}}_e = -j\omega\sigma\hat{\mathbf{A}} + \nabla\hat{\Phi} \quad (3)$$

$$Q = \frac{1}{2\sigma}\hat{\mathbf{J}}_e \cdot \hat{\mathbf{J}}_e^* \quad (4)$$

$$\mathbf{F} = \frac{1}{2}\mathrm{Re}[\hat{\mathbf{J}}_e \times \hat{\mathbf{B}}^*] \quad (5)$$

The symbol (^) denotes a complex number and (*) denotes a complex conjugate number.

After electromagnetic analysis, the structures of the multifunction mold had been designed as shown in Fig.1. The electromagnetic device of the multifunction mold has four iron teeth on wide side of the mold. Each iron tooth has an inner coil and an outer coil is wound over two iron teeth and two inner coils. Fig.2 shows applied current patterns for the multifunction mold. Direct current is applied for EMBr mode, and the two iron teeth wound by the outer coil are magnetized same direction as Fig.2(a). For EMS mode, 3-pahse (U,V,W) alternating current are applied as shown in Fig.2(b) and travelling magnetic field and Lorentz force are generated in the molten steel. It is feature of the multifunction mold of EMS mode that Lorentz force at meniscus of the molten steel is max to obtain uniform stirring at the meniscus.

In the case of EMBr, magnetic flux density at the mold center of thickness needs over 0.3 T during high speed casting. Fig.4 shows dependence of magnetic flux density on the teeth width at the center of mold thickness. It is easy to generate large magnetic flux density by large teeth of iron core, but the large teeth are not suitable for EMS to generate travelling magnetic field. Teeth size were decided by minimum width to generate magnetic flux density over 0.3 T.

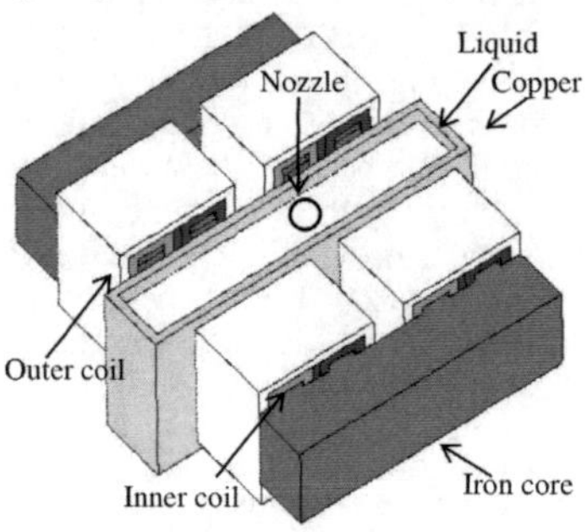

Fig. 1. Schematic illustration of the multifunction mold.

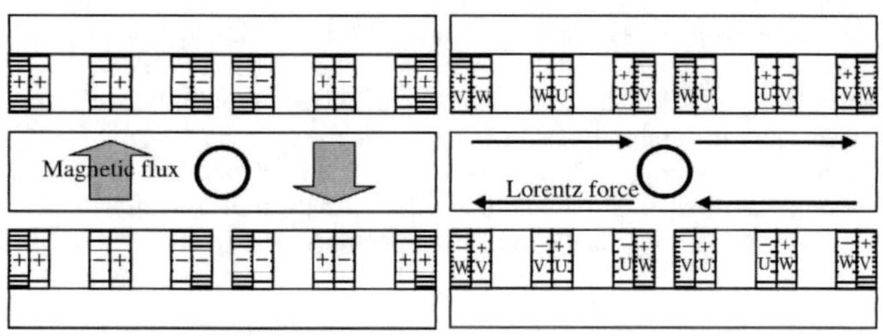

(a) DC for EMBr (b) 3-phase AC for EMS

Fig. 2. Applied current patterns for EMBr and EMS.

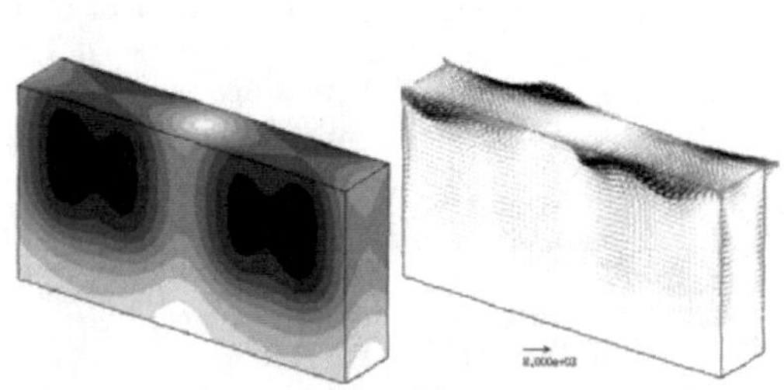

(a) Magnetic flux density for EMBr (b) Lorentz force density for EMS

Fig. 3. Electric magnetic analysis results.

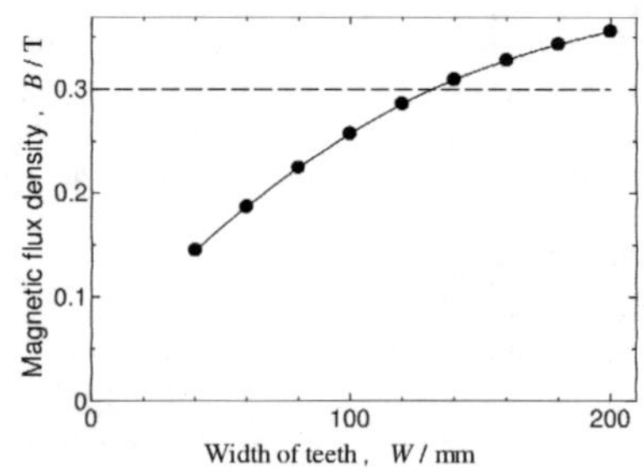

Fig. 4. Dependence of magnetic flux density on width of teeth.

2.2 Heat transfer and Fluid flow analysis

The governing equations of incompressible fluid flow analysis are written by the continuity and Navier-Stokes equations with Boussinesq approximation and LES (Large Eddy Simulation) model [1]:

$$f_g + f_l + f_s = 1 \quad (6)$$

$$\frac{\partial f_g}{\partial t} + \nabla(f_g \mathbf{U}_g) = 0 \quad (7)$$

$$\frac{\partial f_l}{\partial t} + \nabla(f_l \mathbf{U}_l) = -Rs \tag{8}$$

$$\frac{\partial f_s}{\partial t} + \nabla(f_s \mathbf{U}_s) = Rs \tag{9}$$

$$\rho_g \frac{D\mathbf{U}_g}{Dt} = \rho_l \frac{D\mathbf{U}_l}{Dt} + \frac{1}{2}\rho_l \left(\frac{D\mathbf{U}_l}{Dt} - \frac{D\mathbf{U}_g}{Dt} \right) - \frac{3\mu_l}{4{d_g}^2} Re_g C_{Dg} (\mathbf{U}_g - \mathbf{U}_l) + (\rho_l - \rho_g)\mathbf{g} \tag{10}$$

$$\begin{aligned} \frac{\partial (f_l \rho_l \mathbf{U}_l)}{\partial t} + \nabla \cdot (f_l \rho_l \mathbf{U}_l \mathbf{U}_l) = \nabla \cdot \left(\frac{1}{2} f_l (\mu + \mu_t)(\nabla \mathbf{U}_l + \nabla {\mathbf{U}_l}^{t}) \right) - f_l \nabla P \\ + f_l \left\{ \beta (T_0 - T)\rho \mathbf{g} + \gamma (\mathbf{U}_s - \mathbf{U}_l) + \mathbf{F}_{ele} + \mathbf{F}_{gas} \right\} \end{aligned} \tag{11}$$

$$\frac{\partial T}{\partial t} + \nabla \cdot (f_l T \mathbf{U}_l + f_s T \mathbf{U}_s) = \frac{1}{\rho C_p} \left\{ \nabla \cdot \left((\lambda + \lambda_t) \nabla T \right) + \rho \mathrm{Rs}(-\Delta H) \right\} \tag{12}$$

$$f_s = \frac{(T_l - T)/m_l}{(T - T_s)/m_s + (T_l - T)/m_l} \tag{13}$$

$$Re_g = d_g \,|\mathbf{U}_l - \mathbf{U}_g|\, \rho_l / \mu_l \tag{14}$$

$$C_{Dg} = 0.4 + 24/Re_g + 6/(1 + \sqrt{Re_g}) \tag{15}$$

$$\mu_t = \rho_l (C_s \Delta_{\mathrm{cell}})^2 \sqrt{|\mathbf{S}||\mathbf{S}|/2} \tag{16}$$

$$\gamma = \frac{150}{f_l} \frac{\mu_l}{\rho_l} \left(\frac{f_s}{D_2 f_l} \right)^2 \tag{17}$$

$$\mathbf{F}_{gas} = f_g \frac{3\mu_l}{4 d_g} Re_g C_{Dg} (\mathbf{U}_g - \mathbf{U}_l) \tag{18}$$

Here f is volume fraction, **U** is velocity vector, Rs is volumetric solidification rate, ρ is density, μ is viscosity, d is diameter. **g** is gravity vector, β is coefficient of thermal expansion, C_p is specific heat, P is pressure, T is temperature, T_0 is average temperature of the fluid, $\mathbf{F}_{\mathrm{ele}}$ is Lorentz force density vector, λ is thermal conductivity, ΔH is latent heat, m_l is liquidus slope, m_s is solidus slope, T_l is liquidus temperature, T_s is solidus temperature, μ_t is turbulence viscosity of LES, **S** is velocity gradient tensor, Δ_{cell} is cell volume and D_2 is secondary dendrite arm interval. Subscripts l, s and g indicate liquid phase, solid phase and gas phase.

Finite volume method[2] and SMAC method[3] is used to solve these equations and second-order central scheme is used for viscous and convective term with time marching method. The solidification temperature is calculated by lever rule and turbulent Prandtle number is equal to 1. Gas bubble is assumed incompressible and dispersed phase, shape is spherical and diameter is not changed. Momentum equation of gaseous phase is subject to Basset-Boussinesq-Oseen-Tchen equation[4]. How eber Basset term was neglected.

Fig.5 shows velocity and temperature distribution of molten steel in the multifunction mold by fluid flow analysis using LES model. In the case of EMS, velocity in front of initially solidifying shell needs over 0.15 m/s to prevent capturing inclusions and bubbles into the solidification shell [5]. Flow pattern in the multifunction mold of EMBr mode is not much different from its in ordinary EMBr mold.

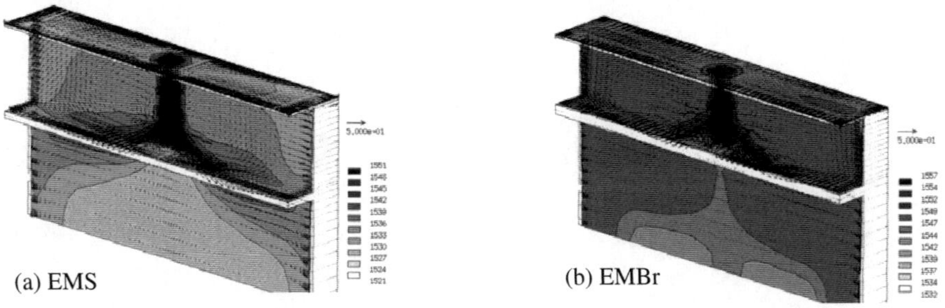

Fig. 5. Velocity (m/s) and temperature (degree Celsius) distribution.

3. Experimental result

Fig.6 shows velocity distribution at solidification interface along wide side of the mold. Experimental velocity was evaluated from deflection angle of dendrite [6]. The stirring velocity in the multifunction mold of EMS mode is over 0.2-0.3 m/s measured from deflection angle of dendrite [5].

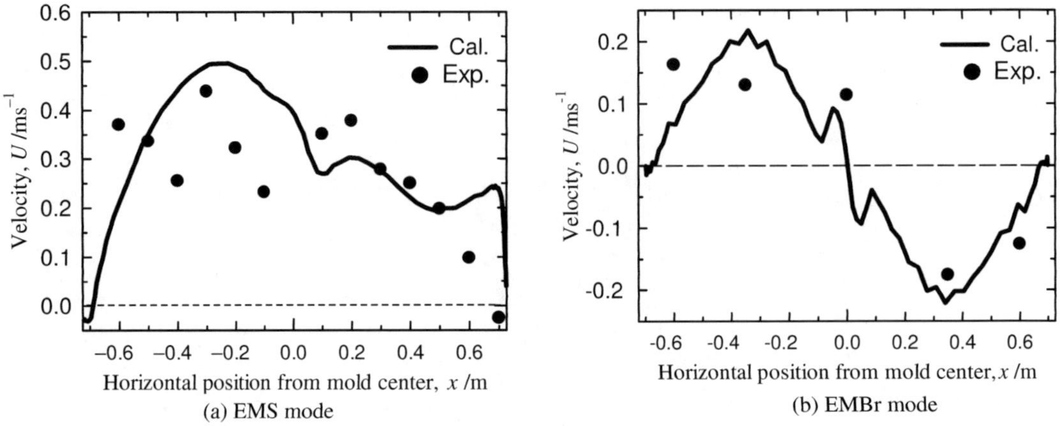

Fig. 6. Velocity distribution of molten steel at solidification interface along wide side of mold.

4. Conclusion

EMBr/EMS multifunction mold has been developed by numerical analysis. It has enough ability to generate magnetic flux density over 0.3 T for EMBr mode as same level as ordinary EMBr mold. And it obtains over 0.2 m/s stirring velocity for EMS mode enough to prevent capturing inclusions and bubbles into the solidifying shell. In this mold, EMBr/EMS mode is could be selected easily during casting for its condition and components. The EMBr/EMS multifunction mold has been applied to Kashima No.3 Caster now.

References

1. K. Takatani : ISIJ International, 43 (2003), 915.
2. H. Jasak: "Error analysis and estimation for the Finite Volume method with applications to fluid flows", PhD. Thesis, Imperial College, University of London (1996)
3. A.A. Amsden, F.H. Harlow, The SMAC method: A New Numerical Technique for Calculating Incompressible Fluid Flow, LA-4370, Los Alamos Scientific Laboratory, (1970).
4. J. O. Hinze : Turbulence, 2nd ed., McGraw Hill, New York, (1975), 460.
5. A. M. Mollinger and F. T. M. Nieuwstadt: J. Fluid. Mech., 316 (1996), 2085.
6. S. Okano, T. Nishimura, H. Ooi and T. Tino : *Tetu-to-Hagane*, 61 (1975), 2982.

Experimental modelling of continuous casting of steel in slab moulds using low melting liquid metals

K. Timmel, T. Wondrak and S. Eckert

HZDR, Bautzner Landstraße 400, 01328 Dresden, Germany

Corresponding author: k.timmel@hzdr.de

Abstract

Electromagnetic actuators are widely used in industry for contactless control of the steel flow in the continuous casting process. However, a real control of the flow structure by those actuators is a challenging task due to the lack of flow monitoring devices. Even a satisfying non real-time characterisation of the melt flow from plant measurements is missing. Beside numerical simulations, only a very few spatially and temporally limited measurements in liquid steel are available to investigate the actual action of the magnetic fields on the fluid.

Therefore, model experiments with low melting point liquid metals are an important tool to investigate the flow structure and related transport processes in the mould of a continuous caster. Their advantage is the availability of a variety of measurement techniques for quantitative flow measurements. The application of the Ultrasonic-Doppler-Velocimetry (UDV) and the Contactless-Inductive-Flow-Tomography (CIFT) allows for a detailed characterization of the mould flow with a reasonable spatial and temporal resolution.

In recent experiments at HZDR, the systematic study on the influence of an electromagnetic brake on the mould flow in a slab caster was continued. The measurements were carried out using a 1:2 scaled model operated with SnBi and a 1:8 scaled model operated with GaInSn, respectively. The melt surface was partly measured by a laser scanner system. In particular, the immersion depth of the submergence entry nozzle (SEN) was varied during the experiments. It became obvious that changes in the mould flow had a strong influence on the free surface of the melt, where strong perturbations can significantly impair the surface quality of the final steel strands. Moreover, effects from the "artificial" clogging of one SEN-port or the injection of Argon gas at the stopper rod were investigated.

Key words: Continuous Casting, Liquid metal models, ultrasonic wave guides

Introduction

The persistent effort to achieve a better product quality and higher productivity of the continuous casting of steel implies the high importance of powerful capabilities to control the flow in tundish and mould, and the initial solidification in the mould. Numerous sophisticated numerical simulations concerned with the metal flow during the casting process need a fundamental experimental validation [1]. The use of water models has the advantage to save expenses and to be able to apply a number of well-proofed measuring methods. However, a generalisation of these results to liquid metal flows has to be considered as questionable because the realistic values of flow parameters (Re, Pr, Gr, Ha, etc.) are difficult to meet. In many cases, for instance liquid metal flows with strong temperature gradients, with an additional gaseous phase or under the influence of electromagnetic fields, the flow phenomena cannot reasonably be modelled by means of water experiments [1].

The application of electromagnetic fields provides a considerable potential to control the fluid flow in the mould and to influence the solidification in the strand. First strategies for EM applications in steel casting were mainly guided by simplified pictures of the magnetic field impact on the global flow field. Many numerical investigations have been reported until now to improve the understanding of the magnetic field influence on the mould flow (see for instance [2-5]). However, the problem has to be considered as challenging because of the complex geometry, the highly turbulent flow, and specific peculiarities occurring in case of MHD turbulence. Obviously, a validation of the numerical predictions by liquid metal experiments is indispensable. However, related experimental studies are rather scarce until now. Several plant trials were carried out [6, 7] to test the efficiencies of electromagnetic brakes in the real casting process. Because of the lack of suitable measuring techniques for liquid steel at 1500 °C such trials can provide only a limited gain of knowledge about the magnetic field effect on the flow. Only rough information might be achieved by visual observations of the surface velocity or by the application of imprecise local mechanical metering devices. First model experiments employing simplified mercury models have been reported [8-11]. For instance, Okazawa et al. [8] used a Vives-type sensor to investigate the effect of an electromagnetic stirrer on the fluid flow in the mercury model.

With our work we want to continue the strategy of cold metal models. The main value of such cold metal laboratory experiments consists in the capabilities to obtain quantitative flow measurements with a reasonable spatial and temporal resolution. New techniques for measuring the velocity in liquid metal flows came up during the last decade allowing for a satisfying characterization of flow quantities in the considered temperature range up to 400°C. The main issue of the experimental program at LIMMCAST concerns investigations about the impact of the magnetic field on the flow in the continuous casting process.

Experimental setup

The experimental programme of the LIMMCAST facility at HZDR aims to model the essential features of the flow field in the continuous casting of steel, namely the flow field in the tundish, in the submerged entry nozzle (SEN) and in the mould cavity. Experimental setups were built up for this purpose and two of them will be introduced in this section: the LIMMCAST and the Mini-LIMMCAST facility. The investigations in both setups will be explicitly focused on the behaviour of the isothermal melt flow. Although the two experiments exhibit different scaling in length, the basic geometric features of submerged entry nozzle and mould are the same. The SEN is a bifurcated one with two sidewise port openings and the mould represents the geometry of a slab caster, i.e. it has a rectangular cross section. The experimental setups were used to investigate intensely the action of electromagnetic fields on the liquid metal flow. An electromagnetic brake (EMBr) of the ruler type was used in the experimental series presented in this paper, i.e. the pole faces of the magnet covered the whole width of the mould.

In principle, it is possible to replace the important parts of the loops in a reasonable amount of time (especially at Mini-LIMMCAST). For example, the action of an electromagnetic stirrer (EMS) on the flow in a round mould at Mini-LIMMCAST can be found in another paper of this proceeding by Schurmann et al.

The bigger experimental facility is called LIMMCAST and an overall view of this facility is shown in Figure 1. All components are made from stainless steel. The low melting point alloy Sn60Bi40 is used as model liquid, whose temperature-dependent material data are reported by Plevachuk et al. [12]. The liquidus temperature of 170°C allows for an operation of the facility in a temperature range between 200 and 350°C. An electromagnetic pump is used for continuously transporting the liquid metal in the experimental loop. The actual mould has a cross section of 500 × 100 mm^2 and offers measurement access for Ultrasonic-Doppler-Velocimetry (UDV) from the top by using ultrasonic wave guides and from the narrow side walls using high temperature ultrasonic transducers. This resulted in a well time resolved measurement of the mould flow. Complementary, the entire mould flow can be also observed with the Contactless Inductive Flow Tomography (CIFT) [13].

Figrue 1: Overview of the LIMMACST facility.

Figure 2: Mini-LIMMCAST with all the main parts.

The second setup is called Mini-LIMMCAST and is illustrated in Figure 2. This setup is been filled with eutectic alloy of GaInSn as model liquid. This alloy is liquid at room temperature and its material properties are reported by Plevachuk et al. [14]. The experiments reported here were conducted in a discontinuous way, i.e. firstly the model tundish was filled by the liquid metal and secondly after opening the stopper rod the model fluid flows just by gravity through the assembly back into the storage tank. The mould was made of acrylic glass and has a rectangular cross section of 140 × 35 mm^2. The mould flow was measured with the UDV using up to 10 transducers in multiplexer mode for a well time and spatial resolved flow mapping in the jet region. Standard transducers can be used due to the operation at room temperature. The global flow in the entire mould can be obtained by CIFT [15-16]. The CIFT-technique and some measurement results are presented in a separate paper in this proceeding, e.g. Ratajczak et al.

The effect of the EMBr was investigated at Mini-LIMMCAST under different electrical boundary conditions in the mould. The standard acrylic glass mould denotes an electrically insulating wall. In some experimental series, brass plates were inserted into the mould and attached to the wide mould walls. The brass plates modelled the solidified steel and represent electrically conducting walls. The thickness was chosen in that way, that it satisfies the wall conductance ratio, as introduced in [17]. The LIMMCAST facility with mould walls made of stainless steel always represents electrically conducting boundary conditions.

Experimental results

Previous experiments showed a dramatic influence of a horizontal, static magnetic field on the liquid metal flow in the mould [17]. Low frequency oscillations of the horizontal jet flow were detected as well as a local acceleration of the flow. The jet is deflected upwards by the static magnetic field resulting in a strong upwards flow near the narrow wall. In this paper, the strong upward flow was measured in the vertical velocities by UDV.

Figure 3 is depicting the mean vertical flow in the upper recirculation zone close to the narrow mould wall for different

IOP Conf. Series: Materials Science and Engineering **424** (2018) 012032 doi:10.1088/1757-899X/424/1/012032

strengths of the electromagnetic brake. It can be clearly seen, that the upwards flow is greatly increased by the application of the static magnetic field. This is in contrast to the otherwise assumed and simplified action of a static magnetic field as an overall brake on the mould flow. Besides some braking, the magnetic field is reorganizing the flow field and can lead locally even to an acceleration of the flow.

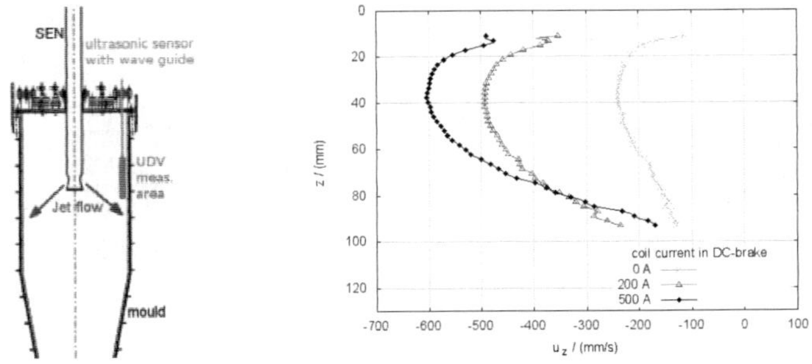

Fig. 3: Vertical velocity along the narrow mould wall (right) and sketch of measurement configuration (left).

Similar experiments were conducted at Mini-LIMMCAST. Two ultrasonic transducers were mounted symmetrically - one in each mould half - and dipped from top through the free surface into the melt. The mean vertical velocity profile recorded with this configuration along the narrow mould wall is shown in Fig. 4. The splitting of the liquid metal jet in a strong downward stream and an upward stream is clearly recognizable by a change in flow direction of vertical velocity and it is in accordance with common picture of the mould flow as a double roll pattern. The static magnetic field is causing a local increase in the upward velocity which is very obvious in Fig. 4 and in accordance with the findings from LIMMCAST. The feature of increased upward flow was found in related numerical simulations of the mould flow, too [18]. The increase in vertical velocity due to the magnetic field is present for insulating and the conducting case as well. However, the electrical boundary conditions have a strong influence on the temporal behaviour of the jet flow (see [17]).

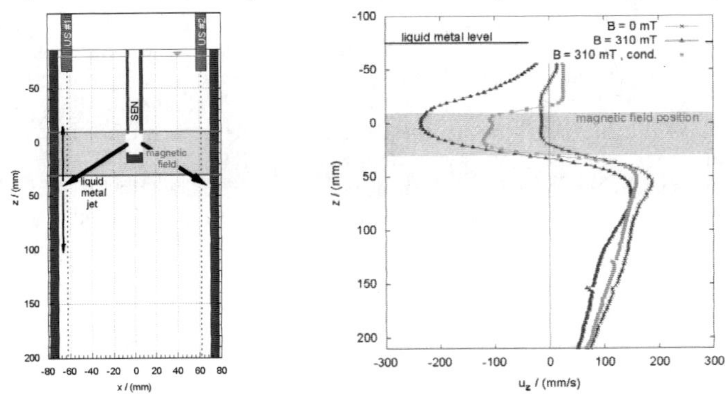

Fig. 4: Measurement configuration for mean vertical velocity along the narrow mould wall at the Mini-LIMMCAST setup (left) and corresponding results of one sensor for different experimental conditions (right).

When the strong upward flow is reaching the free surface, it can affect the shape of the melt surface. This is illustrated in Figure 5. This Figure compares snapshots of the free surface for different boundary conditions. These snapshots show strong differences. The case with EMBr under insulating walls presents the creation of a strong bulge, meanwhile the surface under reference and conducting conditions is rather smooth. Beside the shape, the flow and the magnetic field can also change the temporal and oscillation behaviour of the melt surface.

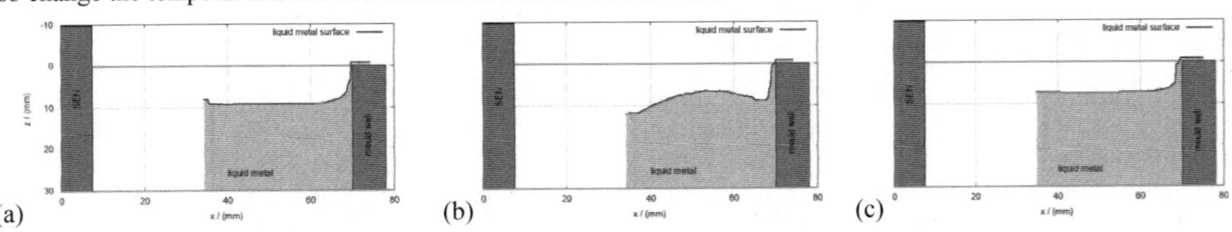

Figure 5: Snapshot of free surface profile near the narrow mould wall measured by a laser scanner system at the Mini-LIMMCAST setup: reference condition without EMBr (a) and applied EMBr with insulating (b) and conducting walls (c).

Figure 6 present the mean vertical velocity recorded at the big LIMMCAST setup using four ultrasonic waveguides in the right mould half. Figure 6(a) is depicting the upward flow of the upper recirculation zone under normal conditions. The blue colour is indicating a melt flow in upwards direction. Figure 6(b) is showing the flow under the condition of artificial clogging of the left nozzle port. The upwards flow close to the narrow mould wall is enforced under this circumstances as the outflow from the nozzle is concentrated on the shown mould half. Additionally, one can see a stronger downward flow for the locations closer to the nozzle.

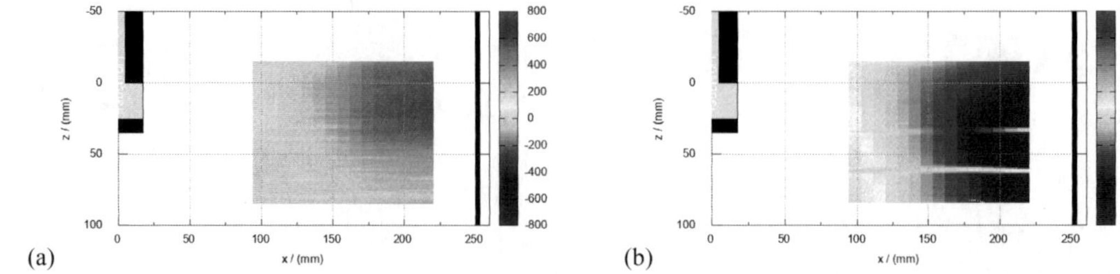

Figure 6: Vertical mould flow at LIMMCAST under normal conditions (a) and with artificial blocking of the left nozzle port (b). Measurements were performed by ultrasonic transducer with wave guides through the top lid of the mould.

Conclusions

For physical modelling of the continuous casting process, it is essential to use liquid metals for conducting the experiments when regarding special effects, e.g. in case of two-phase flows or magnetic fields. To provide an experimental tool for these cases, the experimental facilities of the LIMMCAST-family were built up at HZDR. This paper shortly presented two of these experimental setups.

The experimental setups demanded for an appropriate velocity measurement technique applicable to liquid metals. This technique is the UDV-method. Different kinds of ultrasonic transducers were used in the experiments: standard, high temperature and ultrasonic wave guides. Selected results from each of the different transducer types were presented.

The selected results from velocity measurements showed a strong influence of the static magnetic field on the mould flow near the face of the narrow mould. The flow intensity was locally increased, which is in contradiction to the supposed action of a static magnetic field as contactless brake. Additionally, the detected change in the mould flow becomes recognisable in the free surface shape. The increase in flow intensity can lead to an unstable, fluctuating surface profile. The presented results were achieved from two different facilities with different length scales, but similar scales in dimensionless numbers. The measurement results of the liquid metal experiments represent a valuable data base for the validation of numerical models, e.g. [18-20].

References

1. Thomas, B. G.: Metallurgical and Materials Transactions B 33 (2002), P. 795-812
2. Thomas, B.G.; Zhang, L.: ISIJ Int. 43 (2001), P. 1181-1193.
3. Takatani, K.; Nakai, K.; Watanabe, T.; Nakajima, H.: ISIJ Int. 29 (1989), P. 1063-1068.
4. Li, B.; Tsukahashi, F.: ISIJ Int. 45 (2006), pp. 1833-1838.
5. Cukierski, K.; Thomas, B.G.; Metall. Mater. Trans. B 39 (2008), P. 94-107.
6. Gardin, P.; Galpin, J.-M.; Regnier, M.-C.; Radot, J.-P.: Magnetohydrodynamics, 32 (1996), P. 189-195.
7. Moon, K.H.; Shin, H.K.; Kim, B.J.; Chung, J.Y.; Hwang, Y.S.; Yoon, J.K.: ISIJ Int. 36 (1996), P. S201-203.
8. Okazawa, K.; Toh, T.; Fukuda, J.; Kawase, T.; Toki, M.: ISIJ Int. 41 (2001), P. 851-858.
9. H. Harada, T. Toh, T. Ishii, K. Kaneko, E. Takeuchi (2001), ISIJ Int. 41, P. 1236-1244.
10. J. Etay, Y. Delannoy (2002), PAMIR conference, Proc. Vol. 2, P. 277-284.
11. Yu, Z.; Lei, Z.; Zhang, Z.; Zhang, H.; Ren, Z.; Deng, K.: 6th EPM, 2009, P. 227-230
12. Plevachuk, Yu.; Sklyarchuk, V.; Gerbeth, G.; Eckert, S.: Int. J. Mater. Res. 101 (2010), P. 839-844
13. Ratajczak, M.; Wondrak, T.; Timmel, K.; Stefani, F.; Eckert, S.: J. for Manuf. Sc. and Prod. 15 (2015), 41-48
14. Plevachuk, Y.; Sklyarchuk, V.; Eckert, S.; Gerbeth, G.; Novakovic, R.: J. of Chem. & Eng. Data 59 (2014), 757-763
15. Ratajczak, M.; Gundrum, T.; Stefani, F. & Wondrak, T.: Journal of Sensors 2014 (2014), P. 9
16. Wondrak, T.; Eckert, S.; Gerbeth, G.; Klotsche, K.; Stefani, F.; Timmel, K.; Peyton, A. J.; Terzija, N.; Yin, W.: Metallurgical and Materials Transactions B 42 (2011), P. 1201-1210
17. Timmel, K.; Eckert, S.; Gerbeth, G: Metallurgical and Materials Transactions B 42 (2011), P. 68-80
18. Miao, X.; Timmel, K.; Lucas, D.; Ren, Z.; Eckert, S.; Gerbeth, G.: Metall. Mater. Trans. B 43 (2012), P. 954-972.
19. Chaudhary, R.; Ji, C.; Thomas, B.G.; Vanka, S.P.: Metall. Mater. Trans. B 42 (2011), P. 987–1007.
20. Kratzsch, C.; Timmel, K.; Eckert, S.; Schwarze, R.: steel research int. 86 (2015), P. 400-410.

IOP Conf. Series: Materials Science and Engineering **424** (2018) 012033 doi:10.1088/1757-899X/424/1/012033

NUMERICAL SIMULATION OF TURBULENT STEEL CEM® MOLD UNDER HIGH MASS FLOW CONDITION

J.Y. Hwang[1*], M.J. Cho[1], B.G. Thomas[2] and S.M. Cho[2]

[1] Research Laboratory, POSCO, 20-26, Pokposarang-gil, Kwangyang, Korea
[2] Colorado School of Mines, 1500 Illinois St., Golden, CO 80401, USA

Corresponding author: jyhwang@posco.com

Abstract
The POSCO has invented one of the most innovated steel manufacturing process called CEM® (Compact Endless casting & rolling Mill) which can continuously produce coils without cutting before the finishing mill. In this high throughput process between the casting and rolling processses, a material balance is essential. World-class throughput of CEM® has been achieved with high-speed casting technology (up to 8.0 m/min) and increased slab thickness (up to 100 mm). Many new advanced technologies are required to do this. For instance, instability of the meniscus in the mold is one of the important factors to be solved. Electro Magnetic Braking (EMBr) system has come into the spotlight to control mold surface profile and level fluctuations. Therefore, an accurate evaluation and improvement of EMBr is required. In this work, a numerical and experimental study was carried out to investigate high throughput conditions, to prepare for the next generation of CEM®. In order to effectively stabilize mold flow near the meniscus, a new SEN was designed, combined with a U-shaped EMBr invented by POSCO. The effect of EMBr current on the temperature in the mold is also studied. Numerical results are consistently agreed with plant measurements obtained using nail board tests. Stabilized mold level and temperature are the critical technologies needed to achieve stable casting operation.

Introduction
A new process, CEM® which can produce ultra-thin hot coils was launched by POSCO in 2009 [1]. As reported by Hwang et al., this process features 80~100 mm thick slabs directly going through a roughing mill located just after the caster, being reduced down to 20mm thickness, and then going into a small scale induction heater and finally being manufactured into ultra-thin coils through the hot finishing mill [1]. The coils are continuously produced without cutting before the finishing mill in this process, so is called endless rolling. Thin-gauge coils can be produced with uniform material quality due to consistent material temperature along the whole length of each strip. The mass balance between the caster and rolling mill is very important. Thus, high-throughput casting technology has to be accomplished because the capability of the hot rolling mill is commonly greater than that of the caster. If casting speed is lower than a critical limit, heat loss during rolling part is increased, which leads to excessive rolling loads. Moreover, sufficient FM (Finishing Mill) coiling temperature has to be achieved for the grain structure of the coils not to become mixed. Therefore, the productivity of the caster in CEM® must be much higher than other casters. As shown in Fig. 1, the throughput per unit width of CEM® casters is the highest of all casting processes, including strip casters, thin slab casters, conventional casters and other ultra-thick slab casters. In addition, many producers desire increased production for other casters. Therefore, the stabilization of mold flow during high-speed casting operation is an important topic to be investigated.
Several investigations regarding this subject have been carried out in previous work. In particular, EMBr control of the flow fluctuations at the meniscus is known to improve the quality of the cast steel by reducing the penetration of non-metallic inclusions [2]. Three different EMBr configurations were studied: Local EMBr, EMBr Ruler, and FC Mold [3]. The nail board test is performed by dipping nails down through the slag layer and measuring the meniscus profile and velocity [3, 4]. The shape of the meniscus surface is also measured from oscillation mark profiles. Water modeling has difficulty to show the real flow of molten metal with EMBr in the mold [5]. Vogl et.al carried out not only numerical simulation but also experiments using the liquid metal GaInSn with a length scale of 1:8. It was reported that meniscus speed can be reduced from 0.2 to 0.09 m/sec with EMBr and a single SEN design combined with the installation of EMBr can offer new possibilities for extending the operational range. The transient fluid flow with argon gas was studied by nail board dipping test and Large Eddy Simulation (LES) coupled with the Lagrangian Discrete Phase Model (DPM) [6]. The dominant frequencies of jet flow were reported to be detected as 0.5-2.0Hz. Cho [7] investigated the effect of double-ruler EMBr on transient flow using a Reynolds Averaged Navier-Stokes (RANS) model coupled with Magnetohydrodynamics (MHD) and plant measurement using nail boards. The dominant frequency was observed at 0.03Hz. EMBr lowers the peaks in the level of power spectrum and decreases the level fluctuations near the SEN by 50%. LES is carried out with FC-Mold EMBr, which enables meniscus velocity to be decreased 0.5 to 0.3 m/sec and meniscus level to be reduced 12 to 1 mm [8]. Potential gradients in the Lorentz force term has to be discretized with great care along the boundary. Cho [9] developed a discretized method with second order accuracy, CIM, and showed

good agreement with the analytical solution for plane channel flow.

In this study, numerical analysis of mold flow was performed for high throughput casting at 8.4-8.7 ton / min. This is 45% larger than that of CEM® currently in operation, 6 ton/min. A new SEN for high throughput was developed and applied to quantitatively evaluate the meniscus level and flow velocity.

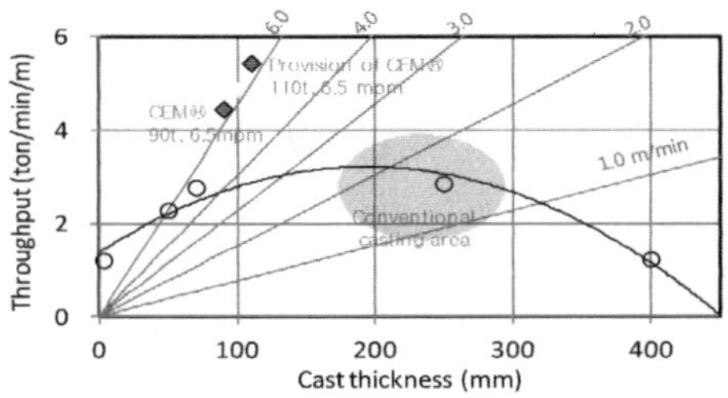

Fig. 1. Productivity related with cast thickness

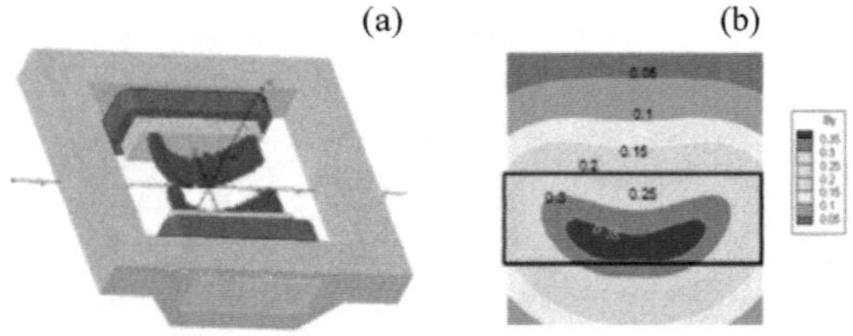

Fig. 2. Geometric configuration of CEM® EMBr, (a) schematic view, (b) magnetic field

Numerical Method, Flow Evaluation Criteria and the effect of EMBr

The main numerical parameters are grid resolution, calculation domain and turbulence model. These parameters are strongly correlated with computational reliability and cost. For turbulence modelling, the k- ε model is used with second order accuracy of spatial discretization. Through long experience with CEM® performance, two design criteria were chosen. That is, the maximum difference of meniscus level (△H) should be less than 8.8 mm to ensure stable operation and quality. The velocity on the meniscus is also important. If the velocity is too low, the local freezing of the meniscus and hook formation [10] may occur. On the other hand, abnormal high surface velocity leading to vortex formation [11] and instability at the interface between the molten steel and slag [12], could entrain slag into the molten steel pool, resulting in both surface and internal defects in the final steel product. Therefore, maximum velocity (V_{max}) range of 0.1-0.38 m/sec is determined to be an optimal velocity range.

Electromagnetic forces are applied to the flow field using EMBr. In order to effectively control the discharge from the SEN, we developed a U-shaped EMBr core as shown in Fig.2. In order to evaluate the present EMBr performance, calculations were carried out under conditions of a slab thickness of 90 mm, an average slab width of 1200 mm, and a casting speed of 6.5 m / min. This condition is conventional for currently performing endless rolling in CEM®. The magnetic field was set to 0.35T, obtained when the maximum current is applied. Figure 3 shows the computational results of the △H and V_{max} for the cases of with and without EMBr at each casting speed. When the EMBr is applied, △H and the speed are controlled to about 50% and 65% compared with the case without EMBr, respectively.

Figure 4 shows velocity vectors and contours on the mid-section in the thickness direction when the magnetic field is applied with 0T and 0.35T. The red solid line represents the position of the EMBr core. The left side of the figure is the case without EMBr, where a secondary vortex with a large momentum is clearly observed compared to the right side of the figure with magnetic field 0.35T. One can notice that magnetic field make strong momentum diffusion happen, resulting in decrease of both △H and V_{max}.

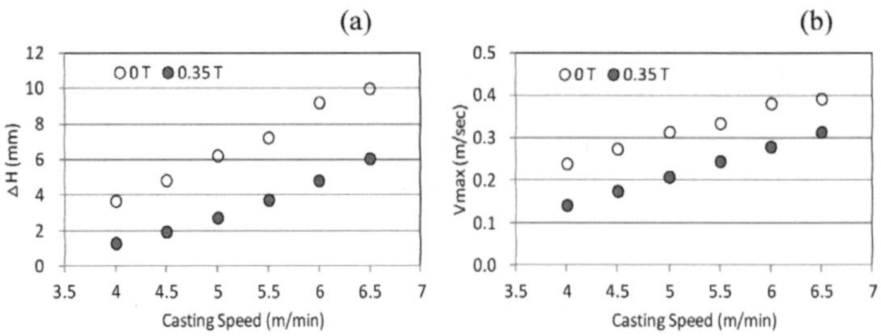

Fig. 3. Effects of casting speed and EMBr on: (a) △H, (b) V_{max}

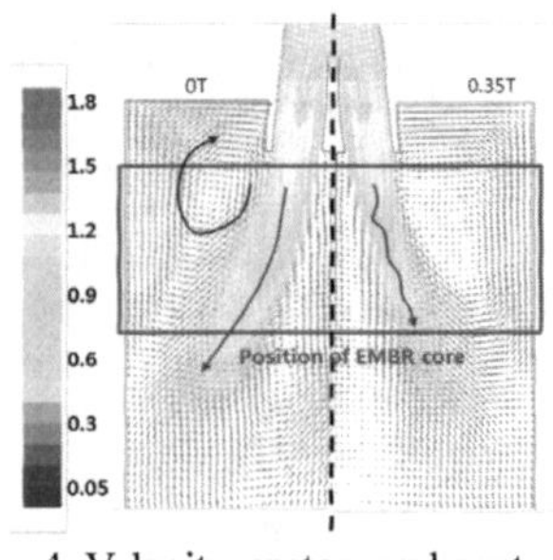

Fig. 4. Velocity vectors and contours

Development of New SEN for High Throughput Operation

A new SEN was required to be developed for high throughput operation. The key technology is to increase the effective exit area of the SEN in order to decrease the velocity exiting the ports. In this study, we developed the new SEN design, which widens the cross sectional area by 50% compared with a conventional nozzle. Figure 5 shows the meniscus level profile from the nozzle center to the edge for the existing and new SEN. Here, the dotted line indicates the case in which EMBr is not applied, and the solid line indicates the case with EMBr 0.35 T. In case of new SEN, it is found that the meniscus is controlled more effectively than the case of existing SEN, and EMBr is also a very useful tool for controlling the meniscus flow. Especially, if the new SEN is with EMBr, $\triangle H$ can be controlled to about 30% of that with the current SEN without EMBr.

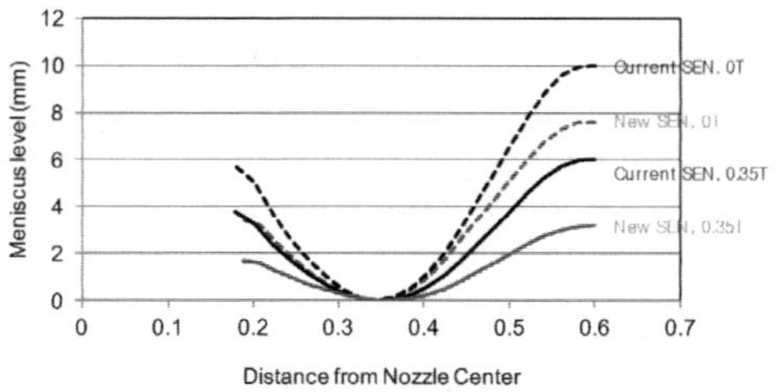

Fig. 5. Meniscus flow level of current SEN compared with new SEN, with and without EMBr

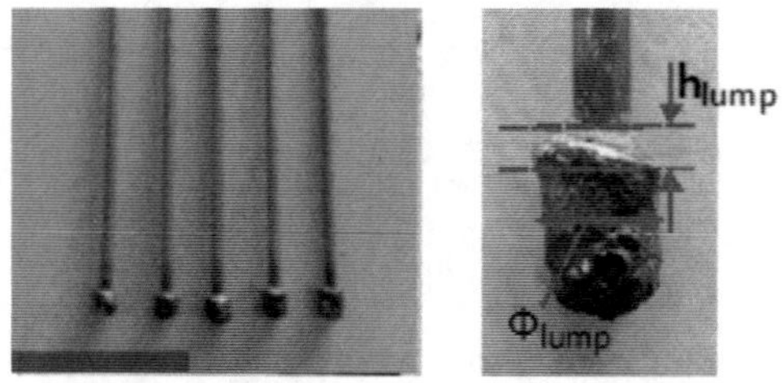

Fig. 6. Photos of the nail board test: overview of nails, lump solidified on a nail

Plant Measurement

Nail board dipping tests, which are commonly used for convenient plant measurement of the meniscus level and velocity at Posco, were carried out to validate the calculation results in this work. Specifically, two sets of measurements using 5 STS nails with 370 mm length and 5 mm diameter were conducted, as shown in Fig. 6. If the lump height difference (h_{lump}) is measured as shown, surface velocity ($V_{surface}$) at each nail can be estimated from the following correlation [13] between h_{lump} and $V_{surface}$ in eq. (1), where and Φ_{lump} means lump diameter.

$$V_{surface} = 0.624 \cdot (\Phi_{lump})^{-0.696} \cdot (h_{lump})^{0.567} \quad (1)$$

Figure 7 and 8 show the meniscus level and meniscus velocity distributions along the width direction from the center of SEN, respectively. The casting conditions are thickness 90 mm, width 1200 mm, casting speed 6.5 m/min and magnetic field 0.23 T using the new SEN. Figure 7 shows that the meniscus level is minimized in the middle region of the width direction. This is because the flow velocity of that part is maximized as shown in Fig. 8. Figure 7 represents the results of calculations are very consistent with those of the nail board tests and the averaged measured value of oscillation mark (OSM). The $\triangle H$ is less than 8mm, indicating a stable meniscus distribution. The meniscus velocity from the nail board tests is obtained using equation (1) and is also in good agreement with the calculated results. Therefore, the numerical analysis used in this study is verified well. In particular, V_{max} is 0.35 m/sec, and we can expect stable casting operation.

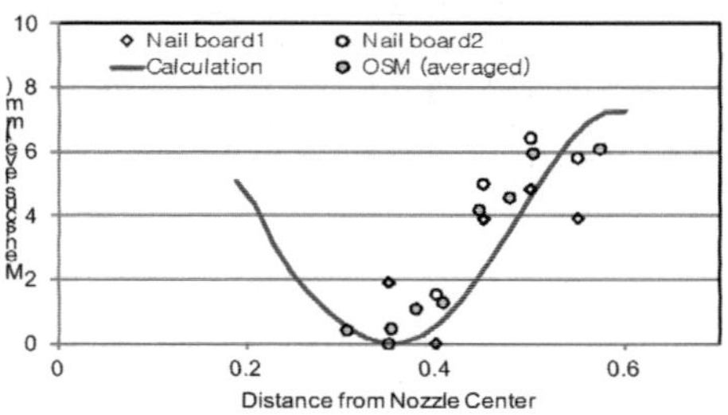

Fig. 7. Comparison of meniscus level : Calculation, nail board test and OSM

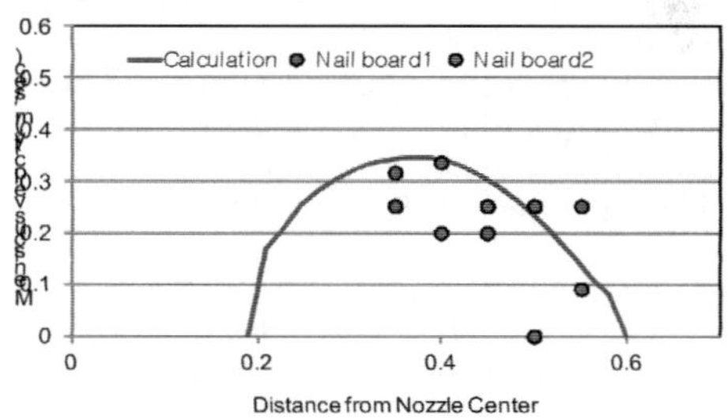

Fig. 8. Comparison of meniscus velocity : Calculation and nail board test

Numerical Simulation for the High Mass Flow Condition

In this section, numerical simulations with high mass flow conditions are conducted. The new SEN combined with EMBr was applied to satisfy the two flow criteria mentioned in previsous section. Figure 9 shows the $\triangle H$ and V_{max} for the highest throughput case of which throughput is 8.7 ton/min, that is slab width of 1,600 mm and thickness of 110 mm with casting speed 6.5 m/min. If the magnetic field is not applied, $\triangle H$ and V_{max} exceed the maximum criteria at 4.5 m/min. The meniscus velocity results are similar to those of $\triangle H$. On the other hand, if EMBr is applied, it can be seen that there is no problem in both meniscus level and velocity.

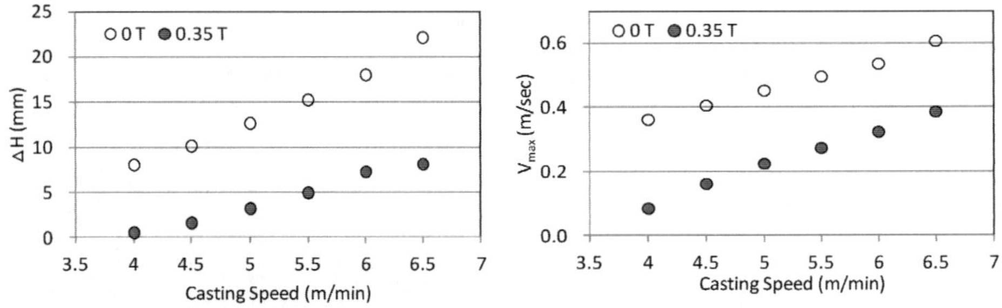

Fig. 9. Meniscus state for high throughput case C, (a) △H, (b) V_{max}

Effect of EMBr on theTemperature on the Meniscus

The effect of EMBr current on the temperature in the mold is also studied in this section. As an temperature index, △T was defined as the tempearature difference between T_{meni} and liquidus temperature, where T_{meni} is average temperature obtained near the corner part on the meniscus. As casting speed and electro magnetic field increase, so does △T. This can be explained from the fact that magnetic field make strong momentum and heat diffusion happen, resulting in increase of △T as shown in Fig. 10.

Table 1. Effect of EMBr and Casting Speed on △T

Casting speed (m/min)	Electro magnetic field (T)	△T (℃)
4.0	0	+6.5
6.5	0	+7.0
	0.25	+7.6
	0.35	+8.0

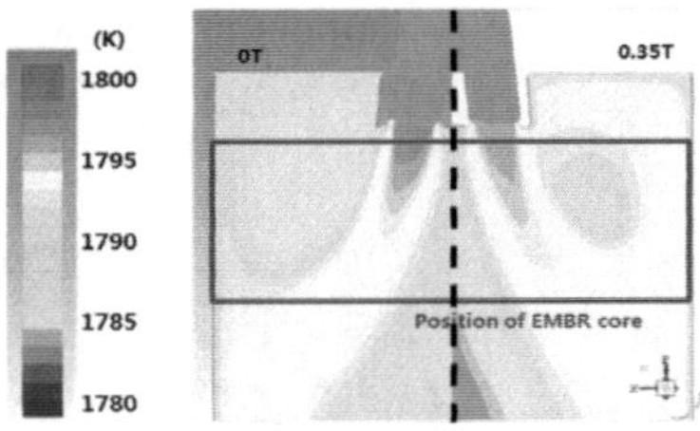

Fig. 10. Temperature Field

Conclusion

In this study, numerical analysis of mold flow was performed for high throughput conditions, which are 45% larger than that used for the CEM® casting process currently in operation. A new SEN designed for high throughput was evaluated for the meniscus level and velocity combined with EMBr. In order to validate the numerical method and results, nail board dipping tests from plant measurements were carried out. Those tests are in good agreement with the calculated results. If electromagnetic field is increased, the meniscus is dramatically stabilized. With this knowledge, numerical simulations were conducted for shigh mass flow case, and it was found that using the new SEN combined with the current U-shaped EMBr design, stable meniscus level and velocity can be expected without any serious problems. The effect of EMBr current on the temperature in the mold is also studied. It was clearly found that themperature on the meniscus decreases as EMBr current increas.

References

1. J.-Y. Hwang, S.-H. Lee, S.-Y. Kim and M.-C. Shin, METEC 2015, (2015, Dusseldof, Germany)
2. A. Lehman, ABB Review 1 (1996), p.4
3. B. G. Thomas and R. Chaudhary, 6th Int. Conf. Elecromagnetic Processing of Materials EPM (2009)
4. B. Rietow and B. G. Thomas, AISTech 2008, (2008, Pittsburgh, USA)
5. V. Norbert, O. Hans-Jurgen and R. Markus, 8th ECCC, Dusseldorf (2014)
6. S.-M. Cho, S.-H. Kim and B. G. Thomas, ISIJ Int. (2014, Vol. 54, No. 4), p. 845
7. S.-M. Cho, S.-H. Kim and B. G. Thomas, ISIJ Int. (2014, Vol. 54, No. 4), p. 855
8. R. Singh, B. G. Thomas and S. P. Vanka, Metallurgical and Materials Trans. B (2014, Vol. 45B), p. 1098
9. M. J. Cho, Numerical Heat Transfer (2006, Part B, 49), p.599
10. J. Sengupta, B. G. Thomas, H. Shin, G. Lee and S. Kim: Metall.Mater. Trans. A, 37A (2006), 1597.
11. S-M. Cho, G-G. Lee, S-H. Kim, R. Chaudhary, O-D. Kwon and B. G. Thomas: Proc. of TMS 2010, (2010), 71.
12. M. Iguchi, J. Yoshida, T. Shimizu and Y. Mizuno: ISIJ Int., 40 (2000), 685.
13. R. Liu, J. Sengupta, D. Crosbie, S. Chung, M. Trinh and B. Thomas: Proc. of TMS 2011, (2010)

Analysis of the Influence of Vertical EMBr on Steel/slag Interface in Continuous Casting

L. Xu[1], E.G. Wang[1]*, C. Karcher[2], A.Y. Deng[1], and X.J. Xu[1]

[1] Key Laboratory of Electromagnetic Processing of Materials (Ministry of Education), Northeastern University, No. 3-11, Wenhua Road, Shenyang 11004, P. R. China)

[2] Institute of Thermodynamics and Fluid Mechanics, Technische Universität Ilmenau, P.O. Box 100565, 98684, Ilmenau, Germany

Corresponding author : egwang@mail.neu.edu.cn.

Abstract

In this article, the impact of an electromagnetic brake (EMBr) on the flow pattern in a continuous casting mold was simulated with an implemented Reynolds-averaged Navier-Stokes (RANS) SST k-ω turbulence model. A new type of EMBr, named vertical electromagnetic brake (V-EMBr) was introduced to investigate its applicability and adaptability. The characteristics of electromagnetic field and flow field inside a 1450 mm × 230 mm mold with V-EMBr have been simulated. The numerical simulation results indicate that the static magnetic field generated by V-EMBr can cover both the vicinity of the mold narrow faces and the impingement region of jet flow, which can effectively control the liquid steel flow in the upper recirculation zone. The parametric study also shows that the braking effect on the surface velocity and steel/slag interface fluctuation is enhanced when the magnetic pole is placed close to the narrow face of the mold.

Key words : electromagnetic brake, continuous casting mold, V-EMBr, steel/slag interface

Introduction

In the continuous casting process, especially under the condition of high casting speed, increased flow rate of the liquid steel at the submerged entry nozzle (SEN) outlet can easily cause disturbances at the steel/slag interface on the top of the mold. Therefore, effectively controlling level fluctuations in the mold is key point in producing high quality steel slabs. Electromagnetic brake is one of the modern continuous casting techniques, the structure of which has been continuously improved. Up to now it has been developed into three typical types: local EMBr, ruler EMBr [1] and double-ruler EMBr [2]. At the present time, there are two widely used EMBr devices: ruler EMBr and double-ruler EMBr. It was found that, in the slab continuous casting process, the braking effects of ruler EMBr on the level fluctuation is not significant and usually ruler EMBr was unable to prevent mold flux entrapment effectively. Compared to ruler EMBr, double-ruler EMBr is good at controlling meniscus velocity and level fluctuation. However, it's not applicable for the continuous casting of thin slabs due to large volume, heavy weight and complicated structure.

Therefore, on the basis of understanding the characteristics of the existing EMBr, a new type of EMBr, named vertical electromagnetic brake (V-EMBr) was proposed by our research group in previous publications [3]. Two pairs of magnetic poles are installed vertically on the mold wide sides close to the narrow faces as shown in Fig. 1. The prominent characteristic of V-EMBr device lies in that the static magnetic field can cover both the vicinity area of the narrow faces of the mold and the impingement region of liquid-metal jet from the SEN, with the aim to suppress the upper recirculation flow and control the fluctuation of meniscus.

In this study, the SST k-ω model was utilized to describe the transient flow in a continuous casting slab mold, and the VOF model was applied for the treatment of the two-phase surface flow. The present work focuses on the effect of a static magnetic field on the fluctuation of steel/slag interface and the flow pattern in the mold, without considering the impacts of solidification and heat transfer on the liquid metal flow. In addition, the influence of the magnet position of V-EMBr on the barking effect was analyzed.

Description of Vertical EMBr

The cross section of the mold is 1450 mm × 230 mm, with a height of 800 mm. The V-EMBr device has two pairs of magnetic poles with a height of 600 mm, which are arranged vertically near the mold narrow face, as shown in Fig. 1. In addition, the sketch representation of different magnet positions with V-EMBr is illustrated in Fig. 2.

Mathematical Model

Under these assumptions the time-averaged governing equations read as follows:

$$\frac{\partial(\rho\phi)}{\partial t}+\nabla\cdot(\rho U\phi)=\nabla\cdot(\Gamma_\phi\cdot\nabla\phi)+S_\phi \quad (1)$$

Where, ϕ stands for various time-averaged variables, Γ_ϕ and S_ϕ represent the diffusion coefficient of the variable ϕ and the source term for various transport equations respectively.

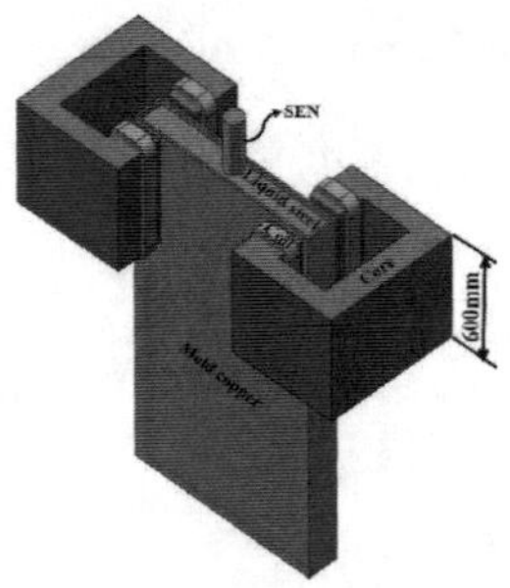

Fig. 1: Schematic of three-dimensional device

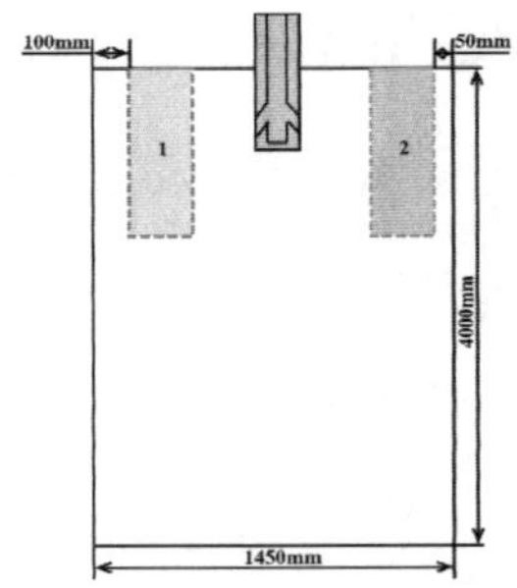

Fig. 2: Sketch representation of magnet positions.

A. Turbulence model

In this research, we utilized the Reynolds-averaged Navier-Stokes (RANS) SST k-ω model to simulate the mean flow characteristics in the mold. Different from the other two-equation eddy-viscosity turbulence models, the SST k-ω model computes the anisotropic turbulent flow field and considers the effect of Reynolds stress. And in the near wall region, where the SST k-ω model is superior to the other two-equation models, an automatic wall function is utilized. And more details for the turbulence model is given in the Reference 4.

B. Electromagnetic force equations

A magnetic induction method was utilized to calculate the induced current and electromagnetic force, which is derived from Ohm's law and Maxwell's equations. The induced current density can be can be found from the Reference 5.

C. VOF model equation

The method of volume of fluid (VOF) was utilized to build two-phase turbulent transient flow model, which could track the shape of fluctuating steel/slag interface in the slab continuous casting mold. The volume fraction of liquid steel should satisfy the following equation.

$$\frac{\partial \varphi_{st}}{\partial t} + \nabla \cdot (\varphi_{st} \boldsymbol{U}_{st}) = 0 \tag{2}$$

Where, φst=1 stands for the liquid steel; φst=0 stands for the mold flux; and between 0 and 1, φst represents the steel/slag interface.

D. Computational domain

Due to the fact that the model is symmetrical, in order to improve the computational efficiency, one half of the mold geometry was considered. The operating parameters and physical properties of the mold are shown in Table I.

Table I. Geometrical parameters and physical properties

Parameter	Value	Parameter	Value
Mold size/mm	1450×230	Total height of SEN outlet/mm	80
Mold length/mm	800	Magnetomotive Force/A·t	0, 31500
Computational domain/mm	1450×230×4000	Liquid steel density/(kg·m-3)	7020
Casting speed/(m·min-1)	1.2	Mold flux density/(kg·m-3)	3500
The height of liquid mold flux/mm	35	Steel viscosity/(kg·m-1·s-1)	0.0062
Angle of nozzle port (°)	-15	Mold flux viscosity/(kg·m-1·s-1)	0.2664
Depth of SEN/mm	150	Electric conductivity/(S·m-1)	7.14×105
Interface tension/(N·m-1)	1.2	Magnetic permeability/(H·m-1)	1.257×10-6
$Re=\rho Ud/\mu$ (d=nozzle diameter)	69500	Hartmann number: $Ha=B_0L\sqrt{\sigma/\mu}$	515
$Re=\rho UL/\mu$ (L=magnetic pole width)	232000	Stuart number: $N=B_0^2L\sigma/\rho U$	1.14

Results and Discussion

Fig. 3, Fig. 4 and Fig. 5 show contour plots of magnetic flux density, vector plots of current density and Lorentz force in the mid-plane at the mold with different positions of V-EMBr. The red dashed lines in the plots indicate the position of V-EMBr. It can been seen from Fig. 3(a) to Fig. 5(a) that the magnetic field is concentrated in the region which is close to the mold narrow face, and it decays significantly from the central zone to the outer area. Fig. 3(b), Fig. 4(b) and Fig.

5(b) indicate that as the magnet position gradually approached to the narrow face of the mold, the induced current is decreased. In addition, the induced current flows clockwise from the region where the jet flow interacts with the higher magnetic field. Fig. 3(c), Fig. 4(c) and Fig. 5(c) represent that Lorentz force produces a braking effect on the ending segment of the jet.

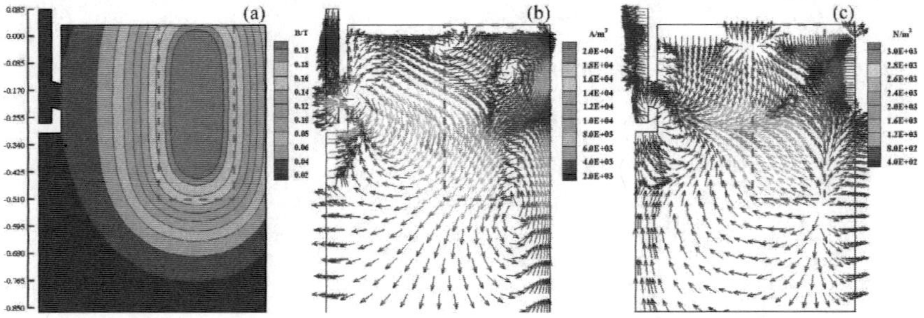

Fig. 3: Distribution of magnetic flux density, induced current density and Lorentz force in the central plane with V-EMBr (P = 100 mm): (a) $B_0 = 0.2$ T, (b) induced current density and (c) Lorentz force.

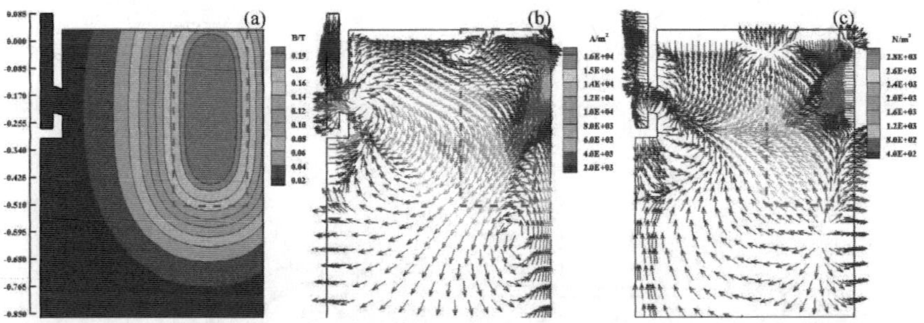

Fig. 4: Distribution of magnetic flux density, induced current density and Lorentz force in the central plane with V-EMBr (P = 50 mm): (a) $B_0 = 0.2$ T, (b) induced current density and (c) Lorentz force.

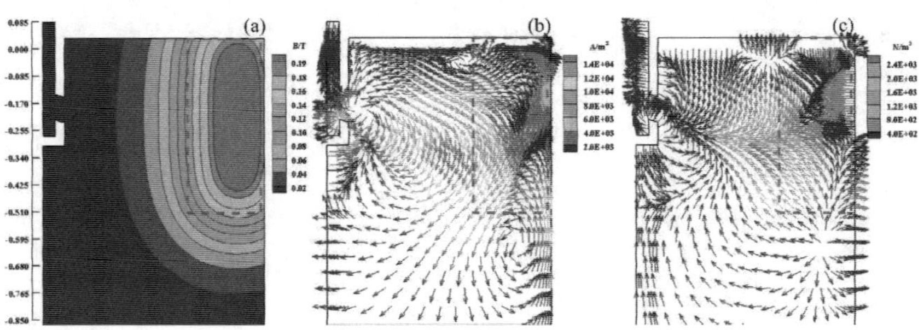

Fig. 5: Distribution of magnetic flux density, induced current density and Lorentz force in the central plane with V-EMBr (P = 0 mm): (a) $B_0 = 0.2$ T, (b) induced current density and (c) Lorentz force.

Fig.6 presents the velocity distribution in the mid-plane at the mold wide face with different magnet positions. Under the condition of without EMBr, a typical double-roll flow pattern as shown in Fig. 6(a), the liquid metal jet discharged from the SEN flows downward and splits into an upward and a downward flow after impacting on the narrow face of the mold. The liquid steel flows upward along the mold narrow face and rushes to the steel/slag interface and then flows back to the center of the mold to form an anticlockwise vortex. The downward liquid steel flows deep into the molten pool and generates a big clockwise vortex. Fig. 6(b), Fig. 6(c) and Fig. 6(d) show that the jet impingement upon the narrow face of the mold is prevented by V-EMBr. In the lower recirculation zone, the core of the vortex is shifted upward as the magnet position approached to the narrow face of the mold. However, an extra vortex is formed below the jet flow, when the corresponding to the magnet position of V-EMBr is 100 mm and 50 mm, respectively.

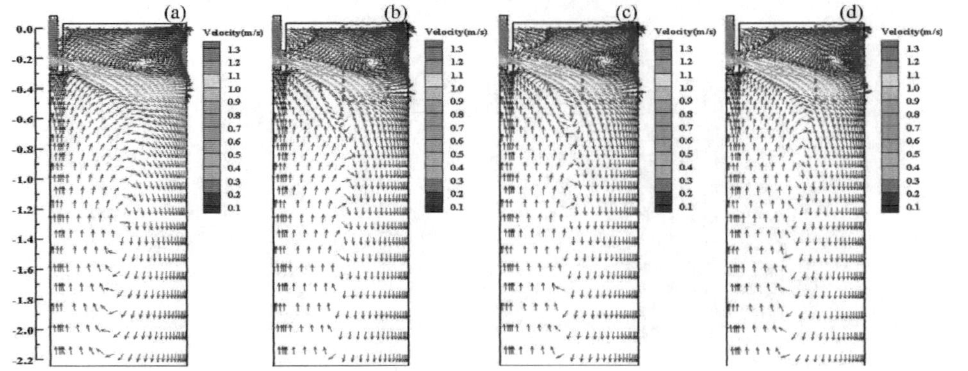

Fig. 6: Velocity distribution under different magnet positions in the central plane: (a) B = 0 T; (b) P = 100 mm, with V-EMBr; (c) P = 50 mm, with V-EMBr and (d) P = 0mm, with V-EMBr.

The fluctuation of steel/slag interface and surface velocity with different magnet positions of V-EMBr is illustrated in Fig. 7. When the V-EMBr is applied, the fluctuation height is reduced to 5.27 mm, 4.33 mm and 2.67 mm, when the corresponding to the magnet position of V-EMBr is 100 mm, 50 mm and 0 mm, respectively. Additionally, the maximum velocity is reduced to -0.11 m/s, -0.10 m/s and -0.09 m/s in turn when the magnet position of V-EMBr is 100 mm, 50 mm and 0 mm, respectively.

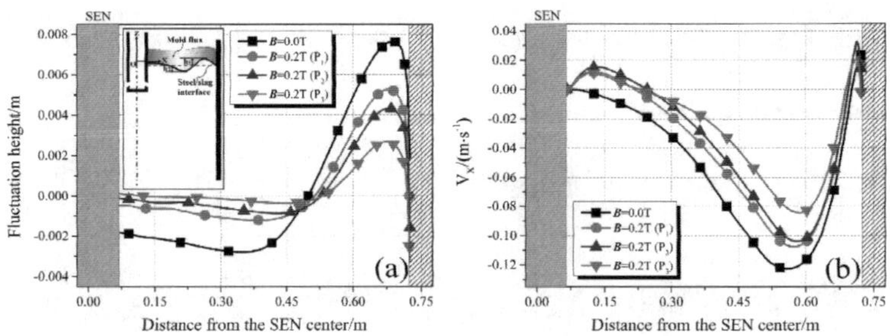

Fig. 7: Fluctuation of steel/slag interface and surface velocity with different magnet positions.

Conclusions

V-EMBr device can not only cover the free surface, but also cover the impact range of the metal jet from the SEN. When the V-EMBr is imposed, the braking effect on the surface flow and steel/slag interface fluctuation is more effective. In addition, the upward deflection of jet flow near the mold narrow face is further suppressed when the magnetic pole is placed close to the narrow face of the mold.

Acknowledgements

This work was financially supported by the National Nature Science Foundation of China (Grant No. 51574083, Grant No. 51474065 and Grant No. U1760206), the Program of Introducing Talents of Discipline to Universities (The 111 Project of China, Grant No. B07015), the Fundamental Research Funds for the Central Universities (Grant No. N160904004), and the Deutsche Forschungsgemeinschaft (DFG) with the Research Training Group Lorentz Force (GRK 1567) at TU Ilmenau.

References

1. R. Singh, B.G. Thomas, and S.P. Vanka, Metall. Mater. Trans. B, 44 (2013), 1201-1221
2. A. Idogawa, M. Sugizawa, S. Takeuchi, K. Sorimachi, and T. Fujii, Mater. Sci. Eng. A, 173 (1993), 293-297
3. Z. Li, E.G. Wang, L.T. Zhang, Y. Xu, and A.Y. Deng, Metall. Mater. Trans. B, 48 (2017), 389-402
4. F.R. Menter, Aiaa Journal, 32 (1994), 1598-1605
5. ANSYS FLUENT user's guide, ANSYS Inc., 2016, ver. 16.2.0 edition

Two Paradigms on Study Slab Continuous Casting Process with Mold Electromagnetic Stirring

Zuosheng Lei[1*], Bin Li[1], Yueming Zhou[2], Xing Wu[3], Yunbo Zhong[1], Zhongming Ren[1]

1. State Key Laboratory of Advanced Special Steel & Shanghai Key Laboratory of Advanced Ferro-metallurgy & School of Materials Science and Engineering, Shanghai University, Shanghai 200072, China
2. Baosteel Central Research Institute, 200072, Shanghai, China.
3. Scool of Computer Engineering and Science, Shanghai University, 200072, Shanghai, China.

* *Corresponding author: lei_zsh@staff.shu.edu.cn*

Abstract: Electromagnetic stirring (EMS) is an efficient technology to control the flow structure in slab continuous casting mold. But there is still an important opened problem to be answered firstly, that is, what kind of flow structure is our goal by using electromagnetic stirring in order to get good casting products, and how to optimize the flow fields in slab continuous casting mold with EMS? Two kinds of research paradigms are proposed in this study. One is based on causal relationship between the flow field and its effects on the mold flux entrapment and solidification process. Two parameters, named Mold Flux Entrapment Index and Velocity Uniform Index were proposed in order to evaluate the flow field in the mold. Based on these two indexes, optimization stirring parameters were proposed under different casting situation. The other research paradigm is machine learning big data model based on correlation analysis of nearly all the input parameters of casting machine and slab quality as an output parameter. Data acquisition from the continuous casting machine, data mining by machine learning algorithms and online automatic control parameters output are three main aspects in the second research frame.

Key words: Slab Continuous Casting, Flow Control, Electromagnetic Stirring, Big Data Machine Learning.

Introduction

Fluid flow in the mold is very important to product quality in the steel slab continuous casting [1]. Electromagnetic stirring [2] is one typical technologies in which Lorentz force can be generated in the up zone of liquid steel to controlling the flow field in the mold and has being widely applied in the industrial process. Mathematical modeling and physical modeling using low melt point metal have been developed to understand the fluid flow phenomena with electromagnetic field. But up to now, most of the work can only tell us ***what it is*** about the flow field in the mold rather than ***what should be***. There is still an important open problem to be answered firstly, that is, what kind of flow structure is our best choice to get the best casting products when we use electromagnetic technologies? How to evaluate the flow field with the imposing of electromagnetic field in the mold is a key problem. In this study, two research paradigms in order to resolve the above-mentioned problems are proposed and examinated. One is based on causal relationship between the flow field and its effects on the mold flux entrapment and solidification process, and the other is machine learning big data model based on correlation analysis of nearly all the input parameters of casting machine and slab quality as an output parameter. Some preliminary results are introduced here.

Mathematical model based on causal relationship

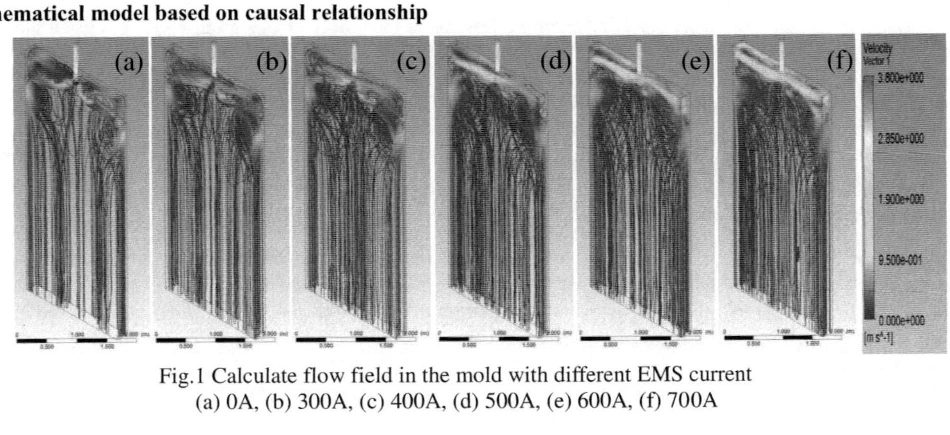

Fig.1 Calculate flow field in the mold with different EMS current
(a) 0A, (b) 300A, (c) 400A, (d) 500A, (e) 600A, (f) 700A

A mathematical model coupled with magnetic fields and flow fields is established to numerical simulate the flow field in the industrial mold under different EMS parameters and continuous casting conditions. This model has been validated by contrast to a physical modlling applying low melt point metal [2]. Fig. 1 shows some typical calculated results when the casting speed, the section of the mold the SEN submerge depth, and the exit port angle are 1.4m/min, 2150mm×230mm, 170mm, and 15°downward, respectively.

It can be seen from Fig.1 that the flow field without EMS is a typical double-roll structure, and a horizontal circulation in the upper part of mold appears with EMS, and the velocity increases with the increasing EMS current. But the problem is which EMS current is the best one in order to improve the slab quality? Two Indexes are proposed in order to evaluate the flow field in the mold.

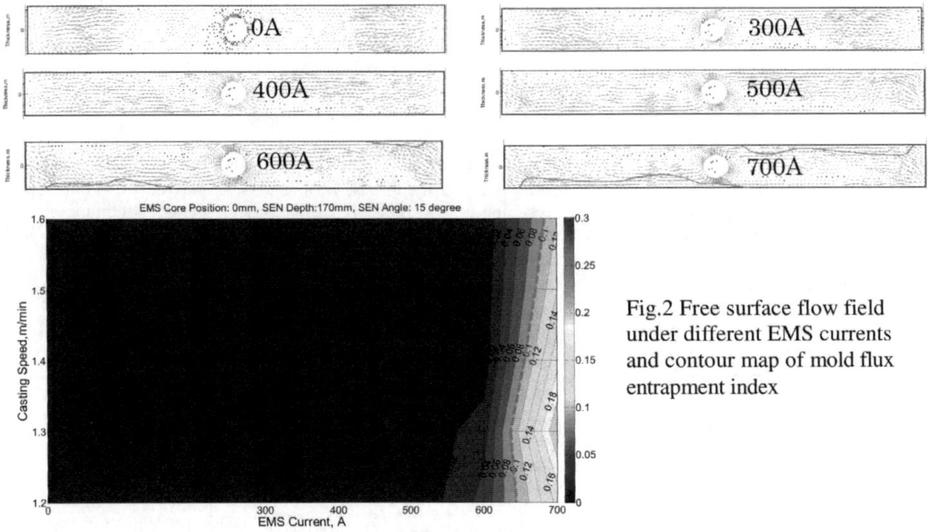

Fig.2 Free surface flow field under different EMS currents and contour map of mold flux entrapment index

In slab continuous casting process, the free surface is very important to control the velocity of the liquid metal flow under the mold flux. If the velocity of this area is too high, the possibility of mold flux entrapment might increase. In order to valuate this effect quantitatively, a new index, named Mold Flux Entrapment Index (MFEI), is introduced and it is defined as the surface $S_{v>v_c}$ in free surface in the mold where the steel velocity is greater than the critical mold flux entrapment velocity v_c to the total surface of free surface S_{total}. The MFEI can be calculated as

$$MFEI = \frac{S_{v>v_c}}{S_{total}} \times 100\% \qquad (1)$$

Where v_c can be obtained by eq.

$$v_c^2 \geq \frac{2(\rho_1+\rho_2)}{\rho_1\rho_2}\left[\gamma(\rho_1-\rho_2)\right]^{1/2} \qquad (2)$$

Where ρ_1 and ρ_2 are the density of liquid steel and mold flux, and γ is the surface tension between liquid steel and mold flux. Eq. (2) is the Kelvin-Helmholtz instability criterion between two immiscible liquid. It suggests that once the velocity difference of liquid steel and mold flux is greater than the critical velocity, the possibility of mold flux entrapment will increase. The critical velocity can be determined by experiments in facts, but here hypothesis was took that the critical velocity is 0.8m/s.

Fig.2 shows the free surface flow field and contour map of mold flux entrapment index under different EMS currents and casting speed. From the map, the region right to the red line is relatively dangerous from the viewpoint of mold flux entrapment. So the maximum EMS current can be selected under different casting speed.

As we know that the growth speed of initial solidifying shell is decided by two aspects. One is the cooling density from the water-cooled copper mold, and the other is the impact of the hot liquid steel flow in front of the initial shell. In order to prevent the crack formation caused by thermal stress, it is very important to keep the shell growing uniformly. It is the flow velocity in front of the initial solidifying shell that determine the convection heat transfer between the interface of the shell and liquid steel, so the uniform of the velocity in front of this interface is applied to valuate the impact of flow field on initial shell growth.

Due to the help of EMS Lorentz force, a horizontal circulation will form in the stirring zone, and it will affect the flow structure of the low part in the mold. An index named, Velocity Uniform Index (VUI), is introduced followed reference [2]. To give a quantitative valuation of the effect of flow field on solidification on every level, The VUI is defined as:

$$VUI = \frac{\oint_\Gamma |V(x)| dx}{\max(|V(x)|)\oint_\Gamma dx} \times 100\% \quad (3)$$

Where Γ is a closed curve 20 mm away from the inner mold wall, see in the Fig.3, and V_x is the absolute velocity of every point on Γ. It is obvious that VUI has a value between 0 and 1. This value greater the better the flow field is. Fig.3 shows the VUI of different levels in the mold when the EMS currents are different. It can be seen that the VUI increases with the EMS current, but decreases a little in the deep part of the mold.

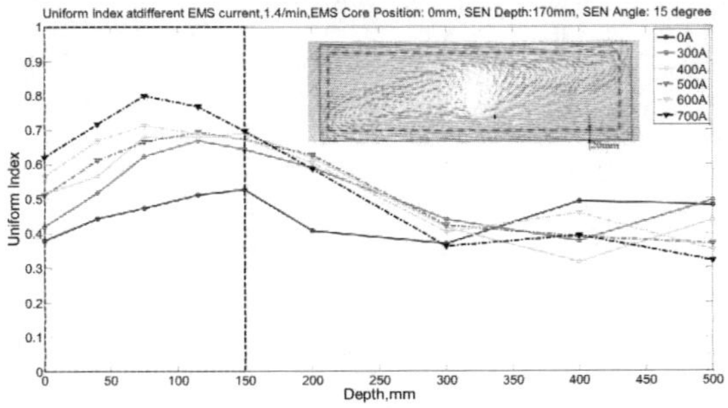

Fig.3 VUI under different EMS currents

By comprehensive evaluation of these two indexes an optimized EMS current can be achieved under a given continuous casting condition. There are still other aspects have not be took into consideration so far, for example, the gas bubble and inclusiong transport behavior in the mold, the stress of solidified shell, and so on.

Big data model based on correlation analysis by machine learning

The study roadmap discribe above is a typical analytical research method in which every detailed mechanism should be clarified before the best operation parameters can be determined. But in fact, it is not only difficult but also time cost and sometimes impossible to studied thoroughly in such a complex continuous casting system with so many input parameters.

With the development of computer technology nowadays, it is possible to deal with massive large data or big data from the industrial process. By artificial intelligence algorithm developed recently the correlation between input parameters and output parameter of an industrial system can be obtained. Continuous casting process, as a complex system, is very suitable for big data analysis in order to find the best opeartion parameters.

The whole research sketch map are showed in Fig.4. There are mainly three parts in this process, data acquisition, data mining and prediction & real-time controlling. So far, we have developed an off-line system including the first two sections. The real-time control sections will be applied after the machine learning system be fully trained and examined.

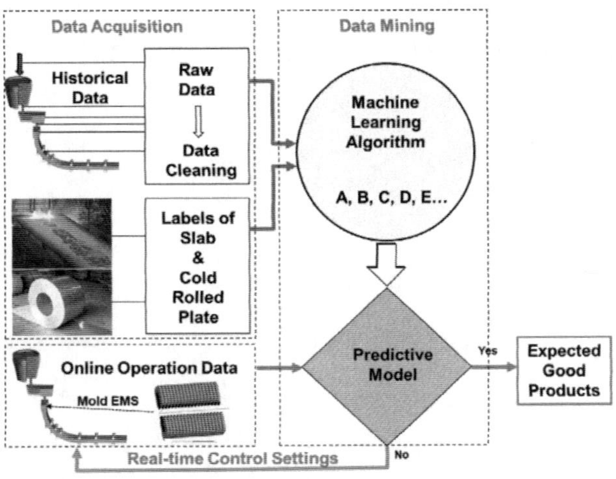

Fig. 4 Sketch map of big data model for continuous casting process

-Data acquisition. Nearly most of the casting operation parameters are collected every one second. These parameters are casting speed, EMS currents and frequency, cooling water flow rate of mold, argon injection flow rate, submerge entry nozzle depth, temperature of the mold copper plate, free surface fluctuation, mold flux addition rate, and so on. The quality of slab and cold rolled plate are labeled corresponding to the operation parameters. All the data must be cleaned and have a good time stamp.

-Date mining. The data collected from the continuous casting machine is divided into two parts. 75% of the data including both the operation parameters and labeled products quality is used as training set and the others 25% data is used as test data. The training set data were input into machine learning software package including more than ten kinds of supervised learning algorithm, then every algorithm can give a predict simulation of the test set data. The algorithm with the highest prediction accuracy is the winner be choosed to obtain ***the model*** of the whole continuous cating system, including the EMS technology. The forecast accuracy can reached 99.5%. Learning from history can predict the future, then the model can be used to predict the product quality when online process parameters were input in it.

So far, the historical production data more than one year duration of No.2 continuous casting machine of Baosteel Zhanjiang works have be collected and stored for the machine learning. Also, the big data model can tell us which parameters affect the slab quality most. For example, from high to low according to weight, surface fluctuation, casting slab width, EMS currents, casting speed and temperature of mold copper plate are the first five important parameters. This give us a possible solution in order to improve the slab quality by taking the most important parameters into account first.

- Intelligent real-time controlling. Based on the big data model, it is possible to realize a totally intelligent control of continuous casting process. When the model give a nagative predition it can intelligently find a parameters settings in the multi-dimension phase space to change the outputs. So far we have not realize this section, it need more profound study under the sincere cooperation of both metallurgical field and computer technology field.

Conclusions

In this study, two kinds of research paradigms are proposed in order to optimize the slab continuous casting process with EMS controlling the flow field in the mold. One is based on causal relationship between the flow field and its effects on the mold flux entrapment and solidification process. The other is machine learning big data model based on correlation analysis of nearly all the input parameters of casting machine and slab quality.

The first research method can give us some mechanism explain on some particular aspect, but can not give a comprehensive undstanding of the whole process. The later reserch method did not give causal explain on every single problem, but can give an online prediction on the product quality.

Different research paradigms provide different thinking method.This two method should rely on each other and explain each other mutually in order to optimize the whole process.

Acknowledgment

This project financially supported by National Science Foundation of China (NO. 51274137) and Shanghai Economic and Information Committee (No.CXY-2016-015).

References

1. Thomas, B. G., Zhang, L. (2001). *ISIJ international, 41*(10), 1181-1193.
2. LI Bin, LEI Zuo-sheng, GE Jiang-meng, ZHONG Yun-bo, REN Zhongming. (2016) *Baosteel technical research,* 10(1), 8-14.

IOP Conf. Series: Materials Science and Engineering **424** (2018) 012036 doi:10.1088/1757-899X/424/1/012036

Resonant pulsed electromagnetic stirring of melt for effective grain fragmentation

D. Köppen[1,2], E. Baake[1], G. Gerstein[1], G. Mrówka-Nowotnik[3], G. Jarczyk[4]

[1]Leibniz University Hannover, Welfengarten 1, Hanover, Germany
[2]Kazan State Power Engineering University, Krasnoselskaya Str. 51, Kazan, Russia
[3]Rzeszów University of Technology, Powstańców Warszawy 8, Rzeszów, Poland
[4]Engineering Consulting, Grosskrotzenburg, Germany

Corresponding author: baake@etp.uni-hannover.de

Abstract
The paper describes the results of an experimental research, demonstrating and explaining the effect of grain fragmentation, caused by pulsed resonant electromagnetic stirring. In the experiments, 6082 aluminium alloy melt was directionally solidified under the influence of continuous (AMF) and pulsed application (PMF) of an alternating magnetic field. The frequency of applied PMF was in accordance to the low-frequency circulation of the melt, causing the resonant increase of a pulsed component of the melt velocity. The structure of electromagnetically stirred specimens was compared to those, solidified without a magnetic field. A strong fragmentation effect (decrease of an average grain size on 51%, comparing with the solidification in natural conditions) for the case of resonant EM stirring was stated. Further, to analyse the influence of the flow, appearing due to the resonant stirring, we observed the formation of solid/liquid interface and a macro-crystalloid structure during solidification of continuously and pulsed stirred melt by applying the novel method of neutron radiography. The results confirmed the strong influence of the pulsed component of velocity on thermal conditions during solidification and, consequently, the metal structure.

Key words: electromagnetic stirring, solidification, grain fragmentation, process imaging, neutron radiography.

Introduction
Since several decades, the electromagnetic (EM) stirring of solidifying materials is known as an advanced technique for approaching metals with desired properties [1 -4]. Yet still, the development of this technique is still based on trial-and-error approach, what makes new solutions applicable only for a limited number of processes and installations. Development of universal and effective method of metallic grain fragmentation and their homogenous distribution remains one of the important challenging problems.
The reason, why this problem was not solved till now, is the enormous role, which electromagnetically driven turbulent flows playing in the process of crystals formation [5], and at the same time an impossibility of performance of in-situ measurements in metallic melts, due to their opacity, chemical aggressiveness, and high melting temperature. These limitations reduce the number of methods, which could be applied for gaining the information about the interaction between turbulent melt flows and the crystallizing material.
The methods, allowing investigating these processes and broadly used nowadays are the numerical simulation and X-ray analysis. Both these methods, being quite effective, have their own disadvantages. All results obtained based on numerical simulation should have to be experimentally verified, what brings investigators back to the problem of measurements in liquid metals. From the other side, the X-ray radiography can be employed to visualize the dendritic growth of metal alloys, though the thickness of the sample under analysis strongly depends to the ray energy and usually does not exceed 1 cm for metals. Due to the high viscous properties of liquid metals development of three-dimensional turbulent flows, which have such a great influence on crystal formation, in such suppressed volumes is practically impossible. These circumstances force the search of technique, which could allow the in-situ observation of metal solidification process for specimens with increased volume and on-time effective correction of EM parameters.
As it has been shown by several recent experimental investigations, the neutron radiography can be applied to solidifying material for such a visualization [6, 7]. Now, when the technology allows employing the method of neutron radiography to close-to-industrial processes, it is more than important to collect the visual information about the interrelated phenomenon in electromagnetically stirred solidifying material with the aim to optimize the technology.
Based on the previous research, related to the optimal application of pulsed EM field (PMF) for effective grain fragmentation and redistribution, this article discusses the visualization results of the metal structure, formed under the EM treatment of the melt. The macro-crystalloid structure of ingots with increased thickness, grown under the influence of turbulent flows is analyzed. Being one of the most promising methods of EM stirring, application of resonant pulsed EM field is in the focus of the present work.

Solidification experiment with EM stirring

On the first stage of the research the solidification of 6082 aluminium alloy in different conditions was experimentally investigated. The detailed description of experimental system, the procedure of specimens post-processing and metallographic analysis can be found in previous publication [8]. In the experiment, we reproduced the operation of a continuous casting device on the limited material volume (500 g) and provided its directional crystallization with and without EM stirring by alternating magnetic field. Furthermore, the magnetic field was applied to the solidifying melt in one case permanently and in the other case with low-frequency pulses. The frequency of applied pulsed AMF was in accordance to the low-frequency circulation of the melt, causing the resonant increase of a pulsed component of the melt velocity. The structure of all obtained specimens was compared and analysed. Two main parameters were taken into account – the average size of metal grains and homogeneity of their distribution. The magnified surface of the samples middle section, grown in natural conditions and stirred by means of continuously applied AMF and resonant pulsed magnetic field is presented in Figure 1.

(a) (b) (c)

Fig. 1: The structure of specimens solidified in natural conditions (a), stirred with continuously applied AMF (b) and pulsed resonant AMF (c). Magnification is ×50.

The strong fragmentation effect was observed in the probe, stirred with resonant magnetic field – the average grain size, calculated for the investigated area was 320 μm for the natural solidification and 155 μm for EM treatment with pulses (reduction on 51%, Fig. 1, a and c). The continuous EM stirring led to grains fragmentation on 40% (187 μm, Fig. 1, b). One can notice the increased homogeneity of the structure, obtained with resonant EM stirring – comparing to other both cases, the variation of grain sizes is much lower. Two more values of pulse frequency, higher than the resonant frequency were tested as well. The comparison of obtained average grain size is provided in the Table 1. It is noticeable, that the fragmentation effect of the pulsed stirring increases with getting closer to the resonant frequency.

The experiment has shown the strong influence of type of the melt motion on solidification conditions. In the further step of the study the model experiment, reproducing the solidification process in comparable conditions, and the reasons, causing the observed fragmentation effect was analyzed. For this purpose, instead of aluminum, the pure gallium was selected as a model melt, due to its much lower melting temperature (approx. 30 °C), which allowed us to simplify the experimental set-up. Visualization of the process enabled us to observe the in-situ formation of metal ingots, and what is more important – to evaluate their macro-crystalloid structure.

Table 1. Comparison of the average grain size, obtained for different stirring methods

Parameters of crystallization process	Average grain size, (μm)
No Force	320
Continuous application of AMF	187
PMF	
f_p=10 Hz	200
f_p=1 Hz	186
f_p=0.1Hz	155

Visualization of a solidification process

This part of the investigation is based on experiments performed at the Swiss spallation neutron source SINQ, Paul Scherrer Institute, Villigen, Switzerland with use of NEUTRA facility [9]. Figure 2 depicts the experimental installation, optimized to work with the beam of thermal neutrons. For this, the setup was designed to have a quasi-two-dimensional form. Here (1) – is the heating system, serving for reduction of heat losses from the melt free surface and for increasing the temperature gradient; (2) – the window-glass vessel with copper bottom for the better heat transfer to the cooling system (7); (3) – pure gallium, preliminary melted and warmed up until 40 °C. A rectangular volume of the liquid had the height of 12 cm, the width of 10 cm and the depth of 2 cm. The induction system (4) produced a travelling magnetic field (TMF), consisted from three-phase coils with a phase shift of 60° and was feed with alternating current at 100 Hz and at 0.89 A of its effective value. The electric current density (j), following the phase shift in a three-phase induction coil, created a three-phase alternating magnetic field (B) and induced a current density (j_{ind}) in the metal. The interaction between the current in the coils and the current induced in the melt gave rise to the Lorentz force (5):

$$\mathbf{F}_{\mathbf{EM}} = \mathbf{j}_{\mathbf{ind}} \times \mathbf{B}, \tag{1}$$

which followed the traveling field and pulled the liquid after. In such conditions, the melt flow had a form of two oval vortices, with streamlines, indicated in Fig. 2 as (6), with the maximum vertical velocity on the vessel's axis (characteristic velocity) $V_{ch} = 18$ mm/s.

In the experiment the application of TMF was realized in two ways – metal solidification in conditions of permanent and resonant pulsed EM stirring. Based on the characteristic velocity of the melt circulation and characteristic length, the resonant frequency of the pulsed TMF was calculated and had a value of $f_{ch} = 0.52$ Hz. Thus the frequency of pulsed applied TMF was chosen to be close to this value – f_p = 0.5 Hz. Here f_p identifies the frequency of connection/disconnection the induction coils form the generator and determined by the sum of the on-time T_{act} and off-time T_0 (2):

$$f_p = 1 / T_P;\; T_P = T_{act} + T_0 \tag{2}$$

The periodic modulations of the magnetic field lead to the appearance of the Lorentz force with pulses. In such a manner the stirring with permanently employed TMF can be described with the value of f_p = 0 Hz ($T_{act} = \infty$). As a reference case, the solidification process in natural conditions (without magnetic field) was considered. In Figure 3 are shown the images of three ingots, solidified in described arrangements. The original frame size of each image is 10 × 12 cm. The analysis of a macro-crystalloid structure was based on the optical local skeletonization of obtained images of ingots.

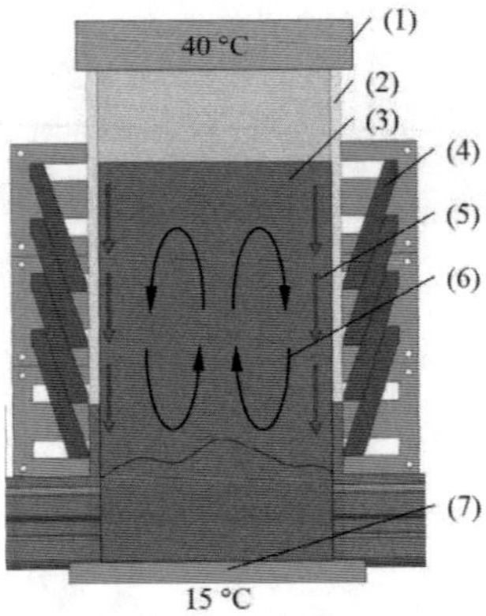

Fig. 2: The experimental system, used for visualization of EM-stirred solidifying pure gallium by means of neutron radiography.

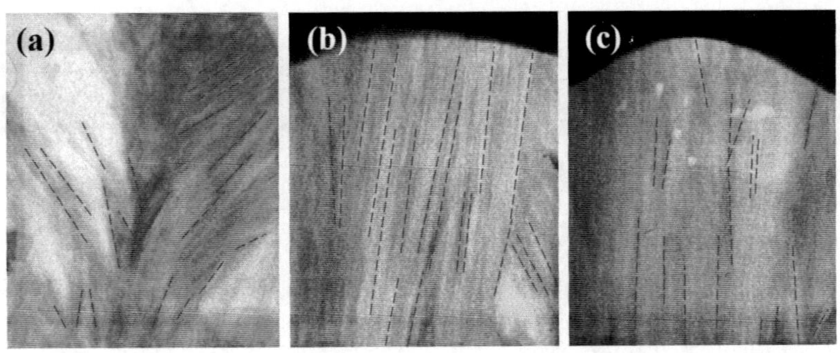

Fig. 3: The visualized macro-crystalloid structure formed in natural conditions (a), stirred with resonant pulsed TMF (b) and with steadily applied TMF (c). Images demonstrate ingots on 150th-minute from the solidification begin.

As it can be seen in Fig. 3, the type of the melt motion, having a direct influence on thermal conditions, led to the formation of different metal macro-structures. The ingot, grown in natural conditions has a big amount of short branches with various directions (some of them are skeletonized in Fig. 3, a). Based on the solidification theory [10], such a structure is assumed to be caused by the existence of more than one (additionally to the artificially applied) temperature gradients. Inhomogeneity in the temperature field provoked a growth of crystal branches, which due to different directions disturbed each other's growth and provided an accumulation of dislocations.

An opposite picture can be observed in Fig. 3, b, where the solidifying melt was stirred with resonant pulsed TMF. Here most of the branches have the extended length and a direction along the main temperature gradient, indicating a uniform heat transfer from the liquid phase. At the same time, the macro-structure of the ingot solidified in conditions of permanent stirring (Fig. 3, c), demonstrates a middle situation – the majority of branches having a growth direction along the main temperature gradient, formed shorter structures. Since all other conditions of the solidification process were the same, one can suggest, that the heat transfer conditions were strongly influenced by the type of melt motion. As it was shown in previous investigations, the pulsed application of magnetic field in accordance with the characteristic frequency of melt circulation is capable to increase the pulse component of the axial velocity considerably [11]. Taking into account the assumptions of other authors [12, 13], that in melt flows of such a configuration an axial component of velocity plays a key role in the heat exchange, one can state better heat transfer conditions and more homogenous temperature distribution, leading to the observed macro-structure. In the case of steady application of TMF, the heat transfer is intensified as well, but into a less extent. Same phenomena took place in the previous solidification experiment and became a reason of Al-alloy grain fragmentation. Although, on the micro-scale, in addition to the intensified heat transfer in the liquid, the intensive interaction between growing crystals and turbulent flows has a great influence as well.

Summary

One of the most important engineering problems is an optimal application of electromagnetic (EM) stirring for processing advanced metallic materials. One of the promising methods of EM stirring is the application of a resonant pulsed EM field to a solidifying melt, which compared to other methods provides the effective grain fragmentation without an increase of the energy consumption on the process. It is shown, that the resonant application of EM field leads to a considerable increase of velocity pulsations in a melt, in their turn changing heat transfer conditions to optimal for a formation of homogenously distributed metallic grains with a reduced size. In the present study, an application of neutron radiography for in-situ observation of a solidification process under the influence of permanent and pulsed EM stirring is considered. An increased volume of a crystallizing melt is investigated, where EM stirring leads to the development of three-dimensional turbulent flows. The principle of beam particles attenuation depending on the material density, used in neutron radiography, allowed us to observe the ingots evolution process and their skeleton structure, influenced by different treatments. The visualization has shown a strong influence of EM stirring, comparing to the solidification in natural conditions – stirred specimens gained a regular macro-structure, evidencing more stable heat transfer conditions. The most pronounced effect was obtained for the application of resonant stirring.

Acknowledgments

This work was partially supported by the Ministry of education and science of Russian Federation (Russian President Grant). Authors are grateful to the Helmholtz-Alliance LIMTECH (Liquid Metal Technology) for financial support of the experiments in PSI and to the colleagues from Paul Scherrer Institute for their help during experiments.

References

1. C. Vives, J. Cryst. Growth, 76-1 (1986), 170-184.
2. K.A. Radjai and K. Miwa, Melt. Mat. Mater. Trans. A., 31A (2000), 755-762.
3. P.A. Nikrityuk, et al., Int. J. Heat and Mass Transfer, 49 (2006), 1501-1511.
4. L. Hachianik, et al., Int. J. of Heat and Mass Transfer, 85 (2015), 438-454.
5. T. Campanella, et al. Metall. Mater. Trans. A35-10 (2004), 3201-3210.
6. D. Musaeva, et al., Magnetohydrodynamics, 53-3 (2017), 583-593.
7. A.S. Tremsin, et al., Sci. Rep., 7 (2017), 46275.
8. D. Musaeva, et al. Materials Science Forum: Materials Engineering and Technologies for Production and Processing II, 870 (2016), 471-476.
9. E.H. Lehmann, et al., Nondestructive Testing and Evaluation, 16:2-6 (2001), 191-202.
10. B. Chalmers Principles of Solidification. – John Wiley & Sons, Inc., New York, London, Sydney, 1967.
11. D.A. Musaeva, et al. St. Petersburg Polytechnical University J.: Physics and Mathematics, 2 (2016), 193–200.
12. M. Kirpo, et al. Magnetohydrodynamics, vol. 42 (2006), no. 2–3, pp. 207–218.
13. A. Umbrashko, Heat and mass transfer in electromagnetically driven recirculated turbulent flows: PhD thesis (Physics) / Andrejs Umbrashko. – Riga (2010) – 108 p.

Permanent magnet dipole stirrer for aluminium furnaces

T. Beinerts[1], A. Bojarevičs[1], R. Baranovskis[1], M. Milgrāvis[1*], I. Kaldre[1]

[1] Institute of Physics University of Latvia, Miera street 32, Salaspils, Latvia

Corresponding author: mikus.milgravis@lu.lv

Abstract
Use of permanent magnet systems for metallurgical applications have been proposed before. We have continued development of such systems by investigation use of single magnetic dipole which is realized as permanent magnet cylinder. Magnet is magnetized orthogonal to its axis and has been mounted to the side of the liquid metal reservoir. System first was modelled using small scale experiment with galinstan melt. Velocity distributions and flow patterns in reservoir using Ultrasound Doppler Anemometer were acquired and will be presented here. Additionally, numerical model using Comsol and Cenos modelling software were developed alongside analytical, non-dimensional estimation calculations. Experimental results were used to adjust and verify calculations. Using these calculations, industrial scale system was modelled and designed, and built. Industrial prototype was installed to 20t aluminium remelting furnace in Norway. It demonstrated that sufficient flow for efficient stirring could be achieved, while power of unit did not exceed 2kW, what is up to 30 times less than for comparable size classical EM stirrers with current windings. Besides system allows unique versatility by changing tilt angle to change flow pattern continuously allowing adaptive melting process, and costs significantly less than classical systems. By the time of the conference another prototype will be commissioned and results will be included.

Key words : Permanent magnet; Metal stirrer; Melt treatment; Magnetohydrodynamics; Aluminum

Introduction

High temperature and aggressive properties of aluminum alloys make it complicated to use different mechanical devices to solve technological problems, such as liquid metal transportation, dosing and stirring. Therefore, even more importance and topicality are contactless electromagnetic methods for processing of molten metals. Along with the traditional electric heating devices in furnaces and mixers, electromagnetic devices for liquid metal transportation, pouring and stirring are being widely applied [1-3]. Liquid metal induction pumps already are used in nuclear applications for reactor cooling systems. Classical electromagnetic (EM) pump with current windings are easy to operate as they are made without moving parts and low service needs. Main disadvantages for these pumps are high energy consumption and necessity for large and powerful transformers. Most of the power consumed by the three-phase induction pumps is spent on the creation of a required induction value of the magnetic field by a set of coils. For more possible applications in metallurgical technologies, the permanent magnet pumps were not effective due to the high temperature of melts ($\geq 700\ ^0C$) and due to the need to produce channels with rather thick walls (200-300 mm) from ceramic refractory materials. As a result, the nonmagnetic gap becomes very big and, hence, the magnetic field strength in the liquid metal zone becomes rather low, not providing sufficient efficiency of the pump.

Nowadays, new technologies for the production of permanent magnets have made them cheaper and available in large quantities. Permanent magnets can be used as magnetic field source instead of current windings [4]. The same as permanent magnet pumps can become competitive in metallurgy as they are already used for low melting temperature metals in nuclear industry and research applications. It is demonstrated that the travelling magnetic field in the induction pump could be generated by a system of permanent magnets with alternating polarity fixed on a rotating ferrous base cylinder or disc [5,6]. Moreover, in permanent magnet systems the only power consumption is needed to rotate magnet along the axis.

In this work we present laboratory scale permanent magnet stirrer and its industrial scale prototype which is built and tested on 20t aluminium remelting furnace. Practical experience is obtained in work with previous permanent magnet pump studies. [7-9]

Experimental setup

Experimental laboratory setup is made as a model to simulate aluminium furnace. Stirrer is single permanent magnet cylinder 50mm in diameter and with height 100mm, magnetized orthogonal to its axis. Magnet is positioned on the side of liquid metal reservoir. For these experiments galinstan melt is used, as it is in liquid state in room temperature and the electrical conductivity is high enough to be used. Permanent magnet rotates around its symmetry axis. The tilt of the stirrer is adjustable so it can be set on horizontal, vertical and tilted (45 degree angle) positions, see Fig. 1. For stirring application use of magnet dipole is more effective than multipole permanent magnet assembly as dipole magnetic field intensity is higher, but rotation frequency has to be low anyway to avoid influence from skin

IOP Conf. Series: Materials Science and Engineering **424** (2018) 012037 doi:10.1088/1757-899X/424/1/012037

depth and complicated technical challenges. Even with 5 to 10 Hz skin effect is significant (3-4 cm) which is comparable to size of liquid metal pool. This sets the maximum operating frequencies both in laboratory and industrial scale stirrer since at that point increase in rotational frequency will not give increase in integral flow.

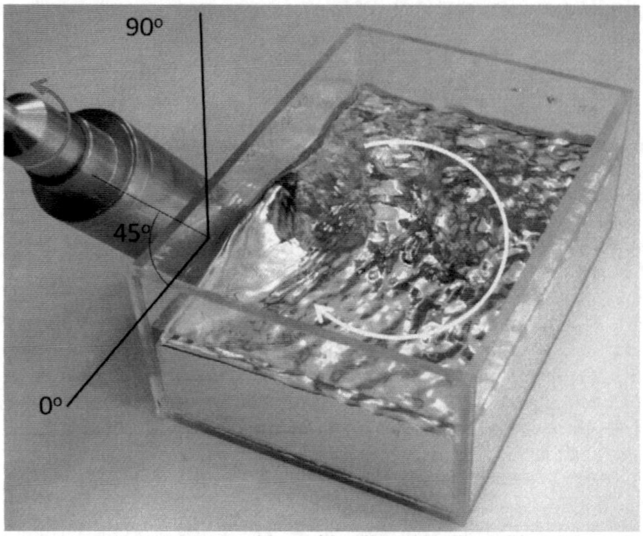

Fig. 1: Laboratory scale setup - permanent magnet stirrer with adjustable tilt position. Magnet creates intense flows and surface deformations in the melt; Yellow arrow shows melt flow direction, red - magnet rotation direction.

Metal flow velocities depending on magnet position and rotation frequency is measured by Ultrasound Doppler Anemometer. Fig. 2 shows comparison of flow characteristic velocities and velocity fluctuation interval close to the reservoir wall with stirrer on a side and on opposite side of the reservoir. Horizontal magnet position showed highest metal surface deformations, but lower characteristic velocities than other configurations, also can be detected with high fluctuation interval. Highest flow intensity in metal volume is reached with magnet tilted in 45 degree angle. In vertical and tilted configuration metal creates 1 vortex instead of 2 or many, as it is in horizontal magnet position. In this setup 1 vortex leads to better steering in whole volume - velocity measurements on opposite side of reservoir are higher.

In these measurements melt depth is 32mm. This value is derived scaling down metal furnace minimum melt depth as simulates the worst case scenario for side stirrer. Even at this level, stirring can be achieved by tilting the magnet. The technology supports the tilt adjustments during operation, which means optimal tilt can be set for varying melt levels during stirring cycle.

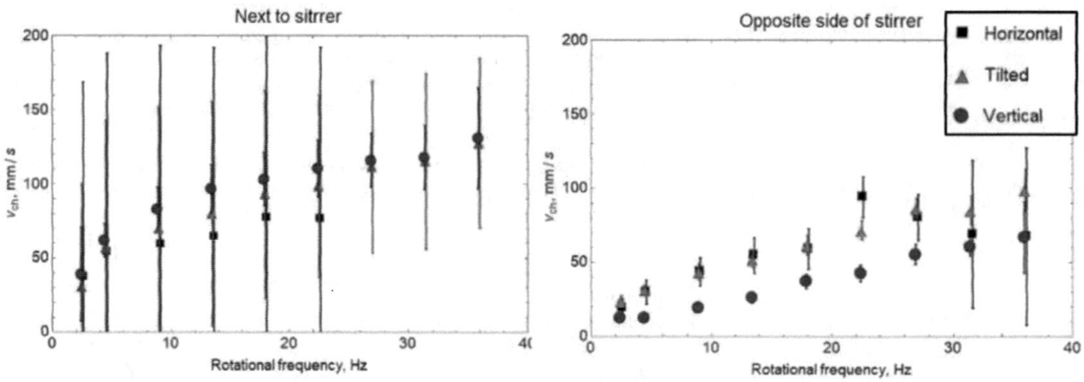

Fig. 2: Liquid metal characteristic velocities and velocity fluctuation interval in various magnet positions: horizontal, vertical and tilted in 45 degree angle. Melt depth is 32mm.

In case of normal melt depth - 52 mm, changing tilt affects the flow type, however, the characteristic velocities are similar. With shallow depth ratio of melt depth and magnet rotor vertical height is too low for effective utilization of magnet flux which affects magnitude of induced current. This reduction of effectiveness is seen Fig. 3, where lower metal depth corresponds to lower velocities on both sides of reservoir. However this effect can be negated by tilting. Our observation and measurements indicate at such low level a 45 degree tilt provide both azimuthal and radial velocity in bath.

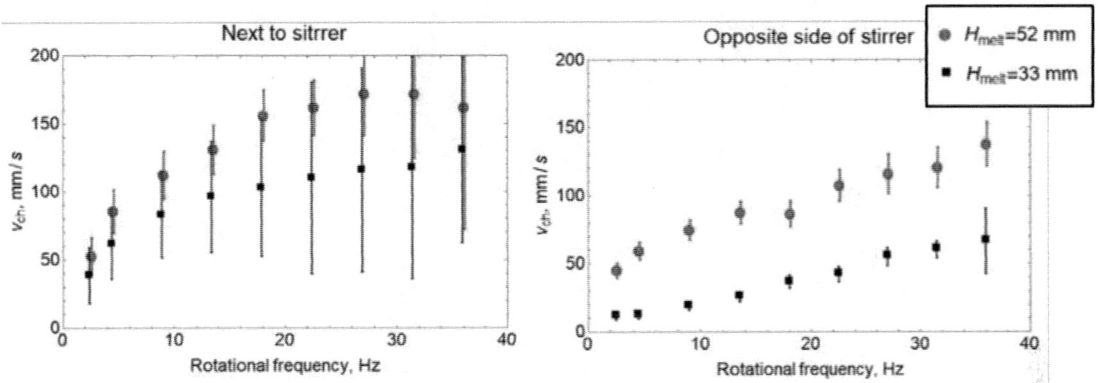

Fig. 3: Liquid metal characteristic velocities on both sides of the reservoir while stirred low and medium depth melt.

Results and discussion

Liquid metal oscillations are detected even with very low magnet rotation frequencies - 1 to 2Hz. Horizontal magnet position for metal stirring seems to be not as effective since metal flows are not intense in whole metal volume. In model experiment there is little increase of stirring intensity when magnet rotation frequency reach around 25 Hz because of skin effect. Scaling that to industrial sizes, gives us optimal frequency of 0.7 Hz. During experiments the most intense stirring is detected in case when magnet is tilted in 45 degree angle, even sharp surface deformations shows up.

For aluminium alloy homogenization purposes industrial scale stirrer is built and tested in industrial environment, see Fig.4. As scaling factors to use symmetry of process conditions we use the same approach as in our previous article [8]. Aluminium furnace for tests is located in Norway, its capacity is 20 tonnes and typical aluminium depth is 80cm. Stirrer is designed to create 22 L/s flow in normal conditions.

This magnet assembly is made from 16 permanent magnet discs, Fig. 4.a. Each disc consists of 12 smaller magnets and is surrounded by stainless steel cover. Steel rods are used to compress the magnet discs in a cylinder. Overall, magnet rotor is 80cm high and 30 cm in diameter. Typical wall thickness for aluminium furnace is 30cm, assembled magnet magnetic field is 40mT on opposite side of the wall. Magnet is built to reach rotation frequency up to 6 Hz but the optimal frequency for the particular furnace geometry is 1-2 Hz.

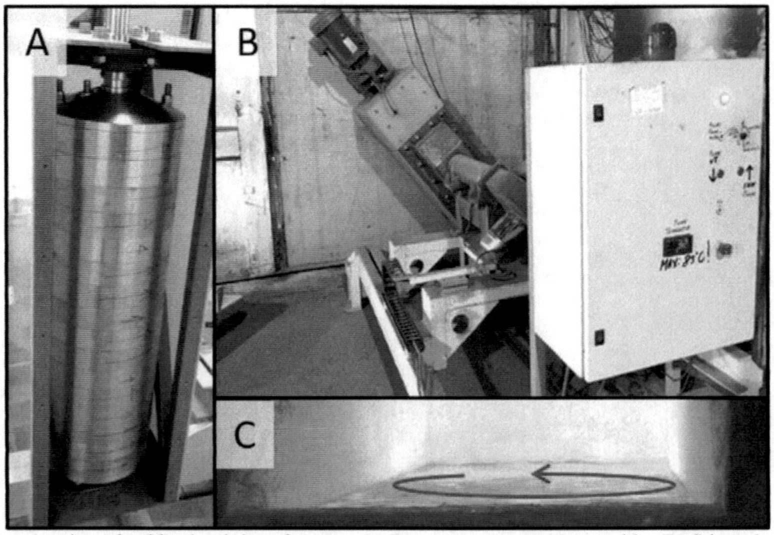

Fig. 4: Industrial scale stirrer for 20t aluminium furnace: A) Permanent magnet assembly; B) Stirrer located next to 20t aluminium furnace wall. Magnet is tilted in 45 degree angle. C) Aluminium furnace is stirred by magnet. Recognizable metal flow on metal surface is detected.

Presented industrial stirrer allows changes of the magnet tilt angle, what results with wide range of flow patterns depending on processes in furnace. Stirrer demonstrated sufficient flow while power unit did not exceed 2kW, what is about 30 times (!) more effective than equivalent EM stirrer with current windings.

Conclusions

This study showed that permanent magnet pumps and stirrers can be used in large scale devices for industrial purposes and have many advantages comparing to other alternative methods - easy operation, low energy consumption, easy to adjust for various low melting temperature alloy furnaces and flow patterns. Model experiments yield high enough velocities throughout whole volume of metal even in shallow depth. The tilt adjustability of magnet allows real-time adaptation during melting cycle for optimal stirring regime. As of now 40 cm rotor is under construction designed for 12t aluminum furnace.

Acknowledgment

This work was funded by European Regional Development Fund under contract "Refinement of metallurgical grade silicon using smart refinement technologies" (No. 1.1.1.1/16/A/097)
ERDF project „Mechanical engineering competence centre", ID Nr.1.2.1.1/16/A/003, "Development of electromagnetic crystallizer for special fine-grained alloy casting"

References

1. L.A. Verte: Magnetohydrodynamics in Metallurgy. Metallurgy Publishing House, 288 p. (1975) (in Russ.).
2. H. Yang: Aluminum Melting Process and Device, ABB: Patent No. WO 2010/094337 Al. (20.02.2009). Applicant: ABB AB.
3. V. N. Timofeev, R. M. Khristinich, S. A. Boyakov, A.A. Temerov: Linear Induction Machine. Patent of Russian Federation No. 2069443. Publication: (20.11.1996)
4. A. Bojarevičs, T. Beinerts: Experiments on liquid metal flow induced by rotating magnetic dipole, Magnetohydrodynamics, Vol.46 (2010), 333-338
5. I. Bucenieks: Magnetohydrodynamics, Vol. 36, Issue 2, (2000), 151-156
6. I. Bucenieks: Magnetohydrodynamics Vol. 39, Issue 4, (2003), 411 – 417
7. A. Bojarevics, T. Beinerts, Y. Gelfgat, I. Kaldre: Permanent magnet centrifugal pump for liquid aluminium stirring, International Journal of Cast Metals Research 29(3), (2016), 154-157
8. T. Beinerts, A. Bojarevičs, I. Bucenieks et al.: Use of Permanent Magnets in Electromagnetic Facilities for the Treatment of Aluminium Alloys, Metall and Materi Trans B, (2016), 47: 1626
9. A. Bojarevičs, R. Baranovskis, I. Kaldre, M. Milgrāvis, T. Beinerts: Two cylinder permanent magnet stirrer for liquid metals, IOP Conf. Ser.: Mater. Sci. Eng. 228 012022 (2017)

IOP Conf. Series: Materials Science and Engineering **424** (2018) 012038 doi:10.1088/1757-899X/424/1/012038

An Experimental and Modeling Study of Al-based Nanocomposites Fabricated by Ultrasonic Cavitation and Solidification Processing

Laurentiu Nastac[1] and Yang Xuan[1]

[1] The University of Alabama (MTE Department, Box 870202, Tuscaloosa, AL, 35487, USA)

Corresponding author: lnastac@eng.ua.edu.

Abstract
In the present investigation, A356 nanocomposites were fabricated via ultrasonic stirring technology (UST) in a coreless induction furnace. SiC nanoparticles were used as the reinforcement. Nanoparticles were added into the molten metal and then dispersed by ultrasonic cavitation and acoustic streaming assisted by electromagnetic stirring. The UST was also applied during the solidification process. The nanocomposite microstructure was investigated by optical microscopy and SEM. The distribution of SiC nanoparticles in the A356 alloy matrix was also analyzed. The SEM and EDS analyses showed that both the matrix microstructure and the SiC particles dispersion into the matrix can be considerably improved when ultrasonic cavitation, induction melting and stirring, and solidification processing techniques are applied together. Molecular dynamics simulations were conducted to analyze the complex interactions between the nanoparticles and the S/L interface.

Keywords:
Al-based nanocomposites; Microstructure refinement; SiC nanoparticle dispersion, Molecular dynamics simulation, Particle pushing and engulfment.

Introduction
Al metal-matrix-composites (MMCs) have been widely used in aerospace, military application, automobile and so on, since their low cost and desirable properties, such as high wear resistance, and high specific strength, *etc.* [1-5]. Because normally the use of micro-size particles could jeopardize the ductility of composites, nano-size ceramic particles has been studied as reinforcement to fabricated metal-matrix-nano-composites (MMNCs). It was shown that MMNCs could significantly improve matrix properties while maintaining good ductility and high temperature creep resistance at the same time [6-8]. Mechanical stirring is widely used to produce MMCs. However, it is not suitable to fabricate MMNCs. Nanoparticles have high specific surface area, which means it is difficult for mechanical stirring to distribute and disperse them uniformly in the melt. In addition, nanoparticles will agglomerate again and float on the liquid surface after stirring due to their poor wettability with the metallic alloy melts [8].
Ultrasonic stirring treatment (UST) is a promising way to fabricate MMNCs, since it will induce nonlinear effects into the melt, such as cavitation and acoustic streaming, which can spread and break the nanoparticle clusters in the melt. Also, extremely high temperature (possibly larger than 5,000K) could help to enhance the wettability between nanoparticles and metal melts, which means the nanoparticles would not gather together or float on liquid surface when the UST processing is stopped [3-5, 9]. It was demonstrated that the application of UST during the melt solidification process refined the matrix microstructure [10]. Previous studies have indicated that the application of UST for A356-Al_2O_3 nanocomposites around liquidus temperature changed the matrix microstructure from dendritic grains to globular grains [3, 5, 11].
In this study, the effect of the UST processing on the microstructure of the A356-SiC nanocomposites has been studied. Additionally, the effect of nanoparticles size on their distribution in the matrix has been discussed. During the solidification process, the particles can be pushed, engulfed or entrapped by the solidification front, among which particle pushing will always lead to particle clustering, which is undesirable as it results in non-homogeneous and lower macroscopic mechanical properties [12]. The particle pushing and engulfment (PEP) is a complicated phenomenon and it is affected by many factors [13] including particle size and shape, interfacial energy between particle, liquid, and solid, as well as a temperature gradient in the melt ahead of the solidification interface.
Molecular dynamics (MD) simulations conducted at the atomic level were performed to study the interface dynamics. In the current study, the open source MD program LAMMPS (Large-scale Atomic/Molecular Massively Parallel Simulator) [14] was applied to conduct MD simulations to analyze the complex interactions between the nanoparticle and the liquid/solid interface.

Experimental Procedure
Aluminum alloy A356 was used as the matrix and SiC nanoparticles (spherical shape, average diameter range is about

45-55 nm) were used as the reinforcement. Jia *et al.* [3] showed that the optimum amount of SiC is about 1.0 wt. %. An Inductotherm induction melting furnace was used to melt the alloy. When the melt temperature reached about 750°C ultrasound cavitation and stirring processing was performed onto the melt. 1.0 wt. % SiC was added into the cavitation area during a 15 min time-frame. During the UST processing, a 40 mm diameter Nb ultrasonic probe was inserted to about 50 mm beneath the melt surface. The sketch of the UST system is shown in Fig. 1. The system parameters are: amplitude is 20 microns, maximum applied power is 1.5kW and frequency is18 kHz. The cast ingots were cylindrical in shape with a diameter of 85 mm and a height of 120 mm. To avoid the effects of different locations (cooling rates) on the microstructure [5], all samples were taken from the middle location of the ingots. Al samples were taken from a section below the probe insertion location. Optical microscopy (OM) (Nikon EPIPHOT 200) photos and scanning electron microscopy (SEM) (JEOL 7000) photos were prepared by grinding with 240, 600, 800, 1200 sand papers followed by polishing with 9 μm, 3 μm, and 1μm water based diamond suspensions.

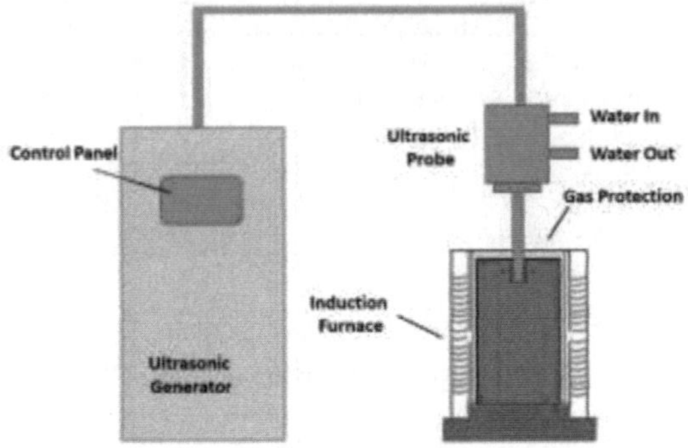

Figure 1. Sketch of the UST system.

Molecular Dynamics Modeling Procedure

The dimensions of the MD geometry in the x, y, and z-direction are 40 nm × 40 nm × 40 nm. The diameter of the SiC nanoparticle is 10 nm. The simulated structure of pure aluminum is a face centered cubic (FCC) structure with a lattice parameter of 4.05 Å and with the <100> directions coincident with the Cartesian coordinates, since planes of looser packing, such as {100}, are better able to accommodate an atom that leaves the liquid to join the solid than a closer packed plane, such as {111}. The nanoparticle consists of a cubic crystalline poly-type of silicon carbide (3C-SiC) with a lattice parameter of 4.36 Å. In total, 3,900,208 atoms are generated. The MD simulation details are provided in [15, 16].

Experimental Results

Figure 2 shows the optical microscopy results in the middle part of the cast ingots [17]. As shown in Figs. 2(a) and 2(b), the A356 samples without UST during solidification have coarser microstructures than the ones with UST. Applying UST during solidification can break up the dendritic structure and thus finer globular grain structure can be obtained. Figure 2(c) shows the microstructure of A356 reinforced with SiC nanoparticles with UST during solidification. They have similar microstructures with those shown in Fig. 2(b), which means adding nanoparticles will not change the microstructure significantly, and the UST is the main cause of grain refinement. Thus, by applying the UST during the solidification (Figs. 2(b) and 2(c)) the matrix microstructure is refined and modified from a dendritic to a globular grain morphology. This can be explained as follows: First, ultrasound will increase the nucleation potential in the melt during solidification. This is because UST will cause the temperature and pressure characteristics of the melt to change periodically at high frequencies. As a result, the local number of nuclei into the melt are increased. Moreover, strong convection produced by UST will promote the diffusion of solute, which will also increase the number of nuclei into the melt. The second reason is because shock waves produced by ultrasonic cavitation will break the dendrite tips during solidification, which will contribute to the microstructure refinement and modification [5].

Figure 3 shows the SEM microstructures at high magnification of the A356 sample reinforced with 1.0 mass % SiC and UST during solidification [17]. Some nanoparticles in Fig. 3(b) were found in the grain boundary area.

Since the particles are in nanometer size, the EDS accuracy is not high enough to identify the particle chemical composition, therefore Fig. 3(c) only shows the EDS mapping results at relatively low magnification. In the mapping photo, the red color represents C (carbon), the green color represents Al (aluminum). The EDS results indicate that some of C is located in the grain boundary area. By combining the EDS results with the SEM photo, it can be determined that some of the SiC nanoparticles are gathered around the grain boundary area.

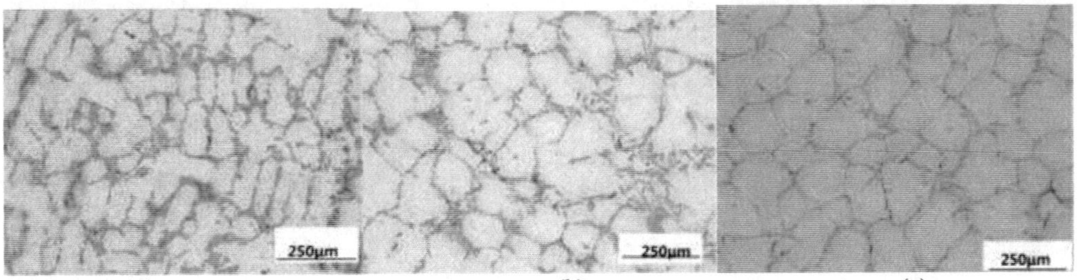

(a) (b) (c)

Figure 2. Optical micrographs in the middle part of the cast samples: (a) A356 without UST;(b) A356 with UST during solidification and (c) A356 sample reinforced with 1.0 mass% SiC and UST during solidification.

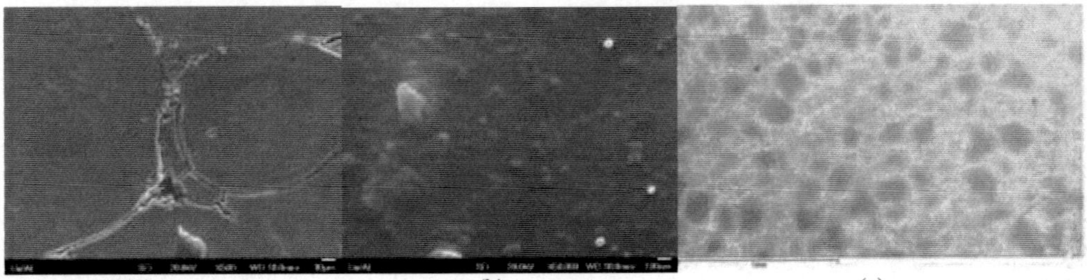

(a) (b) (c)

Fig. 3. SEM results and EDS mapping photo of the sample processed via UST+1.0 mass% SiC. (a) SEM picture at X500 and (b) SEM picture of the red rectangular area in (a) at X50000 and (c) EDS mapping results at X40.

Simulation Results

The two-phase Al system is established by holding the atoms in solid Al region of the original crystalline lattice and melting the liquid Al region by raising the temperature above the melting point. The whole system is equilibrated after 40ps, then the initial solid Al atoms are fixed at some temperature (500 K) below the melting point, and Si, C, and liquid Al atoms which will release heat through the solid atoms. Figure 4 shows the solid/liquid status of the system with a 10 nm SiC nanoparticle after 100 ps, 200 ps, 400 ps, and 600 ps, respectively [15. 16]. As it can be seen from Fig. 4(a), when the liquid/solid interfaces are far away from the SiC nanoparticle, both of the interfaces at the top and bottom are flat. When the interface at the bottom approaches the SiC nanoparticle, a trough is formed below the particle (see Fig. 4(b)). As time goes on, the interface at the bottom passes through the particle and the interface at the top approaches the particle (See Fig. 4(c)). Finally, these two interfaces meet each other, become one, and disappear (see Fig. 4(d)).

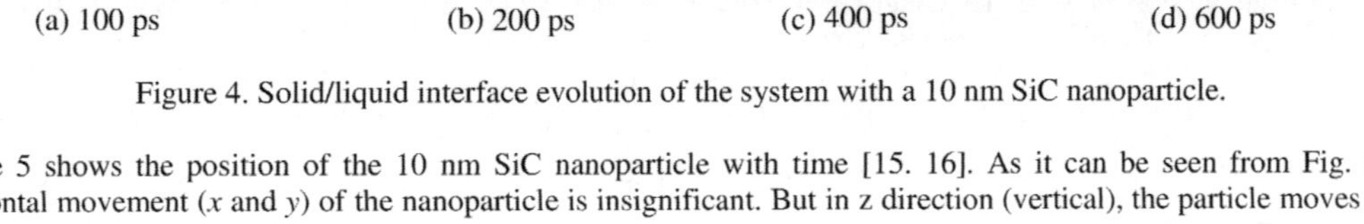

(a) 100 ps (b) 200 ps (c) 400 ps (d) 600 ps

Figure 4. Solid/liquid interface evolution of the system with a 10 nm SiC nanoparticle.

Figure 5 shows the position of the 10 nm SiC nanoparticle with time [15. 16]. As it can be seen from Fig. 5, the horizontal movement (x and y) of the nanoparticle is insignificant. But in z direction (vertical), the particle moves down towards the liquid/solid interface at the bottom in the beginning, and when the interface from the top approaches it, the particle moves up towards that interface again. This MD simulation confirmed that during the solidification process, the SiC nanoparticle will be engulfed by the solidification front instead of being pushed, as predicted by previous PEP models [18]. Similar conclusion was found for the solidification domain with a 20 nm SiC nanoparticle [15, 16].

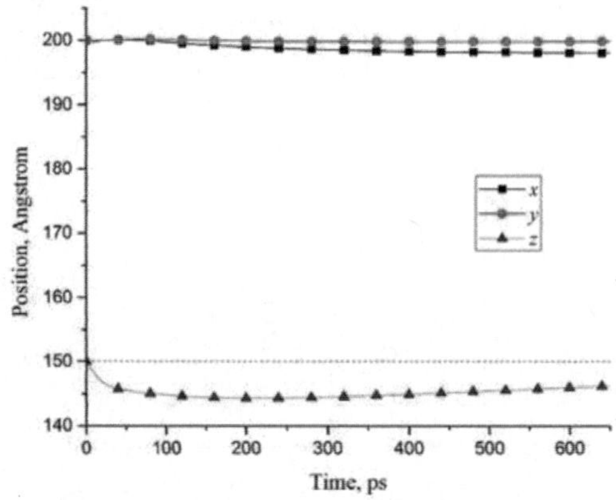

Figure 5. Position evolution of a 10 nm SiC nanoparticle during solidification.

Conclusions

It was determined from this study that the application of UST during solidification plays a key role in the refinement and modification of the microstructure of the A356 cast ingots. Without UST processing, dendritic structures were obtained. By using UST during solidification, fine globular grain structures were obtained. The addition of SiC nanoparticles affects the formation of the microstructure of the A356 cast ingots only marginally.

The SEM and TEM results indicated that the added SiC nanoparticles are dispersed reasonably well into the A356 matrix. However, some SiC nanoparticles are gather around the grain boundary area.

The complex interactions between a SiC nanoparticle and the Al liquid/solid interface were analyzed with a 3D molecular dynamics model using LAMMPS on the High Performance Computing Cluster (HPCC) at the University of Alabama. Molecular dynamics simulation results showed that SiC nanoparticles with diameters less than 20 nm will be engulfed by the solidification front instead of being pushed, which is beneficial for nanodispersion.

References

1. X.D. Liu, S. Jia, L. Nastac, Int J Metalcasting (AFS), 8(1) (2013), 51-58.
2. S. Jia, D. Zhang, L. Nastac. J Mater Eng Perform., 24(6) (2015), 2225-2233.
3. S. Jia, D. Zhang, Y. Xuan, L. Nastac, Appl Acoust., 103 (2016), 226-231.
4. S. Jia, Y. Xuan, L. Nastac, P.G. Allison, T.W. Rushing, Int J Cast Metal Res., 29(5) (2016), 286-289.
5. Y. Xuan, S. Jia, L. Nastac, High Temperature Materials and Processing, 36(4) (2017), 381-388.
6. Akbari M.K., Mirzaee O., Baharvandi H.R. Mater Design. 46 (2013), 199-205.
7. M.K. Akbari, H.R. Baharvandi, O. Mirzaee, Compos Part B-Eng., 55 (2013), 426-432.
8. Y. Yang, J. Lan, X.C. Li, Mat Sci Eng a-Struct., 380(1-2) (2004), 378-383.
9. J.C. Yan, Z.W. Xu, L. Shi, X. Ma, S.Q. Yang, Mater Design. 32(1) (2011), 343-347.
10. J.I. Youn, Y.J. Kim, Jpn J Appl Phys. 48 (2009), 07GM14.
11. Y. Xuan and L Nastac, Ultrasonics, 83 (2018), 94-102.
12. S. Naher, D. Brabazon, L. Looney, Composites Part A: Applied Science and Manufacturing 38 (2007), 719-729.
13. D.M. Stefanescu, B. Dhindaw, S. Kacar, A. Moitra., Metallurgical Transactions A 19 (1988), 2847-2855.
14. S. Plimpton., J. of computational physics 117 (1995), 1-19.
15. D. Zhang and L. Nastac, *JOM*, 68(12) (2016), 3134-3142.
16. D. Zhang, A Ph.D. Dissertation, The University of Alabama, Tuscaloosa, Al, USA (2016).
17. Y. Xuan, A Ph.D. Dissertation, The University of Alabama, Tuscaloosa, Al, USA (2018).
18. D.M. Stefanescu, Science and Engineering of Casting Solidification, Second Edition, Springer (2009).

Effect of time–modulated magnetic fields on the solidification structure and extrusion properties of wrought aluminum alloys

D. Räbiger[1*], C. Resewski[1,2], S. Müller[2], B. Willers[1], W. Reimers[2] and S. Eckert[1]

[1] Helmholtz-Zentrum Dresden-Rossendorf, 01314 Dresden, Germany
[2] Technical University Berlin, Institute of Materials Science and Technologies, 10587 Berlin, Germany

Corresponding author: d.raebiger@hzdr.de

Abstract
Wrought aluminum alloys and in particular extruded products have a large economic importance worldwide. The opportunities to increase the quality or to reduce the production cost have great a strong finance and ecologic incentive. AC magnetic fields provide the possibility to influence the melt flow in a contactless way to influence the temperature and solute distribution in front of the solidification front. Many studies have shown the beneficial effects like a distinct grain refinement or the promotion of a transition from a columnar to an equiaxed dendritic growth (CET). The use of time-modulated magnetic fields are able to avoid segregation effects, just like known from privies studies they work with permanent rotating magnetic fields, for example.
This present study examines the directional solidification of commercial wrought aluminum alloys from a water-cooled copper chill. Rotating time-modulated magnetic fields were used to agitate the melt. Our results demonstrate the potential of time–modulated magnetic fields to control the grain size, the formation of intermetallic phases and extrusion properties.

Key words: wrought aluminum alloy, time modulated magnetic field, bar extrusion

Introduction

The adjustment of monotonic structure and finely dispersed intermetallic phases has been approved to be a crucial issue for bar extrusion to get a homogeneous end product and to reduce the wear of the tool. Small grains are advantage to get this properties in the continuously cast ingot. Standard in industrial applications is the addition on of grain refiners like TiB [1]. Further possibilities to reduce the grain size are rapid cooling conditions [2], mechanical or electromagnetic stirring [3, 4] or ultrasonic treatment [5].
Grain refinement by inoculants is a very effective and well-known method which is widely used in foundry industry. It can be realized easily by addition of Al-Ti or Al-Ti-B master alloys and provides beneficial effects, for instance better feeding, reducing of hot tearing and improvement of strength. Grain refinement of aluminum alloys by inoculants is well reported [6, 7]. AC magnetic fields provide a contactless method to control the flow inside a liquid metal and the grain size of the solidified ingot without any inoculants and additional alloying elements [8, 9]. Many studies have shown that beneficial effects like a distinct grain refinement or the promotion of a transition from a columnar to an equiaxed dendritic growth (CET) can be obtained. However, electromagnetically-driven melt convection may also produce segregation freckles on the macroscale especially in solute-rich alloys showing different equilibrium concentrations of solute in the mixed crystal and the surrounding melt [10, 11]. The achievement of superior casting structures needs a well-aimed control of melt convection during solidification, which in turn requires a detailed knowledge of the flow structures and a profound understanding of the complex interaction between melt flow, temperature and concentration field. Time-modulated rotating magnetic fields have been proposed recently to control the heat and mass transfer at the solidification front and to avoid segregation effects [12-14]. The general positive influence on wrought aluminum alloys is shown in [15].
The main goal of this work is to demonstrate the impact of forced convection on the grain size, microstructure and the extrusion properties of a wrought aluminum alloy. Therefore we compare an unaffected, a TiB grain refined and a electromagnetic stirred alloy. Furthermore, we show the impact of time-modulated magnetic rotating magnetic field on the fluid flow condition.

Experimental Set-up

Fig. 1(a) shows a schematic view of the experimental setup. A cylindrical vessel made of Perspex with an aspect ratio A $= H_0/2\cdot R_0 = 1$ was used. The size of both the inner diameter $D = 2\cdot R_0$ and the height H_0 was chosen to be 67.5 mm. The cylinder was filled with the eutectic alloy GaInSn and closed by a rigid lid. The ultrasound Doppler velocimetry (UDV) was applied to measure the fluid flow inside the cylinder. More details about the measuring principle can be found in [16]. To measure the primary azimuthal flow we employed the DOP2000 instrument (model 2125, Signal Processing SA, Lausanne) equipped with a 4 MHz transducer (TR0405LS, acoustic active diameter 5 mm). To achieve the acoustic

coupling between sensor and fluid we attached the ultrasonic transducer to the wall of the fluid container at two vertical positions. Moreover, a new measuring system of linear sensor arrays was used to realize a flow mapping of the meridional flow. Details of the measuring system are given in [17]. The spatial resolution in lateral direction varied from 5 mm at the sensor to approximately 7.5 mm at the lid of the fluid vessel. In axial direction a spatial resolution of about 1.5 mm was achieved. The velocity data were acquired with sampling frequencies between 0.5 and 6 Hz.
Solidification experiments were performed using two wrought aluminium alloys EN AW 6082 prepared by TRIMET Aluminium SE. The chemical composition is shown in Tab. 1. A third of the samples were treated by TiB using the master alloy AlTi3B1. We give only a brief description here concerning the solidification experiment.
Further details can be found in [14]. The EN AW 6082 samples were solidified directionally from the bottom in a double-walled, cylindrical stainless steel mold. The mold has an internal diameter of 50 mm and a height of 100 mm. The schematic drawing of the experimental set-up can be found in Fig. 1 (b). Preparatory specimens were made with a predetermined weight of 445 g corresponding to a height of 80 mm. After melting the mold was positioned at a water-cooled copper chill which was kept at a constant temperature of about 20°C. In the experiments presented here the cooling of the sample and the influence of the electromagnetic field starts nearly at the same time. Three different conditions were used, solidification without convection or grain refiner, solidification with grain refiner and without forced convection and solidification with forced convection and without grain refiner. From this ingots cylinders with diameters of 28.5 mm was eroded for bar extrusion. A half of these samples were heat-treated. For bar extrusion experiments, both samples with cast structure as well as heat-treated samples were used. After the extrusion the heat-treated samples were aged.

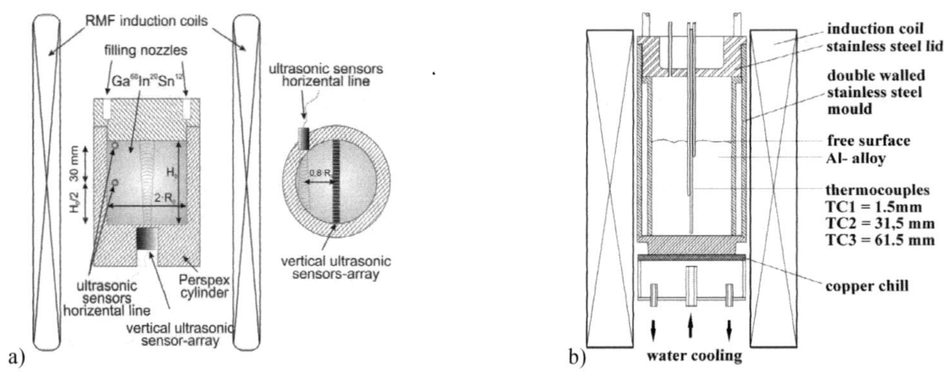

Fig. 1: .Schematic drawing of the experimental setup: (a) experimental setup for the fluid flow measurements and (b) experimental setup of the solidification experiments.

Tab. 1. Chemical analysis of the used alloys (inspection certificate: EN 10204 – 3.1; Method: emission spectrometer OBLF).

Alloy	Si	Cu	Mg	Mn	Zn	Fe	Cr	Ti	Al
EN AW 6082	1.01	0.06	0.92	0.58	0.01	0.28	0.13	0.004	Bal.

Results

In the following description a rotating magnetic field with permanent changes in rotating direction (it means clockwise and counter clockwise) is meant if time-modulated magnetic field is written. The rotating frequency of 50 Hz is not modified. Therefore, the overall energy input in the system is not change compared to a continuously rotating magnetic field.
Fig. 2 presents the distribution of the vertical velocity u_z across the vertical mid-plane for a continuously applied RMF (Fig 2(a) and a time-modulated RMF (Fig 2(b). The amplitude of the vertical velocity is represented by the grey scale value. It becomes apparent that the main secondary flow structure occurs in the form of two toroidal vortices, which is represented by the upward and downward flow components in the plot. For a permanent applied RMF the mean velocity structure is stable in time and this provoke segregation effects in solidified ingots [9]. To overcome this problem we adopted the concept of a periodic change of the rotating direction of the field [12]. We have succeeded to find an optimum change frequency of rotating direction where the amplitude of flow velocity is higher as in the permanent RMF case, as shown in Fig. 2 (b). Additional a permanent change of flow direction can be observed as described in [13]. For this condition high flow velocities during the period can be observes whereas the averaged velocity is strongly reduced. Along this path a grain refining effect of the fluid flow and a reduction of segregation effect can be reached.

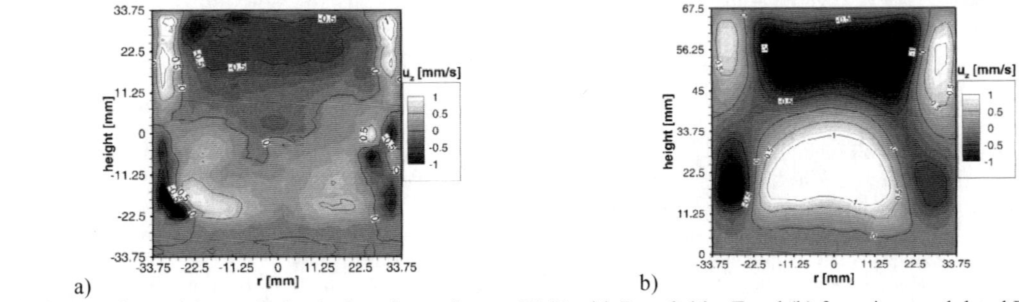

Fig. 2: Snapshots of the vertical velocity of a continuous RMF with $B_0 = 0.46$ mT and (b) for a time modulated RMF with $f_R = 0{,}015$ Hz and $B_0 = 0.46$ mT at the time with the strongest meridional flow.

From the literature it is well known that a forced convection influenced the temperature distribution in the column and the concentration field in front of the solidification front. The forced flow reduces the temperature gradient in the liquid volume and can increase the gradient directly at the solidification front. Furthermore, melt convection in a turbulent regime tends to fluctuations in the temperature and concentration fields both in front of as well inside the mushy zone. This leads to a grain refinement induced by remelting of dendrite arms and the transport of fragments in zones ahead of the solidification front were they can work as new nuclei.

Fig. 3: Macrostructure of cast ingots: (a) without forced convection and without grain refiner, (b) with forced convection and without grain refiner and (c) without forced convection and with grain refiner

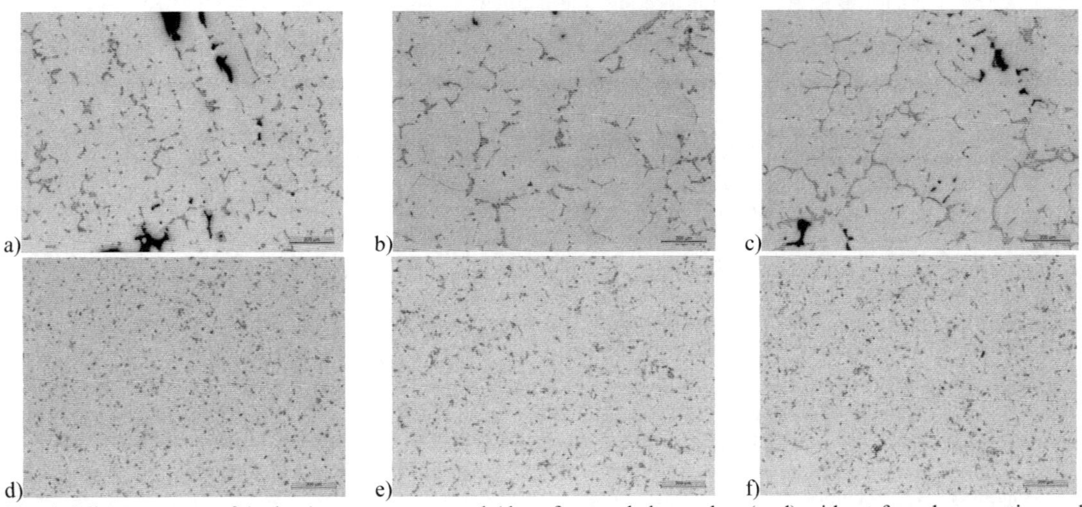

Fig. 4: Microstructure of (a, b, c) cast structure and (d, e, f) extruded samples: (a, d) without forced convection and without grain refiner, (b, e) with forced convection and without grain refiner and (c, f) without forced convection and with grain refiner.

Fig. 3 shows the macrostructure for the three different solidification conditions. In Fig 3 (a) the macrostructure of the alloy without any grain refining method is shown. The grain refining effect of the fluid flow (Fig. 3 (b)) is obviously. The stirred samples show small equiaxed grains as well as the sample with the grain refiner in Fig. 3 (c).

For the mechanical characteristics the size and distribution of the intermetallic phases are just as important as the grain size. In Fig. 4 (a) (b) and (c) the microstructure of the cast ingot is shown. In a qualitative way no significant different of the phase distribution between the solidification conditions is visible. In the case without grain refinement methods (Fig. 4 (a)) the intermetallic phases looks more orientated and the precipitations are alveolar. This is in consequence of the more dendritic growth of the unrefined sample in Fig. 4 (a) and the more globular or rosette-like growth sample in Fig. 4 (b) and (c).
After the extrusion the size of intermetallic phases are reduced and evenly spread across the sample. It becomes obvious in Fig. 4 (a), (b) and (c) that the different solidification conditions have no influence on the distribution of the intermetallic phase after the extrusion. The examples shown here are not heat-treated. But also in the heat-treated case no differences in the phase distribution are visible.
Remarkably enough, there are in principle no differences between the microstructure of the chemical grain refined sample without forced convection, the sample with forced convection and the sample without any grain refiner and without forced convection. The microstructure is more defined by the cooling rate [18]. The impact of stirring on the microstructure is weak because of the small solidification temperature range of wrought aluminum alloys.

Conclusion

The present paper examines the impact of electromagnetic stirring during the solidification of a wrought aluminum alloy on the grain size and microstructure. The results would be compared with an unrefined alloy and an alloy with TiB grain refinement. The application of a forced convection provides a distinct grain refined macrostructure comparable to the grain refining by TiB. The forced convection reveals no significant and direct influence on the nature of intermetallic phases. The phase size and distribution are comparable for the different examined variants. Negative effects like macro segregation are not detectable during the solidification of wrought aluminum alloys.
The results demonstrate the potential of magnetic fields as a powerful tool with prospects for tailoring the microstructure during the continuous casting of wrought aluminum alloys. It looks possible to substitute large portion of TiB grain refiner in the production of extruded aluminum products.

Acknowledgment

The research is supported by the German Helmholtz Association in form of the Helmholtz-Alliance "LIMTECH". Furthermore we thank the TRIMET Aluminium SE for supporting the alloy and fruitful discussions.

References

1 D. G. McCartney, International Materials Reviews, 34 (1989), 247-260.
2 R.K. Trivedi, Materials Science and Engineering A, 178 (1994), 129-135.
3 M.C. Flemings. Metallurgical Transactions A, 22 (1991) 957-981.
4 C. Vives and C. Perry, International Journal of Heat and Mass Transfer, 29 (1986) No. 1, 21-33.
5 G.J. Eskin, Ultrasound treatment of Light Alloy Melt, Gordon and Breach, Amsterdam (1998)
6 T.E. Quested, Materials Science and Technology, 20 (2004) 1357-1369.
7 B.S. Murty, S.A. Kori and M. Chakraborty, International Materials Reviews, 47 (2002) No. 1, 3-29.
8 J.K. Roplekar and J.A. Danzig, International Journal of Cast Metals Research, 14 (2001) No. 2, 79-95.
9 W.D. Griffiths and D.G. McCartney, Materials Science and Engineering A, 216 (1996), 47-60.
10 M. Medina, Y. Du Terrail, F. Durand and Y. Fautrelle, Metallurgical and Materials Transactions B, 35 (2004), 743 - 755.
11 P.A. Nikrityuk, K. Eckert and R. Grundmann, International Journal of Heat and Mass Transfer, 49 (2006), 1501-1515.
12 S. Kojima, T. Ohnishi, T. Mori, K. Shiwaku, I. Wakusagi and M. Ohgarni, Proc. of the 66th Steelmaking Conf., Atlanta, ISS-AIME, Warrendale (1983), 127-131.
13 S. Eckert, P. A. Nikrityuk, D. Räbiger, K. Eckert and G. Gerbeth, Metallurgical and Materials Transactions B, 38 (2007) 977-988.
14 B. Willers, S. Eckert, P.A. Nikrityuk, D. Räbiger, J. Dong, K. Eckert and G. Gerbeth, Metallurgical and Materials Transactions B, 39 (2008), 304-316.
15 D. Räbiger, B. Willers, S. Eckert, M. Rosefort, T. Dang, H. Koch, Proceedings of the XVIII International UIE-Congress (2017), Hannover
16 Y. Takeda, Nuclear Engineering and Design, 126 (1991), 277-284.
17 S. Franke, H. Lieske, A. Fischer, L. Büttner, J. Czarske, D. Räbiger, S. Eckert, Ultrasonics, 53 (2013) No. 3, 691-700.
18 G. Mrówka-Nowotnik, J. Sieniawski, Journal of Materials Processing Technology, 162–163 (2005) 367-372.

Inclusion Removal from Molten Steel Using Electromagnetic Vibrating Force

MARUYAMA Asuka[1] and IWAI Kazuhiko[1]

[1] Hokkaido University, Kita 13, Nishi8, Kita-ku, Sapporo, Japan

Corresponding author : as-maru@es.hokudai.ac.jp

Abstract

An electromagnetic vibrating force is a candidate of a micro-size inclusion removal method [1-5]. Especially, an A.C. magnetic field imposition has an advantage that it enables the inclusion removal and an induction heating of a molten steel in a tundish simultaneously [3-5]. However, appropriate system design and operating conditions for the simultaneous operation of the inclusion removal and the induction heating of the molten steel by the A.C. magnetic field imposition have not been clarified. Therefore, an inclusion removal time from the molten steel under the A.C. magnetic field imposition has been calculated in this study. And then, a minimum channel length for the inclusion removal and the molten steel temperature increase per unit channel length by the induction heating has been evaluated. The A.C. magnetic field frequency should be controlled within a certain frequency range in which the inclusion removal time becomes minimal. Furthermore, the channel length should be longer than the minimum channel length for the inclusion removal calculated by using the minimum inclusion removal time and an assumed maximum molten steel velocity in the channel. The molten steel temperature increase per unit channel length by the induction heating can be controlled by controlling the A.C. magnetic field frequency within the frequency range suitable for the inclusion removal. Especially, a high frequency condition in that frequency range is better for the efficient thermal supply in this calculation condition.

Key words : A.C. magnetic field; induction heating; inclusion removal

Introduction

Demand for high quality steel has been extended in various industries for yield improvement, fabrication of long life time products and so on. Non-metallic inclusions such as alumina particles in the steel causes some troubles in some processes such as rolling and wire drawing, and decrease in fatigue strength of the final products. Therefore, fabrication of high-purity steel without the inclusions is one of important issue in the steel industry today. However, the inclusions formation resulting from control of the molten steel chemical component is inevitable. In general, the inclusions are removed from the molten steel using buoyancy force acting on them in the tundish and a continuous casting machine, and their rising velocity in the molten steel is proportional to the square of their diameter. Thus, removal of micro-size inclusions requires long operating time and it should be improved for high productivity of the high-purity steel. A vibrating electromagnetic force is a candidate of the micro-size inclusions removal method [1-5]. For example, cleanliness factor of the molten steel is improved by the A.C. magnetic field imposition using an induction heating system of the tundish [3]. By imposing the A.C. magnetic field on the molten steel flowing in a channel of the tundish for induction heating, the inclusions move towards the channel inner wall with vibrating motion by the electromagnetic force acting on them and are trapped on the channel inner wall [4]. This method has an advantage that the inclusion removal can be achieved simultaneously with preventing the temperature drop of the molten steel. The A.C. magnetic field with a commercial frequency, for example 50 Hz or 60 Hz in Japan, is generally used in the induction heating system. On the other hands, the authors carried out a numerical calculation of an inclusion motion in the molten steel under the A.C. magnetic field imposition and clarified presence of an optimum magnetic field frequency for the speedy inclusion removal [5]. However, for the simultaneous operation of the inclusion removal and the induction heating of the molten steel, the optimum channel design and operating conditions might be different. From the viewpoint of the inclusion removal, the channel length should be longer than a certain minimum channel length in which the inclusions should be attached to the channel inner wall before it reaches the channel exit. On the other hand, certain level of the heat supply is required from the viewpoint of the induction heating. Both the molten steel velocity in the channel and the A.C. magnetic field frequency are important parameters when the A.C. magnetic field magnitude is fixed.

In order to clarify guidelines of the channel design and the operating conditions, numerical calculation of the inclusion removal time from the molten steel under the A.C. magnetic field imposition based on electromagnetic field theory and fluid dynamics has been carried out at first. Then, the minimum channel length for the inclusion removal and the molten steel temperature increase per unit channel length by the induction heating has been evaluated.

Calculations

Evaluation of the inclusion removal time from the molten steel under the A.C. magnetic field imposition was carried out, in which a couple of parallel horizontal A.C. magnetic fields with frequency of f and half amplitude of B_0 are simultaneously imposed on a horizontally infinite molten steel with vertical thickness of $2w$ from its upper and bottom surfaces. The analytical system is same with the previous report [5]. Because x-axis is in horizontal direction and positive direction of y-axis is vertical upwind direction with the origin of on the center line of the molten steel thickness, the electromagnetic force acting on a spherical insulating inclusion in this molten steel has only y component, and its magnitude per unit volume of the inclusion F_m is given as follows [5];

$$F_m = F_p + F_v \tag{1}$$

$$F_p = C_1(\sinh 2\zeta - \sin 2\zeta) \tag{2}$$

$$F_v = C_1[C_2 \cos(4\pi f t + \alpha)] \tag{3}$$

$$C_1 = 3\left(B_0^2\sqrt{R_\omega}\right)/\left[8\mu w\left(\left(\cosh\sqrt{2R_\omega} + \cos\sqrt{2R_\omega}\right)\right)\right] \tag{4}$$

$$C_2 = \sqrt{(\cosh 4\zeta - \cos 4\zeta)\left[\left(\cosh\sqrt{2R_\omega} + \cos\sqrt{2R_\omega}\right)^2 + sinh^2 2\zeta sin^2 2\zeta\right]} \\ /\left[\sqrt{2}\left(\cosh\sqrt{2R_\omega} + \cos\sqrt{2R_\omega}\right)\right] \tag{5}$$

$$\alpha = \frac{\pi}{4} - arccos\left[\frac{4\sinh\zeta\cosh\sqrt{R_\omega/2}\cos\zeta\cos\sqrt{R_\omega/2} + \cosh\zeta\sinh\sqrt{R_\omega/2}\sin\zeta\sin\sqrt{R_\omega/2}}{2\sqrt{(\cosh 2\zeta - \cos 2\zeta)\left(\cosh\sqrt{2R_\omega} + \cos\sqrt{2R_\omega}\right)}}\right] \tag{6}$$

$$\zeta = (y - w)\sqrt{R_\omega/2} \tag{7}$$

$$R_\omega = 2\pi f \sigma \mu w^2 \tag{8}$$

where t, σ, and μ are time, electrical conductivity of the molten steel (7.2×10^5 Sm^{-1}), and magnetic permeability of the molten steel ($4\pi\times10^{-7}$ Hm^{-1}), respectively. A shielding parameter R_ω defined by Eq. (8) is a ratio of the penetration depth of the A.C. magnetic field to the half thickness of the molten steel w. The Eqs. (1) – (3) indicate that the electromagnetic force acting on the inclusion consists of a pinch force F_p in the direction from the molten steel center line to the molten steel surface and a vibrating force F_v with frequency of $2f$. An alumina inclusion with diameter of D and density of ρ_s (3880 kgm^{-3}) was set at $y = 0$ as the initial condition. The driving forces of the inclusion are a buoyancy and the pinch forces toward the upper direction, and the vibrating force as illustrated in Fig.1. Using inclusion vertical velocity u, density of the molten steel ρ_f (6958 kgm^{-3}), and viscosity of the molten steel η (5.28×10^{-3} Pa·s), an equation of the inclusion motion in the vertical direction can be written as follows [5];

$$\frac{du}{dt} = \frac{1}{(\rho_s + \rho_f/2)}\left[-\frac{18\eta}{D^2}u - \frac{9}{D}\sqrt{\frac{\rho_f\eta}{\pi}}\int_0^t \frac{du/ds}{\sqrt{t-s}}ds + g(\rho_f - \rho_s) + F_p + F_v\right] \tag{9}$$

where, first, second, and third terms in the right-hand side of the Eq. (9) are a drag force, a Basset force, and the buoyancy force, respectively. The inclusion behavior from the initial position ($y = 0$) to the upper surface of the molten steel ($y = w$) was computed by applying the fourth-order Runge-Kutta method for the inclusion with diameter D of 100 μm in the molten steel with half thickness w of 0.075 m under the imposition of the magnetic field with half amplitude B_0 of 0.2 T. And then, the inclusion removal time from the molten steel t_r was evaluated as total moving time of the inclusion from the initial inclusion position to the upper surface of the molten steel.

For evaluation of the minimum channel length for the inclusion removal and corresponding molten steel temperature increase per unit channel length by the induction heating, an induction heating channel model in which the A.C. magnetic field with frequency f and half amplitude B_0 was imposed on a circler cross section channel of radius w and the molten steel flows in the axial direction with average velocity of u_f was adopted. The removal time of an inclusion t_r is obtained using the numerical calculation. Therefore, the minimum channel length for the inclusion removal L_{min} was evaluated by $L_{min} = u_f t_r$. The heating rate per unit channel length Q can be approximated as follows when the shielding parameter R_ω given by the Eq. (8) is much larger than unity (when the A.C. magnetic field frequency f is much higher than 31Hz in this case) [6].

$$Q \approx B_0^2\pi w\sqrt{\pi f/\mu^3\sigma} \tag{10}$$

Thermal balance equation of the molten steel in the channel can be written as follows;

$$Q = \rho_f C_p \Delta T \pi w^2 u_f \tag{11}$$

where, C_p is heat capacity of the molten steel (795 Jkg^{-1}K^{-1}) and ΔT is the molten steel temperature increase per unit channel length by the induction heating. From the Eqs. (10) and (11), the molten steel temperature increase per ui channel length ΔT is given as follows.

$$\Delta T = B_0^2\sqrt{\pi f/\mu^3\sigma}/\rho_f C_p u_f w \tag{12}$$

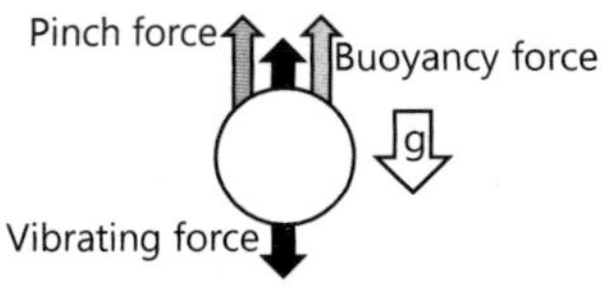

Fig. 1: Driving forces acting on the inclusion

Results and discussions

Table 1 shows the numerically calculated inclusion removal time t_r. The inclusion removal time t_r becomes minimal wh the A.C. magnetic field frequency falls in the range of 156 - 500 Hz. The reason for the presence of the minimal value t_r was discussed in the previous report [5] as follows. The rising velocity of the inclusion in the molten steel can effectively promoted when the pinch force is larger than the buoyancy force. However, the pinch force becomes small than the buoyancy force in most of the region in the molten steel when the A.C. magnetic field frequency is excessive low or high, because the pinch force magnitude becomes small under the excessively low frequency condition and t penetration depth of the A.C. magnetic field becomes thin under the excessively high frequency condition. The regi where the pinch force exceeds the buoyancy force becomes the largest when the A.C. magnetic field frequency falls the range of 156 – 500 Hz in this calculation condition, and therefore the inclusion removal time t_r becomes minimal. F the speedy inclusion removal and minimization of the channel length, the A.C. magnetic field frequency should controlled in this frequency range in which the inclusion removal time t_r becomes minimal. Because the inclusion remov time t_r is about 11.5 s with error of 4.3 % or less when the A.C. magnetic field frequency falls in the range of 156 – 500H the minimum channel length for the inclusion removal L_{min} when the inclusion removal time t_r is 11.5 s will be describ below as a function of the molten steel velocity in the channel. And then, the molten steel temperature increase per ui channel length by the induction heating ΔT of 156 – 500 Hz in the A.C. magnetic field frequency f will be also discusse

Table 1: Inclusion removal time for various A.C. magnetic field frequency

A.C. magnetic field frequency f [Hz]	5	50	156	313	500	5000
Inclusion removal time t_r [s]	23.3	14.9	11.6	11.5	12.0	18.1

Figure 2 shows an effect of the molten steel velocity in the channel u_f on the minimum channel length for the inclusi removal L_{min} when the inclusion removal time t_r equals to 11.5 s. With increase in the molten steel velocity in the chanr u_f, the minimum channel length for the inclusion removal L_{min} increases. In general, the molten steel velocity in t channel u_f is a function of the withdrawal speed in the continuous casting process for keeping certain casting temperatu without consideration of the minimum channel length for the inclusion removal L_{min}. Thus, the inclusion removal effe of the A.C. magnetic field is not enough, when the minimum channel length for the inclusion removal L_{min} is longer th the length for the molten steel heating. From the viewpoint of the inclusion removal, the channel length should be long than the minimum channel length for the inclusion removal L_{min} which is derived from the minimum inclusion remov time t_r and the maximum withdrawal speed in the continuous casting process. For example, the channel length should longer than 1 m or more when the molten steel velocity u_f exceeds about 0.1 ms^{-1} (about 0.7 ton·min^{-1} in flow rate in t channel) in this calculation condition.

Figure 3 shows the effects of the molten steel velocity u_f and the A.C. magnetic field frequency f with the range of 156 500 Hz on the molten steel temperature increase per unit channel length by the induction heating ΔT. The molten ste temperature increase per unit channel length ΔT decreases with increase in the molten steel velocity in the channel u_f. (the other hands, the molten steel temperature increase per unit channel length ΔT increases with increase in the A. magnetic field frequency f. Therefore, the molten steel temperature increase per unit channel length ΔT can be controll by controlling the A.C. magnetic field frequency f. For example, when the molten steel velocity in the channel u_f is (ms^{-1}, the molten steel temperature increase per unit channel length ΔT can be controlled from about 20 K (at 156 Hz) about 36 K (at 500 Hz). And as shown in the Fig. 3, when the molten steel velocity in the channel u_f changes within t range from 0.1 ms^{-1} (about 0.7 ton·min^{-1} in flow rate) to 0.2 ms^{-1} (about 1.5 ton·min^{-1} in flow rate), the molten ste temperature increase per unit channel length ΔT can be kept a certain value of about 36 K by controlling the A.C. magnetic field frequency within the range from 156 Hz (at 0.1 ms^{-1}) to 500 Hz (at 0.2 ms^{-1}) in this calculation condition. Thus, a high frequency is better for thermal energy supply to the molten steel in the frequency range suitable for the inclusion removal. Above discussions provide some guidelines that (1) the magnetic field frequency f should be controlled within the certain frequency range in which the inclusion removal time becomes minimal, (2) the channel length should be longer than the minimum channel length for the inclusion removal L_{min} which is obtained from the minimal inclusion removal time and an assumed maximum molten steel velocity in the channel for industry, and (3) the molten steel temperature increase per unit channel length ΔT can be controlled by controlling the magnetic field frequency f.

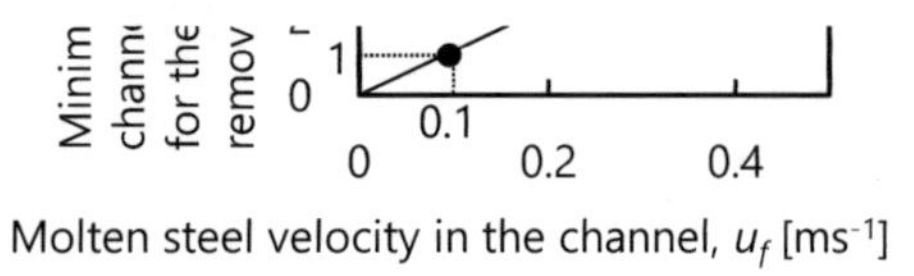

Fig. 2: Effect of the molten steel velocity in the channel on the minimum channel length for the inclusion removal

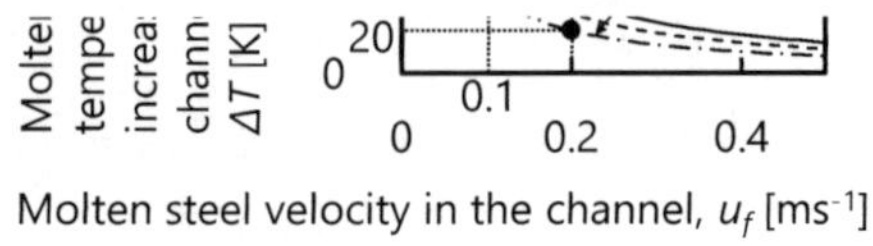

Fig. 3: Effects of the molten steel velocity in the channel and the A.C. magnetic field frequency on the molten steel temperature increase per unit channel length by the induction heating

Conclusions

In order to clarify the guidelines of the channel design and operating conditions for the simultaneous operation of the inclusion removal and the induction heating of the molten steel by the A.C. magnetic field imposition, numerical calculation of the inclusion removal time from the molten steel under the A.C. magnetic field imposition, evaluation of the minimum channel length for the inclusion removal, and evaluation of the molten steel temperature increase per unit channel length by the induction heating were carried out. The results can be summarized as follows.

(1) inclusion removal time becomes minimal when the A.C. magnetic field frequency falls in the range of 156 –500 Hz under the assumptions that the A.C. magnetic field half amplitude B_0 = 0.2 T, the channel radius w = 0.075 m, and inclusion diameter D = 100 μm. The A.C. magnetic field frequency f should be controlled in this frequency range to minimize the minimum channel length for the inclusion removal. In addition, the channel length should be longer than the minimum channel length for the inclusion removal which is calculated by using the minimal inclusion removal time and an assumed maximum molten steel velocity in the channel. The minimum channel length for the inclusion removal becomes 1 m or more when the molten steel velocity exceeds 0.1 ms^{-1} in this calculation condition.

(2) The optimum frequency for the inclusion removal is a relatively wide range in comparison with that for the induction heating in this calculation condition. Thus, the operating frequency should be adjusted for optimum condition for the heat supply in the optimum frequency range of the inclusion removal.

References

1. A. Maruyama and K. Iwai, Tetsu-to-Hagané, 102(2016), 106-112.
2. A. Maruyama and K. Iwai, Tetsu-to-Hagané, 102(2016), 113-118.
3. R. Miura, R. Nishihara, H. Tanaka, Y. Takasaki and T. Imoto, Tetsu-to-Hagané, 81(1995), T30-T33.
4. S. Taniguchi and J. K. Brimacombe, Tetsu-to-Hagané, 80(1994), 24-29.
5. A. Maruyama and K. Iwai, Tetsu-to-Hagané, 103 (2017), 499-507.
6. K. Iwai, Bull. Iron Steel Inst. Jpn., 9 (2004), 705-713.

Microwave Processing of Metallic Materials

N. Yoshikawa

Graduate School of Environmental Studies, Dept. of Materials Science and Enginaeering,Tohoku University
6-6-02, aza-Aoba, Aramaki, Aoba-ku, Sendai, Japan, 980-8579

Corresponding author : yoshin@tohoku.ac.jp

Abstract

The author has performed experimental studies on microwave heating of metallic materials for these twenty years. In this report, it is intended to review the researches and discuss the future prospects on this subject, from the following three points of view: 1. Separated E(Electric)/H(magnetic) field heating in a single mode cavity contributed to understanding of the microwave heating mechanisms of various materials. It was demonstrated that heating behavior of metal film depends on the substrate normal direction with respect to the microwave oscillating magnetic field. 2. In heating of metal/ceramic composite materials, its effective permittivity values have to be determined. In relation with occurrence of conductivity percolation, some measurement data are presented exhibiting negative permittivity (real part), and possible interpretation of the composites' permittivity was discussed. 3. In microwave processing of materials, "non-thermal effect" has been proposed, so far. In this report, the non-thermal effect observed in annealing the Au thin film will be presented. And possible effect of magnetic resonance heating is proposed.

Key words : microwave, metal heating, electric field, magnetic field, permittivity, non-thermal effect

Introduction

Ever since discovery of microwave application to heating materials in 1940s [1], there are many researches and attempts performed for processing materials. One of the major applications is as a heat source of domestic microwave oven for cooking. It has been also applied for some industrial purposes, such as for drying leaves, timbers and roasting coffee beans etc. Because of its specific characteristics of rapid, internal and selective heating, its usage has been widely spread. For example, it has advantages of heating a cold food once stored in a refrigerator, immediately. Thick wet timbers can be dried from the internal area, within a shorter time than the general processes by hot air drying.

Different from the above heating applications, microwave application to metals has been less common, because metals reflects microwave, and a bulk metal cannot be heated effectively. This is related with a fact that the microwave penetration (skin) depth to metals is about or less than 1 microns, therefore heat generated on a surface thin layer is dissipated by heat conduction to the internal area and temperature of the whole metal bulk cannot be raised very much. Microwave heating can be applied to metal powders or films, having their scales of an order of their skin depth. Researches on microwave heating application to metal sintering has been paid attention at the end of '90s [2], and several attempts have been undertaken since then, such as for preparation of nano-grained metal sintered compacts [3], together with some theoretical studies. One of the another contribution to microwave metal heating researches is by the method of a separated electric (E-) and magnetic (H-) field heating in a microwave cavity [4]. In measurement of electron spin resonance (ESR), it is common to place a specimen in a magnetic field maximum position in a microwave cavity and use the microwave H-field, however, specific utilization of separated E- and H- field for heating materials is a relatively new method. In this study, the author's recent findings in separated E- and H- heating of metals is presented.

In an analysis of microwave heating of ceramics/metal composite materials, averaged (effective) physical properties of the composites are important parameters. In this study, the authors' measurement of average permittivity and conductivity of the composites will be presented, and it is intended to discuss their significance related with the heating mechanisms.

The authors have investigated the microwave heating of metal films and demonstrated its dependence of heating efficiency on directions of microwave magnetic field [5]. The detailed study on the metal films has revealed some specific phenomena of microstructural evolution in the film annealing processes. This findings are to be demonstrated as one on the non-thermal characteristics of microwave heating. And lastly, possibility of ferro magnetic resonance (FMR) heating application will be proposed.

Experimental

Specimens used for the experiments are reagent grade powder particles of metals, graphite and ceramics (cordierite, magnetite) having grain size between several hundred nanometer to ten microns. Au film is sputter-coated on SiO_2 substrates with variation of sputtering time for controlling the film thickness in a range of several ten to hundred nanometers, the substrates were cleansed with en ethyl alcohol under ultrasonic irradiation in advance. For the experiments of FMR heating, Fe foil (Johnson-Matthey, 130μm in thickness) was prepared. Microwave heating of the specimens was conducted using a single mode cavity (5.8GHz, TE103, P_{max} 700W(magnetron), 100W(solid state oscilator)) or a single mode wave guide applicator (2.45GHz, TE10, P_{max}2kW). Detail of the single mode applicator can be found elsewhere [5]. Temperature of the heating objects was measured by an optical method using a sapphire rod (Luxtron, PhotoriX system), the pyrometer has a lower temperature detection limit of 350°C. Measurement of complex permittivity of powder mixture was conducted using a vector network analyzer (VNA, Agilent E5062A) in a frequency range between 100MHz and 3GHz by a co-axial transmission method. External static magnetic field less than $4x10^5$[A/m] was imposed to the microwave cavity with an electromagnet.

Results and Discussion

1. Microwave heating of metals in separated E- and H- field

In a microwave cavity, standing wave is generated and the distributions of E- and H- field are formed in different manners (Fig. 1). Namely, E-field maximum position is different from that of H-field by a quarter of the wave length (TE10 mode). Therefore, if a specimen (heating object) is small enough with respect to the field distribution, separated E- and H- field heating is possible. This heating method enables rapid and energy effective heating, because of the concentrated irradiation of either E- or H- field. In addition, it contributes to academic purposes of clarifying the heating mechanisms of materials, according to an equation of the microwave energy loss : P [J/m^3/s] (Eq. 1), where **E** and **H** are the electric and magnetic field vector, ω is an angular frequency and is equal to 2πf (f : frequency [s^{-1}]), ε" and μ" are imaginary parts of complex permittivity and magnetic permeability, respectively. And σ is an electric conductivity of material.

$$P = \left(\omega\varepsilon''\left|\boldsymbol{E}\right|^2 + \omega\mu''\left|\boldsymbol{H}\right|^2 + \sigma\left|\boldsymbol{E}\right|^2\right)/2 \tag{1}$$

Dielectrics such as water, paper and some ceramics are heated well in E- field maximum position. On the other hand, metallic materials are better heated in H-field maximum position than in E- field (metal powder, for example [4]). It is rather natural, considering the induction current generated by an alternating magnetic field. However, the third term in right hand side of Eq. 1, which corresponds to the ohmic loss due to induction current, is not described as the contribution of alternating H-field. It is interpreted that there is a relationship between E- and H- field on metal surface. And the electromagnetic analysis has been performed by setting some boundary conditions.

Metal thin film can be heated well in H-field, however, our previous study has shown that heating behaviors are different when films are placed in different ways with respect to the oscillating microwave magnetic field [5]. It was pointed out that the best heating was obtained in the cases of film normal, parallel to the magnetic field direction (the Au film thickness is less than 1 micron, and microwave penetrates the film, and induction current is effectively generated in the whole volume of films, comparing with the other two configurations.). Therefore, H2 in Fig. 2(a) is the most effective configuration for generating the electric current on the surface.

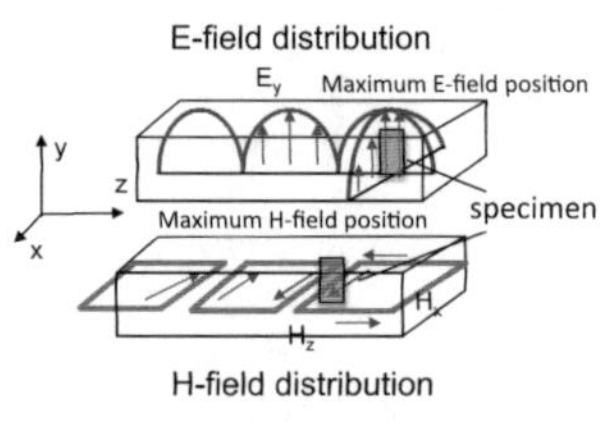

Fig. 1: Schematics of E- and H- field distributions in TE103 microwave cavity.

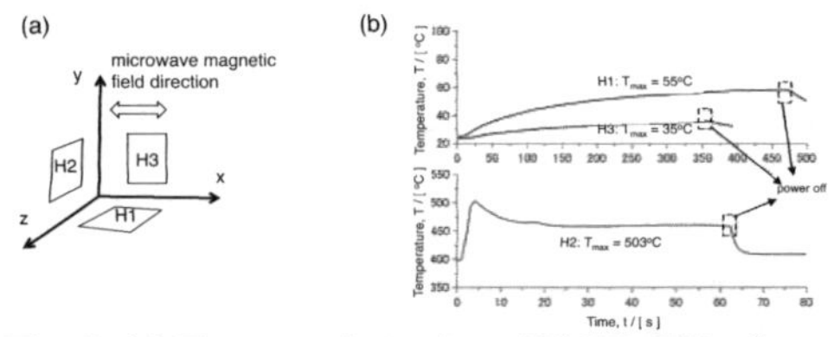

Fig. 2: (a) Three configurations (H1,H2, H3) of metal film on SiO_2 substrate with respect to the oscillating microwave magnetic field and (b) heating curves of Au films placed in the three configurations [5].

2. Physical properties of metal/ceramics composites

In analyzing microwave heating processes for fabricating metal/ceramics composite materials, it is required to measure their physical properties related with microwave energy loss. They are electric conductivity, permittivity and magnetic permeability. The latter two are complex values, but it is possible to use DC electric conductivity for the metals at microwave frequency. Specific features are demonstrated in the composites' permittivity.

Measured permittivity of graphite (C) and goethite (FeO(OH)) mixtures are shown in Fig. 3, as a function of frequency for different graphite compositions. When graphite content is low (Fig. 3(a)), both real and imaginary parts of permittivity exhibited positive values, however, the real part becomes negative (Fig. 3(b)) when graphite content increases and exceed the conductivity percolation value [6]. The same tendency was observed in the cases of Mo/cordierite ($Mg_2Al_4Si_5O_{18}$) mixtures. Fig. 4 indicates the Mo volume fraction dependence of (a) real and (b) imaginary parts of permittivity. The negative real part values appear in the conditions of largest grain size. As previously discussed [7], occurrence of negative real part permittivity is related with the formation of conduction path or percolation, namely at a composition when the whole composite body becomes conductive. However, it has to be noted that the permittivity measurement condition is sometimes out of the possible range of measurement, because of the conductivity. Therefore, the only positive values of graphite / FeO(OH) mixtures are plotted in Fig. 5, with fitting curves by a mixing law expressed by Eq. 2, which is based on the EMA (Effective Medium Approximation, symmetric composite configuration [7]). ε_1, ε_2 and ε_m are permittivity of phase 1, 2 (inclusion) and average (effective) permittivity of composite, respectively. f is a fraction of phase 2, and f_c is a fraction of percolation threshold. In the fitting procedure, complex permittivity of graphite has to be assumed. Two complex numbers having positive and negative real part values were input, for the trial fittings. It looks both of them fits the data well. The real part value influences the microwave penetration distance. And the imaginary part is closely related with the microwave heating characteristics. All the values of imaginary parts increased as an increase of both graphite and Mo fractions. Nevertheless, it has to be noted that when conductor fraction is low, large heating rate is not attained, because of low loss, and in conditions of large fraction of conductor conditions, in which whole body becomes metallic, and heating rate decreased, as well. This is because of decrease in penetration distance (δ) of the composite body according to Eq. 3.

$$f\frac{\varepsilon_2-\varepsilon_m}{\varepsilon_2+A\varepsilon_m}+(1-f)\frac{\varepsilon_1-\varepsilon_m}{\varepsilon_1+A\varepsilon_m}=0\ ,\quad A=(1-f_c)/f_c \tag{2}$$

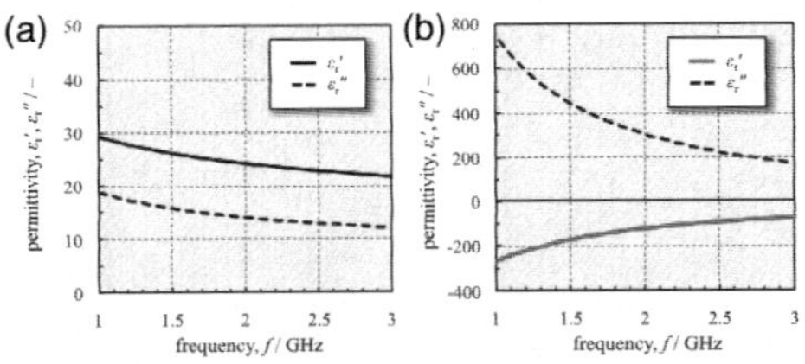

Fig. 3: Real and imaginary parts of permittivity as a function of frequency for graphite/goethite mixtures of graphite (a) 8.2%, and (b) 11.2% [6].

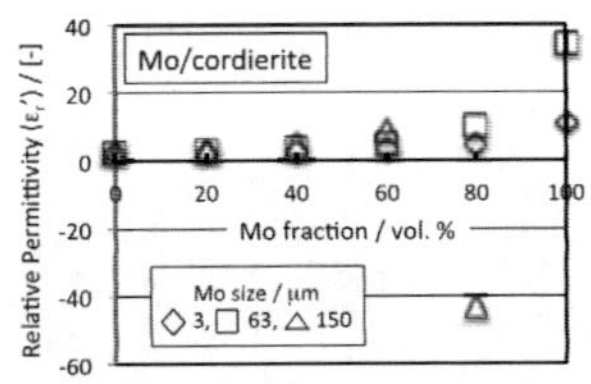

Fig. 4: Permittivity of Mo / cordierite powder mixtures having various particle sizes, plotted as a function of Mo volume fraction,

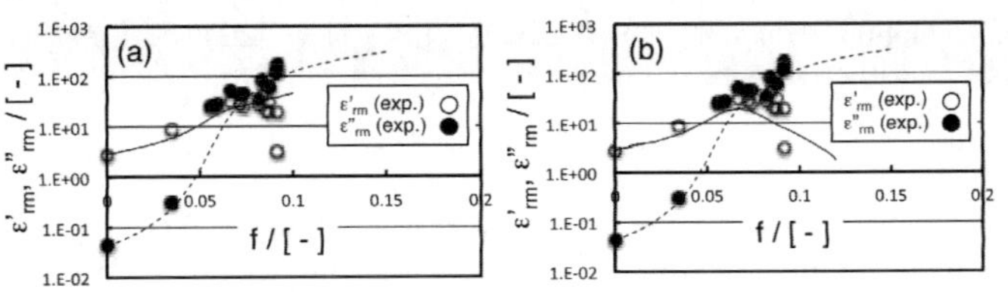

$$\delta=\sqrt{\frac{2}{\sigma\omega\mu}} \tag{3}$$

Fig. 5: Measured complex permittivity of FeO(OH) / C powder mixtures plotted as a function of C volume fraction, real (solid line) and imaginary (dotted line) are fitted by (a) 1060-3568i and (b) -73-3350i using Eq. 2.

3. Specific phenomena by microwave heating of metallic materials

There are many reports, proposing so-called "microwave non-thermal effects". Some of them are the phenomena, which are not interpreted in terms of temperature at which they would occur. For example, they claim enhancement of kinetics of reaction, sintering and some mass transport etc. However, some of reports may contain experimental results due to erroneous temperature reading. For example, temperature measurement under microwave irradiation is not necessarily easy, such as a thermocouple being heated and playing a role of antenna, and thus an erroneous reading may occur. The optical method is the most effective for the temperature measurements, however, setting of the emissivity is not necessarily straightforward.

In this report, microwave annealing process of Au thin film (~20nm) deposited on SiO_2 substrates is presented as an example. Same optical method (pyrometer using a sapphire rod, because of no influence by microwave irradiation) was employed for the measurement of temperature both in microwave and conventional (electric furnace) heating. Because microwave heating in separated H-field takes place very rapidly, microwave power was switched off immediately, after temperature reached the predetermined value. In the conventional heating, electric furnace temperature was controlled at the predetermined value, and the specimen film (on the substrate) and pyrometer rod was inserted into the furnace, together and then removed, imitating the same temperature history as that of microwave heating. AFM photographs of the film surface are depicted in Fig. 6 for comparing the two heating methods. It can be seen that the microwave heating brought about more flat surface and connecting of grains. It is anticipated that the surface diffusion is enhanced by microwave irradiation, not by thermal effect but by some effect due to inhomogeneous distribution of surface electric current, which could function together with microwave magnetic field. This phenomenon is under investigation for clarifying as an electromagnetic field effect.

Lastly, it would be of significance to introduce one of our findings [8] on the microwave magnetic heating mechanism. Fig. 7 demonstrates a heating curve, specimen temperature is indicated by a left vertical axis. A Fe sheet is placed in a magnetic field maximum position, and started at 480°C by an ordinary microwave heating. When a slowly varying external magnetic field (H^{ext}) is applied after 200s (the magnitude of H^{ext} is expressed by a right vertical axis) perpendicular to the microwave (oscillating) magnetic field, temperature increased and experienced a temperature peak, at about 230s, then decreased even when H^{ext} was kept increasing, however, after about 450s when H^{ext} is decreased, the temperature peak appeared again at 650s, the two peaks corresponding to H^{ext} equal to 0.08[T] (6.4×10^4 [A/m]). This phenomenon is due to ferro-magnetic resonance (FMR). So far, the authors examined not only iron but also other ferro (ferri) magnetic materials, no particular difference has been observed in microstructural difference from the ordinary microwave heating or conventional heating. However, the temperature increase caused by FMR is due to an acquisition of microwave energy by electron spins of their simultaneous precession motion (precession frequency equal to microwave frequency), then the energy transferred to the lattice of the materials and was detected as a temperature increase. It is expected that the microwave energy once received by spin motions could cause some effect before transfer to the lattice or could transfer in different ways to lattice. Further investigation is being carried out.

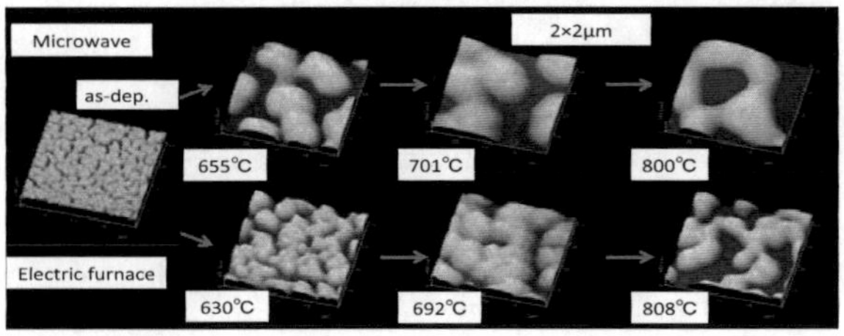

Fig. 6: Comparison of microstructural evolution between microwave and conventional heating in annealing of Au thin film, observed with AFM.

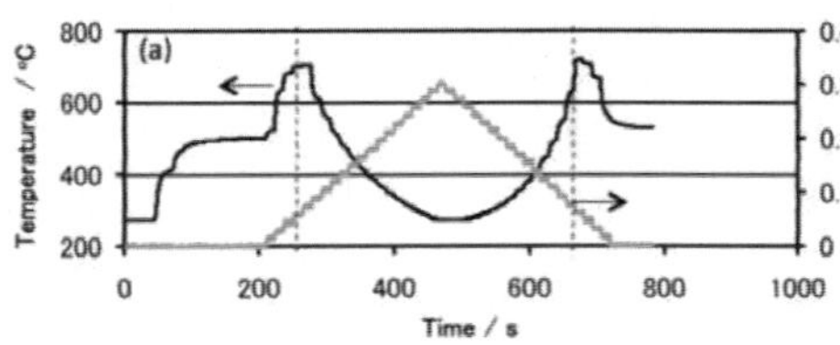

Fig. 7: Heating curve (black line) under imposition of slowly varying external magnetic field (glay line) [8].

Conclusions

Microwave heating of metallic materials has not been a major processes applied to their fabrication. However, since the beginning of this century, its applications to sintering of metal powder and to annealing metal thin films are being performed. In addition, method of separated E/H field microwave heating contributed to understanding the role of microwave magnetic field for metallic materials. This report started with describing the present author's recent studies and introduced some topics on negative permittivity of metallic materials and some specific phenomena related with the so-called "microwave non-thermal effect".

Acknowledgment

Last part of this study was financially supported by JSPS, Grant in Aid of Challenging Exploratory Research (# 16K14433) and is acknowledged.

References

1. H.Sobol, IEEE Trans. Microwave Theory Tech. **50** (2002)594-611.
2. R.Roy, D.Agrawal, J.Cheng and S. Gedevanishvili, Nature, **399** (1999) 668-670.
3. K.S.Tun and M.Gupta, Composite Sci. technol., **67** (2007) 2657-2664.
4. J.Cheng, R.Roy and D.Agrawal, J.Mater.Sci. Lett. **20** (2001) 1561-1563.
5. Zi.Cao, N.Yoshikawa, S.Taniguchi, Mater. Chem. Phys., **117** (2009) 14-17.
6. K.Kawahira, Y.Saito, N.Yoshikawa, H.Todoroki and S.Taniguchi, Metal. Mater. Trans., **45B**, (2014) 212-220.
7. N.Yoshikawa, K.Kawahira, Y.Saito, H.Todoroki and S.Taniguchi, J.Appl. Phys., **117** (2015) 084105.
8. N.Yoshikawa and T.Kato, J. Phys. D: Appl. Phys., 43 (2010) 425403-425408.

Coupling of Magnetite Particles with Microwaves at Temperatures lower than the Curie Point

A. Amini[1], K. Ohno[2], T. Maeda[2] and K. Kunitomo[2]

[1] Department of Materials Process Engineering, Kyushu University, Fukuoka, Japan
[2] Department of Materials Science and Engineering, Faculty of Engineering, Kyushu University, Fukuoka, Japan

Corresponding author: Ahmadreza.Amini.ir@gmail.com

Abstract

Effect of the particle size (*d*) and apparent density on the coupling of microwaves with Fe_3O_4 was investigated at temperatures lower than the Curie point, $T_C \approx 585$ °C. Two samples in the form of tablet with particle sizes of 45-75 μm (MT75) and <45 μm (MT45) and one sample in the form of powder with a particle size of <45 μm (MP45) were heated in multi-mode and maximum E- and H-field modes using a microwave generator at a frequency of 2.45 GHz. According to the results, an earlier temperature increase and also a higher temperature was achieved in the sample heated in the maximum H-filed mode. Regarding the incubation time in all samples, the particle size of Fe_3O_4 has no significant effect on the time required for the initial temperature increase in the presence of the H-field. In the maximum E-field mode, a shorter time was required for the temperature increase in the MT75 sample than MT45. At $T \leq T_C$, magnetic loss and Joule loss are the dominant heating mechanisms in the presence and absence of the H-field, respectively. Magnetic loss is independent of the particle size whereas Joule loss which is influenced by electrical conductivity, affected by particle size. Therefore, above-mentioned effect of the particle size is attributed to the dominant heating mechanism. Also, some of the small particles seems to be transparent owing to a greater penetration depth (δ), ca. 80 μm, at room temperature causing an earlier onset of temperature increase in sample with larger particle size, MT75. Moreover, microwave absorption in a sample with higher apparent density, MT45, was lower because of a higher electrical conductivity of sample in tablet form, MT45, than powder form, MP45.

Key words: Magnetite; Microwave absorption; Particle size; Apparent density; Curie point;

1. Introduction

Iron production with microwave heating is considered a possible solution to mitigate the CO_2 emission. Reduction of iron oxide using highly efficient microwave heating is reported as a new method for energy conservation and also for decreasing the CO_2 emission [1–3]. For a more efficient use of the microwave irradiation, it is important to understand the heating mechanism and also find effective parameters for the microwave heating of powdery materials. Microwave power absorption by materials, P (W/m^3), can be evaluated by Eq. (1) which is directly related to the penetration depth, δ (m). Penetration depth presents the distance from the surface to a point inside the materials where the power of the exposed electromagnetic waves decreases to 1/e, 36.8%, of the surface value [4,5] which can be calculated according to Eq (2) [4]:

$$P = \frac{1}{2}\sigma|E|^2 + \pi f \varepsilon_0 \varepsilon_r''|E|^2 + \pi f \mu_0 \mu_r''|H|^2 \quad (1)$$

$$\delta = \left[\frac{\omega^2 \varepsilon' \mu'}{2}\left(\sqrt{1+\frac{\sigma^2}{\omega^2 \varepsilon'^2}} - 1\right)\right]^{-\frac{1}{2}} \quad (2)$$

where σ (S/m) is the electrical conductivity, E (V/m) is the electric field amplitude, H (A/m) is the magnetic field amplitude, f (Hz) is the microwave frequency, ε_0 (F/m) and μ_0 (N/A^2) are the permittivity and permeability of vacuum, respectively. ε_r'' (dimensionless) and μ_r'' (dimensionless) are the imaginary part of permittivity and permeability, respectively. ε' and μ' are the real part of permittivity and permeability, respectively. According to Eq. (1), three different mechanisms including Joule loss (first term, W/m^3), dielectric loss (second term, W/m^3), and magnetic loss (third term, W/m^3) would contribute to microwave heating. Hayashi *et al.* [4] observed that microwave absorption of magnetite at the steady state hardly depends on the particle size and theorized that it is because of the great penetration depth of Fe_3O_4, ca. 80 μm, at room temperature at 2.45 GHz. However, such large penetration depth should be considered in the case of interaction between Fe_3O_4 and microwaves at temperatures lower than the Curie point, ca.

585 °C [4]. At the steady state, the penetration depth would be significantly shallower than 80 μm due to an increase in the conductivity with temperature. Moreover, no sufficient information is available about initial stages of temperature increase during microwave heating. In fact, microwave absorption at the steady state is related to higher temperatures whereas initial stages of temperature increase depend on the low-temperature (T< 585 °C) interaction between the microwaves and materials. In this study, the effect of the apparent density and particle size on the microwave absorptivity of Fe_3O_4 is investigated by focusing on the initial stages of temperature increase for a better understanding of the interaction between the microwaves and materials at temperatures lower than the Curie point.

2. Experimental Procedure

2.1. Sample preparation

Three types of magnetite samples were prepared as follows:

Magnetite powder (ca. 1 μm, purity of 99%) was pressed into a briquette shape (diameter, 30 mm; thickness, 20 mm) at a pressure of 42 MPa and then heated under a flow of Ar at 1350 °C for 1 h in an electric resistance furnace. The heat-treated briquettes were crushed and ground to a certain grain size, as mentioned above. To prepare the tablet samples, 0.02 g of a 5 mass% solution of polyvinyl alcohol in water was added as a binder to 3 g of the crushed magnetite powder. Cylindrical tablet shape samples (diameter, 15 mm and thickness, 5 mm) were formed by cold-pressing followed by drying at 120 °C for 10 h. The properties of the samples are presented in Table 1. The apparent density of the tablet shape sample was calculated by dividing the weight of the sample by its apparent volume (height×π×[radius]^2) at 25 °C and pressure of 1 atm. To evaluate the density of the powder sample, 3 g of the powder was loaded into a 2.5 mL graduated cylinder followed by vibrating for 3 min using an ultrasonic cleaner (USM, AS ONE, JAPAN). Then, the bulk density of the powder sample was calculated by dividing the mass (g) of the powder, 3 g, by the volume occupied in the cylinder [6]. The electrical conductivity of the tablet shape samples was evaluated using an Analog tester (CX-270N, Custom) at 25 °C. The average electrical conductivity of the sample was calculated from the electrical conductivity along two directions: Along the pressure direction and orthogonal to the pressure direction. To evaluate the electrical conductivity of the powder samples, 3 g of each sample was loaded into a silica tube (20.6 mm outer diameter (OD), 15.4 mm inner diameter (ID), and 35.5 mm height). Then, the conductivity was measured between the two end of the tube occupied by the sample using an Analog tester (CX-270N, Custom) at 25 °C.

2.2. Microwave heating

The samples were subjected to microwave irradiation under multi- (whereby both the magnetic field (H) and the electric field (E) perpendicular to it contribute to heating) and single-mode operation using a microwave generator with a maximum output power of 1.5 kW at 2.45 GHz under N_2 atmosphere. For microwave heating under each mode (multi- and single-), a specific cavity was designed by the manufacturer. Typically, the sample was loaded in an alumina crucible (41 mm OD, 36.5 mm ID, and 49 mm height) surrounded by ceramic wool and covered with an insulating ceramic brick. The crucible was placed in a quartz chamber (95 mm OD, 85 mm ID, 96 mm height) and the sample was protected from oxidation by flowing N_2 during the experiments. The ceramic wool and insulation fibre, which do not absorb microwaves, served as thermal insulators preventing heat loss during heating as well as protectors of the quartz chamber. In case of the single mode, the H- or E-field has maximum amplitude, whereas the amplitude of the other is ca. 0. For heating in the single-mode, a cavity was used different from that of the multi-mode one. An adjusting plunger was used to define the maximum magnetic or electric field position. The sample was placed on a piece of ceramic fibre and placed inside a quartz tube, and it was protected from oxidation by flowing N_2 during the experiments. A constant power of 1050 W was manually set using three tuning stubs until the end of the heating process in all the experiments. The temperature of the sample was measured through a window above the sample using an infrared radiation thermometer. The measurement temperature range is 330–1500 °C.

3. Results and Discussion

3.1. Effect of particle size

According to the heating patterns of the samples in form of tablet shape shown in Fig. 1, an earlier temperature increase and a higher temperature during treatment was observed in the sample heated in the maximum H-filed mode compared to the multi-mode and maximum E-filed mode. In case of heating in the maximum H-field mode, the temperature increased rapidly to ca. 690 °C via the magnetic loss mechanism and then increased gradually at a lower rate to ca. 1020 °C. The magnetic loss does not contribute to the heating at temperatures higher than the Curie point [4,7] and an increase in the electrical conductivity due to the temperature increase [4] would be the reason for the increase in the temperature of the sample via Joule loss, regarding Eq. (1). The incubation time for temperature increase in samples MT75 and MT45, during treatment in the multi-mode, maximum E- and H-field modes are shown in Fig. 2. This result indicates that in the presence of H-field, where the magnetic loss is the dominant mechanism for heating at temperatures lower than the Curie point, ca. 585 °C [4], microwave heating is not affected by particle size of the sample. This is possibly due to the permeability, which is independent of the particle size, as reported by Hotta *et al.* [8].

Table 1: Properties of magnetite samples.

Sample No.	MP45	MT45	MT75
Particle size (μm)	< 45	< 45	45-75
Form	Powder	Tablet	Tablet
Apparent density (g/cm³)	2.1	3.3	3.3
Average σ at 25 °C (S/cm)	[a]	1.6×10^{-4}	2.0×10^{-4}

[a] The value is lower than the measurement limitation of ca. 8.3×10^{-8} S/cm.

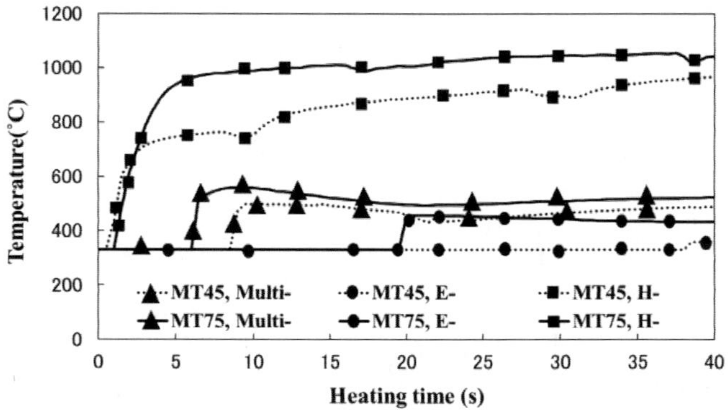

Fig. 1 Heating pattern of samples MT45 and MT75.

In the absence of a high H-field at temperatures lower than the Curie point, neither magnetic loss nor dielectric loss can contribute to microwave heating and only Joule loss would be an effective mechanism for heating, according to Eq. (1). It seems that the Joule loss of sample MT75 is higher than that of MT45 because of the higher electrical conductivity of the sample with a larger particle size, as presented in Table 1. In addition, a greater penetration depth is expected owing to a significant decrease in μ' in the maximum E-field mode, according to Eq. (2), which would affect its microwave absorption. It has been reported that the ratio of the particle size (d) to the penetration depth (δ) is a critical factor for optimum microwave absorption by conductive materials [5,9]. For a very low d/δ ratio, the particle acts as a transparent material and cannot couple with the microwaves. For a very high d/δ ratio, most of the microwaves will be reflected and the sample cannot heat well. For instance, a d/δ ratio of ca. 2.4 is suggested for maximum microwave absorption in nonmagnetic metal particles [5,9]. Thus, larger particles can absorb more of the microwave power owing to their higher d/δ ratios whereas some of the smaller particles cannot couple with the microwaves and behave like a transparent material, owing to their low d/δ ratio.

It seems that most of the particles in sample MT45 fall in the transparent region at temperatures lower than the Curie point, whereas the particles in samples MT75 fall in the absorption region for the same temperature range. Therefore, an early onset of temperature increase in sample MT75 compared to the onset of temperature increase of sample MT45 would be due to both higher Joule loss and higher d/δ ratio of the sample with a larger particle size, MT75. This result proves the considerable impact of the particle size on microwave absorption at temperatures lower than the Curie point.

3.2. Effect of the apparent density

To investigate the effect of the apparent density on the initial stages of temperature increase during the microwave heating, each experiment was conducted at least twice and the average time required for the initial temperature increase for each sample is plotted in Fig. 3, with the standard deviation represented as error bars.

A shorter time was required for the initial temperature increase in sample MP45, indicating better microwave absorption by this sample compared to sample MT45 even at low temperatures.

For acicular magnetite powders at room temperature and a frequency of 2.45 GHz, δ has been calculated to be ~80 μm by Hayashi *et al.* [4], with $\varepsilon' = \sim40$, $\mu' = \sim1.7$, and $\sigma = \sim1.0 \times 10^2$ (S/cm). According to Eq. (2), an increase in the electrical conductivity causes a decrease in the penetration depth, whereas an increase in the permittivity has an inverse effect. If the values of both σ and ε' increase to the same extent, the effect of the electrical conductivity would be dominant, causing a decrease in δ. This explanation is in excellent agreement with that provided by Peng *et al.* [10]; they reported a decrease in the penetration depth of Fe_3O_4 from 0.0153 to 0.0023 m despite a significant increase in ε'' from 1.0976 to 83.3109 with an increase in the temperature from 24 to 880 °C. The penetration depth of a compacted magnetite sample would be shallower than 80 μm owing to the increases in both the permittivity and electrical conductivity with an increase in the relative density, as reported by other researchers [4,5,7,8,11,12].

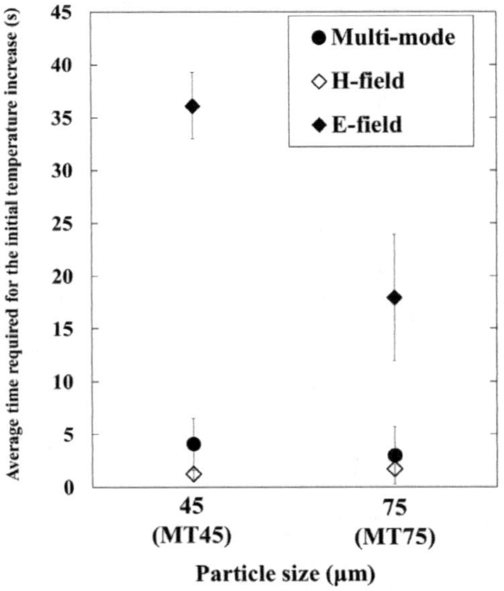

Fig. 2: Average time required for the initial temperature increase in samples MT45 and MT75.

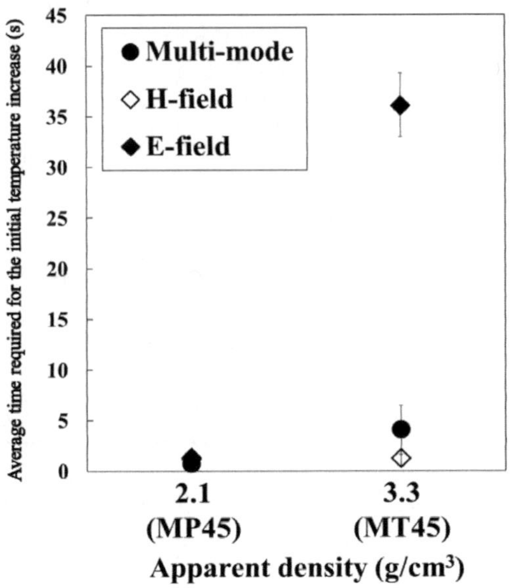

Fig. 3: Average time required for the initial temperature increase in samples MP45 and MT45.

At the initial stages of temperature increase, the dielectric loss could be neglected because of the very low permittivity of magnetite ($\varepsilon'' \simeq 0$) [11]. Moreover, Hotta *et al.* [8] reported an increase in the permeability of Fe_3O_4 with an increase in the relative density. An increase in the permeability leads to a decrease in δ; moreover, an increase in the electrical conductivity also causes a decrease in the penetration depth (Eq. (2)). Therefore, an early onset of temperature increase in sample MP45 compared to the onset of temperature increase in sample MT45 could be related to the shallower penetration depth of the latter even at low temperatures.

4. Conclusions

To expand our understanding of the interaction between the microwaves and magnetite particles temperatures lower than the Curie point, the effect of the apparent density and particle size of Fe_3O_4 on their microwave absorption was studied using a microwave generator at 2.45 GHz at 1050 W. The results can be summarized as follows:

1. A lower apparent density leads to a shorter time required for the initial temperature increase owing to higher microwave absorption caused by the higher electrical resistivity of the material.
2. At temperatures lower than the Curie point, the particle size of Fe_3O_4 has no effect on the time required for the initial temperature increase in the presence of the H-field. In the maximum E-field mode, Joule loss is the dominant mechanism for heating, leading to a higher temperature in a sample with larger particles, because of their higher electrical conductivity.

References

1. K. Kashimura, M. Sato, M. Hotta, D. Kumar Agrawal, K. Nagata, M. Hayashi, T. Mitani, N. Shinohara, *Mater. Sci. Eng. A*. 556 (2012) 977–979.
2. K. Hara, M. Hayashi, *J. Microw. Power Electromagn. Energy*. 45 (2011) 137–147.
3. M. Hayashi, K. Takeda, K. Kashimura, T. Watanabe, K. Nagata, *ISIJ Int*. 53 (2013) 1125–1130.
4. M. Hayashi, Y. Yokoyama, K. Nagata, *J. Microw. Power Electromagn. Energy*. 44 (2010) 198–206.
5. C.A. Crane, M.L. Pantoya, B.L. Weeks, M. Saed, *Powder Technol*. 256 (2014) 113–117.
6. S.M. Jafari, M. Ghalegi Ghalenoei, D. Dehnad, *Powder Technol*. 311 (2017) 59–65.
7. Z. Peng, J.-Y. Hwang, J. Mouris, R. Hutcheon, X. Huang, *ISIJ Int*. 50 (2010) 1590–1596.
8. M. Hotta, M. Hayashi, A. Nishikata, K. Nagata, *ISIJ Int*. 49 (2009) 1443–1448.
9. M. Ignatenko, M. Tanaka, M. Sato, *Jpn. J. Appl. Phys*. 48 (2009).
10. Z. Peng, Z. Li, X. Lin, M. Yang, J.Y. Hwang, Y. Zhang, G. Li, T. Jiang, *JOM*. 69 (2017) 178–183.
11. M. Hotta, M. Hayashi, K. Nagata, *ISIJ Int*. 51 (2011) 491–497.
12. M. Hotta, M. Hayashi, K. Nagata, *ISIJ Int*. 50 (2010) 1514–1516.

Relevance of Low-Rm MHD for Surface Viscosimetry of Liquid Metals

K. Patouillet[1,2], L. Davoust[1], O. Doche[1], V. Ebrahimian[2]

[1] SIMaP/EPM, 1340 rue de la piscine, 38402 St. Martin d'Hères, France
[2] Montupet, Techcenter 3 rue de Nogent, 60290 Laigneville, France

kevin.patouillet@simap.grenoble-inp.fr

Abstract
This paper presents a theoretical background which justifies the benefits of MHD using a magnetic field for surface viscosimetry of liquid/melted metals. The system considered is based on an annular shallow-channel layout filled with a melted metal covered by an oxide layer, and pervaded by a uniform static magnetic field. The latter permits to monitor the coupling between surface rheology and a supporting flow due to the well-known 2-D tendency of low-Rm magnetohydrodynamics. The impact of a non-planar interface and/or a non-uniformly oxidized interface can be investigated and the surface viscosity better identified.

Key words: Liquid/melted metal, MHD, Surface viscosimetry, Capillary length, Non-uniform surface viscosity.

Introduction

During casting processes, melted metals tend to oxidize quickly in contact with ambient air. A thin oxide layer of about 1 μm thick develops at the air/metal interface. From a modeling point of view, the oxide layer can be described from a revised boundary condition.
Through the Gibbs approach, the physicochemical parameters of the oxidized surface can be taken into account with a macroscopic point of view. This approach requires the introduction of in-excess quantities assigned to the interface: the surface shear viscosity η_s, the surface dilatational viscosity κ_s and the surface tension γ.
To identify the rheological parameters of the oxidized layer, a surface viscometer is used. The concept is firstly introduced in 1970 by Mannheimer and Schechter [1] and has shown a good sensitivity in an annular deep-channel viscometer. More recently, many authors [2] worked on this kind of configuration to determine the rheological properties of a given fluid covered by a monolayer of surfactants.
The use of MHD surface viscosimetry has been successfully proposed for investigating the case of liquid metals, as the galinstan, covered by a thin oxide layer [3]. Indeed, since the liquid metals are electroconductive, the application of a uniform magnetic field B is relevant for several reasons. It permits to monitor the coupling between the viscous surface and the carrier sub-phase by activating Hartmann layers [4]. Moreover, it promotes the 2-D diffusion of the input shearing from the rotating floor up to the liquid surface, and can promote the magnetic extinction of the swirl component of the centrifugally-induced overturning flow [5]. Thus, for a large enough magnetic field, only the surface shear viscosity governs the behavior of the oxide layer avoiding the classical artifact due to surface dilatation.
In this study, two assumptions usually made in the literature are thrown into question. First, the liquid surface is no more considered as planar, which is particularly required for a capillary length, l_c, as large as the one of aluminum. Second, the radially-inwards oxide packing [6] induced by centrifugation of the underlying bulk, is no more considered negligible since it causes a non-uniform surface viscosity along the oxide layer.

Theoretical ingredients

Geometry:

The system considered is an annular MHD shallow-channel where a fluid is end-driven by a rotating floor at angular speed Ω between an inner side wall (radius r_i = 3 cm) and an outer side wall (radius r_o = 7 cm). The shape of the surface, no more considered as planar, results from the resolution of Young-Laplace equation expressed in cylindrical coordinate $\{O, \boldsymbol{e}_r, \boldsymbol{e}_\theta, \boldsymbol{e}_z\}$,

$$\frac{1}{(1+h'^2)^{3/2}}\left(h''+\frac{h'}{r}+\frac{h'^3}{r}\right)=\frac{h}{l_c^2}, \tag{1}$$

where h' and h'' are respectively the first and the second radial derivatives of the height h. The capillary length l_c is defined from the surface tension γ, the density ρ and the acceleration of gravity g:

$$l_c^2=\frac{\gamma}{\rho g}. \tag{2}$$

The wetting angle β is imposed on both walls through the following Neumann boundary condition:

$$h' = -\coth\beta\,. \tag{3}$$

For this special layout, the average liquid depth, h_0, is chosen as small as possible compared to the radial gap of the channel. In other word, the depth-to-width ratio, ε, is much smaller than unity. Indeed, it can be demonstrated that it tends to magnify the coupling terms between surface and subphase flows due to stronger velocity gradients involved along the vertical direction.

Bulk flow:

The annular MHD flow is supposed to be axisymmetric, permanent, laminar, and incompressible. The fluid is Newtonian and electroconductive with a dynamic viscosity η and an electric conductivity σ. The conventional dimensionless number governing the bulk flow dynamics, the Reynolds and the Hartmann numbers, are respectively defined by:

$$Re = \frac{\rho \Omega r_o^2}{\eta}, \qquad Ha = B r_o \sqrt{\sigma / \eta}. \tag{4}$$

As usual in MHD of liquid metals, the magnetic Reynolds number is small so that the imposed magnetic field is not convected by the annular flow. The weakly coupled MHD model is then formulated with the following electromotive density current $\boldsymbol{j}$ and Lorentz force $\boldsymbol{F}$:

$$\boldsymbol{j} = \sigma(-\boldsymbol{\nabla}\phi + \boldsymbol{v} \times \boldsymbol{B}), \qquad \boldsymbol{F} = \boldsymbol{j} \times \boldsymbol{B}, \tag{5}$$

where ϕ is the electric potential, $\boldsymbol{v}$ is the velocity field and $\boldsymbol{B}$ is the magnetic field approximated by:

$$\boldsymbol{B} = B\boldsymbol{e_z}. \tag{6}$$

The electromagnetic part is described by a formulation based on Maxwell equations and the continuity equation for the total magnetic field, completed by the generalized Ohm's law. The flow is governed by Navier-Stokes equations with the Lorentz force as a body force.
The boundary conditions are summarized in Fig. 1. The constant vertical magnetic field is imposed in the surrounding atmosphere with an electrical insulation at the melted metal boundaries. No-slip boundary conditions are applied at the outer and inner walls together with lubrication conditions in the gaps of δ thickness. Finally, the boundary condition at the liquid/gas interface is found by solving surface rheology equations in next section.

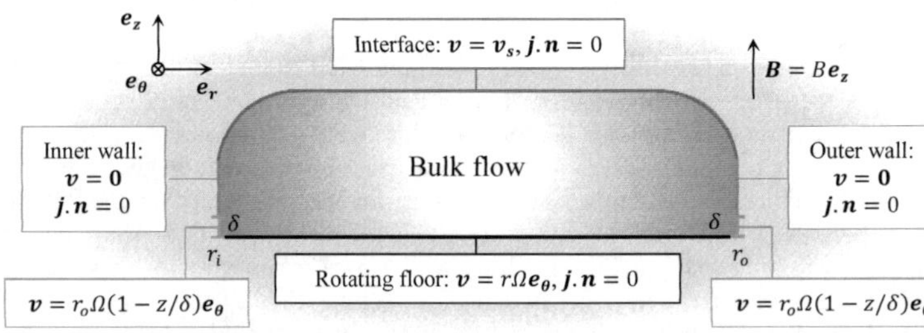

Fig. 1: Geometry and boundary conditions of the channel cross section

Surface flow:

Concerning the mechanics of the oxidized interface, as previously mentioned, the macroscopic point of view is adopted with a zero-thickness interface. This leads to the introduction of a (in-excess) surface stress tensor $\bar{\bar{T}}_s$ linearly related to a (in-excess) surface rate-of-deformation tensor (Boussinesq-Scriven law):

$$\bar{\bar{T}}_s = [\gamma + (\kappa_s - \eta_s)\boldsymbol{\nabla_s}.\boldsymbol{v_s}]\bar{\bar{P}}_s + \eta_s\left[\bar{\bar{P}}_s.(\boldsymbol{\nabla_s v_s}) + (\boldsymbol{\nabla_s v_s})^t.\bar{\bar{P}}_s\right]. \tag{7}$$

The surface projection tensor $\bar{\bar{P}}_s$ is defined from normal unit vector, $\boldsymbol{n}$, and the "s" subscript refers to the variables at the surface. This law involves the three rheological parameters (γ, η_s, κ_s), each of them involves the following dimensionless numbers: the Marangoni number, the surface shear and surface dilatational Boussinesq numbers :

$$Ma = \varepsilon\frac{\gamma}{r_o\eta\Omega}, \qquad Bo_{\eta s} = \varepsilon\frac{\eta_s}{r_o\eta}, \qquad Bo_{\kappa s} = \varepsilon\frac{\kappa_s}{r_o\eta}. \tag{8}$$

It is assumed that these quantities depend on the level of oxidation, which itself is related to the oxygen rate of the surrounding atmosphere. The governing equation of surface mechanics is a jump momentum balance (JMB) derived from an elementary volume straddling the liquid surface. With no surface body force, the JMB writes according to a balance between the surface stresses and the jump in bulk stresses normally applied on both sides of the oxide layer [7]:

$$\boldsymbol{\nabla_s}.\bar{\bar{T}}_s = [\![\bar{\bar{T}}.\boldsymbol{n}]\!]. \tag{9}$$

Magnetic damping:

If the following condition is satisfied,

$$\frac{Ha}{\sqrt{Re}}\frac{r_o}{h_0} \gg 1, \tag{10}$$

the swirl component of the centrifugally-induced overturning flow is damped [3], and the MHD flow is purely 2D-azimuthal. Therefore, the surface shear viscosity can be selectively studied, since the surface dilatational viscosity is no longer solicited without any significant radial component of the surface flow. The eventual residual impact of a radially-inwards oxide packing can be taken into account through a non-uniform surface shear viscosity.

Under these conditions, the dimensionless azimuthal component of the JMB can be explicitly written as:

$$\underbrace{Bo_{\eta s}\frac{\partial}{\partial s^*}\left(\frac{1}{r^*}\frac{\partial}{\partial s^*}(r^* v^*_{\theta s})\right)}_{\text{2-D viscous shear}} + \underbrace{\frac{\partial Bo_{\eta s}}{\partial s^*}\left(r^*\frac{\partial}{\partial s^*}\left(\frac{v^*_{\theta s}}{r^*}\right)\right)}_{\text{2-D non-uniformity in shear visosity}} + \underbrace{Bo_{\eta s}\frac{v^*_{\theta s}}{r^*}\frac{2\varepsilon^2 h'^* h''^*}{\left(1+\varepsilon^2 h'^{*2}\right)^2}}_{\text{Surface curvature}} = \underbrace{r^*\frac{\partial}{\partial n^*}\left(\frac{v^*_\theta}{r^*}\right)}_{\text{Liquid shear from underlying sub-phase}}, \tag{11}$$

where the dimensionless tangential and normal derivatives are respectively defined by:

$$\frac{\partial .}{\partial s^*} = \frac{1}{\sqrt{1+\varepsilon^2 h'^{*2}}}\left(\frac{\partial .}{\partial r^*} + h'^*\frac{\partial .}{\partial z^*}\right), \quad \frac{\partial .}{\partial n^*} = \frac{1}{\sqrt{1+\varepsilon^2 h'^{*2}}}\left(\frac{\partial .}{\partial z^*} - \varepsilon^2 h'^*\frac{\partial .}{\partial r^*}\right). \tag{12}$$

It is worthy to note that the normal component of the JMB is nothing but the Young-Laplace equation.

Surface shear viscosity model

The radially-inwards oxide packing leads to a non-uniform surface shear viscosity. Then, the Boussinesq number $Bo_{\eta s}$ is defined along the dimensionless curvilinear abscissa, s^*, according to the model proposed in [6]:

$$Bo_{\eta s}(s^*) = Bo_m\left(1 + \frac{1-R}{1+R}\tanh\left(\frac{s^* - s^*_f}{\delta^*_f}\right)\right). \tag{13}$$

The latter involves the median Boussinesq number Bo_m, the location of the segregation front s^*_f, the thickness of the segregation front δ^*_f, and the ratio R between the inner and outer Boussinesq numbers. All these quantities are normalized by the radial extent of the liquid/gas interface or by the average liquid depth .

Results and interpretations

Numerical simulations are performed with Comsol Multiphysics which is a software based on finite element method. A fully-coupled approach is solved with a linear stationary direct solver (MUMPS). The internal mathematical functions are used to model the bulk MHD flow and weak formulations for surface equations (azimuthal JMB and Young-Laplace equations). The mesh is composed of triangular elements with a rectangular boundary layer mesh at the walls.

With a planar interface and uniform surface properties, a good agreement is found between this numerical model and existing analytical results for MHD flow [4]. The hydrodynamic flow with a radially-dependent surface shear viscosity is also well validated based on an asymptotic analytical model [6].

Physical properties of pure aluminum at 973 K, used in numerical simulations, are summarized in Tab. 1.

ρ [kg.m^{-3}]	η [Pa.s]	σ [Ω^{-1}.m^{-1}]	γ [N.m^{-1}]	l_c [m]
2357	1.286×10^{-3}	4.034×10^{6}	0.703	5.514×10^{-3}

Tab. 1: Physical properties of aluminum at 973 K used for numerical simulations

First results presented in this study are dedicated to surface velocity profiles. It can be seen that the impact of the wetting angle is significant close to the side walls and that the tangential surface shear tends to increase with the wetting angle (see Fig. 2). Indeed, a dead zone arises near the side walls for the wetting case, which causes a decrease in the surface velocity, while the bulk flow drives the side parts of the oxide layer for the non-wetting case. These effects are weakened for a high Hartman number due to the 2-D tendency.

Typical Boussinesq profiles used to evaluate the impact of the radially-inwards oxide packing are shown in Fig. 4. Surface velocity profiles are mainly affected at the outer part of the viscometer where a bump is created (Fig. 3 and 5). As expected, a higher jump between the inner and outer Boussinesq numbers, which can be obtained with a larger median Boussinesq number Bo_m or Boussinesq ratio R, leads to greater effects. As previously, the impact of a non-uniform contamination along the surface is reduced for a large enough Hartmann number.

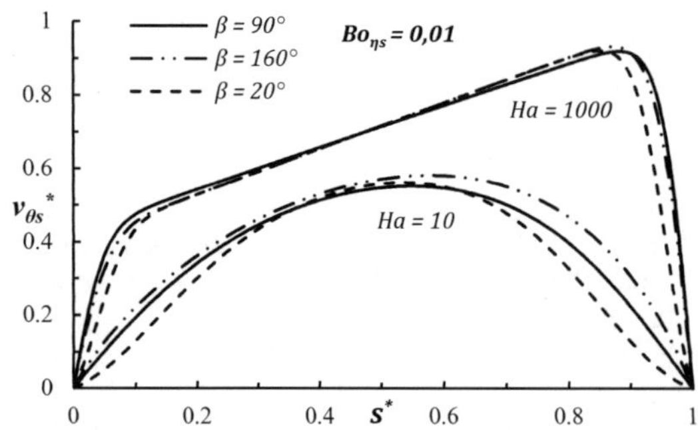

Fig. 2: $v_{\theta s}^{*}$ profiles for $R = 1$ and $Bo_m = 0.01$.

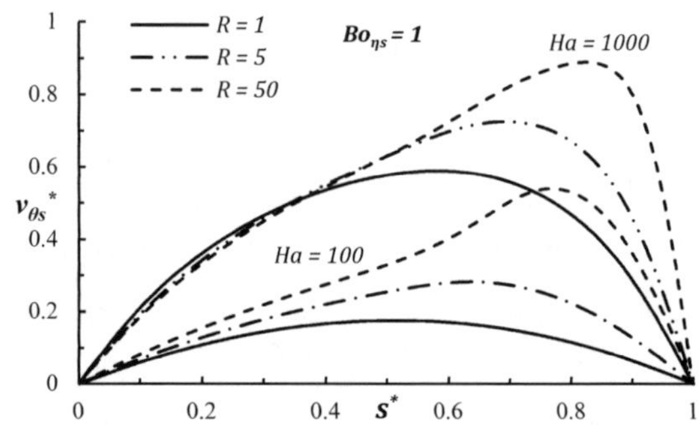

Fig. 3: $v_{\theta s}^{*}$ profiles for $\beta = 160°$ and $Bo_m = 1$.

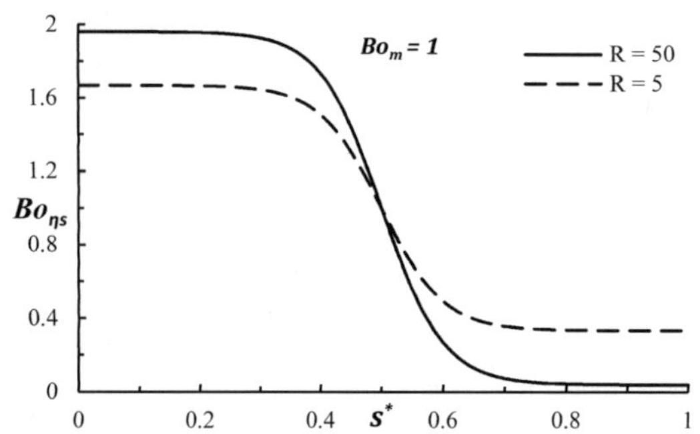

Fig. 4: $Bo_{\eta s}$ profiles for $Bo_m = 1$, $s_f^{*} = 0.5$ and $\delta_f^{*} = 0.1$.

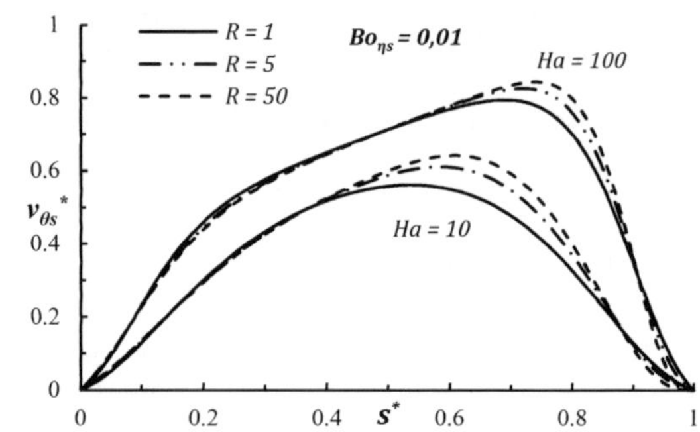

Fig. 5: $v_{\theta s}^{*}$ profiles for $\beta = 20°$ and $Bo_m = 0.01$.

Thus, the magnetic field can be helpful to detect the possible presence of defects usually neglected, due to the 2-D tendency. However, the surface shear stress and the bulk shear stress at the interface corrected by the magnetic effects should be balanced.

To compare these quantities, a revised Boussinesq number [4] is introduced:

$$\widetilde{Bo}_{\eta s} = \frac{2\,Bo_m}{\varepsilon^3 Ha\big(1 + \exp(-\varepsilon Ha)\big)}. \tag{14}$$

For a Hartmann number significantly larger than the Boussinesq number, the surface velocity is almost perfectly aligned with the rotating floor, whatever the interface shape or the surface viscosity profile. In other word, a revised Boussinesq number smaller than unity does not allow surface flow to be sensitive to the wetting effect or to surface viscosity non-uniformities.

Conclusion

By comparing the experimental surface velocity field along an oxide layer with numerical simulations presented in this study, it is now possible to evaluate the surface physicochemical parameters of oxidized melted metals with a better accuracy. Indeed, artifacts due to the wetting effect and to surface viscosity non-uniformities are taken into account as long as the revised Boussinesq number is greater than or equal to unity.

Acknowledgments

The authors are grateful to C. Bozonnet for his contribution to modeling during his master thesis. The laboratory SIMaP is part of the LabEx Tec 21 (Investissements d'avenenir - Grant Agreement No. ANR-11-LABX-0030).

References

1. R.J. Mannheimer & R.S. Schechter, *J. Fluid Mech*, **470** (2002) 135-149
2. J.M. Lopez, R. Miraghaie & A.H. Hirsa, *J. Colloid Interf. Sci.* **248**:1, (2002) 103-110
3. J. Delacroix, L. Davoust & K. Patouillet, *Rev. Sci. Instrum.* **89**:1 (2018) 015103
4. J. Delacroix & L. Davoust, *Phys. Fluids*, **26**:3 (2014) 037102
5. J. Delacroix & L. Davoust, *Phys. Fluids*, **27**:6 (2015) 062104
6. L. Davoust, Y.L. Huang, S.H. Chang, Surface Science, **603**:17 (2009) 2777-2788
7. J.C. Slattery, L. Sagis, & E. Oh, New York: Springer-Verlag, (2007) 22,360,712.

Optimization of Amplitude and Frequency Modulated Magnetic Field Parameters in a Square Mold Wall

A. Kapusta[1,2], B. Mikhailovich[2*], M. Havkin[1], B. Tilman[1], K. Smirnov[1], A. Kudriavtcev[1]

[1]Nord Israel Research and Development Ltd, Beer-Sheva, Israel
[2]Department of Mechanical Engineering,
Ben-Gurion University of the Negev, Beer-Sheva, Israel

[*] *Correspoing author: borismic@bgu.ac.il*

Abstract
The possibility of optimization of parameters of amplitude and frequency modulated rotating magnetic field (AFM RMF) acting on the copper wall of a mold with a square cross-section aimed at an increase of the efficiency of electromagnetic impact on the mold in the process of continuous casting of a steel billet has been theoretically substantiated.

Key words: continuous casting of steel billet, mold, rotating magnetic field, amplitude and frequency modulation, resonance.

Introduction
To improve the continuous ingot quality, a variety of methods of action on liquid metal are used and constantly modernized [1]. In the presented study, we analyze the possibility of optimizing the non-stationary electromagnetic forces excited in the wall of the crystallizer by means of an amplitude-frequency modulated rotating magnetic field (AFM RMF) [2], which ensures a maximal amplitude of electro-magnetic fields (EMF) in it. The authors suggest a method of optimization of non-stationary electro-magnetic forces excited in the mold wall using AFM RMF. The conditions for the resonant energy transfer by a part of the Fourier-pack harmonics to the mold wall are determined.

Problem Formulation and Solution
A problem of excitation of oscillations in the wall of a square copper mold using an AFM RMF source is examined for the optimization (determining the maximal amplitude of the vectorial potential) of this influence.
Electrodynamic processes in the mold wall arising under the action of a circularly polarized AFM RMF excited by an ideal inductor are described in rectangular coordinates (x, y, z) by a dimensionless equation for the z – component of the vectorial potential a_z ($\mathbf{b} = \text{rot}\mathbf{a}$)

$$\frac{1}{\pi^2}\left(\frac{\partial^2 a_z}{\partial x^2} + \frac{\partial^2 a_z}{\partial y^2}\right) - \frac{\overline{\omega}}{\pi}\frac{\partial a_z}{\partial t} = U(a_z) = 0, \tag{1}$$

where $\overline{\omega} = \frac{\mu_0 \sigma_m \omega_0 x_0^2}{\pi^2}$ - dimensionless frequency of EMF in a mold wall with specific electrical conductivity σ_m and with the half of external mold side length x_0, μ_0 – magnetic permeability in the vacuum, ω_0 - carrier angular frequency of AFM RMF, t – dimensionless time.

Boundary conditions for Equation (1) are determined in the first approximation by the magnetic induction **b** of the RMF excited by an ideal inductor in whose circular boring the mold is located (Fig.1).

In rectangular Cartesian coordinates, these conditions have the following forms:

$$\left.\frac{\partial a_z}{\partial x}\right|_{x=1} = \frac{-(1 - iy)\phi(t)}{\sqrt{1 + y^2}}; \left.\frac{\partial a_z}{\partial x}\right|_{x=x_1} = -\frac{(x_1 - iy)\phi(t)}{\sqrt{x_1^2 + y^2}}e^{-d} \tag{2}$$

$$\left.\frac{\partial a_z}{\partial y}\right|_{y=1} = \frac{(x - i)\phi(t)}{\sqrt{1 + x^2}}; \quad \left.\frac{\partial a_z}{\partial y}\right|_{y=-y_1} = \frac{(x + iy_1)\phi(t)e^{-d}}{\sqrt{x^2 + y_1^2}} \tag{3}$$

where $d = \frac{\langle\delta\rangle}{\Delta_p}$, $\langle\delta\rangle$ is a dimensional averaged mold wall thickness which is determined by the following formula:

$$\langle\delta\rangle = \frac{\delta_0}{2}\int_0^1 \sqrt{1+y^2}dy = 1.148\delta_0$$

δ_0 – dimensional thickness of the mold wall, $\Delta_p = \sqrt{\frac{2}{\mu_0\sigma_m\omega_0}}$ - depth of magnetic field penetration in copper mold.

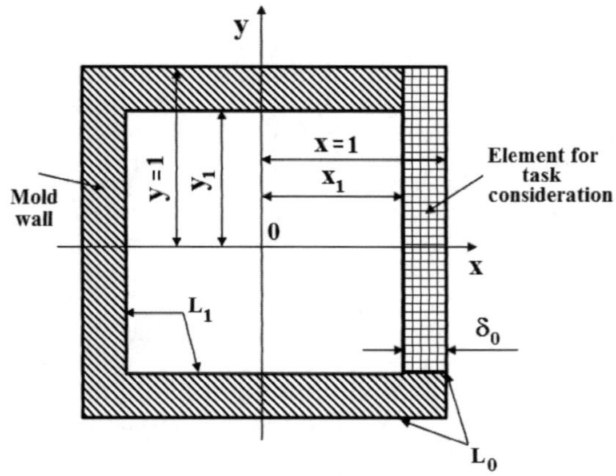

Fig.1: Scheme for calculation of electrodynamical processes in a square copper mold section; x and y are dimensionless Cartesian coordinates.

According to [3], the function of time $\phi(t)$ can be presented in the following form:

$$\phi(t) = (1+\chi e^{i\pi\omega_a t})\sum_{n=-\infty}^{\infty} J_n(\xi)e^{i\pi\Omega_n t} \quad (4)$$

where χ is the depth of amplitude modulation, ω_a – dimensionless angular velocity of amplitude modulation, $J_n(\xi)$ – Bessel function of the n order, Ω_n - dimensionless angular velocity of the n - th Fourier harmonic, $\xi = \frac{\Delta\omega}{\omega_f}$ – frequency modulation index, $\Delta\omega$ – angular frequency deviation, ω_f – angular frequency of frequency modulation.
To solve the problem (1) with boundary conditions (2), we represent vectorial potential a_z in the form

$$a_z = \alpha_z - F(x,y)\phi(t) \quad (5)$$

where

$$F(x,y) = \iint_{x_1-y_1}^{11} \frac{(x-iy)}{\sqrt{x^2+y^2}}\left(e^{-f_1(x)} + e^{-f_1^*(y)}\right)dxdy\,;$$

$$f_1(x) = \frac{1-x}{1-x_1}d; \quad f_1^*(y) = \frac{1-y}{1+y_1}d.$$

The transformation (5) reforms a semi-uniform problem (1) into the following semi-uniform problem:

$$U(\alpha_z) = U[(F(x,y))\phi(t)], \quad \left.\frac{\partial\alpha_z}{\partial x}\right|_{L_0} = \left.\frac{\partial\alpha_z}{\partial y}\right|_{L_1} = 0 \quad (6)$$

where L_0, L_1 are external and internal contours of the mold cross-section (Fig. 1).

We look for the solution of the problem (5) by Galerkin's method in the form

$$\alpha_z = \sum_{k=1}^{\infty}\sum_{l=1}^{\infty} T_{kl}(t)u_k(x)\,v_l(y) \tag{7}$$

where $u_k(x) = \sin\left(k\pi\frac{1-x}{1-x_1}\right)$, $v_l(y) = \sin\left(l\pi\frac{1-y}{1+y_1}\right)$, $x_1 = 1-\bar{\delta}$, $y_1 = 1-\bar{\delta}$, $\bar{\delta} = \frac{\delta}{x_0}$.

Substituting (7) into (6) and accomplishing the procedures of Galerkin's method, we obtain

$$\frac{1}{\pi}\frac{\partial T_{kl}}{\partial t} + P_{kl}T_{kl} = \frac{\frac{P_{kl}}{\pi^2}(\mathrm{Int}_1 - i\mathrm{Int}_2)\phi(t) - (\mathrm{Int}_3 - i\mathrm{Int}_4)\frac{1}{\pi}\frac{\partial\phi(t)}{\partial t}}{\mathrm{Int}_0} \tag{8}$$

where

$$P_{kl} = \frac{1}{\bar{\omega}}\left(\frac{k^2}{\bar{\delta}^2} + \frac{l^2}{\left(2-\bar{\delta}\right)^2}\right);\ \mathrm{Int}_0 = \iint_{x_1-y_1}^{1\,1} u_k^2(x)v_l^2(y)dxdy$$

$$\mathrm{Int}_1 = \iint_{x_1-y_1}^{11} \Delta F_R u_k(x)v_l(y)\,dxdy;\ \mathrm{Int}_2 = \iint_{x_1-y_1}^{11} \Delta F_I u_k(x)v_l(y)\,dxdy$$

$$\mathrm{Int}_3 = \iint_{x_1-y_1}^{11} F_R u_k(x)v_l(y)\,dxdy;\ \mathrm{Int}_4 = \iint_{x_1-y_1}^{11} F_I u_k(x)v_l(y)\,dxdy$$

$$F(x,y) = F_R - iF_I,\ \Delta = \frac{\partial^2}{\partial x^2} + \frac{\partial^2}{\partial y^2}$$

We look for the solution of Equation (8) in the form

$$T_{kl} = (A_{Rkl} + iA_{Ikl})e^{i\pi\Omega_n t} \tag{9}$$

Substituting (9) into (8), we obtain

$$A_{Rkl} = \frac{-1}{\mathrm{Int}_0}\frac{P_{kl}\left[\left(\frac{P_{kl}\mathrm{Int}_1}{\pi^2} - \Omega_n\mathrm{Int}_4\right) + \Omega_n\left(\frac{P_{kl}\mathrm{Int}_2}{\pi^2} + \Omega_n\mathrm{Int}_3\right)\right]}{P_{kl}^2 + \Omega_n^2} \tag{10}$$

$$A_{Ikl} = \frac{1}{\mathrm{Int}_0}\frac{\left[P_{kl}\left(\frac{P_{kl}\mathrm{Int}_2}{\pi^2} + \Omega_n\mathrm{Int}_3\right) + \Omega_n\left(\frac{P_{kl}\mathrm{Int}_1}{\pi^2} - \Omega_n\mathrm{Int}_4\right)\right]}{P_{kl}^2 + \Omega_n^2} \tag{11}$$

Exploring (11) for the presence of extremes using the condition

$$\frac{\partial A_{Ikl}}{\partial\Omega_n} = 0$$

we obtain

$$\Omega_n^2 + 2P_{kl}\left(\frac{P_{kl}\mathrm{Int}_4 + \mathrm{Int}_2}{P_{kl}\mathrm{Int}_3 + \mathrm{Int}_1}\right)\Omega_n - {P_{kl}}^2 = 0 \tag{12}$$

Owing to the symmetry of the problem with respect to variables x and y,

$$\mathrm{Int}_1 = \mathrm{Int}_2;\ \mathrm{Int}_3 = \mathrm{Int}_4$$

The solution of Equation (12) looks as follows

$$\Omega_{nr} = P_{kl}(\sqrt{2}-1) = 0.414P_{kl} \tag{13}$$

Since

$$\Omega_{nr} = 1 + n\omega_f \quad (14)$$

we obtain

$$\omega_{fr} = \frac{(\Omega_{nr} - 1)}{n} \quad (15)$$

where ω_{fr} is the resonance frequency of the RMF frequency modulation.

As an example, we present a plot of the dependence of the vectorial potential amplitude A_{Ikl_n} on the Ω_n frequency harmonic in the copper mold wall with a square cross-section (Fig.1). Parameters for calculation: carrier frequency of the current $f_0 = 10$ Hz; billet cross-section 150x150 mm^2; $\delta_0 = 12$ mm; $\Omega_0 = 0.015$, dimensionless resonance frequency $\Omega_{nr} = 14.1$.

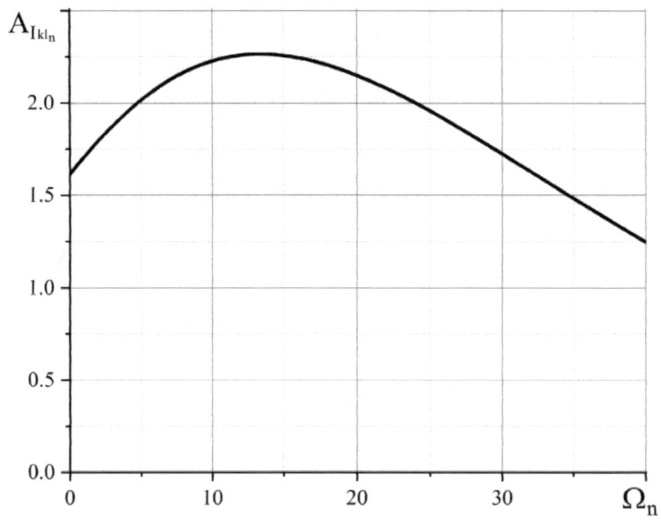

Fig. 2: Dependence of the vectorial potential amplitude A_{Ikl_n} on the frequency of Ω_n Fourier-pack harmonics in the square cross-section mold.

In view of the fact that the AFM RMF is characterized by a superposition of two Fourier-packs (see Equation (4)), the total amplitude of resonance harmonics of the vector potential is more than twice greater than when using a harmonic field.

Conclusion

Results of the analysis show that the amplitudes of resonance values of the vectorial potential exceed more than twice the amplitudes of the vectorial potential corresponding to harmonic currents used at the existing continuous casting machine. Since the electromagnetic body forces are proportional to a squared vectorial potential, the effect of force impact grows more than four times. This point indicates to the possibility of development the highly-efficient continuous casting technology using the amplitude-and-frequency modulation of rotating magnetic field acting on the copper wall of the mold.

References

1. E. Herrmann. Handbuch des Stranggießens. Düsseldorf: Aluminium-Verlag, 1958.
2. M. Havkin, B.Tilman, K. Smirnov, A. Kudriavtcev. Method of optimizing electromagnetic stirring in metallurgical technologies, Foreign License Granted US 62/642,617 March 14, 2018.
3. A. Angot, Compléments de Mathématiques. Paris, 1957.

IOP Conf. Series: Materials Science and Engineering **424** (2018) 012045 doi:10.1088/1757-899X/424/1/012045

3D Numerical Modeling for Inductive Processes

A. Gagnoud[1], Y. Du Terrail-Couvat[1], O. Budenkova[2]

[1] Univ. Grenoble Alpes, CNRS, Grenoble INP, SIMAP, 38000 Grenoble, France

Corresponding author : annie.gagnoud@simap.grenoble-inp.fr

Abstract
In order to analyze the performance and to optimize inductive processes we develop a dedicated software coupling integral method and finite element method. The integral method is dedicated to model electromagnetic phenomenon and the finite element method is used to model heat transfer and fluid mechanic. This software is developed for 2D and 3D configurations. In the article some results are given for various inductive processes.

Key words : induction processes, numerical modeling, couped phenomena, integral method, finite element method

Introduction

Inductive processes are used in many industrial processes and applied to various materials such as metals, semi-conductors and oxides within a large frequency domain from Hz to MHz. These applications involve coupled phenomena: electromagnetism, heat exchange and fluid mechanics. The phenomena manifest themselves and interact depending on the properties of the material, frequency and geometry of the process. Our aim was to develop dedicated software to model a large diversity of electromagnetic processes. Usually, electromagnetic phenomenon is modeled with the finite element method (FEM). Thus, various 3D potential formulations are available[1]. To take into account thin electromagnetic skin depth, boundary impedance formulation can be used [2]. The management of these formulations is complex. We choose to model electromagnetic phenomenon with Integral Method (IM) which is better adapted to inductive processes [3] [4] compared to the FEM. The main advantage of this method is that the calculating mesh is limited to conductive parts that allows one to perform calculations for significantly different scales. For example: some parts of installation can be very large whereas the space between them can be extremely small, or a thin electromagnetic skin depth can be easily modeled. For numerical modeling of heat exchange and fluid mechanic we choose Finite Element Method (FEM) with Galerkin's projection method which allows for describing various formulations. In the software IM and FEM use a discretization with finite elements and Lagrange polynomial interpolation. IM and FEM are coupled using interpolation between the mesh of each method. Depending on the physical process under consideration, different algorithms can be adopted: for strong coupling phenomena one non-linear system for all physical variables is solved [5] and for weak coupling the IM and FEM are used sequentially.

Electromagnetic model

Usually, the different parts of the inductive processes are electrical conductors which can be metal, semi-conductors or melt oxides. In this case, the integral method is based on two equations : the local Ohm's law (1) and the conservation of current equation (2). A sinusoidal alternating potential difference is imposed at the edges of the coil, the equations are written in their complex form:

$$\boldsymbol{J} = -\sigma \boldsymbol{\nabla} V - i\sigma\omega \boldsymbol{A} \tag{1}$$

$$\boldsymbol{\nabla} \cdot \boldsymbol{J} = 0 \tag{2}$$

where $\boldsymbol{J}$ is the complex current density, σ is the electrical conductivity of the material, V is the electrical scalar potential, i is the imaginary number, ω is the angular frequency and $\boldsymbol{A}$ is the magnetic vector potential. The magnetic vector potential is expressed by Biot and Savart law (3) :

$$\boldsymbol{A} = \frac{\mu_0}{4\pi} \iiint_{\Omega_c} \frac{\boldsymbol{J}}{r} dv \tag{3}$$

where μ_0 is the magnetic permeability of the vacuum, Ω_c is the volume of the conductors, r is the distance between a point of the conductor and the calculation point of the magnetic potential. By introducing the expression of the vector potential (3) in the relation (1) an integral formulation is obtained. A first order finite elements mesh is constructed for the electrical conductors of the system, i.e. for the inductor coil, crucible, load, etc. If the electromagnetic skin depth of a conductor is thin compare to the size of the object, a surface mesh is done, otherwise a volume mesh is done. Equation (1) is written on each node of the mesh and equation (2) is integrated on each element of the mesh. We impose zero normal current density at the surface of the conductors except at the boundary of the inductor where electrical potential is imposed. Finally, a linear system is obtained where unknowns are the complex current density and the complex electrical potential. The coefficients of the matrix are integrals which are calculated numerically by Gauss method, like in FEM.

Heat exchange and fluid mechanic model
To model the heat exchange and fluid mechanic a finite element method formulation is obtained by Galerkin's projection. For finite element method various algorithms are developed : linear, non linear, transient and non linear transient.
As an example, we consider static heat transfer equation without phase change (4).

$$\rho C_p \frac{\partial T}{\partial t} + \rho \boldsymbol{U} \cdot \boldsymbol{\nabla} T + \boldsymbol{\nabla} \cdot (-k \boldsymbol{\nabla} T) = Q_{th} \tag{4}$$

where T is the temperature, ρ is the density, C_p is the calorific capacity, $\boldsymbol{U}$ is the velocity, k is the thermal conductivity, t is the time and Q_{th} is the source term. For the temperature, we use second order elements, so the temperature is interpolated by second order Lagrange polynomial functions. For the projection of heat exchange equation second order Lagrange functions are used as test functions.

To model fluid mechanic for an incompressible fluid we consider the Navier-Stokes equation (5) and the continuity equation (6). The density is considered constant.

$$\rho \frac{\partial \boldsymbol{U}}{\partial t} + \rho (\boldsymbol{\nabla} \cdot \boldsymbol{U}) \boldsymbol{U} = \rho \boldsymbol{g} + \boldsymbol{F} + \boldsymbol{\nabla} \cdot \boldsymbol{\Sigma} \tag{5}$$

where $\boldsymbol{g}$ is the gravity vector, $\boldsymbol{F}$ is the external force and $\boldsymbol{\Sigma}$ is the stress tensor.

$$\boldsymbol{\nabla} \cdot \boldsymbol{U} = 0 \tag{6}$$

The velocity is interpolated by second order Lagrange polynomial functions and the pressure is interpolated by first order Lagrange polynomial functions. The Navier-Stokes equation is projected on second order Lagrange functions. The continuity equation is projected on first order Lagrange functions. To establish the fluid mechanic formulation we use P2P1 polynomial functions.

Coupling algorithms
In the case of inductive systems the source term of heat exchange is equal to time average Joule power density (7), and the external force of fluid mechanic is equal to time average electromagnetic force density (8). So we have :

$$Q_{th} = \frac{\boldsymbol{J} \cdot conj(\boldsymbol{J})}{2} \tag{7}$$

$$\boldsymbol{F} = \frac{1}{2} real(\boldsymbol{J} \times conj(\boldsymbol{B})) \tag{8}$$

where $\boldsymbol{B}$ is thee complex magnetic induction, and $conj(\boldsymbol{B})$ is conjugate of $\boldsymbol{B}$. To model the interaction between phenomena three kinds of algorithm are developed :
- sequential algorithm with IM in the first step and FEM in the second step. In this case coupling terms are source terms of thermo-hydrodynamic equation: Joule power density and electromagnetic force density. These quantities are interpolated at the integration points of the elements of the finite element mesh on the IM mesh.
- iterative algorithm between the two methods IM and FEM. In this case coupling terms are the source terms of the thermo-hydrodynamic equation and the electrical conductivity in the electromagnetic formulation which is interpolated at the integration point of the elements of the integral method on the FEM mesh.
- strong coupling : one system of equation is constituted by the formulations of the two methods.

Results
Three cases of inductive cases are presented. We consider first an axisymmetrical configuration of the electromagnetic stirring at 100kHz. The inductor coil is constituted of two wires. The radius of the load is equal to 4cm and its eight is 6cm. A sequential algorithm is used. The thin skin depth model is applied, so the electromagnetic quantities are calculated on the boundary of the electrical conductors : load and wires. A laminar fluid flow is calculated with the FEM formulation. The electromagnetic force is presented in the Fig. 1, and the velocity in the Fig. 2. Secondly we model the electromagnetic levitation at 164kHz. The radius of the load is equal to 0.494cm. The configuration and the electromagnetic force density are presented in Fig. 3 and the fluid flow in the Fig. 4. In the last example a 3D case is presented : inductor coil and cylindrical load at 30kHz frequency. The radius of the load is 2.6cm and its eight is 4cm. The thin skin depth model is applied. A sequential algorithm is used to solve electromagnetism and thermal problem. In the Fig. 5 the configuration and the density power is presented. In the Fig. 6 the temperature at the surface of the load is presented.

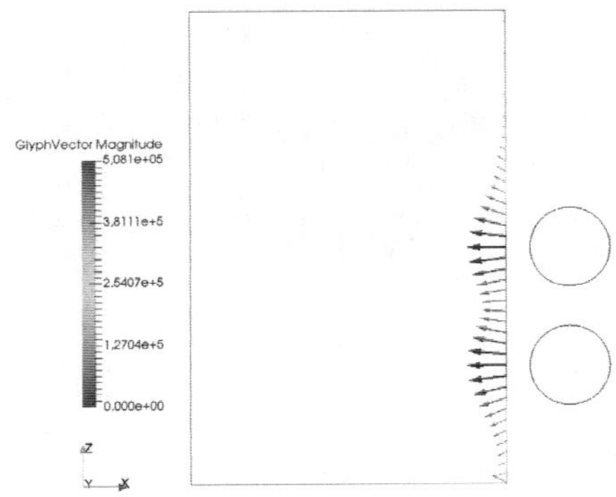

Fig. 1: Vectors of the electromagnetic force density on the boundary of the melt load

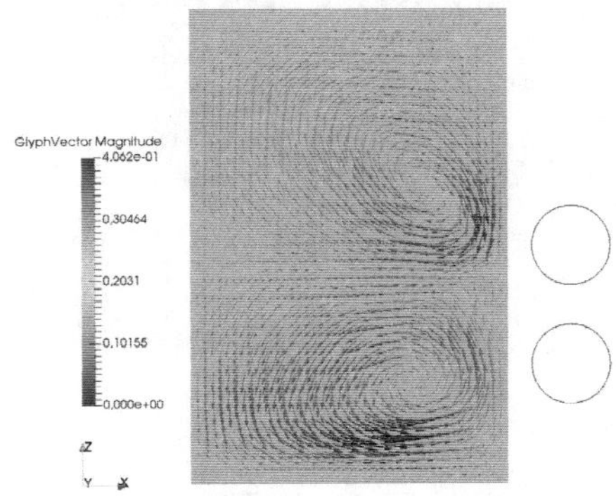

Fig. 2: Fluid flow on the melt load, maximum velocity : 4.2 10^{-1} m/s

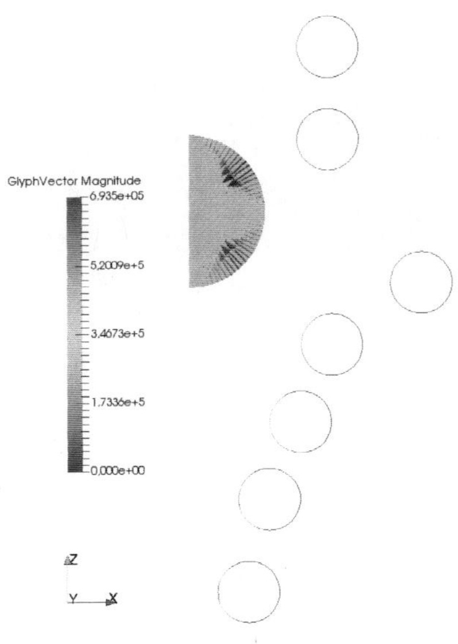

Fig. 3: Electromagnetic levitation : geometry and electromagnetic force density on the melt load

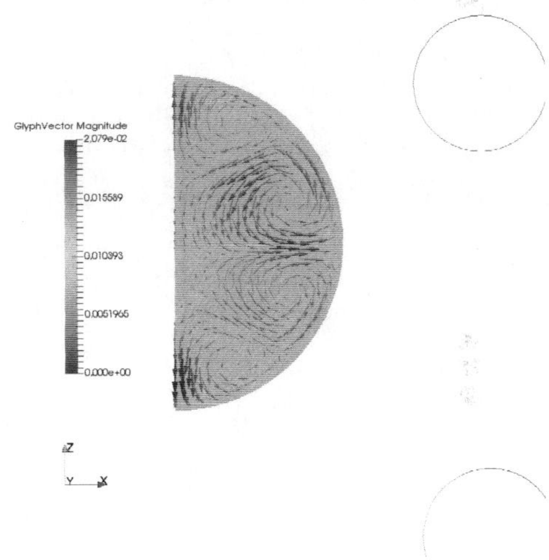

Fig. 4: Fluid flow on the melt load, maximum velocity : 2.03 10^{-2} m/s

Conclusion

A software dedicated to the inductive problem has been developed based on the coupling between the integral method and the finite element method. The integral method is well adapted to model inductive systems because the mesh is limited to active part of the installation. So even a small air gap between the objects of the installation can be modeled. Furthermore, a model adapted to the case of thin skin depth is developed. The use of finite element method permits one to implement various formulations for heat exchange and fluid mechanic. A developed approach allows use of different coupling algorithms.

IOP Conf. Series: Materials Science and Engineering **424** (2018) 012045 doi:10.1088/1757-899X/424/1/012045

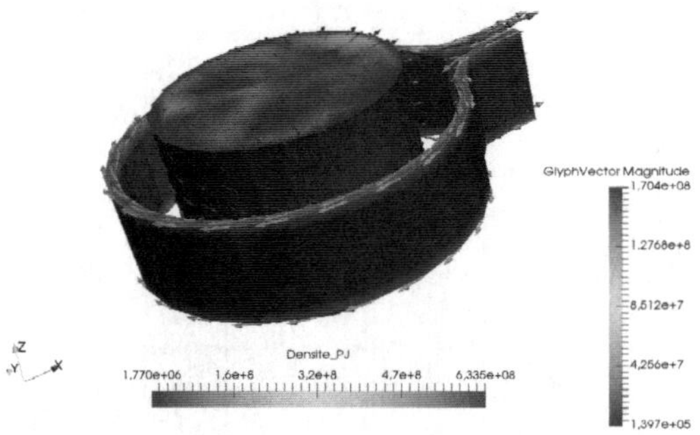

Fig. 5: 3D configuration : geometry and Joule power density and real part of the current density vectors

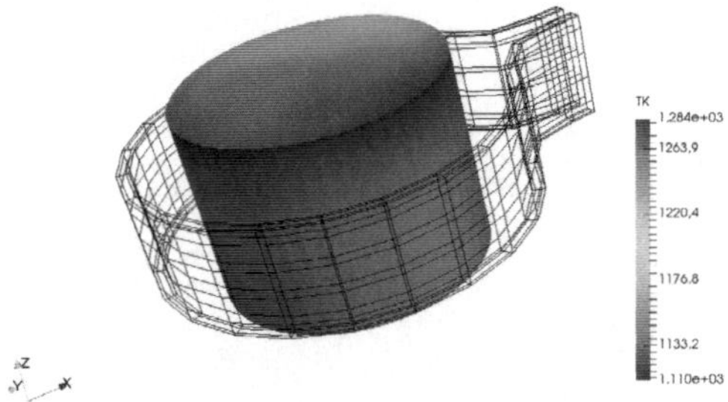

Fig. 6: Temperature on the surface of the load

References

1. O. Biro, K. Preis , IEEE Trans. Magn. 25 (4) (1989), 3145-3159.
2. S. Yuferev, N. Ida, IEEE Trans. Magn. 35 (3) (1999) , 1486-89.
3. R. Scapolan, A. Gagnoud, Y. Du Terrail, IEEE Trans. Magn. 50 (2) (2014), 7023504
4. A. Gagnoud, IEEE Trans. Magn. 40 (1) (2004), 29-36
5. P. Triwong, A. Gagnoud, Serbian Journal of Electrical Engineering 5 (1) (2008), 87-98

IOP Conf. Series: Materials Science and Engineering **424** (2018) 012046 doi:10.1088/1757-899X/424/1/012046

Numerical modelling of liquid metal electromagnetic pump with rotating permanent magnets

V. Dzelme[1], A. Jakovics[1], I. Bucenieks[2]

[1] Laboratory for Mathematical Modelling of Environmental and Technological Processes, Department of Physics, University of Latvia, 23 Zellu street, Riga, Latvia
[2] Institute of Physics, University of Latvia, 32 Miera street, Salaspils, Latvia

Corresponding author : valters.dzelme@lu.lv

Abstract
In this work, we study rotating permanent magnet based liquid metal pump. Full 3D model of the pump is created and different approaches to electromagnetic force computation are investigated. First, we seek a possibility to model the permanent magnet block consisting of many separate magnets as a single block with equivalent analytically prescribed magnetization. Second, different approaches regarding coupling of electromagnetic field to fluid dynamics simulation are considered. Main results show that for pump integral parameter calculation, time-dependent coupling is not necessary, although it can lead to different velocity and turbulent structure distribution in the channel.

Keywords: magnetohydrodynamics, liquid metal pump, simulation

Introduction

For liquid metal transportation, electromagnetic pumps (EMP) have an advantage over mechanical pumps in channel hermetization and the absence of moving parts in the melt. Among EMPs, permanent magnet based ones have an advantage over, for example, linear induction or conduction pumps regarding setup flexibility, the absence of electrodes or windings and overall pump efficiency [1].

Generally, pump performance is described by *p-Q* characteristics, which are curves showing the relation between developed pressure difference Δp and flow-rate Q. Previous studies of electromagnetic pumps are mostly analytical estimations and experimental measurements of those integral parameters [1, 2]. Such approaches, although usually giving reasonably correct Δp and Q, do not give specific information about velocity distribution, turbulence and instabilities. Numerical modelling, on the other hand, can give information about pump operation and flow field, leading to greater understanding of various processes and instabilities and consequently to improvements of pump design and operation.

There have been efforts in EMP numerical modelling, but, as pump setups are usually quite complex, simulations are heavily limited by computer resources, requiring many simplifications to be made. Rotating permanent magnet EMP simulation in 2D was performed in [3], which could correspond to pumps with very high aspect ratios of channel width to height. Liquid metal flow driven by travelling magnetic field in a straight channel was investigated numerically in 3D in [4], showing good developed pressure agreement to analytical estimations. More realistic 3D simulations of a disk-type permanent magnet pump are found in [5], using an iterative electromagnetics – fluid dynamics coupling algorithm to acquire time-average velocity distribution. The iterative algorithm is quite time consuming and it would be virtually impossible to apply it to time-dependent coupling.

In this work, we study rotating permanent magnet based liquid metal pump numerically in 3D. First, we seek a possibility to model the permanent magnet block consisting of many separate magnets as a single block with equivalent analytically prescribed magnetization. Second, different approaches regarding coupling of electromagnetic field to fluid dynamics simulation are considered.

System description

The pump is located at the Institute of Physics, University of Latvia in Salaspils, Latvia. It is used for research purposes and smaller commercial pump testing. Picture of the active part is shown in Fig. 1.

Melt in the channel is Pb-Bi eutectic at 523K (relevant material properties are given in Tab.1). Channel walls, magnet rotor core, shell around the rotor and other parts are made of stainless steel. Below the channel is a laminated iron yoke with relative magnetic permeability $\mu \approx 100$. The permanent magnet rotor consists of 20 permanent magnets arranged in a Halbach array (see Fig.3). Each permanent magnet remanence is $B_r = (1.39 \pm 0.03)$ T.

Fig. 1: Pump picture

Table 1. PbBi properties at 523K

Density, ρ	10430 kg/m^3
El. conductivity, σ	$8.6 \cdot 10^5$ S/m
Viscosity, μ	2.20 mPa·s

Numerical model

The electromagnetic pump is a complex multiphysical system that includes electrically conductive fluid flow, time-varying magnetic field and electric currents and the mutual interaction of those phenomena, mainly in the form of Lorentz force density (fluid driving force) and Joule heat. The electromagnetic part of the simulation is performed in Elmer [6], which solves the Maxwell's equations in potential formulation using the finite element method, but fluid dynamics in OpenFOAM [7], which solves the Navier-Stokes and, in this work, k-ω SST turbulence model equations using the finite volume method. The coupling is achieved with the EOF-Library [8], which effectively couples Elmer and OpenFOAM and is an open-source library itself.

Generally, the current density and magnetic field in this system is, respectively,

$$\vec{j} = \sigma\left(-\frac{\partial \vec{A}}{\partial t} + \vec{v} \times \vec{B}\right) \tag{1}$$

$$B = B_0 - B_i - B_v, \tag{2}$$

where $\vec{j}$ – current density, $\vec{B}$ – magnetic flux density, B_0 is magnetic field of the permanent magnets, B_i is secondary magnetic field created by the $\sigma\frac{\partial \vec{A}}{\partial t}$ current ($\vec{A}$ – vector potential) and B_v is magnetic field created by the $\sigma(\vec{v} \times \vec{B})$ term. B_i and B_v oppose the primary field B_0, hence they are subtracted from it.

There are two ways of taking the $\vec{v} \times \vec{B}$ term into account. The simplest is to compute electromagnetic problem for static fluid, then transfer current density and magnetic field to flow simulation adding $\sigma(\vec{v} \times \vec{B})$ to the imported current density. This approach leads to neglecting of B_v in formula (2) and to underestimation of Lorentz force, as explained in [9]. The correct approach is to solve the electromagnetic problem in its entirety importing velocity from fluid dyamics simulation and iteratively (or every time-step) updating the Lorentz force in fluid flow simulation. Approaches used in this work are summarized in Tab. 2.

Table 2. Coupling strategies used in this work

A	Constant instantaneous $\vec{j}$ and $\vec{B}$; $\vec{v} \times \vec{B}$ computed in fluid simulation
B	Time-dependent coupling; $\vec{v} \times \vec{B}$ computed in fluid simulation
C	Time-dependent coupling; $\vec{v} \times \vec{B}$ computed in electromagnetic simulation

Numerical model corresponding to active part of the pump was created as shown in Fig. 2, with cross-section in Fig. 3. In the beginning of the results section it will be shown that the rotating magnet block consisting of many magnets can be equivalently modelled as a single piece with analytically prescribed remanence components:

$$B_{rx} = B_r \sin\left(\omega t + (n-1)arctg\left(\frac{y}{x}\right)\right) \tag{3}$$

$$B_{ry} = B_r \cos\left(\omega t + (n-1)arctg\left(\frac{y}{x}\right)\right) \tag{4}$$

where n – periodicity (in this case n=5), ω – magnet block rotation frequency, t – time.

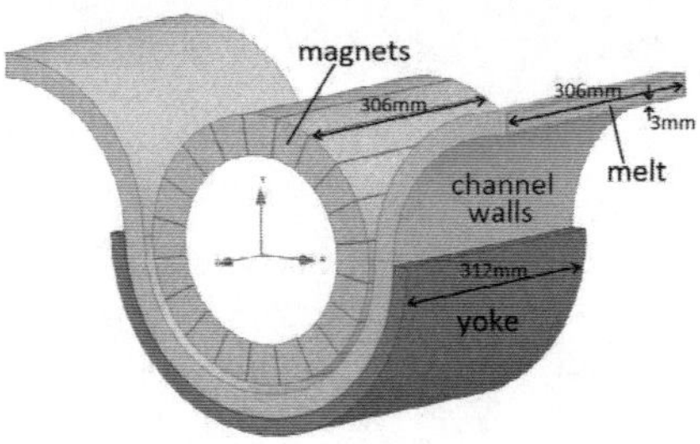

Fig. 2: Numerical model scheme

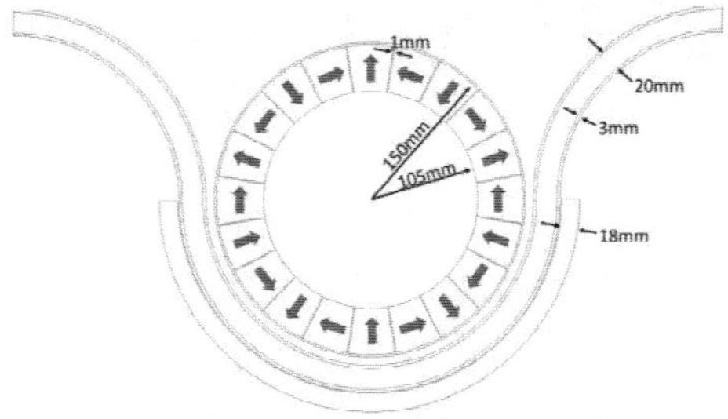

Fig. 3: Numerical model cross-section

Electromagnetics mesh consists of about 3M finite elements in 3D and 200k in 2D simulation. For simplicity channel walls are set the same electric conductivity as the melt (see Tab.1). Fluid dynamics mesh consists of about 220k elements in 3D and 20k in 2D. No-slip boundary condition is applied to all walls, inlet and outlet are openings with set relative pressure difference (the developed pressure) and standard wall functions are used for boundary layer resolution. Fluid density and viscosity as in Table 1.

The main problem with 3D simulations, especially of such coupled type, is computation time. For example, 1 second flow simulation with coupling approach B and C (which recalculate electromagnetics every time-step) takes about 4 days on computer with Intel Xeon E3-1240 3.7GHz 4-core processor with 32Gb RAM. However, the open-source solutions used in this work can be efficiently used on computer clusters, if available.

Results

First of all, Fig. 4 shows instantaneous magnetic flux density vectors in the cross-section of the system, revealing the Halbach array. The instantaneous normal magnetic field component along the channel between magnets and yoke is shown in Fig. 5. The analytical magnetizations are multiplied by a factor of 0.91 to acquire the same amplitude as in the separate magnet case (this is because in the real case there are small gaps between magnets, decreasing magnetic field). Clearly, the magnetic field distribution is completely identical, so further we use only the case of single piece rotor with analytical magnetization.

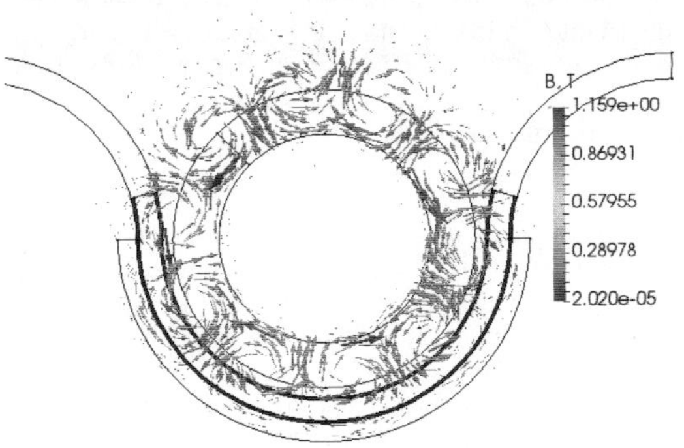

Fig. 4: Magnetic field distribution

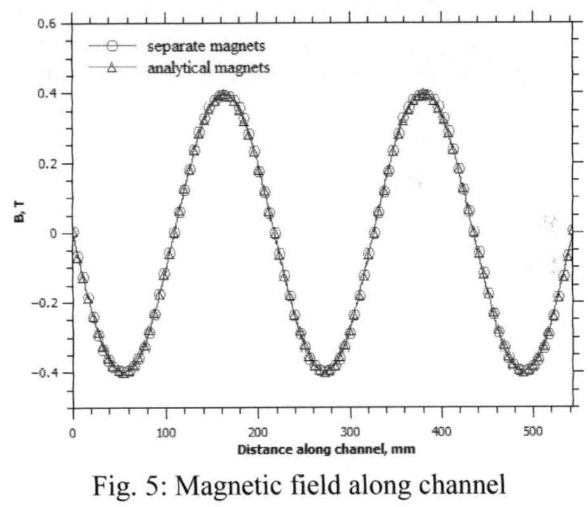

Fig. 5: Magnetic field along channel

An example of instantaneous Lorentz force and velocity distribution is shown in Fig. 6 and 7, respectively. Note the nonhomogeneity of force density along the channel width, which causes fluid recirculation and vortices near the channel sides as seen in Fig.7. This is known as channel end effects and is the main reason why 2D models are incapable of giving reasonable results without introducing some corrections.

Magnet block rotation frequency is taken $f = 6$ Hz. We calculate the *p-Q* characteristics using approaches A, B and C (see section *Numerical model*, Tab. 2) and, to show the necessity of 3D simulations, approach A is also applied to 2D model. Results, including experimental measurements, are shown in Fig. 8.

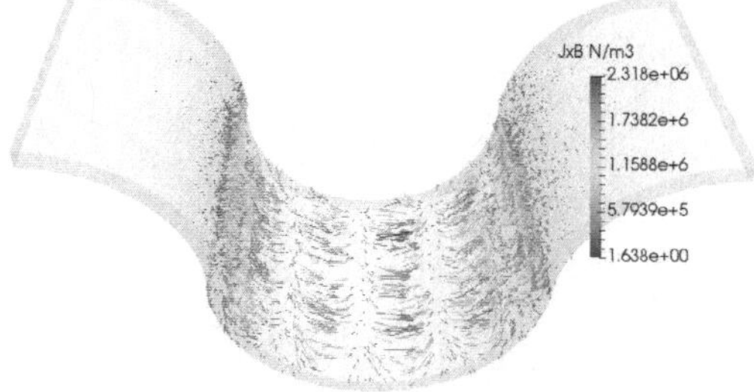

Fig. 6: Lorentz force density

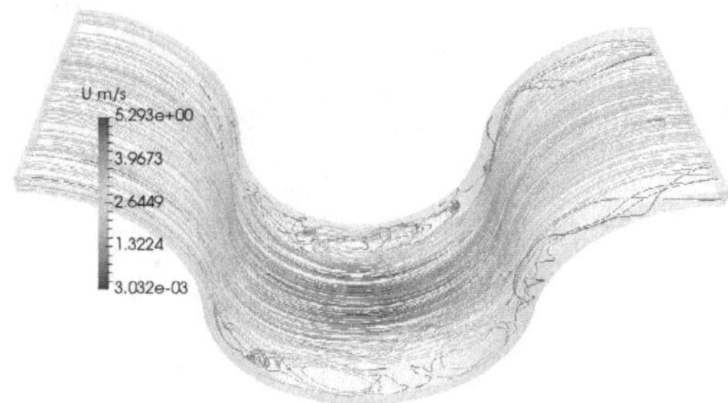

Fig. 7: Velocity streamlines

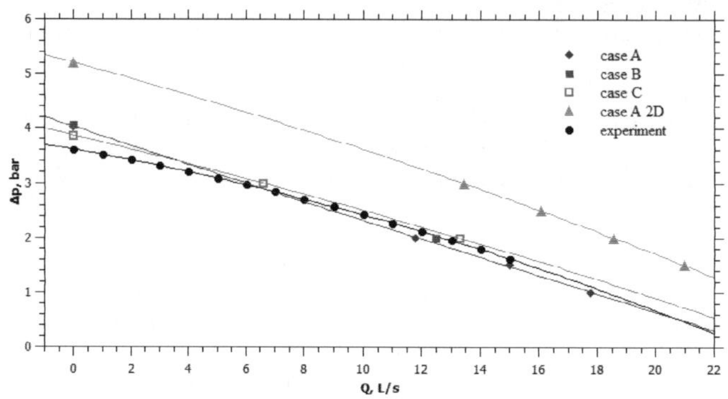

Fig. 8: *p-Q* characteristics

The simple approach A of computing electromagnetic field in static fluid and including $\vec{v} \times \vec{B}$ term only in flow simulation gives results reasonably close to experimental data, with slightly underestimated flow-rate at a given pressure difference. Approach B gives integral results very close to those of case A. As also explained in [9], this is so because the integral force acting on the fluid is constant during the magnet block rotation. The correct coupling approach is C, which is also reflected in the *p-Q* characteristics – the results are closest to experimental data.

Conclusions

Realistic 3D model of permanent magnet based electromagnetic induction pump was created and simulated using open-source software and different coupling approaches. From *p-Q* curves the necessity of 3D simulation is obvious, since 2D model remarkably overestimates induced currents and consequently Lorentz force. The 3D model clearly reveals channel side or end effects – electromagnetic force decreases and changes direction near the channel sides, which leads to change of flow direction and vortices in those regions.

Overall, if only integral parameters are of interest, transient coupling is not necessary and approach A of using constant instantaneous forces is as good as transient approaches B and C. Transient coupling is necessary when studying particular flow structures and instabilities, although in that case the choice of turbulence model plays a big part and approach C is recommended, as the forces can be very sensitive to neglecting some part of magnetic field.

The open-source package EOF-Library for coupling Elmer and OpenFOAM proves to be a powerful tool in MHD applications where transient coupling is needed.

References

1. I. Bucenieks et al., Magnetohydrodynamics, Vol. 47 (2011), No. 1, pp. 97-104
2. I.E. Bucenieks, Magnetohydrodynamics, Vol. 47 (2003), No. 4, pp. 411-417
3. X. Wang, Y. Kolesnikov, Magnetohydrodynamics, Vol. 50 (2014), No. 2, pp. 139-156
4. L. Goldšteins, L. Buligins, Magnetohydrodynamics, Vol. 48 (2012), No. 4, pp. 623-635
5. E. Koroteeva, et al., Fusion Engineering and Design, Vol. 106 (2016), pp. 85-92
6. ElmerFEM, https://www.csc.fi/web/elmer
7. OpenFOAM, https://openfoam.org
8. EOF-Library, https://eof-library.com
9. V. Dzelme, et al., Magnetohydrodynamics, Vol. 53 (2017), No. 4, pp. 731-737

IOP Conf. Series: Materials Science and Engineering **424** (2018) 012047 doi:10.1088/1757-899X/424/1/012047

Numerical and experimental study of liquid metal stirring by rotating permanent magnets

V. Dzelme[1], A. Jakovics[1], J. Vencels[1], D. Köppen[2], E. Baake[2]

[1] Laboratory for Mathematical Modelling of Environmental and Technological Processes, University of Latvia, 23 Zellu street, Riga, Latvia

[2] Institute of Electrotechnology, Leibniz University of Hannover, 4 Wilhelm Busch street, Hannover, Germany

Corresponding author: valters.dzelme@lu.lv

Abstract

In this work, we study liquid gallium stirring by rotating permanent magnets. We demonstrate possibility of easily creating different flow patterns by rotating permanent magnets, which can be industrially important for controlling heat and mass transfer processes in the system. Unlike the typical approach of simulating magnet rotation as a transient problem and time-averaging the Lorentz forces, we solve the magnet rotation as a harmonic (frequency domain) problem, which leads to forces equal to time-averaged ones and decreases the simulation time considerably. Numerical results are validated using qualitative flow structure results from the neutron radiography visualization of tracer particles and quantitative data from Ultrasound Doppler velocimetry.

Keywords: magnetohydrodynamics, liquid metal stirring, neutron radiography, UDV, numerical modelling

Introduction

In metallurgy, melt stirring is used to intensify heat and mass transfer processes, as well as chemical reactions, to improve final product quality. Different methods of stirring can be employed, such as gas bubble flow [1] or electromagnetic stirring [2]. Among electromagnetic stirring techniques, moving permanent magnet systems have advantages over, for example, travelling magnetic fields provided by AC inductors, since no heat is lost to coil heating and different field configurations can be achieved with simpler and smaller setups.

Nowadays, numerical modelling is an important part of industrial process design and improvement, as it can help understanding phenomena and system characteristics that are difficult to study experimentally or theoretically, such as melt flow instabilities. However, numerical models usually require experimental validation. Recently, application of neutron radiography for liquid metal flow visualization has been reported [3]. This opens up new possibilities of numerical model validation, as two-dimensional (or three dimensional with specific configurations of neutron beams) images of flow structures can be obtained, unlike with conventional flow measurement methods, such as Ultrasound Doppler velocimetry (UDV).

In this work, we concentrate mostly on numerical modelling of liquid metal stirring by rotating permanent magnets, trying to obtain flow structures as seen in the neutron experiments. Different possible flow configurations are demonstrated. Since Particle Image Velocimetry (PIV) analysis of tracer particles is not in a good agreement with numerical results [4], we perform Ultrasound Doppler Velocimetry measurements for quantitative validation.

System description and numerical model

The scheme of main liquid gallium stirrer under investigation is shown in Fig.1. It consists of 10x10x3cm glass container with gallium (density 6080 kg/m^3, viscosity 1.97 mPa·s, conductivity 3.7 MS/m) and four cylindrical permanent magnets (remanence 1.4 T) attached to electric motor. The magnets are cylinders magnetized perpendicular to their axes. Magnet positions are fixed relative to the vessel, but their direction of rotation and gallium level in the vessel can be changed. This system was used in the neutron radiography experiments. Basically, the experimental procedure was neutron beam striking the vessel perpendicular to its plane and transmitted neutrons being detected behind the vessel. The acquired data was shadow images of contrast particles entrained in the gallium flow. Further details of the experiments and methodology can be found in [3].

The system shown in Fig.2 is used only for quantitative validation against UDV measurements.

The numerical model contains liquid gallium volume, the magnets and air around them. The fluid is driven by Lorentz force:

$$\vec{f}_L = \vec{j} \times \vec{B} = \sigma\left(-\frac{\partial \vec{A}}{\partial t} + \vec{v} \times \vec{B}\right) \times \vec{B} \tag{1}$$

where $\vec{j}$ – current density, $\vec{B}$ – magnetic flux density, $\vec{A}$ – magnetic vector potential, $\vec{v}$ – fluid velocity.

The electromagnetic part of the simulation is performed in *Elmer* [5], which solves the Maxwell's equations in potential formulation using the finite element method, but fluid dynamics in *OpenFOAM* [6], which solves the Navier-Stokes and k-ω SST turbulence model equations using the finite volume method. The coupling is achieved with the an open-source code *EOF*-Library [7], which effectively transfers data between *Elmer* and *OpenFOAM*.

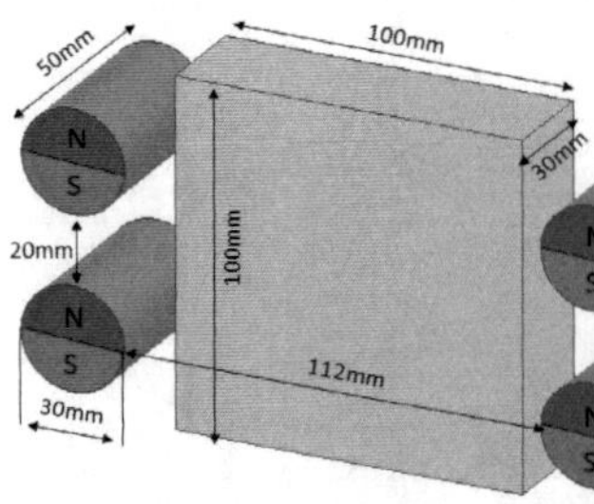

Fig. 1: Main stirrer under investigation

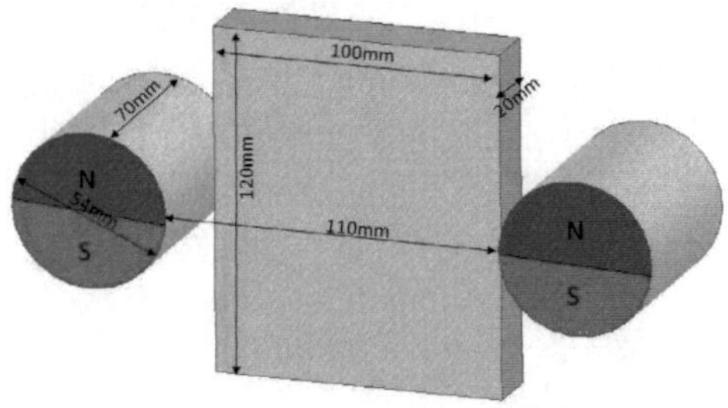

Fig. 2: Stirrer for UDV validation

In this work, we consider the magnet rotation in quasi-stationary approximation (frequency domain), where permanent magnet rotation is modelled by setting real and imaginary parts for magnetization components. It was tested that in such a way obtained Lorentz force is equal to time-averaged one. This approach considerably decreases simulation time compared to transient magnet rotation computation and performing time-averaging.
The $\vec{v} \times \vec{B}$ term is computed directly in *Elmer*, with velocity field imported from *OpenFOAM*. The coupling algorithm used to obtain velocity coupled Lorentz force is shown in Fig.3, where superscript *i* designates coupling iteration. Usually, no more than three coupling iterations are needed to achieve converged force distribution.

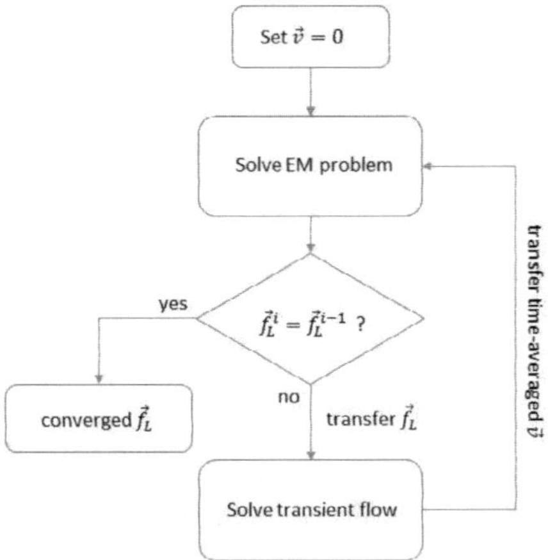

Fig. 3: Algorithm to compute velocity coupled Lorentz force

Electromagnetics mesh consists of 240422 elements, but fluid dynamics of 37500 elements. No-slip boundary condition is applied to all walls, free surface is fixed (modelled as a wall) and standard wall functions are used for boundary layer resolution. For the main system under investigation (Fig.1) two gallium filling levels are considered – 100mm and 75mm with two magnet rotation configurations as shown in Fig.4. The second system (Fig.2) is used only for comparison to UDV measurements, in which case the gallium level is 120mm and the magnets are counter-rotating (left one clockwise, the right one counter-clockwise).

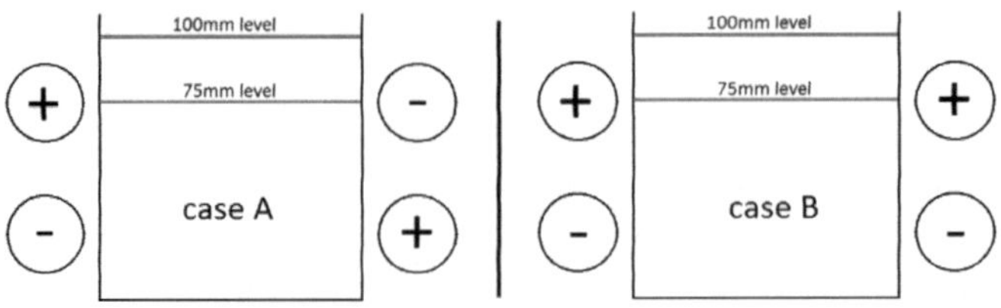

Fig. 4: Considered cases; "+" means clockwise, "-"means counter-clockwise

Results

One of the goals of the current study is to validate numerical models, both qualitatively and quantitatively. Quantitative validation cannot be achieved using PIV analysis of neutron radiography of tracer particle movement [5], therefore we resorted to traditional methods of measurement. The measurements were performed in system shown in Fig.2. Simulation results in comparison to UDV measurements on a vertical line near the vessel side wall (where velocity has maximum) for magnet rotation frequency 3.18Hz are shown in Fig.5. Quantitatively, the results agree well. The curve shape disagreement can be due to instabilities – vortex transitions in permanent magnet stirred flows have been reported in the past [8] and the lifetime of some specific vortical configuration can be longer than the time UDV measurements were averaged over.

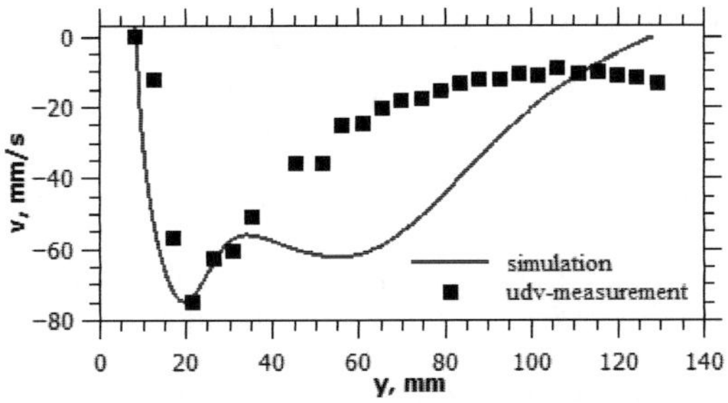

Fig. 5: Velocity on vertical line near vessel side wall

Another goal of the current work is demonstrating, by neutron radiography and numerical simulation, different stirring regimes achieved by rotating permanent magnets. The following figures contain simulated velocity vectors in vertical cross section in the middle of the vessel and still frames from the neutron radiography. Since the still neutron images don't reveal flow structures, vortex directions are indicated as arrows (the flow can be seen in videos shown in presentation), although the flow direction near the surface can be deduced from particle distribution (darker areas).

Fig.6 and 7 are results of 100mm filling level for magnet rotation case A, Fig.8 and 9 is with 75mm filling for case A. To demonstrate other possible flow structures, Fig.10 and 11 shows simulated velocity results for case B (Fig. 4) with 100mm and 75mm gallium level, respectively. Magnet rotation frequency for all cases is 21Hz.

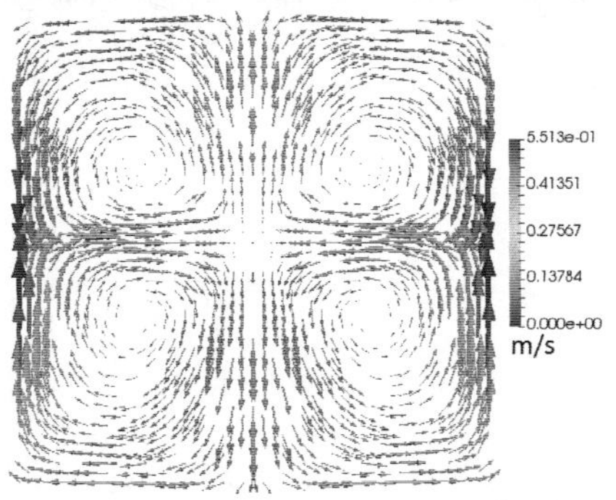

Fig. 6: Case A, gallium level 100mm, simulation

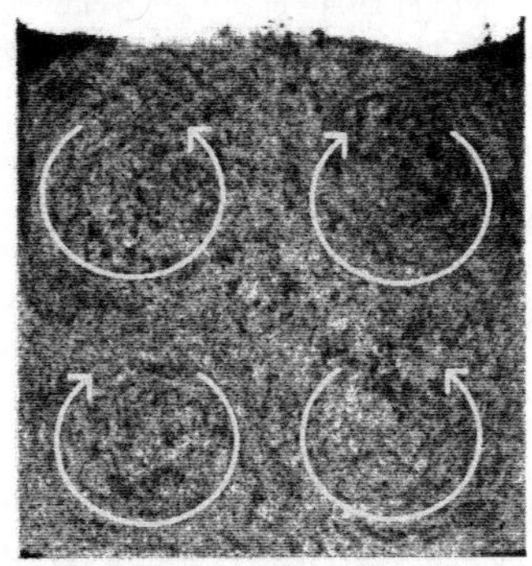

Fig. 7: Case A, gallium level 100mm, neutron radiography

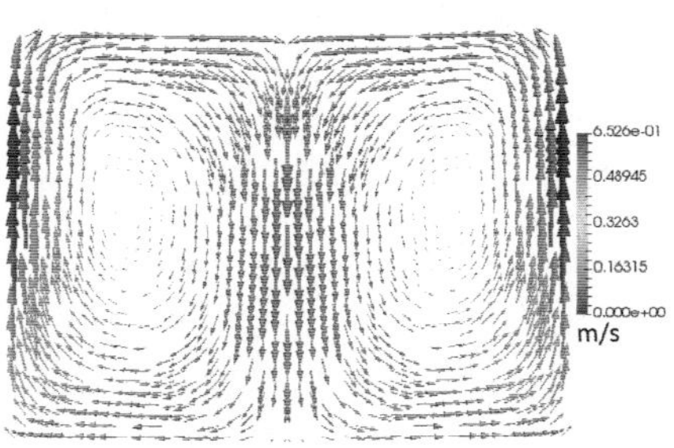

Fig. 8: Case A, gallium level 75mm, simulation

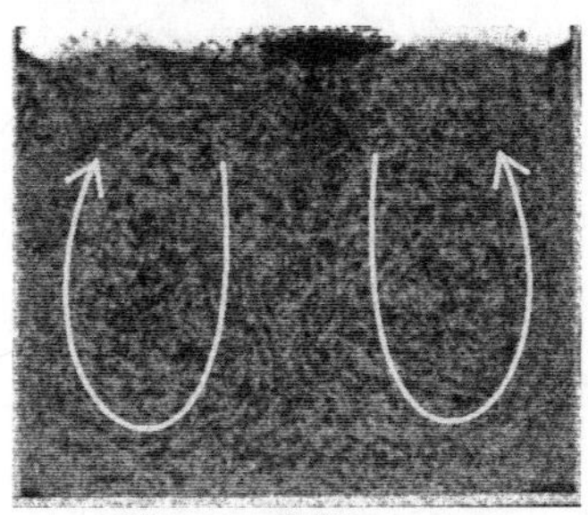

Fig. 9: Case A, gallium level 75mm, neutron radiography

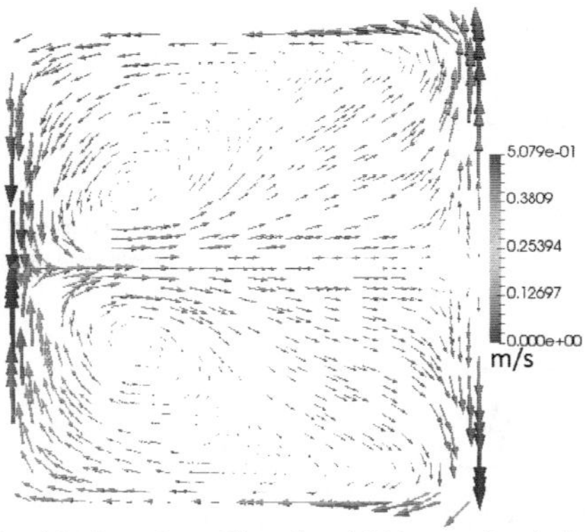

Fig. 10: Case B, gallium level 100mm, simulation

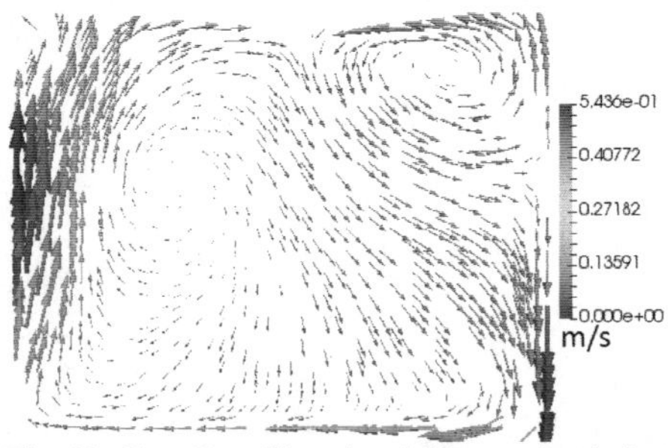

Fig. 11: Case B, gallium level 75mm, simulation

Such simple four magnet system can obviously provide many different flow patterns. Compared to EM stirring by fields produced by inductor coils, permanent magnet systems are more energy efficient and flexible regarding setup and desirable magnetic field distribution. For industrial applications, it means that if flow field can be easily controlled, so can other parameters, such as temperature and inclusion concentration distribution. However, although high Curie point materials exist, usage of permanent magnets for stirring high temperature melts is still limited.

Conclusions

This paper is concerned with numerical simulation of liquid metal stirring by means of rotating permanent magnets. Qualitative validation was done using images from neutron radiography of tracer particles entrained in gallium flow and velocity magnitude validation was achieved using UDV measurements.

An improvement of simulation time was achieved by modelling the magnet rotation in quasi-stationary approximation, that is, solving the EM problem in frequency domain, which leads to Lorentz force density identical to time-averaged one. Even more advantages are provided by using free open-source software, which does not have any license limitations and therefore can be run on large computer clusters.

Different flow configurations easily achievable by rotating permanent magnet system were demonstrated numerically and by neutron radiography visualization of tracer particles. For industrial applications, the ability to easily and controllably create different flow patterns can be useful in controlling process parameters, for example, temperature and concentration distributions.

References

1. J. Wannasin et al., Scripta Materialia 55 (2006) pp. 115–118
2. S. Eckert et al., Metallurgical and Materials Transactions B, Vol. 39 (2008), pp. 374–386
3. M. Scepanskis et al., Metallurgical and Materials Transactions B, Vol. 48 (2017), pp. 1045-1054
4. V. Dzelme, et al., Proceedings of XVIII International UIE-Congress, Hannover, Germany, 2017, pp. 362-367
5. ElmerFEM, https://www.csc.fi/web/elmer
6. OpenFOAM, https://openfoam.org
7. EOF-Library, https://eof-library.com
8. V. Dzelme et al., Proceedings of the 10th PAMIR International Conference, Cagliari, Italy, 2016, pp. 252-256

Fast 3D simulation of various applications of induction heating

H.K. Bui, A. Ba, G. Berthiau, D. Trichet

Laboratory IREENA
University of Nantes, France

Corresponding author: Gerard.Berthiau@univ-nantes.fr

Abstract
In this paper, we present some advances in 3D finite element modeling of the induction heating process. Lighten finite element models are used for fast simulation of Carbone Fiber Reinforced Polymers and massive metals at high frequencies. The lightening methods include the use of degenerated hexahedral Whitney elements for very thin anisotropic layers and the impedance boundary condition with external electrical circuit constraint.

Keywords: 3D FEM, Degenerated Whitney Elements, Impedance Boundary Condition, Induction Thermography NDT, Induction Heating.

Introduction

In this work, various applications of induction heating including welding and nondestructive testing (NDT) are presented. For metal material and carbon fiber reinforced polymers (CFRP), induction welding has already proved its efficiency whereas induction thermography may provide a very promising NDT method [1]. In induction thermography, the presence of defect leads to abnormal temperature distribution on the surface of the piece which can be revealed on the thermal image measured by an infrared thermal camera [2].
Simulation software using finite element method (FEM) provides a powerful tool for the design and optimization of the techniques [3]. However, simulations at high frequencies with external circuit coupling using classical FEM are very time consuming. For CFRP, classical FEM must deal with the numerical issues of very thin and anisotropic multi-layers region which also leads to high computation time.
In this work, we present fast and accurate 3D finite element models for coupled electromagnetic thermal simulation of induction heating processes. For CFRP, the electromagnetic model is lightened by using degenerated hexahedral Whitney elements. For strong skin effect simulations, the electromagnetic model is lightened by means of the Surface Impedance Boundary Condition (SIBC) [4] applied to both the massive induction coil surface taking into account the coupled external electrical circuit and the surface of the conductor work-piece. In this case, the thermal diffusion in the work-piece is solved using surface electromagnetic power density as thermal source. Applications on induction welding and induction thermography NDT are shown.

Classical mathematical formulation

The weak electromagnetic $\boldsymbol{A}-\psi$ and thermal formulations (1) and (2) can be used to calculate induced electromagnetic fields and temperature distribution in the conductor body:

$$\int_{\Omega}\left(\boldsymbol{curlN}_e\frac{1}{[\boldsymbol{\mu}]}\boldsymbol{curlA}+[\boldsymbol{\sigma}](\boldsymbol{N}_e+\boldsymbol{grad}N_n)\,j\omega(\boldsymbol{A}+\boldsymbol{grad}\psi)\right)d\Omega-\int_{\Gamma}\boldsymbol{N}_e\left(\boldsymbol{n}\wedge\frac{1}{[\boldsymbol{\mu}]}\boldsymbol{curlA}\right)=\int_{\Omega}\boldsymbol{N}_e\boldsymbol{J}_s d\Omega \quad (1)$$

$$\int_{\Omega_c}\left(\rho C_p\frac{\partial T}{\partial t}-div[\boldsymbol{\lambda}]\boldsymbol{grad}T\right)N_n\,d\Omega=\int_{\Omega_c}PN_n\,d\Omega \quad (2)$$

where $\boldsymbol{A}$ and ψ are respectively the magnetic vector potential and the primitive in time of electric scalar potential, J_s the source current density, P the electromagnetic induced power density, T the temperature, C_p the specific heat, ρ the specific mass of the material, $[\boldsymbol{\sigma}]$ and $[\boldsymbol{\lambda}]$ respectively the tensors of electrical and thermal conductivity which are function of fiber orientation θ^i. The edge shape function and nodal shape function are denoted by $\boldsymbol{N}_e$ and N_n respectively.

Degenerated hexahedral Whitney elements

The degenerated nodal elements are determined by:

$$(N_n^2 \quad , N_n^2 \quad) \tag{3}$$

where N_n^{2D} denotes the nodal elements defined on the quadrilateral mesh of the upper and lower surfaces of the thin region which are:

$$\begin{aligned} &= (1-u)(1-v)/4, \quad &= (1+u)(1-v)/ \\ &= (1+u)(1+v)/4, \quad &= (1-u)(1+v)/ \end{aligned} \tag{4}$$

and $= (1 \pm w)/2$ are the interpolation functions along the thickness of the region. The degenerated edge elements are written as:

$$(\boldsymbol{N}_e^2 \quad , \boldsymbol{N}_e^2 \quad , N_n^{2D}\boldsymbol{n}/\varepsilon) \tag{5}$$

$\boldsymbol{N}_e^{2D}$ is the set of edge elements defined on a quadrilateral, $\boldsymbol{n}$ the normal vector of median surface of the element, ε its thickness. In the case of a multi-layer region, double layer quadrilateral meshes are created to represent each layer of the region. The coefficients associated with the common edges and common nodes between two double layer meshes are the contribution of those calculated in each mesh. The left-hand side of the $\boldsymbol{A} - \psi$ formulation in the CFRP laminate domain Ω_c can be determined by [5]:

$$\begin{aligned} \int_{\Omega_c} \left(\boldsymbol{curlN}_e \frac{1}{[\boldsymbol{\mu}]} \boldsymbol{curlA}\right) d\Omega &= \sum_{i=1,k} \left(\int_S \frac{1}{\mu_{z,i}\varepsilon_i} [\boldsymbol{n} \wedge \boldsymbol{N}_e^{2D}][\boldsymbol{n} \wedge \boldsymbol{A}]\, dS \right) \\ &+ \sum_{i=1,k} \left(\int_S \frac{i}{[\boldsymbol{\mu}]_{xy,i}} \left(\int \langle \boldsymbol{rotN}_e^{2D} \rangle \langle \boldsymbol{rotA} \rangle dw \right) dS \right) \end{aligned} \tag{6}$$

$$\begin{aligned} \int_{\Omega_c} ([\sigma](\boldsymbol{N}_e + \boldsymbol{grad}N_n)\, j\omega(\boldsymbol{A} + \boldsymbol{grad}\psi))\, d\Omega &= \sum_{i=1,k} \left(\int_S \frac{\sigma_{z,i}}{i} (\varepsilon_i \boldsymbol{nN}_e + [N_n^{2D}])(\varepsilon_i \boldsymbol{nA} + [\psi])\, dS \right) \\ &+ \sum_{i=1,k} \left(\int_S [\sigma]_{xy,i}\varepsilon_i \left(\int \langle \boldsymbol{N}_e^{2D} + \boldsymbol{grad}N_n^{2D} \rangle \langle \boldsymbol{A} + \boldsymbol{grad}\psi \rangle dw \right) dS \right) \end{aligned} \tag{7}$$

where k is the number of degenerated elements, $_i$ the thickness of element i, S the median surface of the CFRP laminate, $\mu_{z,i}$ and $\sigma_{z,i}$ the permeability and the electrical conductivity in the normal direction to the median surface of the material, $[\boldsymbol{\mu}]_{xy,i}$ and $[\boldsymbol{\sigma}]_{xy,i}$ the physical tensors of the material defined in the plan of its median surface. The symbols $\langle a \rangle = a \quad + a$ and $[a] = a \quad - a$ denote respectively the weighted average of a function a over the thickness of the layer and the jump of a across the layer.

Case study 1

An induction welding case study is shown in this section. The piece to be heated is a 16 plies composite plate with the lay-up sequence $[135°/90°/45°/0°/135°/90°/45°/0°]_S$. The thickness of a ply is 140μm. An U-shaped coil is used in this study. It is placed right above the piece and centered on it. The excitation current intensity is of 200A RMS at 2MHz. The geometry of the composite plate and the inductor are shown in the Fig. 1. In the same figure, the simulated distribution of eddy current inside the plate is shown. The temperature rise after 5s of heating is shown in the Fig. 2. The induced power distribution depends strongly on the lay-up sequence of the composite. The temperature distribution can be controlled by means of inductor design and operational parameters choice.

Surface impedance boundary condition with external electrical circuit constraint

In this method, only the surface meshes of the conductors are required. The magnetic vector potential $\boldsymbol{A}$ is defined everywhere whereas electric scalar potential ψ is defined only on the surface of the conductors (inductor and pipe). On the surface of the conductors, one applies the SIBC:

$$\boldsymbol{n} \wedge \frac{1}{\mu} \boldsymbol{rotA} = \frac{1}{Z_c} \boldsymbol{n} \wedge \left(\boldsymbol{n} \wedge j\omega(\boldsymbol{A} + \boldsymbol{grad}\psi)\right) \tag{8}$$

where $\boldsymbol{n}$ is the outward normal vector of the surfaces of the conductors, Z_c the surface impedance which depends on the conductivity of the conductor. The $\boldsymbol{A} - \psi$ weak formulation with a voltage V imposed as external circuit condition [6][7] reads:

$$\int_{\Omega} \boldsymbol{curlN}_e \frac{1}{\mu} \boldsymbol{curlA}\, d\Omega + \int_S \frac{1}{Z_c}(\boldsymbol{n} \wedge j\omega(\boldsymbol{A} + \boldsymbol{grad}\psi))(\boldsymbol{N}_{es} \wedge \boldsymbol{n})dS = -\int_S \frac{1}{Z_c}(\boldsymbol{n} \wedge \boldsymbol{grad}V\alpha)(\boldsymbol{N}_{es} \wedge \boldsymbol{n})dS \quad (9)$$

$$\int_S \frac{1}{Z_c}(\boldsymbol{n} \wedge j\omega(\boldsymbol{A} + \boldsymbol{grad}\psi))(\boldsymbol{grad}N_{ns} \wedge \boldsymbol{n})dS = -\int_S \frac{1}{Z_c}(\boldsymbol{n} \wedge \boldsymbol{grad}V\alpha)(\boldsymbol{grad}N_{ns} \wedge \boldsymbol{n})dS \quad (10)$$

where $\alpha = \sum_{C_S} N_{ns}$ is the sum of all the nodal shape function defined on the contour of the electrodes (C_S) where the voltage V is imposed. S is the surface of the conductors (inductor and pipe). The case study 2 is presented as an application of this method to the NDT by induction thermography.

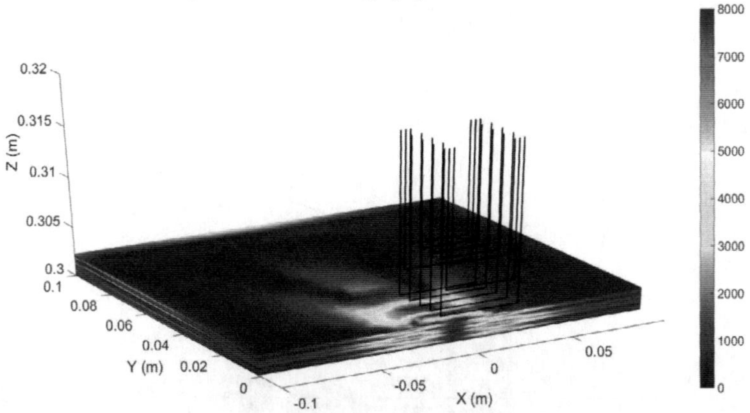

Fig. 1. Case study 1: Geometry of the composite plate and the U-shaped inductor. ½ of the plate ($x > 0, y > 0$) is shown.

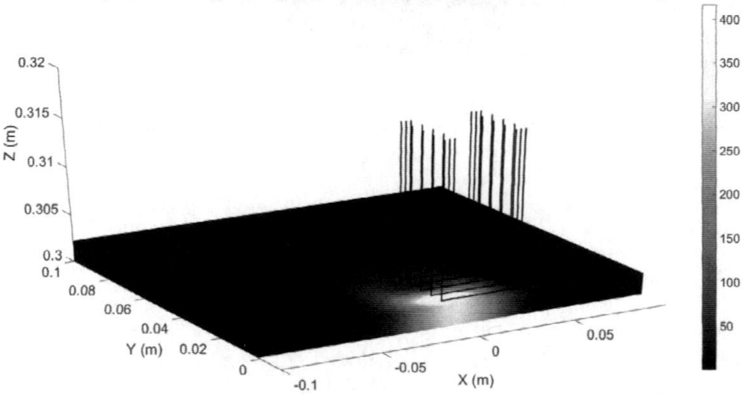

Fig. 2. Case study 1: Eddy current distribution in the plate. ½ of the plate ($x > 0, y > 0$) is shown.

Case study 2

An induction thermography study case is shown in this section. The piece to be inspected is a metal pipe with a vertical crack. A massive inductor is used to carry a high intensity current. The electrical conductivities of the pipe and the inductor are 2×10^6 S/m and 10^7 S/m respectively. The excitation electromagnetic frequency is 5kHz. The surface impedance boundary conditions are applied to both the surface of the pipe and the inductor.

The Fig. 3 shows the current density distribution in the inductor and eddy current in the pipe. The presence of the crack disturbs the circulation of eddy current and leads to a high current density around the crack tips due to the edge effect. This effect is highly observed on the temperature distribution after the heating as shown in the Fig. 4. In the practice, this image can be observed and recorded using an infrared thermal camera. The crack defect can be then detected by this principal.

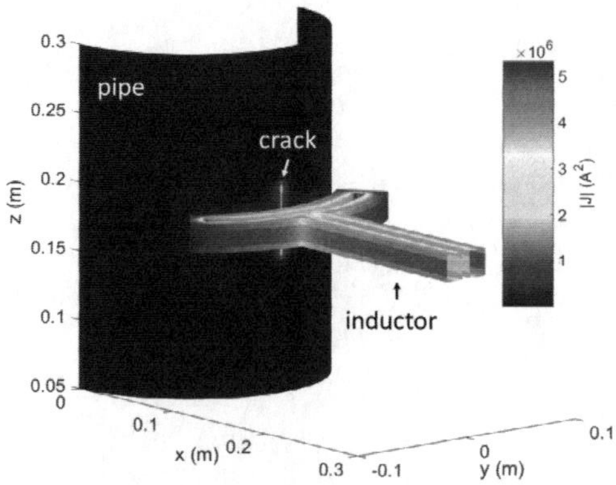

Fig. 3. Case study 2: Current density distribution in the inductor and in the pipe with a crack defect.

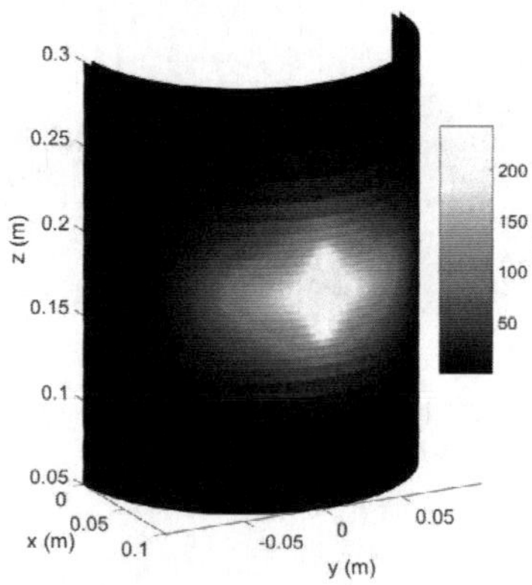

Fig. 4. Case study 2: Temperature distribution showing the crack tips with higher temperature.

Conclusion

Degenerated hexahedral elements can be used to deal with thin and anisotropic layered materials such as carbon fiber reinforced polymers. The use of these elements can achieve the same accuracy as classical methods with reduced computation time. The surface impedance boundary condition can be implemented with external electrical circuit condition and provides a fast method to deal with simulation with strong skin effect.

Acknowledgment

This project has received funding from the European Union's Horizon 2020 research and innovation programme under the Marie Skłodowska-Curie grant agreement No 722134 – NDTonAIR.

References

1. J. Fouladgar (Edt.) (2012), Electrothermics, Wiley.
2. H.K. Bui, G. Wasselynck, D. Trichet, and G. Berthiau (2013), IEEE Trans. Magn., vol. 49, no. 5, 1949-1952.
3. H.K. Bui, G. Berthiau, D. Trichet and G. Wasselynck (2015), 8th Int. Conference on Electromagnetic Processing of Materials, Cannes (France), 403-406.
4. S. Yuferev, N. Ida (1999), IEEE Trans. Magn., vol. 35, no. 3,1486 - 1489.
5. H. K. Bui, G. Wasselynck, D. Trichet and G. Berthiau (2016), IEEE Trans. on Magn., vol. 52, no. 3.
6. S. M. Mimoune, J. Fouladgar, A. Chentouf, and G. Develey (1996), IEEE Trans. on Magn., vol. 32, no. 3, 1605.
7. A. Desmoort, Z. D. Greve, P. Dular, C. Geuzaine, and O. Deblecker (2017), IEEE Trans. on Magn., vol. 53, no. 6.

IOP Conf. Series: Materials Science and Engineering **424** (2018) 012049 doi:10.1088/1757-899X/424/1/012049
***** Formatting Issue - Best Available Paper/Graphic *****

Numerical study of surface waves generated by low frequency EM field for silicon refinement

V. Geza[1], J. Vencels[1], G. Zageris[1] and S. Pavlovs[1]

[1] University of Latvia, LV-1002, Riga, Latvia

Corresponding author: vadims.geza@lu.lv.

Abstract

One of the most perspective methods to produce solar grade silicon is refinement via metallurgical route. The most critical part of this route is refinement from boron and phosphorus due to high segregation coefficients. One possible approach to remove boron is use of reactive gas on surface of silicon melt. An approach of creating surface waves on silicon melt's surface is proposed in order to enlarge its area and accelerate removal of boron via chemical reactions.

This paper focuses on numerical analysis of surface wave creation by means of low frequency magnetic field. Frequency of magnetic field and its amplitude significantly change the character of surface waves with most changes occurring when waves become nonlinear. Mechanism of impurity removal from silicon surface to the gaseous phase is directly connected with surface area enlargement. It was found that two effects are responsible for enhancement of refinement rate – enlargement of surface area and wave-assisted diffusion close to the surface of silicon.

Keywords : silicon refinement, surface waves, electromagnetic stirring

Introduction

The exhaustion of fossil fuel resources, global warming, and increased energy consumption all make finding alternative energy sources important. In terms of environmental impact, solar energy is one of the most perspective and attractive sources. However, development of solar power is limited by the solar power costs in comparison to other power sources. As shown in [1], by 2020 the energy cost for solar power is still expected to be higher than for hydro, onshore wind, geothermal, nuclear and even biomass energy sources. Reduction of solar cell production costs is necessary for this technology to develop, and one solution is to reduce the cost of raw materials like polysilicon. Solar cell costs are directly linked to polysilicon production costs, which can be reduced by lowering energy consumption during solar grade silicon (SoG-Si) production.

In many investigations phosphorus removal is found to be the most challenging task, since the purity threshold (< 1 ppm) was not reached [2]. Although, oxidation and slag refinement is simple process for removing phosphorus from silicon, it is insufficient to reach low purity threshold. Mostly used approach for phosphorus removal is evacuation, it is dependent on free surface area and bulk melt stirring. Refinement enhancement can be achieved by superposing DC and AC magnetic fields [3]. Furthermore, this impact creates also mixing effect in the bulk melt, which is also important for boron removal. In comparison, classical induction stirring can enhance only bulk melt mixing, leaving surface area size unchanged.

Known approach for boron removal is blowing of oxidizing gas on free surface [4], and with additional creation of surface waves by means of electromagnetic impact thus creating capillary waves and increasing surface area. Previously mentioned electromagnetic impact technology (superposition of DC and AC fields) can be used for boron evaporation, because it enhances surface area and stirring.

Highest boron refining rates can be achieved with plasma refining, but it has high energy demand. In reactive gas refining, concentration of boron changes according to first order rate law

$$ln\left[\frac{C_{B,0}}{C_{B,t}}\right] = k_B \frac{A}{V} t \qquad (1)$$

Here C_B – boron concentration with indices after comma indicating time, A is the area of interfacial surface between oxidizing gas and silicon melt, V is the volume of silicon melt, t is time, k_B is boron mass transfer coefficient at the surface, measured in μm/s. k_B is determined experimentally and is dependent on oxidizing gas supply rate, type of oxidizing gas and melt temperature. Creation of surface waves would increase A/V ratio, thus speeding up boron removal. Typical k_B values found in literature are 6 – 40 μm/s, and k_P are 2-5 μm/s for phosphorus.

Modelling of free surface waves

A 3D problem of mercury with free surface is considered. The mercury tank is surrounded by inductor, which creates alternating magnetic field, see Fig. 1 (a) for illustration. A similar device and experiment was described in [5]. This mercury experiment is only used for model validation purposes.

In this model, mercury properties were set in the melt domain – electrical conductivity $\sigma_{melt}=1.04\cdot10^6$ S/m, kinematic

***** Formatting Issue - Best Available Paper/Graphic *****

viscosity $\nu_{melt} = 1.14 \cdot 10^{-7}$ m²/s, density $\rho = 13534$ kg/m³, surface tension $\gamma = 0.5$ N/m and contact angle $\theta = 90°$. The inductor has homogeneous current distribution with density $J_{ampl} = 6.5 \cdot 10^6$ A/m² and frequency f = 14 Hz. Turbulence model k-ε is used, and maximal Courant number is set to 0.4. The transient simulation reached flow time 4 sec.
For numerical simulations we use two open-source software packages – the finite element multi-physics simulation code Elmer and computational fluid dynamics code OpenFOAM. Efficient parallel coupling between both codes is done using EOF-Library [6].

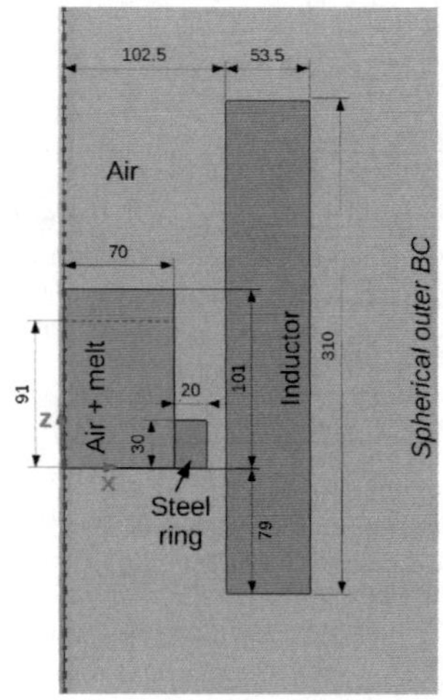

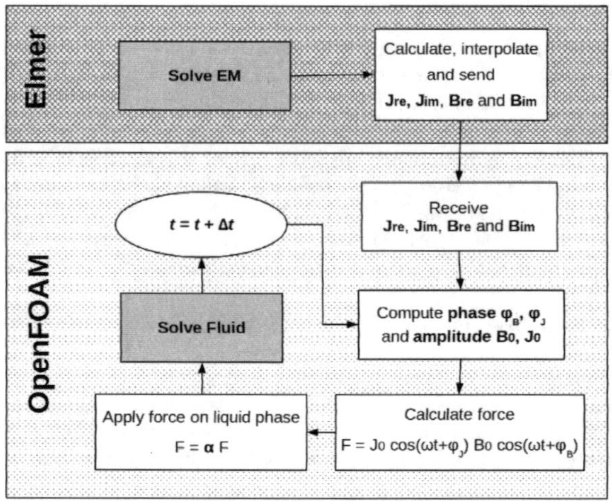

(a) (b)

Fig. 1: (a) Schematic drawing for 2D axisymmetric model, where vertical dashed line is symmetry axis and horizontal dashed line is melt's initial filling level. (b) Computational scheme for coupled OpenFOAM and Elmer simulation

Computation scheme is shown in Fig. 1 (b). OpenFOAM package solves electromagnetic quantities in Elmer package and fluid dynamics with free surface using volume of fluid (VOF) method in OpenFOAM. Elmer solves the complex problem for the time-harmonic electromagnetic (EM) field, then J and B amplitudes and the corresponding phases are computed. Here electromagnetic field is not recalculated when free surface is changed. This effect is neglected to speed up the simulation time. This approach can be used because penetration depth of EM field is much larger than free surface wave amplitude.
Results of 3D surface simulation show different wave structure depending on the strength of magnetic field. Three different regimes were obtained (see Fig. 2):

- Regular axisymmetric waves. Wave structure similar to Fig. 2 (upper-left), but without azimuthal waves. This is most regular pattern and is obtained at weak magnetic field values (below 0.15 T).
- Regular axisymmetric waves with azimuthal waves, seen in Fig. 2 (upper-left and bottom-left). This regime is obtained at stronger magnetic fields (0.15-0.21 T)
- Azimuthal waves Fig. 2 (upper-right). Here axisymmetric have disappeared. This regime appears at magnetic fields above 0.21 T.

These results are in agreement with findings of Galpin and Fautrelle [35]. However, they found fourth regime, highly chaotic surface motion with jets of liquid metal appearing on the surface. This regime appears at even higher magnetic field values. In this research, such magnetic field values were never used.
For refinement of silicon, important parameter is surface area enlargement S(t)/S(0). Here S(t) is surface area at some time, but S(0) is area of surface at rest. Fig. 2 (bottom-right) shows surface enlargement for different magnetic field values. For lowest magnetic field strength, B=0.18 T, the area enlargement is slightly above 1%, for B=0.21 T – about 2.5%, for B=0.25 T – about 4.5%.

Modelling of silicon refining process

For silicon refinement, a system consisting of gas domain with incoming jet (see scheme in Fig. 3) and molten silicon domain is modelled. The computations of silicon melt flow in crucible and gas flow injected with lance are performed using ANSYS FLUENT package with *k–ε* turbulence model. Boron concentration both in melt and in gas is traced with UDS (User Defined Scalar) option in ANSYS FLUENT. Lorentz force, induced by coil is modelled with EMAG module of ANSYS Multiphysics package. Parameters of simulations are following:

- radius of crucible with silicon melt is 0.031 m;
- argon jet lance diameter is 0.0062 m;
- argon flow rate is 4 l_N/min;
- boron-carrying-molecule diffusivity in argon is $3.4 \cdot 10^{-4}\ m^2/s$;
- boron diffusivity in silicon is $2.4 \cdot 10^{-8}\ m^2/s$;
- amplitude value of electrical current in coil is 10 kA.

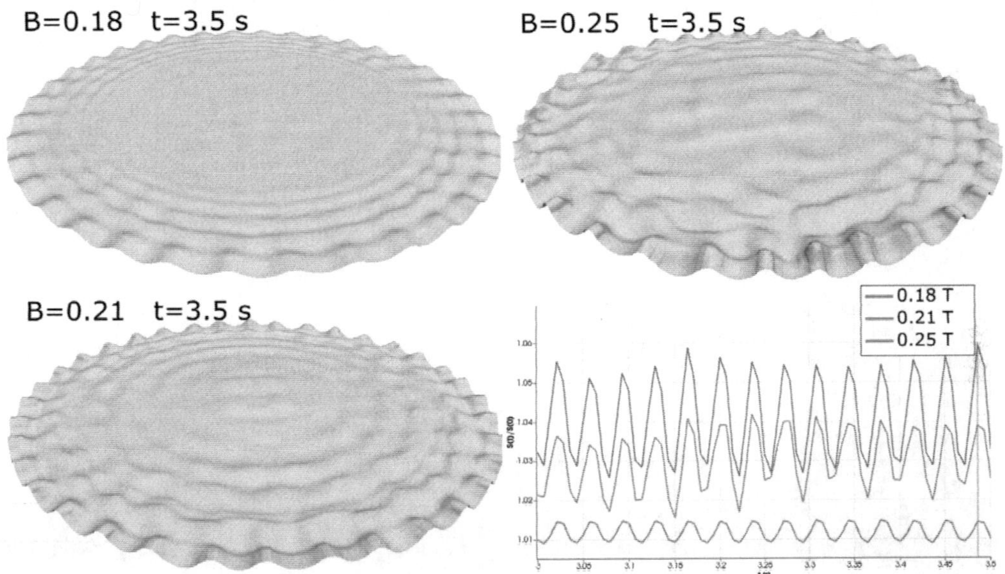

Fig. 2: Free surface shape after 3.49 seconds for different magnetic field strength values and corresponding surface area enlargement (bottom right)

In reality there is a chemical reaction taking action on the surface, where reacting gas oxidizes boron and new molecules are formed which carry boron away from the melt. In this simulation idealized system is used, where one scalar variable was used in both melt and gas domain. In the beginning, fixed concentration of boron in melt is assumed. Then, gas with boron concentration 0 is blown on the surface of silicon, and part of boron is removed.

This simulation was done to estimate the boron mass transfer coefficient k_B at gas–silicon interface (see equation 1). The mean boron concentration in melt during this refinement process simulation is shown in Fig 4. The estimated kB value based on simulation of 5 hours of refinement is 12.7 μm/s. This value is in good agreement with measurement results in [4].

To account the influence of surface waves, a simulation with only gas domain was performed. Gas domain had moving bottom boundary – harmonic wave motion was assigned to it. To measure the influence of different wave amplitudes and frequencies, a boron flux through outlet was observed.

Surface waves with realistic amplitudes did not show any enhancement in boron flux in comparison to static gas-silicon interface. The reason for that is strong diffusion, which leads to diffusion length scale comparable to wave amplitude. It turned out that refinement enhancement can be observed only if following criterion is met:

$$A > 4\sqrt{\frac{D}{f}}$$

Here A – wave amplitude, D – diffusion coefficient of boron-carrying molecule in gas, f – frequency of waves. Wave amplitude in this criterion is, however, linked to frequency, therefore it cannot be directly used to estimate required frequency.

Boron flux through outlet of the gas domain was obtained in simulations with different wave amplitudes (Fig. 5). There is obvious increase of boron flux with increase of amplitude. However, values normalized on surface area lay on the horizontal line, matching values for flux with static surface. This clearly indicates that boron flux increase is connected

IOP Conf. Series: Materials Science and Engineering **424** (2018) 012049 doi:10.1088/1757-899X/424/1/012049

with surface area enlargement.

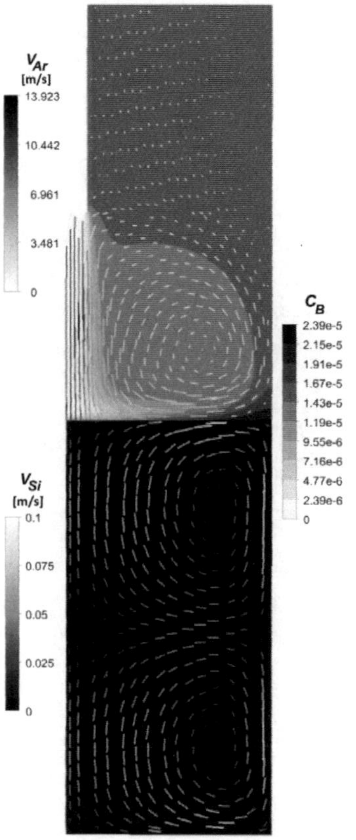

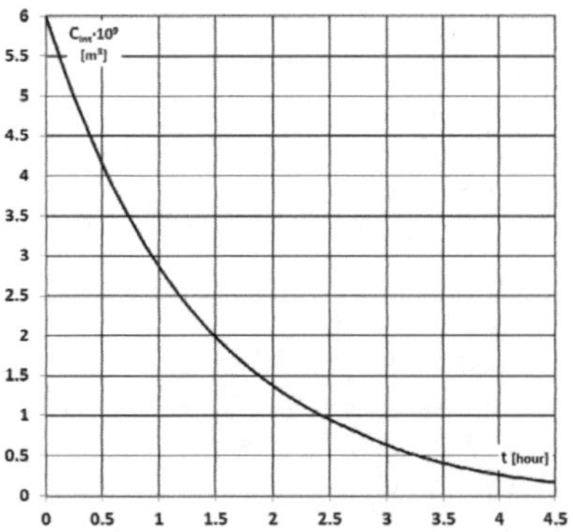

Fig. 4: Boron integral concentration C_{int} in silicon melt as function of silicon refining process time t

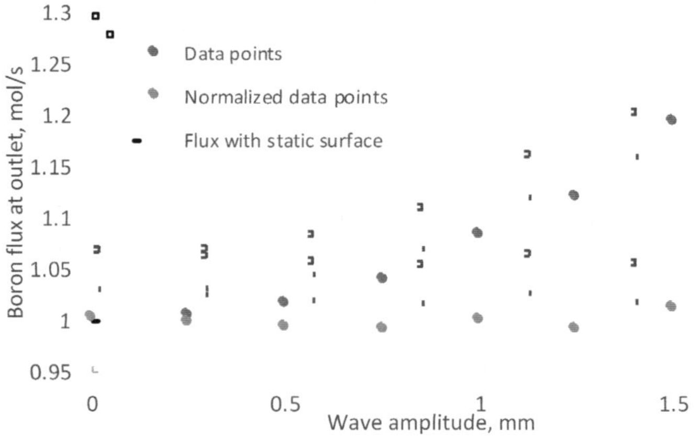

Fig. 3: Modelling of boron removal from silicon with oxidizing gas jet: C_B – boron concentration, v_{Si} and v_{Ar} – silicon melt flow and argon flow velocities for $t = 0.41$ h

Fig 5: Integral boron flux through outlet dependence on wave amplitude. Error bars – 4% of the value. Data are obtained when steady state is reached

Conclusions

Surface waves on liquid metal surface can be captured using VOF approach. Simulations showed three wave regimes, with less structured waves appearing at higher magnetic field values. Increase of magnetic field amplitude also leads to increased surface area, which is important for refinement. Simulations show clear dependence of refinement rate on surface area.

Acknowledgment

This work was funded by European Regional Development Fund under contract "Refinement of metallurgical grade silicon using smart refinement technologies" (No. 1.1.1.1/16/A/097).

References

1. U.S. Energy Information Administration (EIA) – Source // www.eia.gov (retrieved 2016-06-07)
2. Khattak C P, Joyce D B, Schmidt F 2002 Solar Energy Materials and Solar Cells 74 77
3. Bojarevičs A, Beinerts T, Grants I, Kaldre I, Šivars A, Gerbeth G, Gelfgat Yu 2015 Magnetohydrodynamics 51 437
4. Sortland Ø S, Tangstad M 2014 Metallurgical and Materials Transactions E 1 211

***** Formatting Issue - Best Available Paper/Graphic *****

5. Saadi B, Bojarevics A, Fautrelle Y, Etay J 2005 Récents Progrès en Génie des Procédés, Numéro 92 (Paris)
6. Vencels J, Jakovics A, Geza V, Scepanskis M 2017 XVIII Int. UIE-Congress (Hannover) p 312

IOP Conf. Series: Materials Science and Engineering **424** (2018) 012050 doi:10.1088/1757-899X/424/1/012050

Electromechanical Milling - Conception and Design of the Excitation System

B. Halbedel

Technische Universität Ilmenau, Institute of Materials Engineering,
Department of Inorganic-Nonmetallic Materials, Group of Materials research in or with Magnetic Fields
Gustav-Kirchhoff-Straße 6, D-98693 Ilmenau, Germany

Corresponding author: bernd.halbedel@tu-ilmenau.de.

Abstract
The novel electromechanical milling principle is a solution providing reduced energy consumption with milling beads directly moved with a time- and local-variable magnetic field, which is generated in a working chamber by an external, electrical excitation system. Hence, the main design parameters of such machines for electromechanical milling (so-called EMZ plants) must be derived from the time- and local-variable magnetic flux density, the resulting vector gradient and the electromagnetic force distribution in the process room. For the calculation and optimization of these distributions we developed a numerical model on the basis of the commercial software tool ANSYS Maxwell®. We verified this model with experimental measurements of the electromagnetic field in the process room. Furthermore, we have studied the influence of the excitation system design on the vector gradient distribution and discuss selected results in this paper. The study identifies potentials for the further improvement of electromechanical units built in the past.

Key words: milling, electromagnetic field, simulation, optimization, design

1 Introduction

Milling is an important but very complicated and energy consuming process step in numerous material productions such as the preparation of raw or starting materials for the glass, ceramic and the building industry as well as the recycling of waste materials. State of the art is the mechanical grinding with different type of mills. Most used mills are ball mills e.g. in the field of cements production with an input power of up to 28 MW with pull rates of up to 350 t/h, lengths of up to 15 m and diameters of up to 12,8 m [1]. However, the energy efficiency of such mills is disastrous, e.g. the energy efficiency of ball mills is < 9 % or of jet mills even < 2 % [2]. Thus, about 91 % up to 98 % of the input power are losses. These large power parts do not contribute to particle size reduction and are wasted. Therefore, the energy efficiency improvement of the milling machines is an important goal of current research and development activities and a necessary component of environmental strategies [3-6]. The low energy efficiency is caused by the indirect motion of the milling beads connected with high energy losses. Therefore, pivotal questions for the improvement of a novel mill conception are:

+ Can we move the milling beads directly, e.g. with magnet fields?
+ Is the movement of the milling beads enough intense and relative together?

The last requirement is important for the generation of mechanical stresses at the material to be milled in the process room.

In *Section 2* the basics for electromagnetic movement of milling beads and the conception as well as the necessary base design of electromechanical mills as a result of the deduced conditions are explained. In *Section 3* an electromagnetic model based on the commercially available software tool ANSYS Maxwell® for the design of the excitation system – the most important component of electromechanical milling machines – is introduced. *Section 4* presents the simulated vector gradient distributions in the process room and how they are influenced by the design of the excitation system. Finally, *Section 5* summarizes the main findings and ends with a perspective on further development of the electromechanical milling system.

2 The electromechanical milling principle (EMZ)

2.1 Basics

A milling bead moves in a magnetic field if

a) the milling bead is hard magnetic, i.e. milling beads have a fixed magnetization and
b) the magnet field is nonhomogeneous, i.e. magnetic field exhibits a vector gradient

then an electromagnetic force $\vec{F}_{\nabla\vec{B}}$ on a milling bead is generated as shown in Fig. 1. The generated force of one milling bead is given by equation (1) and (2):

$$\vec{F}_{\nabla\vec{B}} = (\vec{p}_M \cdot \nabla)\vec{B}, \tag{1}$$

where $\vec{p}_M$ is the vector of the magnetic moment, given for milling beads by

$$\vec{p}_M = \int_V \vec{M} dV = \frac{1}{\mu_0} \int_V \vec{J} dV \,, \tag{2}$$

with $\vec{M}$ – the fixed permanent magnetization of the milling beads, $\vec{J}$ – its magnetic polarization, V – the volume of a milling bead and $\mu_0 = 4\pi \cdot 10^{-7}$ Vs/Am – the absolute permeability.

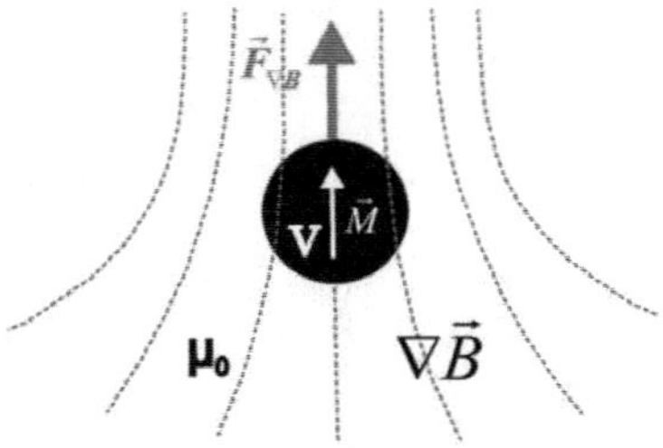

Fig. 1. Magnetic force $\vec{F}_{\nabla \vec{B}}$ on a milling bead with a magnetization $\vec{M}$ and the volume V in a magnetic field $\vec{B}$ with a vector gradient $\nabla \vec{B}$

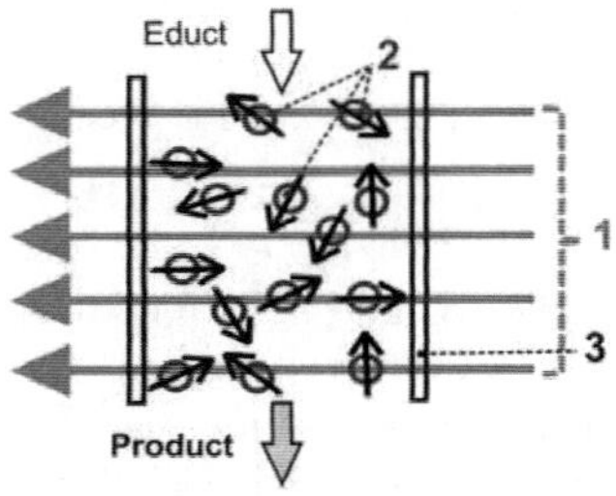

Fig. 2. Conception of an electromechanical mill (EMZ)
1 - time- and local-variable magnet field, 2 - hard magnetic milling beads, 3 - process room

With the simplifying assumptions that the magnetization of the milling bead is $\vec{M} = M = const.$ in its volume and directed in the same direction as $\nabla \vec{B}$ we can deduce equation (3) for description of the electromagnetic force density $\vec{f}_{\nabla \vec{B}}$ on one milling bead:

$$\vec{f}_{\nabla \vec{B}} = \frac{\vec{F}_{\nabla \vec{B}}}{V} = M \cdot (\nabla \vec{B}) = \frac{1}{\mu_0} J \cdot (\nabla \vec{B}) \,, \tag{3}$$

with M – magnitude of the magnetization $\vec{M}$ and J – magnitude of magnetic polarization $\vec{J}$.
As revealed in equation (3) the vector gradient of the magnetic field $\nabla \vec{B}$ dictates the direction of the force density $\vec{f}_{\nabla \vec{B}}$ and the product $J \cdot (\nabla \vec{B})$ determines its magnitude.
Additionally, if the magnetic vector gradient is time- and local-dependent, then a time- and local-dependent force density distribution $\vec{f}_{\nabla \vec{B}}(r, \varphi, z, t)$ is generated in the process room, consequently intensive relative movements between several milling beads are induced with different types of mechanical stresses so that the novel electromechanical milling principle can be used for disintegration of biomass, dry or wet grinding of raw materials and autogenous grinding of ferrites [7, 8].

2.2 Base Design

For the technical implementation of such requirements it is necessary to build a process room (cf. Fig. 2) – the best layout is an annular gap - with non-ferromagnetic walls, which is

+ penetrated from a time and local variable electromagnetic field $\vec{B}$ - generated from a multi-phase electrical exciter system surrounding the process room and
+ filled with numerous and freely movable hard magnetic milling beads.

Then the magnetic field in the process room is nonhomogeneous and exhibits a time and local variable vector gradient distribution, which generates a time and local variable force distribution on the milling beads, whereof

+ intensive relative motions of the milling beads in the process room result and
+ different types of mechanical stresses on powders or in suspensions develop (e.g. friction, shear, impact) which different types of preparation effects of powders or in suspensions generate (e.g. deagglomeration, disintegration, dispersion, grinding).

All these are mainly dependent on the design of the excitation system and its operating parameter. Therefore, we need a model for the determination of the optimal design and for adaption of the operating parameter of the EMZ machine.

3. Model for electromagnetic design

Our developed electromagnetic model for the design of the exciter system of the EMZ machine is based on the following important steps:

1) Simulation of the magnet flux density distribution $\vec{B}(r, \varphi, z, t)$ in the draft of the exciter system of the EMZ machine,

2) Determination of the vector gradient distribution $\nabla\vec{B}(r,\varphi,z,t)$ in the process room dependent on design parameters (e.g. width of the process room, type of winding, number of poles and of phases) as well as on operating parameters (e.g. mean magnetic field strength, phase number, frequency of electric current) and
3) Selection of the design and operating parameter with the adequate vector gradient distribution in the process room.

The simulation of the magnetic field distribution $\vec{B}(r,\varphi,z,t)$ in the exciter system is carried out in ANSYS Maxwell® by iterative solution of the partial differential equations (4) and (5)

$$\nabla\times(\frac{1}{\mu}\nabla\times\vec{A})=\vec{j} \qquad \text{and} \qquad \vec{B}=\nabla\times\vec{A} \qquad \text{(4) and (5)}$$

with $\vec{A}$ as the magnetic vector potential resulting from the design of the windings and the electric current distribution $\vec{j}$ in the exciter system and with μ as the magnetic permeability in the solution domains.

For the determination and evaluation of the vector gradient distribution in the process room, we must discretize the process room again, in 2D only (see Fig. 3) because the z component of the magnet flux density is negligibly small.

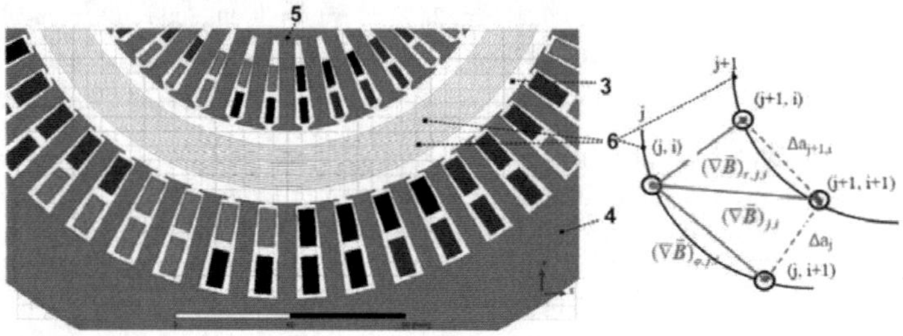

Fig. 3. Extract from a 2D model of an excitation system (*left*) with the details of the discretization of the process room for the calculation of the magnet vector gradient components $(\nabla B)_{r,j,i}$ and $(\nabla B)_{\varphi,j,i}$ in the points (*i, j*) (*right*)

3 - process room, 4 - outer part and 5 -inner part of the excitation system, 6 – *j* circles with the radii R_j and distance Δa_j

Taking wall thickness of the process chamber and sizes of the used milling beads into account, we define *j* circles with radii R_j and n_j points with the distances Δa_j and $\Delta a_{j,i}$. Then we calculate the r and φ components of the vector gradient $(\nabla\vec{B})_{j,i}$ in each point (j, i) on the base of the simulated magnetic flux density distribution $\nabla\vec{B}(r,\varphi)$ to a fixed time, e.g. $\omega t = 0$, with the equations (6):

$$(\nabla\vec{B})_{j,i}=(\nabla B)_{r,j,i}\cdot\vec{e}_r+(\nabla B)_{\varphi,j,i}\cdot\vec{e}_\varphi=(\frac{\partial B_{r,j,i}}{\partial r}+\frac{1}{r_j}\frac{\partial B_{r,j,i}}{\partial\varphi}-\frac{B_{\varphi,j,i}}{r_j})\vec{e}_r+(\frac{\partial B_{\varphi,j,i}}{\partial r}+\frac{1}{r_j}\frac{\partial B_{\varphi,j,i}}{\partial\varphi}+\frac{B_{r,j,i}}{r_j})\vec{e}_\varphi \qquad (6)$$

For the evaluation of different design layouts as well as operating parameter we use the RMS value of the vector gradient components of selected circles *j* with equation (7) and (8).

$$(\nabla B)_{r,j,eff}=\sqrt{\frac{1}{n_j}\cdot\sum_{i=1}^{j}((\nabla B)_{r,j,i})^2} \quad \text{and} \quad (\nabla B)_{\varphi,j,eff}=\sqrt{\frac{1}{n_j}\cdot\sum_{i=1}^{j}((\nabla B)_{\varphi,j,i})^2} \qquad \text{(7) and (8)}$$

with n_j as the number of points (j, i) on the circle *j*.

The electromagnetic model was evaluated with experimental measurements of the magnetic flux densities in the process room and comparison simulations in COMSOL Multiphysics [7, 9].

4. Results and discussion

Fig. 4 presents the simulated vector gradient distributions of the radial and azimuthal component in the process room of an electromechanical lab mill on three selected circles, which are located in the near of the inner part of the excitation system (R1), in the middle of the process room (R8) and close to the outer part (R16) of the excitation system at a mean magnetic field strength of H_δ = 70 kA/m and a frequency of f = 50 Hz.

The components of the vector gradient of the magnetic field have the same order of up to about 10 T/m, but their changes are larger close to the exciter systems (circle R1 and R16) than in the center (circle R8) of the process room. Consequently, the milling beads are more accelerated or decelerated close to the excitation system. Here the movement changes are maximal.

In the center of the process room the distributions are approximately sinusoidal along the circle R8. However, they are shifted, so that the azimuthal component $(\nabla B)_{\varphi j}$ is larger where the radial component $(\nabla B)_{r,j}$ is smaller or vice versa.

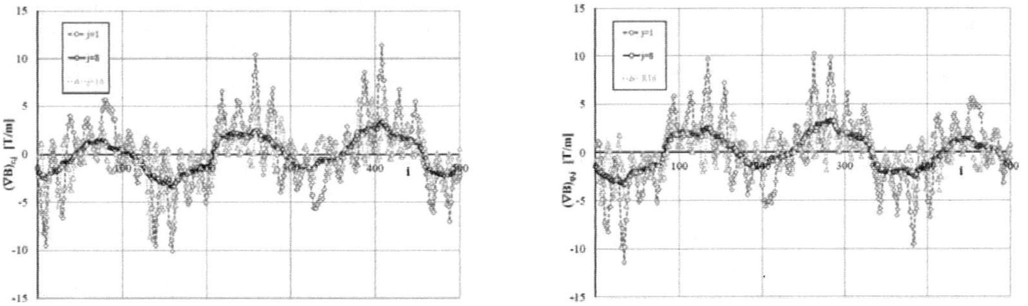

Fig. 4. Radial *$(\nabla B)_{r,j}(i)$ (left)* and azimuthal *$(\nabla B)_{\varphi,j}(i)$ (right)* components of the vector gradient of the magnetic field in the process room of a electromechanical lab mill on three circles *j*, which are located in the near inner part of the excitation system (R1), in the middle (R8) of the process room, and close to the outer part of the excitation system (R16) dependent on *i* at H_δ = 70 kA/m, f = 50 Hz, $\omega t = 0$

Other simulations show that the RMS values of the vector gradient components increase with the enhancement of the mean magnetic field strength H_δ in the process room (see Fig. 5). The mean magnetic field strength H_δ in the process room results from the magnitude of electric currents in the coils and their layout.

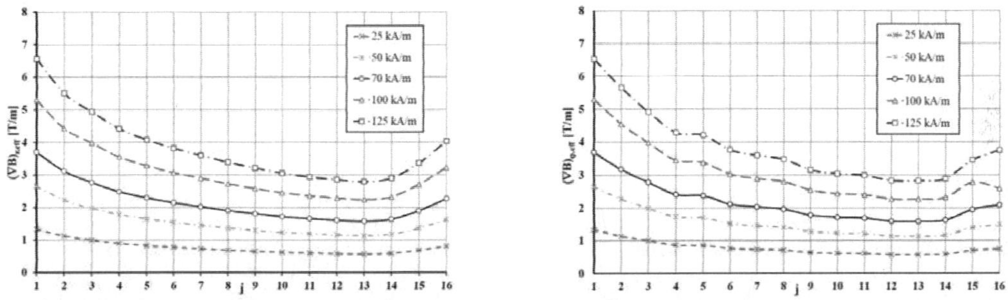

Fig. 5. Distributions of the RMS values of the radial *$(\nabla B)_{r,eff}$ (left)* and azimuthal *$(\nabla B)_{\varphi,eff}$ (right)* components of the vector gradient over the width of the process room dependent on the mean magnetic field strength H_δ for f = 50 Hz and $\omega t = 0$

The graphs show that the RMS values of the vector gradient components on the circles *j* are about same. Hence, the milling beads are approx. uniformly distributed in the *r* - φ plane of the process room [7]. But, how we have seen in the diagrams in Fig. 4, the RMS values of the vector gradient components are larger close to the exciter systems than in the vicinity of the outer excitation system. This disparity increases with larger magnet field strengths in the process room. The reason is the realized ampere turns in the inner and outer parts of the excitation system.

Further crucial influencing factors on the vector gradient components are the windings layout and the number of poles. It is possible to perform the windings of the excitation system as distributed windings or as concentrated two-layer and single-layer windings. The simulations prove that concentrated windings generate larger vector gradients in the process room at the same operating parameters (see Fig. 6).

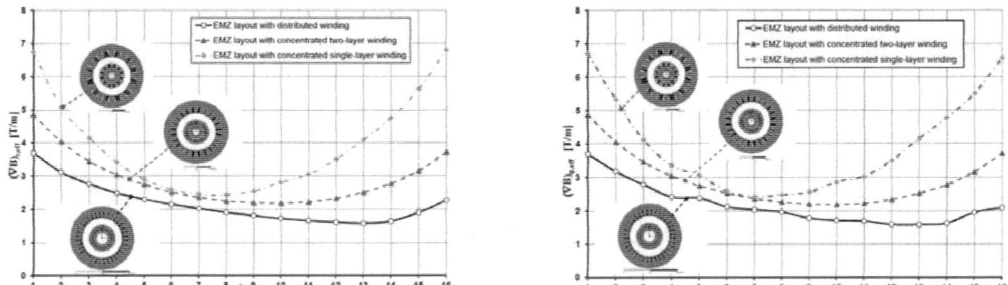

Fig. 6. RMS value distribution of the radial *$(\nabla B)_{r,eff}$ (left)* and azimuthal *$(\nabla B)_{\varphi,eff}$ (right)* components of the vector gradient in the process room for different windings layouts of the excitation system at H_δ = 70 kA/m, f = 50 Hz, $\omega t = 0$

When the magnetic vector gradient is larger, the milling beads move more intensively und the mill produces a finer product at the same operating parameters. Moreover, the heights of its end turns are smaller and consequently there is less cupper required for the fabrication of the windings. This is an important additional advantage for design of EMZ devices [10].
Also the adjustment of number of poles $2p$ dependent on the dimension of the electromechanical mill as well as the reduction of the number of slots per pole per phase q and of the chording of the windings ε are potentials for further improvements of the performance of mills based on the electromechanical principle [10].

5. Summary and Outlook

The theoretical investigation of the function of the novel electromechanical milling principle and the study of the design of the excitation system – the most important component of electromechanical milling machines – together with the developed electromagnetic model based on the commercially available software tool ANSYS Maxwell® show that the distribution of the magnetic vector gradient in the process room of electromechanical mills determine the movement of the milling beads and thereby also the milling result [7, 8]. The study of the influence of the excitation system design on the vector gradient distribution identifies potentials for the further improvement of electromechanical lab units built in the past [12, 13]. In particular, the design of the windings of the excitation system provides great potential for further improvement. The next steps are the connection of developed electromagnetic model with DEM-simulations [14] and a stress model [15] to determine the required magnet field structure in the air gap for an efficient milling process.

Acknowledgment
The authors acknowledge financial support by the German Research Foundation (DFG) (Ha 2338/1) and by the Federal Ministry for Economic Affairs and Energy (BMWi) in ZIM-Cooperation Project EMZ-W (KF2184744KO4) as well as support by several industrial partners and former co-workers for their major commitment.

References

1. http://www.at-minerals.com/de/artikel/at_2010-10_Groesste_Muehle_der_Welt_1003987.html (23.05.2018)
2. Pahl, M. H.: Zerkleinerungstechnik. 2. Auflage, Fachbuchverlag Leipzig/Verlag TÜV Rheinland, 1993.
3. Schäfer, St. and Hoenig, V.: Chancen und Grenzen der energetischen Optimierung und CO_2-Reduktion in der Zementindustrie. 2. Energiepanel Energieeffizienz und CO_2-Reduktion in der Industrie Mainz, 25.03.2015.
4. Shi, F., Morrison, R, Cervellin, A., Burns, F., Musa, F.: Comparison of energy efficiency between ball mills and stirred mills in coarse grinding. Minerals Engineering, Volume 22, Issues 7–8, pp 673-680, June–July 2009.
5. Schubert, G.: How to Make Milling More Energy Efficient. PROCESS Worldwide, Volume 13, No. 3, pp 19-21, June 2011.
6. Bakker, J.: Energy Use of Fine Grinding in Mineral Processing. METALLURGICAL AND MATERIALS TRANSACTIONS E, Vol. 1E, pp 9-19, March 2014, DOI: 10.1007/s40553-013-0001-6.
7. Halbedel, B., Kazak, O.: Potenziale eines neuen Mahlverfahrens – Die elektromechanische Zerkleinerung (EMZ) (Potential of a New Milling Principle – The Electromechanical Milling). Chem.-Ing. Tech., Vol. 90, Issue 4, pp 540-551, April 2018, DOI: 10.1002/cite.201700103.
8. Halbedel, B., Kazak, O.: Development of electromechanical principle for wet and dry milling. IOP Conf. Ser.: Mater. Sci. Eng. 355 012021, 8 pages, 2018, DOI 10.1088/1757-899X/355/1/012021.
9. Kazak, O.; Halbedel, B.: Approaches for modelling electromagnetic parameters in milling demonstration plant of organic agents. In: Proceedings 8th International Conference Electromagnetic Processing of Materials, Cannes/ France, 12 – 16 October, 2015.
10. Halbedel, B.; Kazak, O.: Vorrichtung und Verfahren zum Zerkleinern, Desagglomerieren, Dispergieren und Mischen von dispersen Stoffen und pumpfähigen Mehrphasengemischen. DE 10 2017 008 513.7, registered: 06.09.2017.
12. Halbedel, B.; Müller, W.; Baudrich, R.; Hülsenberg, D.: Process and apparatus for reducing, dispersing, wetting and mixing pumpable, non-magnetic multiphase mixtures. US 5,348,237, 19.12.1991.
13. Halbedel, B.; Hülsenberg, D.: Elektromechanische Zerkleinerung. In: Technische Keramische Werkstoffe, John von Freyend GmbH, Fachverlag für Wirtschaft und Außenhandel, Köln, Nov. 1995.
14. https://www.cfdem.com/liggghts-open-source-discrete-element-method-particle-simulation-code, 28.05.2018
15. Kwade, A.: A Stressing Model for the Description and Optimization of Grinding Processes. Chemical Engineering & Technology 26 (2) 199 – 205, 2003, DOI: 10.1002/ceat.200390029.

IOP Conf. Series: Materials Science and Engineering **424** (2018) 012051 doi:10.1088/1757-899X/424/1/012051

→

→

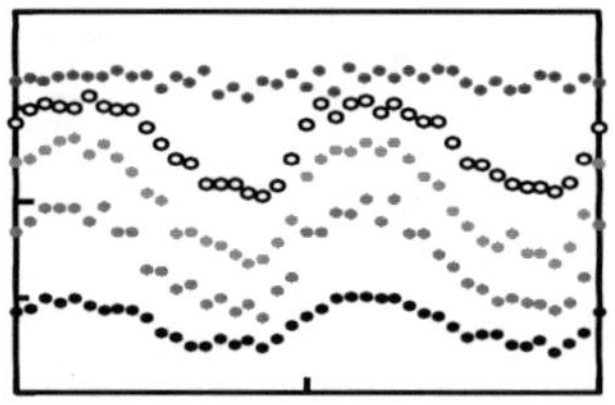

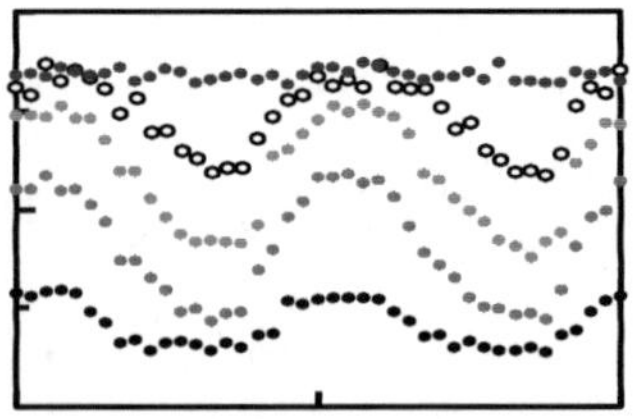

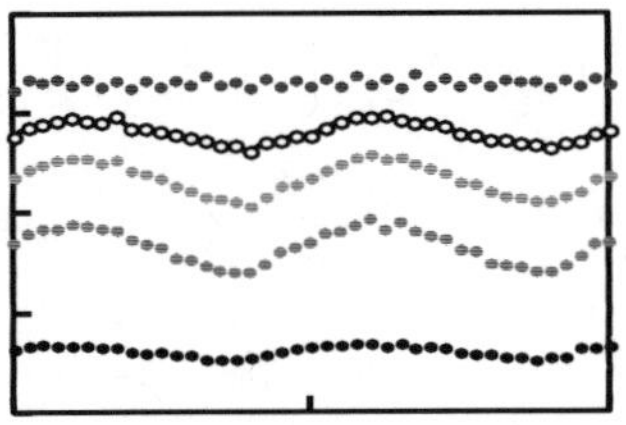

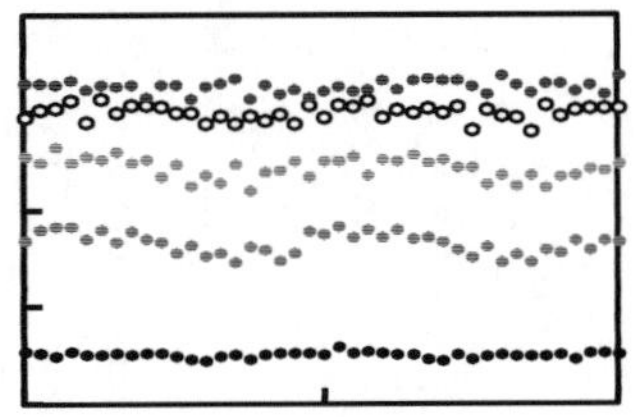

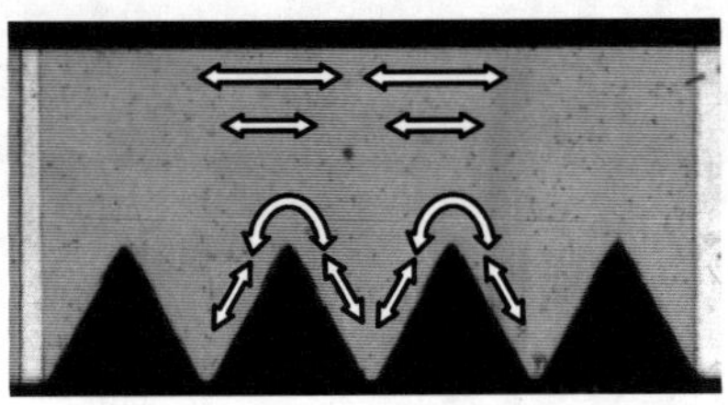

Experimental investigation of the effect of travelling magnetic field on the CET in Sn-10*wt.*%Pb alloy

K. Zaidat[1*], I. Sari[2], A. Boumaaza[2], A. Abdelhakem[2], L. Hachani[2], and Y. Fautrelle[1]

[1] Laboratoire SIMaP-EPM, BP 75, 38402 Saint Martin d'Hères, Grenoble, France
[2] Laboratoire Physique des matériaux, BP 37 G Boulevard des martyrs, Laghouat, Algeria

* *Corresponding author: kader.zaidat@simap.grenoble-inp.fr*

Abstract
In order to eliminate internal and surface defects, which appear in the ingots during solidification process of metallic alloys, homogeneous and fine structures are often required. Therefore, when the adjunction of inoculant substances is not possible, the electromagnetic stirring (EMS) process thanks to the use of travelling magnetic fields has technically shown its effectiveness. The present experiments consist to create convective movements in a Sn-10wt.%Pb liquid metal, by the action of an electromagnetic field during its solidification. A significant improvement in the columnar-equiaxed transition (CET) mechanism has been observed and consequently a fine equiaxed grain structure can then be detected. The main goal of this experimental study is to examine the effect of the EMS process on grain structures for that alloy as well as the corresponding segregations. In particular, it included the determination of the optimal experimental parameters to obtain a grain refinement, the characterization of their effect on the solidification and the understanding of the physical mechanisms driving grain refinement.

1. Introduction

The phase transition study related to the melting and solidification processes is the basis of many numerical and experimental researches, as well as of various engineering applications. In particular, the use of traveling magnetic fields (TMF) makes possible to stir liquid metal and control characteristics of its flow [1-3] and, hence, to affect heat and mass transfer in the melt (including the phase transition boundary) in the process of solidification [4-5]. The study of interaction between the parameters of electromagnetic effect and the structure of metal, its homogenization level, mechanical properties etc., as well as the study of the influence of these parameters on the metal melting rate at high frequency heating and solidification, remain extremely relevant until today. Solidification and melting, like many other thermal processes in industry, require strict thermal conditions during the process. For example, homogeneous temperature, solidification rate and temperature gradients during directional solidification exert a significant influence on a convective flow of liquid during solidification, and therefore have a great effect on the final product. For that purpose, a solidification benchmark experiment was designed and developed at SIMAP Laboratory in Grenoble in order to investigate the structure formation as well as solute macro-mesosegregation by means of a well-controlled meso-scale solidification experiment. Indeed, as the industrial continuous casting process is rigid and complex, the investigations were easier to carry out on the Benchmark installation at laboratory scale, enabling us to define the influence of each experimental parameter: thermal gradient, cooling rate, stirring direction, temperature of the liquid and composition of the alloy.

2. Experimental process

The benchmark experiment developed in the SIMAP/EPM laboratory is a key tool for the validation of the numerical models by comparison with detailed measurements obtained with thoroughly controlled initial and thermal boundary conditions. In its principle, the experiment is similar to the well-known Hebditch and Hunt experiment [7], with a special emphasis on obtaining reproducible quantitative measurements. The Sn-Pb binary alloy pre-sample is enclosed in a rectangular cavity of 100 mm in length, 60 mm in height and 10 mm in width. All walls except for two narrow vertical ones are held at approximately insulating conditions during the whole process via a Kirchhoff box in vacuum conditions. The alloy is solidified by two temperature controllable heat exchangers. An array of fifty K-type thermocouples, is used to record the temperature field, and therefore the evolution of solidification. A linear stirrer is placed beneath the sample able to generate a TFM in order to homogenize the sample after the melting process until the end of the solidification stage. The reader is referred to

IOP Conf. Series: Materials Science and Engineering **424** (2018) 012052 doi:10.1088/1757-899X/424/1/012052

the previous works [8-9] for further details concerning the description of the experimental setup. The experimentis equipped with a three-phase coil, which produces a pulsating electromagnetic force. Its amplitude and direction depend on the instantaneous value of the current flowing through the coil. Figure (1-a) illustrates the configuration of the resulting Lorentz forces distribution in space at chosen moment as well as the corresponding mean fluid flow. The behaviour of the stirred metal in the cavity can be described as follows. At the beginning the melt is under the effect of natural convection only. When the inductor is switched on, the resulting electromagnetic body force accelerates the fluid (see Fig.1-b). Analysis of velocity profiles seems to indicate that the flow configuration consists of a main vortex located in the central part with two small vortices at each upper corner. The present work is focused on the experimental investigation of the effect of the TMF, in particularly the stirring direction on the solidification process in terms of thermal behavior of the melt, grain structure and finally inhomogeneity of concentration (segregation).

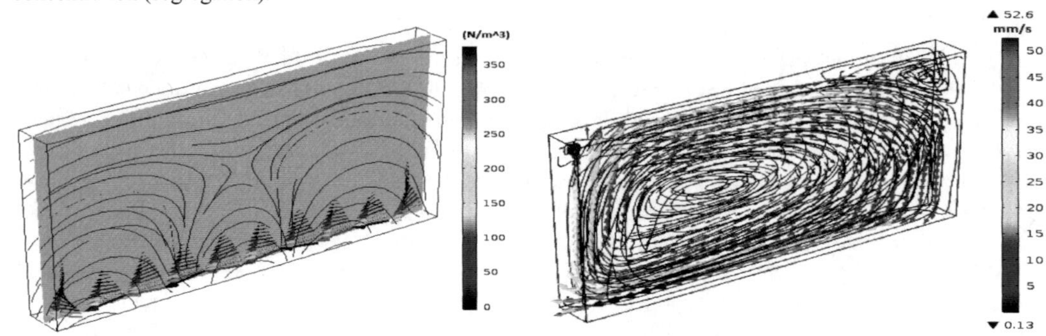

(a) (b)

Fig.1. 3D spatial evolution obtained from numerical simulation of: (a) electromagnetic forces witch the black lines correspond to the magnetic field streamlines and volume vectors are the Lorentz forces, (b) the flow configuration in the melt. The black lines correspond to the streamlines and volume vectors are the velocity field. Results obtained for $I = 8\,A$, and $f = 50Hz$.

3. Results and discussions

3.1. Effect on the temperature field evolution

The experimental temperature maps provided in Fig. 2, are coupled with vectors indicating the local thermal gradients for two solidification cases (Case I: without any electromagnetic stirring and Case II: under EM stirring in the opposite direction of natural convection) at selected instant (t = 9 mn after the beginning of the solidification phase). In the stirred case, coil intensity is set to I = 8 A. The temperature gradients were calculated at each thermocouple location from the temperature difference between two adjacent points by means of a centred difference numerical scheme. The arrows correspond to the $-\vec{\nabla}T$ vector, which is proportional to local heat flux density.

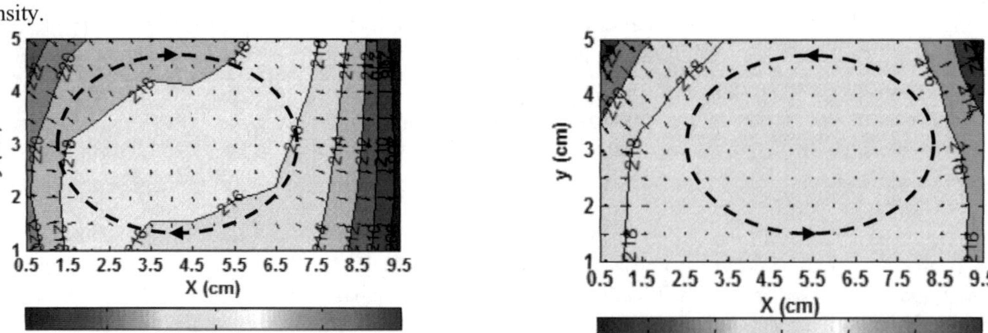

(a) Case I : non-stirred **(b) Case II : e.m. stirred**

T_{max} = 221.42°C; T_{min} = 202.72°C **T_{max} = 222.43 °C; T_{min} = 208.73 °C**

Fig. 2. Instantaneous temperature maps with superimposed local temperature gradient vectors ($-\vec{\nabla}T$) for the four solidification experiments at t = 9 min. The experimental conditions are: Sn-10wt.%Pb alloy, ΔT = 40 K, CR = 0.03 K/s, inductor current I = 8 A.

Indeed the figure (2-b) show clearly the effect of electromagnetic stirring on temperature distribution, namely homogenization of temperature in the melt as well as modification of flow configuration. A comparative study shows that stirring in the opposite direction to convection modifies flow configuration significantly. This is shown by deformation of the isotherms, meaning that a vortex appears in the opposite direction with respect thermosolutal convection.

3.2. Effect on the grain structure

To evaluate the effect of the electromagnetic stirring in the opposite direction of natural convection on the final metallographic structure and grain size, macrostructures of ingots obtained from two solidification experiments were revealed by the procedure detailed in the works of Hachani et al. [6-8]. Fig. 3(a) shows the macrostructure of solidification of the Sn-10wt.%Pb under the effect of thermosolutal convection only. We observe that the dominant morphology of the macrostructure in this case is columnar with an equiaxed zone corresponding to the last liquid at the end of solidification. The process of columnar-to-equiaxed transition can be accounted for by the drop in temperature gradient in the liquid bath as shown by temperature maps illustrated in Fig. 2(a) (Case I). The upwind tilting of the columnar grains is consistent with the direction of natural convection. At the base of the ingot, we notice a growth of a second columnar structure of different orientation. A possible explanation is the effect of convection, which supplies ''cold'' liquid to the bottom part of the ingot, as indicated by Fig. 2(a). Note that the orientation of the columns in that region supports the aforementioned statement.

(a) Case I: without electromagnetic stirring

(b) Case II: stirring in the opposite direction to natural convection

Fig. 3. Macrostructure on the lateral plane of different ingots of Sn-10.*wt*.%Pb. The experimental conditions are: temperature difference ΔT = 40 K, cooling rate CR = 0.03 K/s and the applied current intensity I = 8.2 A. Two cases are presented: (a) Case I solidification under the effect of natural convection only. (b) Case II solidification under the effect of electromagnetic stirring in the opposite direction to natural convection.

The case of stirring in the opposite direction to thermosolutal convection is shown in Fig. 3(b). We notice that the area of the columnar region has regressed. Moreover, the columnar zone of the lower part has almost disappeared, while the equiaxed area has increased significantly. Indeed, the structure mainly consists of fine size equiaxed grains. A small columnar region is located in the bottom right angle of the ingot. It is important to note that the efficiency of electromagnetic stirring in refining the structure is improved significantly when we reversed the direction of stirring against natural convection. Details of the various cases are provided in [6]. This result can be attributed to electromagnetic stirring that exerts two main effects. First, the Lorentz force generates liquid metal convection that homogenizes the pool and lowers the temperature gradients, thereby creating favourable conditions for appearance of a CET. Second, a hot liquid impinging the solid-liquid interface can enhance fragmentation/remelting of the secondary arms. Transported by convection, under certain conditions these fragments can give birth to equiaxed grains. In parallel with the mechanical fragmentation effect, this stirring mode seems to have a significant consequence on the dendrite remelting process. Indeed, EM stirring generates an upward motion of lead-enriched liquid along the columnar front.

3.3. Effect on solute distribution

A post-mortem analysis of solute distribution in the sample was performed. A chemical method coupled with the ICP (Inductive Coupled Plasma) technique was used to determine quantitative analysis of solute distribution. X-ray analysis of the local lead composition (averaged in the sample thickness) were also performed. The results illustrated in Fig. 4-a shows the segregation distribution for the case of solidification under the effect of thermosolutal convection (Case I). Solute rejection starts from the right side of the sample (along with channeling

effects) so that, at the end of solidification, a lead-enriched area appears on the bottom left. This zone which corresponds
to the last solidified liquid has a concentration level exceeding 17%. However, a macrosegregation zone marked by the highest concentration (>18%) was observed in the bottom-right corner of the sample. This concentration peak corresponds to the formation of segregated channels at this location. We checked that the measured average concentration over the whole ingot was close to nominal concentration (10%).

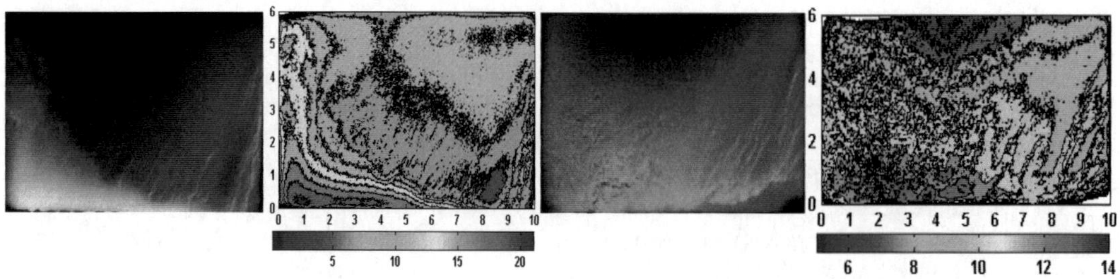

(a) Case I :non-stirred ingot **(b) case II : e.m. stirred ingot**

Fig. 4. X-ray photos (left) and digital processing (right) of two solidified ingots showing the different types of meso and macrosegregation. The alloy is Sn-10 wt.%Pb. The light grey color corresponds to the lead-rich zones. The experimental conditions are: temperature difference DT = 40 K, cooling rate CR = 0.03 K/s, and intensity of the applied current I = 8 A. The composition scales in wt.% are indicated in the scale of the left-hand figures.

Fig. 4-b shows the case of solidification under the effect of electromagnetic stirring in the opposite direction to natural convection (Case II). We noticed a significant displacement of the macrosegregation area with a level of concentration ranging between 11% and 15%, which corresponds to the last liquid solidified. These results can be accounted for by development of the second solidification front (hot side), which distinguishes this case (see temperature maps shown in Fig. 2. The segregation pattern still consists of an upward plume resulting from the effect of two active contra-rotating vortices at the end of solidification. However, this case remains inactive on the elimination of the segregated channels in the right of the ingot for the same reasons cited in the previous case. Note that stirring slightly lowers the segregation level but does not suppress them.

4. Conclusion

In the light of the experimental comparative study presented in this work, we can deduce some findings and general conclusions about the effect of the electromagnetic stirring in the opposite direction to natural convection modes applied to the solidification process, particularly concerning the CET process. Indeed, this mode of stirring has shown its effectiveness in achieving the finest equiaxed structures. Also this stirring mode combines both the remelting and the mechanical fragmentation mechanisms. However, Electromagnetic stirring does not remove the segregated channels. Stirring can even create and promote segregation.

References

1. Niemietz, K., Galindo, V., Patzold, O., Gerbeth, G., Stelter, M., 2011. *Flow modelling with relevance to vertical gradient freeze crystal growth under the influence of a travelling magnetic field.* J. Cryst. Growth 318, 150–155.
2. Oborin, P., Khripchenko, S., Golbraikh, E., 2014. *Influence of conventional and reverse travelling magnetic fields on liquid metal stirring in an asymetric cavity.* Magnetohydrodynamics 50 (3), 291–301.
3. Räbiger, D., Leonhardt, M., Eckert, S., Gerbeth, G., 2011. *Flow control during solidification of SnPb-alloys using time modulated AC magnetic fields.* IOP Conf. Series 27.
4. Ramachandran, N., Mazuruk, K., Volz, M.P., 2000. *Use of traveling magnetic fields to control melt convection.* Jpn. Soc. Microgravity Appl. 17 (2), 98–103.
5. Schwesig, P., Hainke, M., Friedrich, J., Mueller, G., 2004. *Comparative numerical study of the effects of rotating and travelling magnetic fields on the interface shape and thermal stress in the VGF growth of InP crystals.* J. Cryst. Growth 266, 224–228.
6. L. Hachani, K. Zaidat, Y. Fautrelle, *Experimental study of the solidification of Sn–10 wt.%Pb alloy under different forced convection in benchmark experiment*, International Journal of Heat and Mass Transfer 85, 438-454.
7. L. Hachani, K. Zaidat, B. Saadi, X-D. Wang, Y. Fautrelle, *Solidification of Sn-Pb alloys: Experiments on the influence of the initial concentration*, International Journal of Thermal Sciences 91, 34-48.
8. Lakhdar Hachani, Kader Zaidat, Yves Fautrelle, *Multiscale statistical analysis of the tin-lead alloy solidification process,* International Journal of Thermal Sciences 110, 186-205.

IOP Conf. Series: Materials Science and Engineering **424** (2018) 012053 doi:10.1088/1757-899X/424/1/012053

Investigation of Si content on the grain refinement of Al-Si alloy under pulsed magnetic field

J.C. Jie *, S.P. Yue, Z.L. Zheng, Z.K. Guo, T.J. Li

Key Laboratory of solidification Control and Digital Preparation Technology (Liaoning Province), School of Materials Science and Engineering, Dalian University of Technology, Dalian 116024, China

Corresponding author: jiejc@dlut.edu.cn (J.C. Jie)

Abstract
The influence of Si content and pulsed magnetic filed parameters on the solidification structure of the Al-Si binary alloy was investigated. Experimental results indicated that the addition of Si has a significant refinement on the macrostructure of the commercial pure aluminum. The grain size of the Al-Si binary alloy decreases first and then increases with increasing of Si content which could be caused by the constitutional-supercooling, but for the case where a PMF was applied, the grain size was further refined compared with as-cast state.

Keywords: Al-Si binary alloy; Si content; Grain refinement; PMF

Introduction
Aluminium alloys are considered the most abundant, thus important, after steel due to their light weight, physical and mechanical characteristics[1].Among the aluminum alloys, the Al–Si alloys have shown good performance mainly for aerospace,transportation and automotive applications[1-3]. The grain refinement can reduce the hot tearing susceptibility and to improve the structure homogeneity and mechanical properties of aluminum castings. Various methods have been proposed to refine the solidification structure such as the common addition of inoculation particles in industry[4, 5]. However, the inoculation of the melt may produce undesired particle agglomerate which will contaminate the melt, the ultrasonic melt treatment [6]. Recently, the electromagnetic fields provide a contactless method to control the fluid flow in the liquid melt and achieve the fine equiaxed grains in the alloys. A large number of studies have already been reported concerning the impact of rotating magnetic fields (RMF) [7], travelling magnetic fields (TMF) [8] and Pulsed magnetic field (PMF) [9] in solidifying metals. However, few studies have been reported until now the coupling influence of the alloy composition and forced convection on the grain refinement.

In the present study, the effect of both the alloy composition and the forced convection on the grain refinement was investigated through the experimental and numerical simulation methods, and the results were discussed.

Experiment
Fig. 1 shows a schematic view of the experimental setup which consists of a straight solenoid connected with a self-prepared power which includes a capacitor bank of 1000 μF, temperature measurement system, cylindrical graphite mold. All experiments used a pulse frequency of 10 Hz and a half sinusoidal waveform with a peak voltage 700 V. Commercial-purity (CP) aluminum (99.7 wt. %) was melted in a resistance furnace first and then added different contents of Si to the melt and held for 30 min at temperature 100 K higher than the liquidus temperature. Furthermore, the pouring temperature for each experiment was fixed at 60 K higher than the liquidus temperature, and the mold temperature was maintained at 973 K before casting. The solidification process was rendered a near isothermal condition by use of thermal insulating blanket on the lateral and bottom surfaces of the mold. The liquid melt was poured into a graphite mold with an inner diameter of 45 mm and height of 100 mm and then transferred into the PMF immediately. The K-type thermocouples were arranged vertically along the axis of the specimen to monitor the distribution of temperature during solidification process.

The specimens were sectioned longitudinally along the mid-plane and then grounded on SiC papers and polished to remove any scratches for metallographic analysis. The etching reagent used to reveal the macrostructure was a solution of 75 ml HCl, 25 ml HNO3 and 5 ml HF. A CanoScan (LiDE120) was employed to obtain the photographs of the macrostructure. Quantitative measurements of the grain size were determined by the mean linear intercept method using Nano Measurer software package.

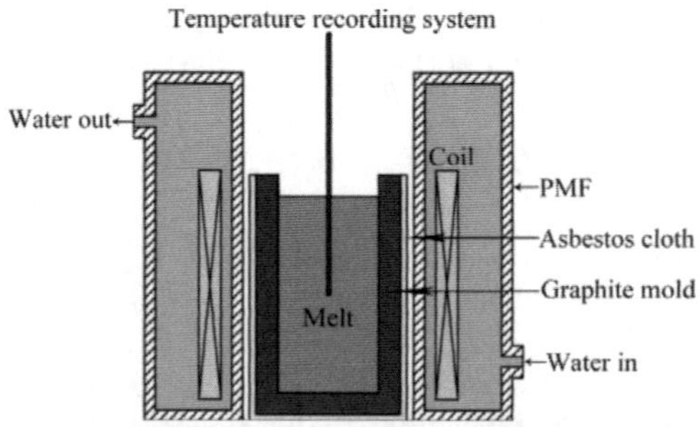

Fig. 1 The schematic view of the experimental setup

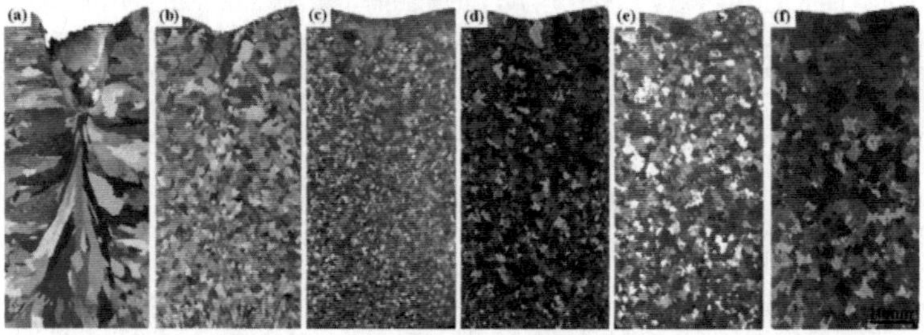

Fig. 2 As-cast macrostructure of Al-Si alloys with different Si contents: (a) Commercial-purity aluminum. (b) Al-1%Si. (c) Al-2%Si. (d) Al-3%Si. (e) Al-5%Si. (f) Al-7%Si

Result and discussion

Fig. 2 shows the as-cast macrostructures of Al-Si alloys with different Si contents. The Si contents have an obvious influence on solidification macrostructure of commercial-purity aluminum. The reference experiment conducted with commercial-purity aluminum (Fig. 2a) shows that a dominance of columnar grains growing opposite to the direction of heat flow. However, the solidification structure changes to fined equiaxed grains with addition of Si element. The corresponding average grains size is shown in Fig. 4. It can be seen that when the addition of Si content ranges from 1 % to 7 % , the average grains size of Al-Si alloys is not reduced monotonously with increasing of Si content, but decreases first and then increases. It is obvious that when the mass fraction of Si in Al-Si alloys is equal to 2 %, the average grains size is the smallest.

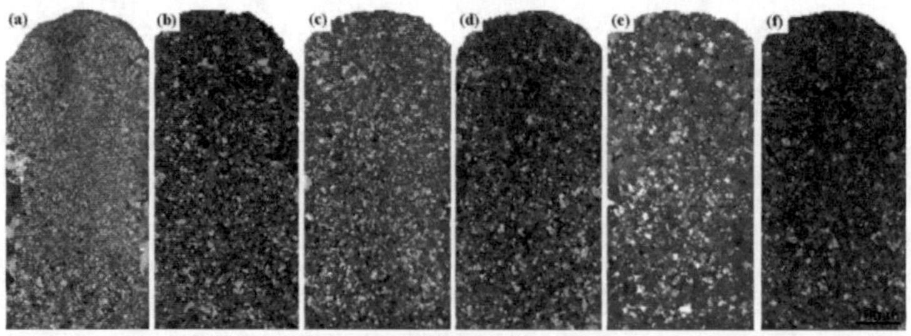

Fig. 3 The macrostructure of Al-Si alloys with different Si contents under PMF: (a) Commercial-purity aluminum. (b) Al-1%Si. (c) Al-2%Si. (d) Al-3%Si. (e) Al-5%Si. (f) Al-7%Si

The representative macrostructures of Al-Si alloys with different Si contents under PMF are shown in Fig. 3. The average grains size under PMF is much smaller compared with the macrostructure as cast but it should be mentioned that the grains size of commercial-purity aluminum is the smallest which is different from the as-cast condition.

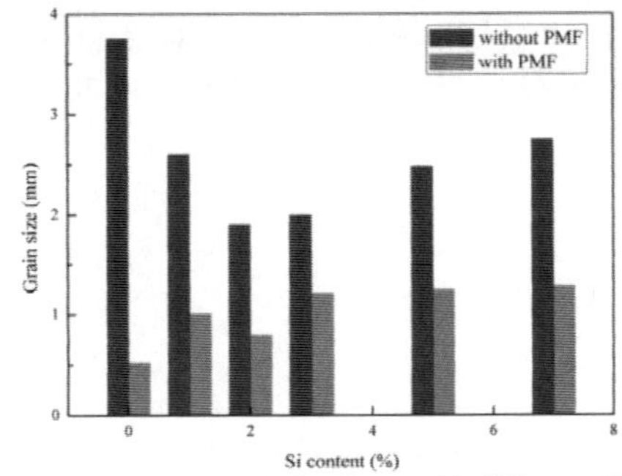

Fig. 4 Grain size of Al-Si alloys with different Si content

In order to yield a better understanding of the refinement mechanism of both Si content and the treatment with PMF, a numerical simulation method was performed to obtain the distribution of Lorentz force in the liquid melt.

According to Maxwell's equations, the eddy current J is induced in case of applying PMF in the liquid melt. The equation can be expressed as follows:

$$\nabla \times E = -\frac{\partial B}{\partial t} \tag{1}$$

$$J = \sigma E \tag{2}$$

where B is magnetic flux density, E is intensity of electric field, J is eddy current density, σ is electrical conductivity.

The above equations are solved by the magnetic vector potential method:

$$B = \nabla \times A \tag{3}$$

$$E = -\frac{\partial A}{\partial t} \tag{4}$$

where A is magnetic vector potential.

As a consequence, the Lorentz force acting on the liquid melt during early stage of solidification can be determined by:

$$F = J \times B \tag{5}$$

Where F is Lorentz force.

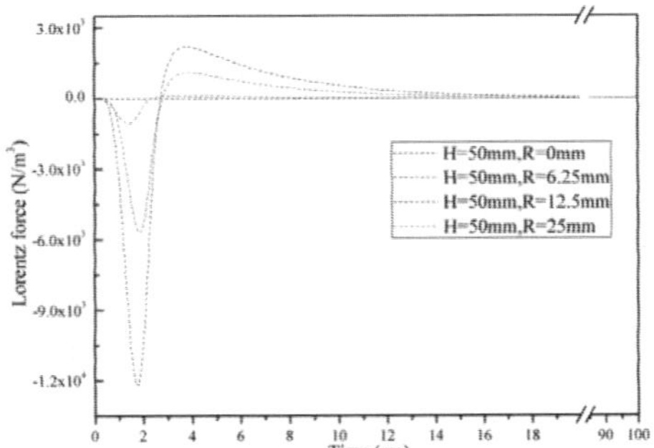

Fig. 5 Variations of the radial Lorentz force with time

Fig. 5 shows the radial Lorentz force induced by the PMF at five points along the radius of the middle section. All of those forces arise as a pressure force, and then turn into a pull force, and finally become negligible before 0.002 s. Due to the inhomogeneous distribution of Lorenz force, there exists a forced convection in the melt. The fluid flow in the bulk of the unsolidified melt promotes an effective homogenization of temperature distribution in the liquid phase, which is beneficial for the formation of the fine equiaxed grain. The refinement effect of PMF can be attributed to the nuclei dissociation from the wall by the melt variation.

Conclusions

In the present study, the coupling effect of Si content and forced convection induced by the PMF on the refinement was investigated. The results show that the average grains size of Al-Si alloys are not reduced monotonously with increasing of Si content, but decreases first and then increases. Furthermore, due to the inhomogeneous distribution of Lorenz force, there exists a forced convection in the melt.The refinement effect of PMF can be attributed to the nuclei dissociation from the wall by the melt variation.

Acknowledgment
The authors gratefully acknowledge the supports of National Key Research and Development Program of China (Nos. 2017YFA0403800), the National Natural Science Foundation of China (Nos. 5177104 0, 51501028, 51690163, 51525401), and Fundamental Research Funds for the Central Universities of China (No. DUT17JC44).

References

1. L.F. Mondolfo, Aluminum Alloys : Structure and Properties, Butterworths, 1976.
2. J. Campbell, Castings, (2003) 320–330.
3. F.C.R. Hernández, J.H. Sokolowski, Journal of Alloys & Compounds, 426 (2006) 205-212.
4. A.L. Greer, Journal of Chemical Physics, 145 (2016) 211704.
5. A.L. Greer, A.M. Bunn, A. Tronche, P.V. Evans, D.J. Bristow, Acta Materialia, 48 (2000) 2823-2835.
6. W. Zhai, Z.Y. Hong, X.L. Wen, D.L. Geng, B. Wei, Materials & Design, 72 (2015) 43-50.
7. Z. Yan, X. Li, Z. Cao, X. Zhang, T. Li, Materials Letters, 62 (2008) 4389-4392.
8. V. Metan, K. Eigenfeld, D. Räbiger, M. Leonhardt, S. Eckert, Journal of Alloys & Compounds, 487 (2009) 163-172.
9. Q. Li, C. Song, H. Li, Q. Zhai, Materials Science & Engineering A, 466 (2007) 101-105.

The electromagnetic stirring apparatus was attached under the mould of continuous casting machine. Three phase alternating current flows through the coils of 12 poles, rotating magnetic field is induced and aluminium alloy melt is stirred and its intensity was changed by applying current and frequency. The mould has 180mm inner diameter and its magnetic intensity inside the mould is simply controlled by adjusting coil current. So, in this study ac current value is directly related to stirring action. Four samples were fabricated by different electromagnetic stirring conditions during continuous casting which are direct chill casting(DC), electromagnetic stirring one way with 80A current(EMS 80A), electromagnetic stirring two way of 5 and 10 seconds switching and 40A (EMS 40A 5s, EMS 40A 10s). The cooling rate of aluminium billets was controlled by adjusting the cooling water and fixed amount of cooling water is used to determine the effect of electromagnetic stirring.

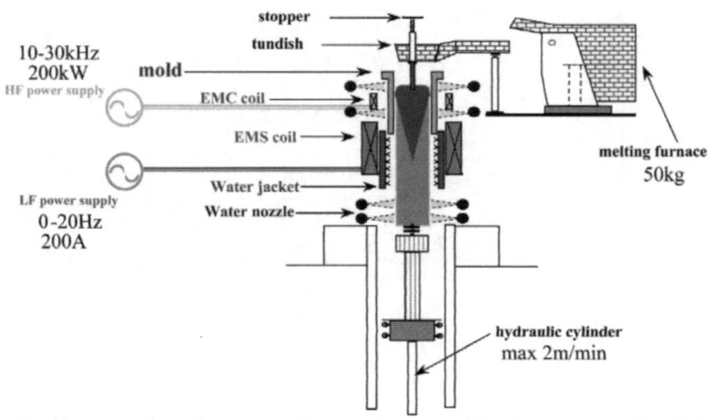

Fig. 1: schematic diagram of continuous casting equipment with electromagnetic stirring equipment

Microstructural survey of billets were conducted on optical microscopy and image analysis and the size and distribution of primary silicon particles were determined for each condition. As following steps, heat treatment of homogenization was conducted at 530 °C for 15 hours. Cast billets was subjected to extrusion process. Initial temperature of billet was 340 °C and extrusion ratio was 30. After extrusion, the samples was subjected to further heat treatment of age hardening. Mechanical properties of samples were tested via tensile test and wear test.

Results and Discussion

Continuous casting of hypereutectic aluminium silicon alloy billets were successfully carried out by adjusting experimental conditions. Hot cracking or tear-out was not occurred during the casting experiments. Figure 2 shows photograph of cross sectional view of cast billets. The diameter of billet is 180 mm and primary silicon is segregated in outer 20mm region. In case of EMS one way with 80A casting, primary silicon segregation becomes more severe and primary silicon is rarely found in the centre region. However, EMS is applied as two way 5s to suppress segregation problem, which means clockwise direction for 5s and counter clockwise direction for 5s subsequently. In case of EMS 5s two casting, primary silicon segregates as island shape and its distribution becomes more uniform. If EMS is applied as two way 10s, it is clear that segregation of primary silicon is less and more uniform distribution of primary silicon in cast billet is obtained.

(a) Direct Chill Casting (b)EMS Casting 80A (one way) (c)EMS Casting (two way, 5s) (d)EMS Casting (two way, 10s)

Fig. 2: Cross sectional views of cast billets, (a) Direct Chill, (b) EMS one way 80A, (c) EMS Casting 40A two way 5s and (d) EMS Casting 40A two way 10s

The image analysis of the micrographs was conducted to calculate the primary silicon aerial fraction. The result is shown in figure 3. In case DC casting, primary silicon is segregated at the region of 20 mm from billet surface. Primary silicon move outward in case of EMS one way 80 A casting samples and the region of 10 mm from billet surface shows high aerial fraction of primary silicon. However, primary silicon is well distributed at EMS two way 10s casting sample. The segregation of primary silicon is mainly due to the flow melt. In case of DC casting, melt flow from nozzle move outward and primary silicon sticks to mush zone region. If EMS is applied to melt pool in the billet, the movement of primary silicon toward outer surface is accelerated and segregation of primary silicon becomes more severe than direct chill casting. The phenomena is also studied by other group and it says high current value of EMS induces congregation of primary silicon. Previous study reports that low current value of EMS is helpful to reduce the congregation phenomena. In this study, EMS two way casting method is applied to reduce the congregation of primary silicon. Two way magnetic field generation with time interval generates the turbulent flow of melt in the casting billet. The turbulent flow of melt does not push primary silicon into outer surface so it is possible to distribute primary silicon uniformly through the billet. From the experiments, 10 seconds is more effective than 5 seconds alterations. The frequency and interval of EMS two way casting is needed to improve by adjusting its value.

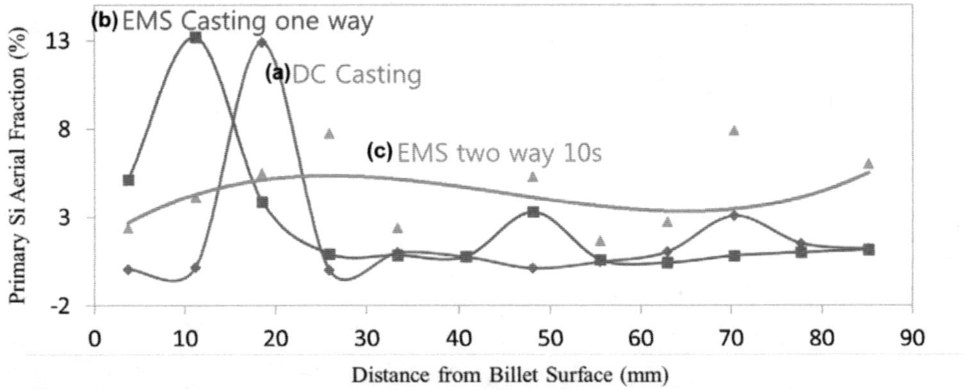

Fig. 3: Primary silicon aerial fraction of cast billet with different casting conditions (a) direct chill casting, (b) EMS casting one way 80A and (c) EMS casting two way 10s 40A

The billets of direct chill casting and EMS two way 10s were subjected to extrusion process and bar type samples were fabricated. Also heat treatment of T6 was applied to enhance the tensile strength of samples because this alloy have some precipitates. Three samples for each conditions were tested and the results are shown in figure 4. The yield strength of samples were 349.9 and 347.8 MPa and tensile strength of samples were 394.2 and 394.7 MPa. The strength of each sample shows no significant difference between DC and EMS two way samples. However the elongation of DC casting is 4.8% but elongation of EMS two way 10s is 7.1%. EMS casting samples shows excellent elongation property than DC casting samples. Primary silicon congregation is main source of failure because it acts as a crack initiation defect. The matrix of used alloy in this study shows excellent yield and tensile strength but the crack initiates from the primary silicon congregation. The low value of elongation from direct chill casting samples is explained by this phenomena and the observation of fracture surface is also good evidence.

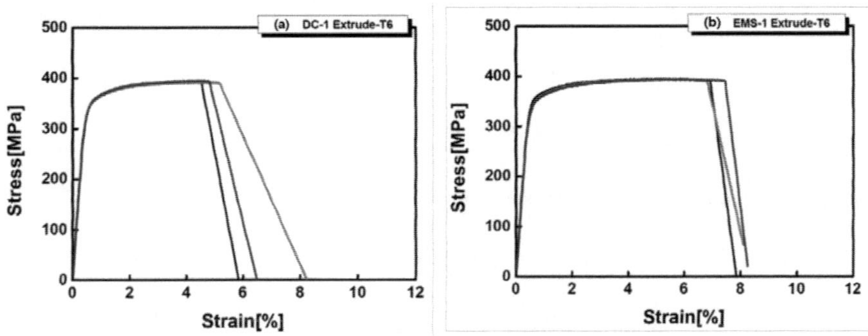

Fig. 4: Stress strain curve of tensile test of (a) direct chill casting-extruded-T6 heat treated and (b) EMS two 10s 40A casting-extruded-T6 heat treated of 3 samples

IOP Conf. Series: Materials Science and Engineering **424** (2018) 012054 doi:10.1088/1757-899X/424/1/012054

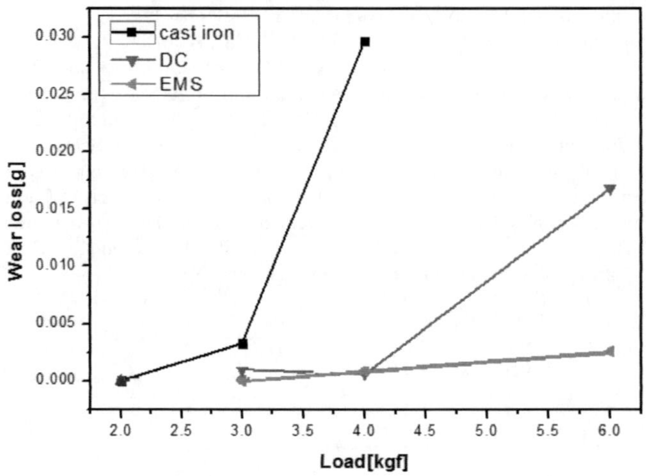

IOP Conf. Series: Materials Science and Engineering **424** (2018) 012055 doi:10.1088/1757-899X/424/1/012055

Direct chill casting with reversing rotational electromagnetic field

S. Shimasaki[1] and A. Minagawa[2]

[1] National Institute of Technology, Kagawa College, 355 Chokushi-cho, Takamatsu, Japan
[2] UACJ Corporation, 3-1-12, Chitose, Minato-ku, Nagoya, Japan

Corresponding author : shimasaki@t.kagawa-nct.ac.jp

Abstract
In order to achieve higher qualities of aluminum metal products, fine grain structures of materials are required. It is well known that agitation of a liquid metal during solidification process is effective to promote grain size refinement. In this research, the reversing RMF, i.e., the RMF that reverses rotation direction of an electromagnetic field at regular intervals, was applied to a molten metal pool in a cylindrical container of a round billet of DC casting. Electromagnetic fields generated by the reversing RMF stirrer were measured and compared with simulation results of electromagnetic analysis. The molten metal in the pool caused a clockwise and a counter-clockwise flow by applying the reversing RMF, and the flow field was simulated by a CFD calculation. It was found that the liquid metal was effectively stirred with the reversing RMF, resulting a higher turbulent intensity of the liquid metal. Grain sizes of a billet were measured and a relation between the flows and the sizes were investigated.

Key words: DC casting, electromagnetic stirring, reversing RMF, grain refinement

Introduction

Grain size and structure have a large impact on qualities of aluminum metal products. Many techniques are proposed and used to obtain finer and equiaxed grain structure, such as grain refiner (TiB_2 particle etc.) [1, 2], control of melt flow in a casting pool [3, 4], oscillating of melt [5], and so on. Among them, electromagnetic stirring is very attractive due to its nature of contactless, it is fitting well to solidification process in a DC casting mold.

Eckert et al. [4] reviewed effects of electromagnetic melt flow control on solidified structures. They showed that Rotating Magnetic Fields (RMF) and/or Travelling Magnetic Fields (TMF) have good effects of grain refinement and an earlier transition from columnar to equiaxed structure (CET). On the other hand, it was pointed out that electromagnetic stirring during solidification of metallic alloy might have frequently a drawback of macrosegregation as a result of strong convection. In order to overcome the flow-induced macrosegregation, Eckert et al. [6] and Willers et al. [7] proposed a pulsed RMF stirring with a constant or alternating direction. They concluded that the time-regulated RMF stirring have a potential to be a more suitable one than conventional continuous RMF stirring methods.

In this research, the reversing RMF, i.e., the RMF that reverses rotation direction of an electromagnetic field at regular intervals[a], was applied to a molten metal pool in a cylindrical container of a round billet of DC casting. Spatial distribution of magnetic flux density, which was generated by the reversing RMF stirrer, was measured by a gauss meter. The experimental result was compared with simulation results of electromagnetic analysis. The molten metal in the casting pool caused a clockwise and a counter-clockwise flow alternately by applying the reversing RMF, and the flow field was simulated by a CFD calculation. A pilot test of DC casting of aluminum alloy with the reversing RMF was carried out and grain sizes of a billet were measured.

Electromagnetic Stirrer

Fig. 1 shows a photograph of the electromagnetic stirrer used in this study. The stirrer is an AC 3-phase 6-pole induction motor without a rotor. It generates a RMF to stir molten metal rotationally inside its bore. An inner diameter of the centre bore is 140 mm and a height of the stirring coil is 210 mm. An AC coil current of 35.7 A at a frequency 50 Hz was applied to the stirrer. Schematic diagram of a concept of the reversing RMF stirring is shown in Fig. 2. It changes alternately the rotation direction with a regular reversing interval.

[a] This stirring method is referred as "RMF Pulse Sequence of Alternation Direction (RMF-PSAD)" in the literature mentioned above [6, 7].

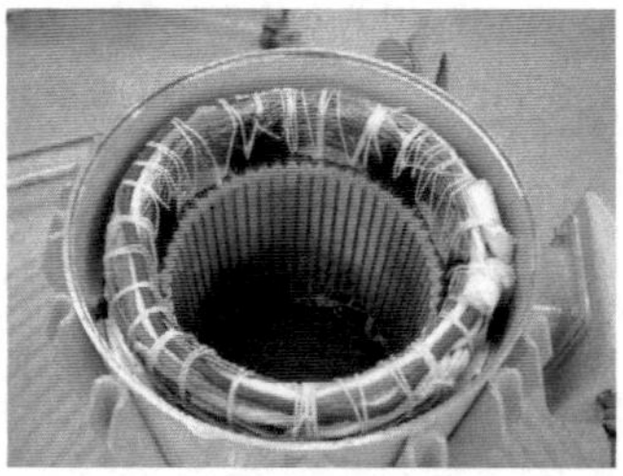

Fig. 1: Photograph of the electromagnetic stirrer.

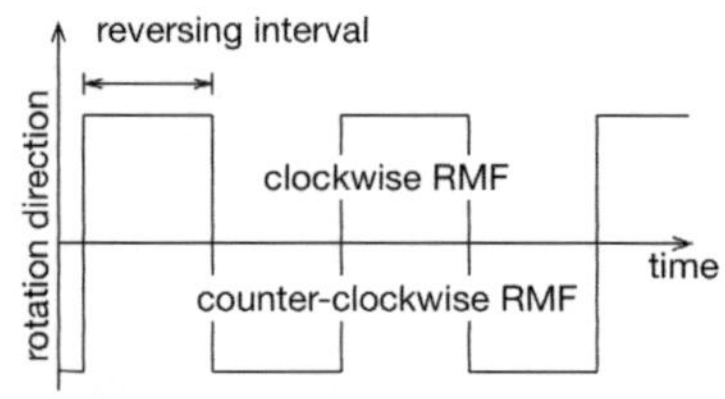

Fig. 2: Schematic diagram of the reversing RMF stirring.

In order to estimate Lorentz force applied to the molten metal by the stirrer, an electromagnetic field analysis was carried out by using a commercial code JMAG (JSOL Corp.). Fig. 3 shows the result of vector field of magnetic flux density on a horizontal cross section at height of coil centre. The figure clearly shows that the stirrer has 6 poles along a circumferential direction. Radial distribution of tangential component of magnetic flux density inside the stirrer's bore was measured by a gauss meter and compered with the electromagnetic analysis. The results are shown in Fig. 4 and the agreement between the experimental data and the numerical simulation is quite well. Spatial distribution of Lorentz force induced by the stirrer is shown in Fig. 5. The force is applied only to the limited area near the RMF coil. The tangential component of the force is much larger than the radial and the vertical components.

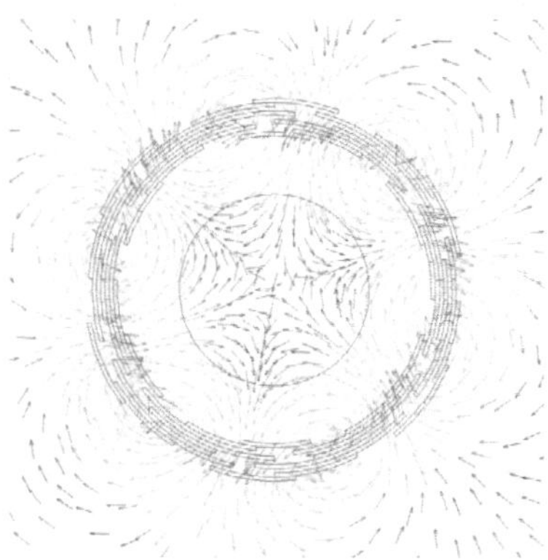

Fig. 3: Vector field of magnetic flux density.

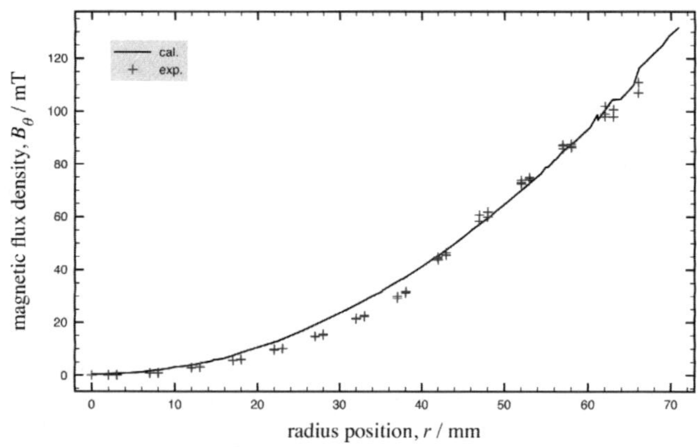

Fig. 4: Radial distribution of magnetic flux density in the bore.

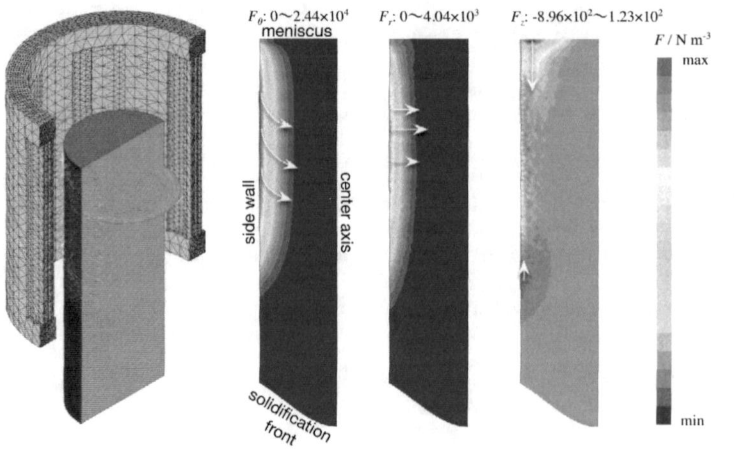

Fig. 5: Spatial distribution of Lorentz force acting on molten aluminum in the casting pool.

Flow Field driven by reversing RMF

Fluid flow of the molten aluminium in the casting pool was conducted by using a commercial CFD code FLUENT (ANSYS, Inc.). Lorentz force estimated by the electromagnetic analysis (shown in Fig. 5) was implemented as a source term of momentum equation. A diameter of the round billet is 92 mm and a height of the pool is from 202 mm (at side wall position) to 219 mm (at centre axis position), the shape of the solidification front (namely, the bottom of the pool) is determined experimentally. For the sake of reducing calculation cost, the following assumptions were used:

- ✓ A flow filed did not affect an electromagnetic field.
- ✓ No energy transportation and no solidification phenomena. Properties of melt were uniform and constant.
- ✓ No feed of molten metal and no casting withdrawal.
- ✓ 2-dimensional axisymmetric flow with swirl.

Temporal change of swirl velocity at the point of $r = 0.87\ R$ (near side wall) and coil centre height is shown in Fig. 6. A reversing time is 0.3 s and the flow direction alternating periodically with the reversing time. Vector field of fluid flow is shown in Fig. 7. From the figure, it can be seen that it is not a simple swirl flow but a 3-dimensional complex flow. Due to the complexity, a turbulent intensity of a melt flow induced by the reversing RMF is much higher than that of a continuous RMF stirring.

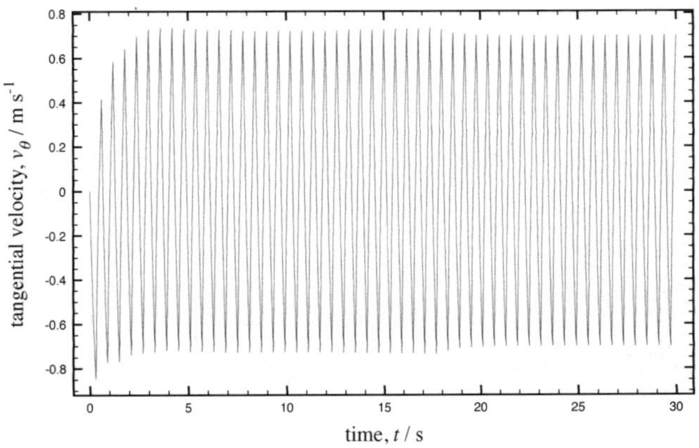

Fig. 6: Temporal change of tangential velocity.

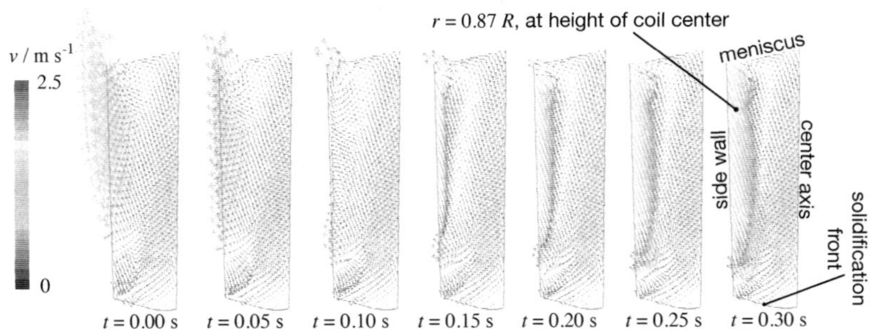

Fig. 7: Flow field generated by the reversing RMF stirrer during reversing interval of 0.3 sec.

Microstructure

Optical micrographs of grain structure of the billet obtained by the pilot test of DC casting of 7000 series aluminium alloy are shown in Fig. 8. Grain sizes are summarized in Fig. 9. The grain size became finer over the entire radius position by applying the reversing RMF stirring, it was about two thirds or less. The smallest grain size was less than 50 μm without addition of grain refiner.

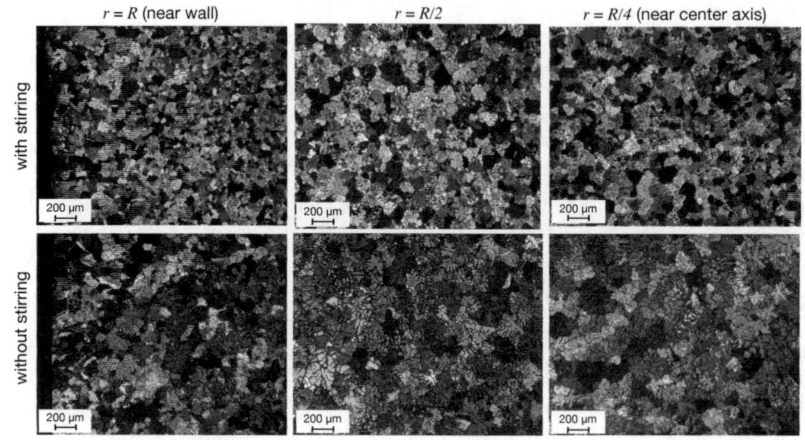

Fig. 8: Polarized optical micrographs of grain structure of 7000 series aluminum alloy with/without stirring.

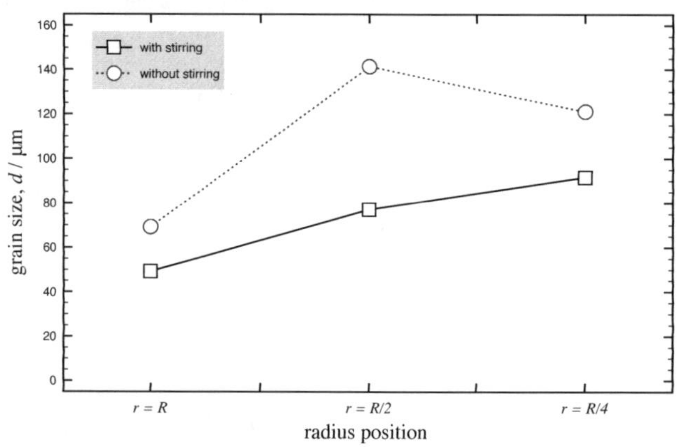

Fig. 9: Grain size of aluminum alloy with/without electromagnetic stirring corresponding to figure 8.

Conclusion

The reversing RMF stirring was investigated through numerical simulation of electromagnetic and CFD analysis. A 3-dimensional complex flow was generated by the alternation of rotation direction of RMF. A pilot test of DC casting was conducted and grain refinement effect by the reversing RMF was confirmed.

Acknowledgment

This paper is based on experimental results obtained from a project commissioned by the New Energy and Industrial Technology Development Organization (NEDO).

References

1. D. G. McCartney, Int. Mater. Rev., 34 (1989), 247–260.
2. B. S. Murty, S. A. Kori, and M. Chakraborty, Int. Mater. Rev., 47 (2002), 3–29.
3. V. Metan, K. Eigenfeld, D. Räbiger, M. Leonhardt, and S. Eckert, J. Alloy Compd., 487 (2009), 163–172.
4. S. Eckert, P. A. Nikrityuk, B. Willers, D. Räbiger, N. Shevchenko, H. Neumann-Heyme, V. Travnikov, S. Odenbach, A. Voigt, and K. Eckert, Eur. Phys. J. Special Topics, 220 (2013) 123–137.
5. K. Iwai and T. Kohama, Tetsu-to-Hagané, 96 (2010), 637–640.
6. S. Eckert, P. A. Nikrityuk, D. Räbiger, K. Eckert, and G. Gerbeth, Metall. Mater. Trans. B, 39B (2008), 374–386.
7. B. Willers, S. Eckert, P.A. Nikrityuk, D. Räbiger, J. Dong, K. Eckert, and G. Gerbeth, Metall. Mater. Trans. B, 39B (2008), 304–16.

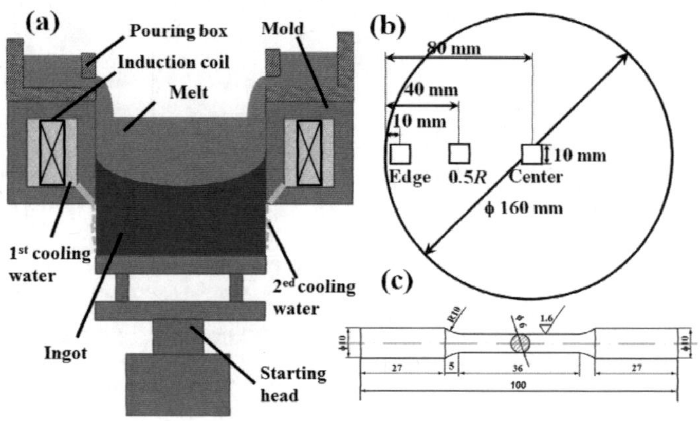
(a)
Pouring box
Mold
Induction coil
Melt
1st cooling water
2ed cooling water
Ingot
Starting head
(b)
80 mm
40 mm
10 mm
10 mm
Edge
0.5R
Center
ϕ 160 mm
(c)
27
5
36
27
100

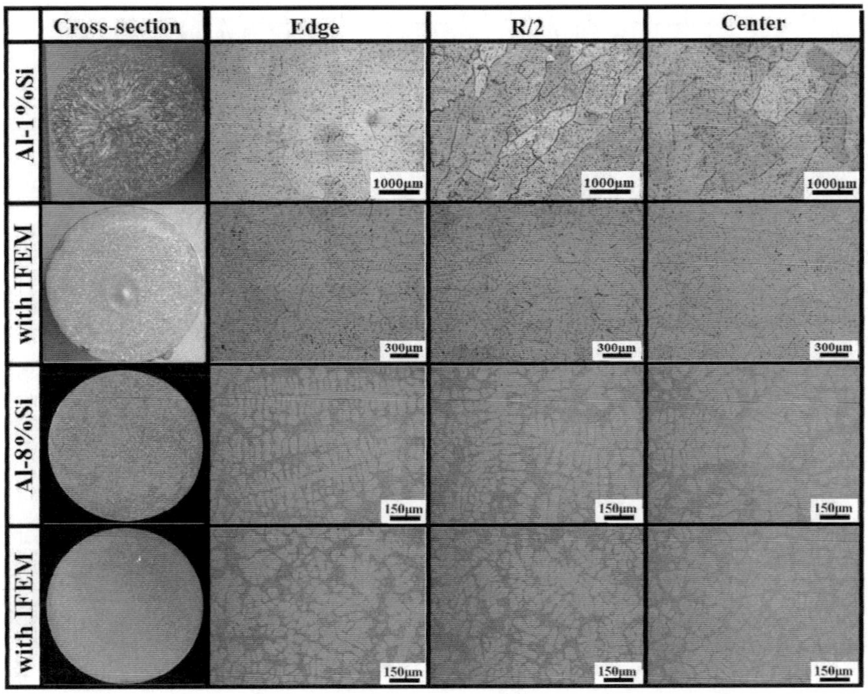
Cross-section
Edge
R/2
Center
Al-1%Si
with IFEM
Al-8%Si
with IFEM
1000μm
300μm
150μm

IOP Conf. Series: Materials Science and Engineering **424** (2018) 012056 doi:10.1088/1757-899X/424/1/012056

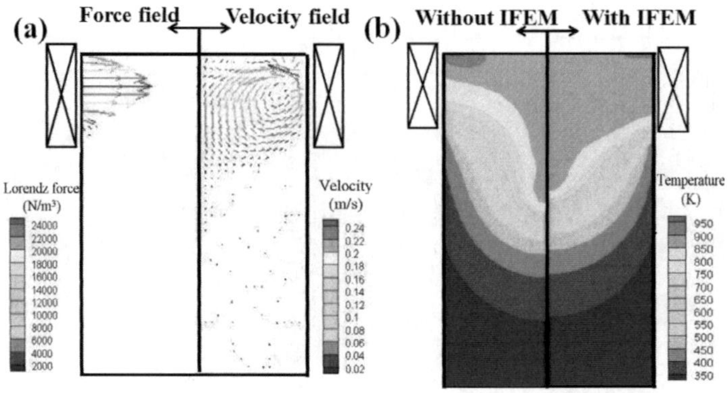

227

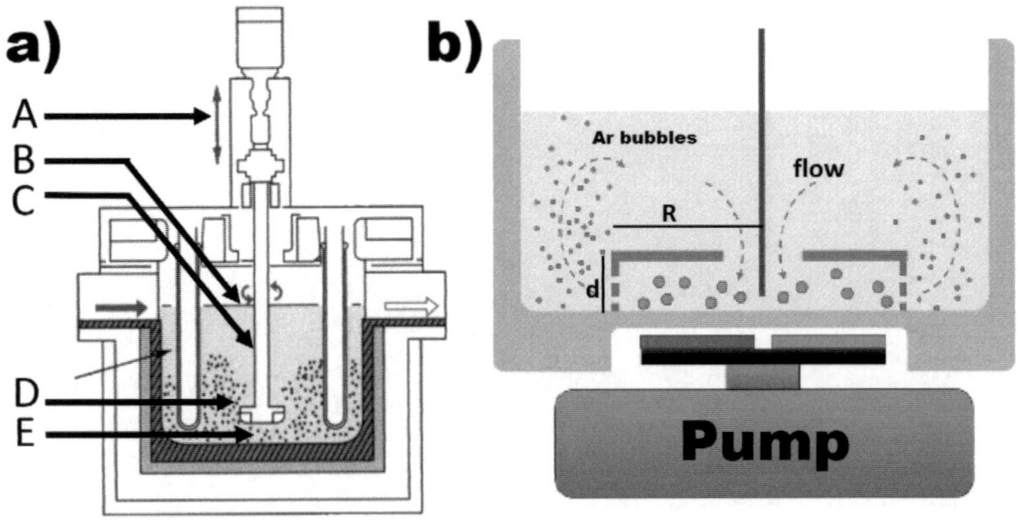

Fig. 1. A) The current degassing process using rotary gas intakes; b) proposed degasser design using permanent magnet rotor to refine Ar bubbles

Results and discussion

To demonstrate the working principle of the proposed setup experimental facility is designed and series of experiments are planned. One of the biggest challenges is the fact that liquid aluminium is contained in refractory material containers which normally have thick walls for better thermal insulation. This is a problem because the magnetic field created by permanent magnets or electric windings decays rapidly with the distance. In this article we are looking at the permanent magnet assembly, which can be rotated, thus setup for flow induction in liquid metal is as simple as possible. Fig. 2 compares two different designs of permanent magnet assemblies which could be used. Fig. 2(a) shows the rotor of a cylindrical electromagnetic pump [6], where several magnet pole pairs are placed on a ferrous plate. Fig. 2(b) shows the concept by using single permanent magnet cylinder. Ferrous yokes in magnetization direction may be used to increase the field penetration distance. Numerical simulations and analytical solution shows that for the current task it is better to use a single permanent magnet cylinder, because liquid metal chamber is placed far away from the magnet surface and magnet diameter should be as small as possible because of difficulties to make large thin wall in aluminium chamber.

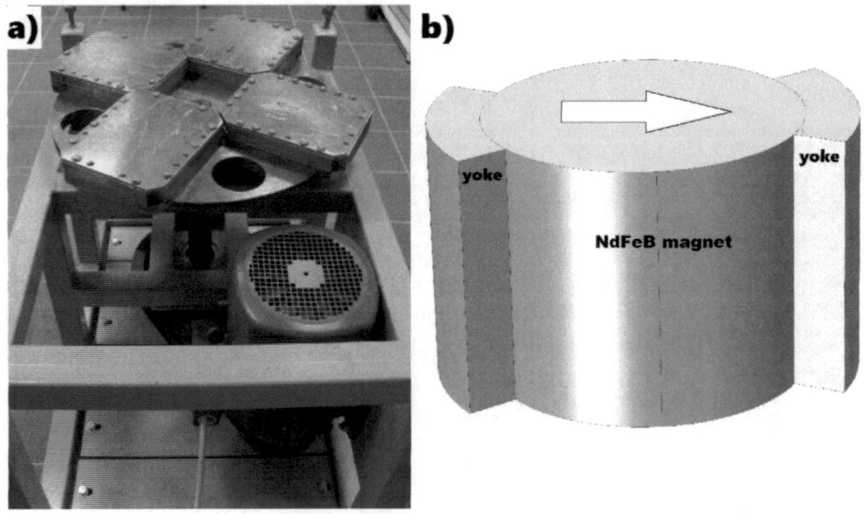

Fig. 2. Comparison of two different permanent magnet rotor designs: a) four rectangular magnet poles on the ferrous plate; b) Cylindrical magnet with yokes for magnetic flux optimization

IOP Conf. Series: Materials Science and Engineering **424** (2018) 012057 doi:10.1088/1757-899X/424/1/012057

$\sqrt{}$

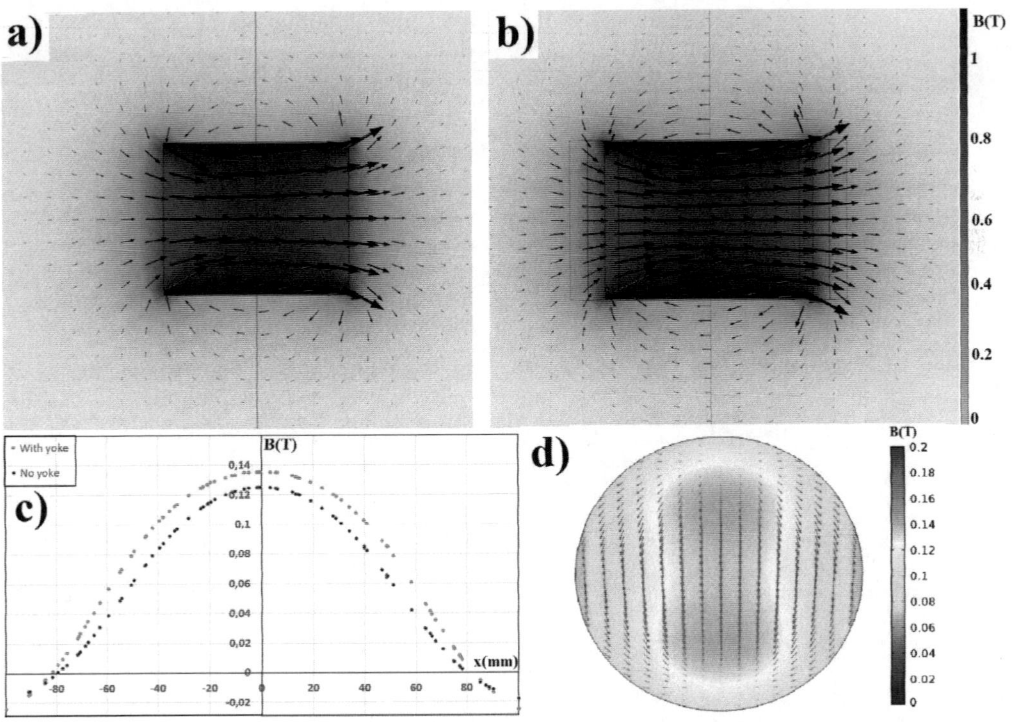

IOP Conf. Series: Materials Science and Engineering **424** (2018) 012058 doi:10.1088/1757-899X/424/1/012058

Potentials of single stage induction heating for press hardening of steel blanks

B. Nacke, A. Dietrich

Institute of Electrotechnology, Leibniz University Hannover, Wilhelm-Busch-Str. 4, 30167 Hannover, Germany

Corresponding author: nacke@etp.uni-hannover.de

Abstract
Due to growing challenges regarding crash-performance, CO_2 emission as well as increasing demand for lightweight construction, hot metal forming of car body parts has risen to one of the most important technologies for saving weight of a car body. During hot forming shaped blanks of steel are heated and austenitized at around 950°C and subsequently quenched for martensitic formation. Currently the heating is realized in roller hearth furnaces which allow only a slow heating and, therefore, a limited production rate. Induction heating of the blanks offers a big potential to increase the production rate dramatically and also to improve the energy efficiency. Only due to the fact, that the heated blanks typically are already pre-shaped with holes and cut-outs induction heating becomes a very complex task. The paper compares different possible induction heating methods (longitudinal heating, transverse flux heating, single stage induction heating, hybrid heating by induction and conventional heating). For the case of single stage induction heating a detailed numerical and experimental investigation in the frame of a recently completed research project is presented.

Keywords: induction heat treatment, hot metal forming, induction heating of blanks, strip heating

Introduction

Nowadays conventional heaters with oil or gas are mainly used in industry for the heating of high strength steel work-pieces. All of them are working by the principle of heat transfer through the surface. This principle has significant disadvantages and restrictions: limited power density, low speed of motion, large size of installations, originally intensive scale formation on the material surface because of long heating time, environmental pollution.

Beside all of these disadvantages the benefit of conventional heaters is the homogeneous heating of the work-pieces. At the same time, this big advantage limits the process variability. An inhomogeneous heating, which is necessary for applying different mechanical properties inside one work-piece, cannot usually be realized.

In comparison with conventional heating methods induction heating offers a lot of advantages, which are mainly based on the principle of direct heating within the heated material. The advantages are theoretically unlimited power density, short heating times, high efficiency, low floor space needs, high flexibility in processing, etc.

Recently different induction heating concepts for hot metal forming were investigated. The potential of continuous heating with longitudinal and transversal flux concepts is analysed in [1,2]. With those heating concepts it is possible to heat blanks up to the Curie temperature. The potential to heat blanks above the Curie temperature is described in [3,4] using single side induction heaters.

Induction heating concepts

Induction heaters can be used for a single-stage heating of work-pieces. The main advantage of induction heating is the fast and energy efficient heating, for holding they are less efficient. Therefore, induction heating for heating up the work-piece can be combined with conventional heating methods if holding or equalization of temperature is necessary.

There are two induction heating concepts for flat material. They differ by the direction of resulting electromagnetic flux provided by the induction coil - longitudinal and transversal flux. Beside the two variants for homogeneous heating also adapted induction heaters for local or inhomogeneous heating of the work-piece are available.

Longitudinal flux heating

For longitudinal flux heating an induction coil encloses the material to be heated, like a strip in Fig. 1. A current in the coil provides the electromagnetic field with a flux orthogonal to the cross section of the strip. By the electromagnetic field eddy currents are induced close to the surface of the strip which are closed in the cross section of the strip (shown as "S" in Fig. 1).

The depth of this current flow is influenced by different material properties and mainly by the frequency of the coil current. This depth is called penetration depth and is given by

$$\delta = \sqrt{\frac{\rho}{\pi \cdot f \cdot \mu}}$$

with ρ - specific electrical resistance, μ - permeability and f -frequency. Consequently, the induced current heats up the material by Joule heat.

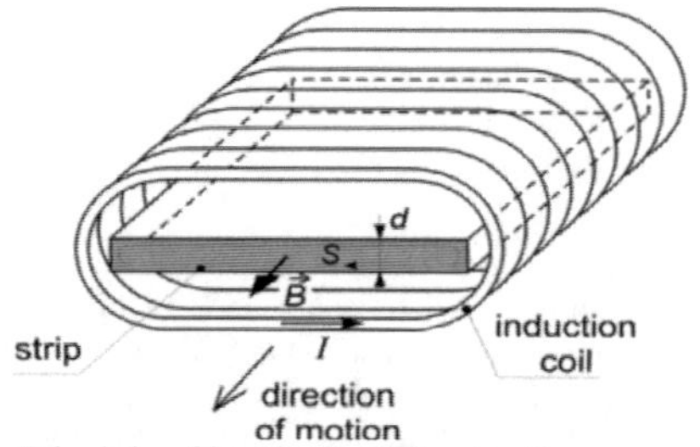

Fig. 1: Principle of longitudinal flux heating

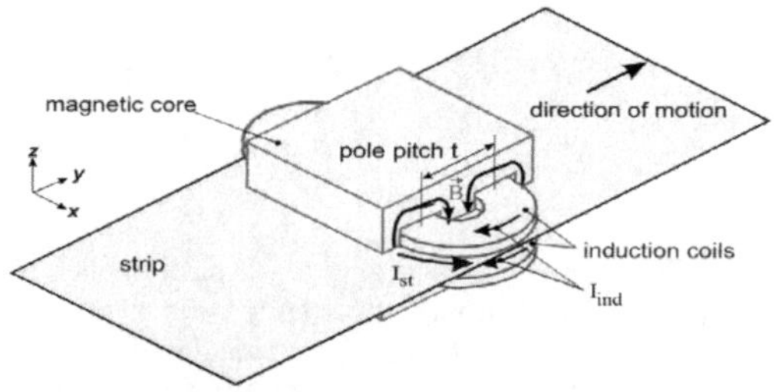

Fig. 2: Principle of transverse flux heating

For nearly flat materials with constant cross-section the longitudinal induction heating concept can be applied, ensuring a homogenous temperature profile. Unfortunately, the application range is limited. If the needed temperature of steel material is above the Curie point (around 760 °C depending on the material), the minimal material thickness which can be heated sufficiently is limited to approximately 1 mm. This is caused by maximum achievable heating frequencies of the power supplies which are below 1 MHz.

Transverse flux heating

The principle of transverse flux heating (TFH) is shown in Fig. 2. The material to be heated is positioned between two inductors located above and below the strip. The electromagnetic field created by a certain frequency current in the induction coils is mainly oriented orthogonally to the material surface. Due to the direction of the field, eddy currents are induced in the plane of the strip and produce Joule heat in the strip. Depending on particular requirements, TFH installations can be accomplished by magnetic flux guiding elements. By an appropriate design of the induction coil it is possible to influence the currents in the strip and herewith the temperature for a concerted non-uniform heating. This opportunity of influencing the temperature distribution is not given by longitudinal flux heating. However, to reach homogeneous temperature profiles with transverse flux heating, the system has to be accurately designed.

Transverse flux heating has advantages even against the induction longitudinal heating principle. On one hand this is indicated by working conditions such as two decades lower frequency, higher efficiency and the induction coils which do not enclose the heated strip.

Comparison of both induction heating concepts for the use in hot metal forming

For hot metal forming where the work-pieces typically are blanks and not strips both induction methods feature advantages in energy consumption and in the ability of achieving inhomogeneous temperatures. But the task to get finally a homogeneous temperature distribution on a high temperature level is difficult for transverse induction heating, therefore, longitudinal flux heating is the preferable concept.

Induction heating is not predestined for the task of temperature holding, which can be required e.g. for metallurgical aspects. It is rather a method for fast heating up. Conventional furnaces with adapted radiation temperature are more effective to realize holding and equalization.

Furthermore, holes or notches in the sheets are in general a problem for induction heating, because the induced current in the material will be concentrated at borders of holes. This can lead to local overheating at the borders. This fact has to be considered by the design of the induction heating process.

Currently a lot of investigations are carried out for the optimal heating process for hot forming of metal blanks in automotive industry.

Fig. 3 illustrates four possible concepts to heat a pre-shaped sheet or endless strip [5]. The concepts are shortly described in the following.

For thin moving sheets longitudinal flux heating can be used as a booster till Curie temperature (Fig. 3-1). The final temperature above 760 °C is reached by conventional heating such as hot air or radiation heating. In batch mode (the work-piece is placed without movement in the heating stations) it is also possible to heat thin sheets by longitudinal flux heating (Fig. 3-2). But higher temperatures need also an additional conventional heating. The design of the heater must be adapted for each work-piece. Changes in the cross-section can be considered by progressive coil design. This concept can be reasonable for small production lots. Transverse flux heating is beneficial for endless strip. Therefore, a concept could be the heating of endless strip from ambient to final temperature by transverse flux heating with very good efficiency (Fig. 3-3). High flexibility can be reached using several small transverse flux heaters with separate power control (Fig. 3-

4). Depending on the work-piece an optimal selection of heaters and individual power is chosen. But it has very high requirements on the design.

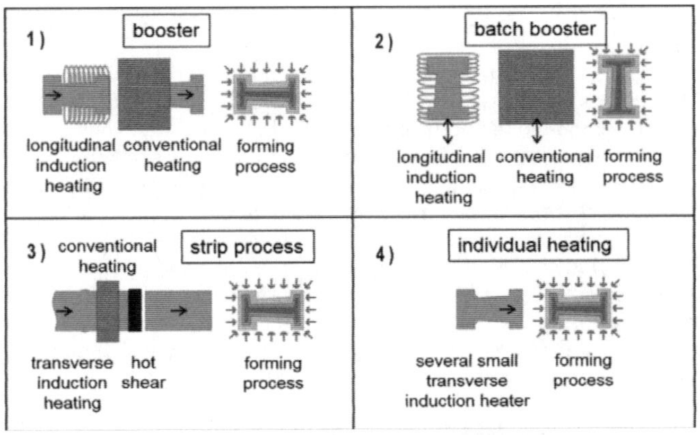

Fig. 3: Heating concepts for hot forming of high strength steel: 1) combined longitudinal flux heating and conventional furnace for continuous mode; 2) batch mode of the combined process; 3) transverse flux heating of endless strip with hot shear; 4) heating by a combination of several small heaters

Investigation of a single-stage induction heating concept regarding discontinuities in the blanks

In the frame of a research project funded by the German Government an induction heating system was investigated which allows a single-stage induction heating up to temperatures of 950 °C. Detailed studies are described in [6-8]. In this paper the most important results are summarized.

The advantage of single-stage induction heating is that it can work without conventional heating systems and, therefore, is most energy effective. In the hot metal forming process, the blanks are typically heated up to a temperature between 930 °C and 950 °C for AlSi coated and uncoated material in order to reach the austenitization temperature.

The blanks typically have irregular forms. Irregularities are holes, cut-outs or changes in the cross-section. In this investigation particularly holes are taken into account and the electromagnetic and thermal behavior due to these irregularities are shown. Furthermore, different geometrical parameters of the system of the induction coil and the work-piece are investigated and from this investigation design rules for the induction coil are derived.
The investigations are mainly carried out by numerical simulations followed by experimental tests for verification of the numerical results.

The numerical model

For the calculation of the following results a 3-dimensional FEM model is used. The material properties such as specific electrical conductivity, relative permeability, specific thermal conductivity and specific heat capacity are adjusted to the transient temperature distribution. Because of symmetry only one half of the system is modeled in order to reduce the calculation time. A sketch of the investigated heating system is shown in Fig. 4. The induction coil has a rectangle cross-section. For an exact calculation it is important to mesh the edges of the blank very fine. The calculations are based on the material properties of 22MnB5 steel [2]. The reference thickness of the blank is 1,5 mm, because in many industrial applications blanks with a thickness in this range are used. The width of the blank is 200 mm. The width of one winding is 30 mm and the distance between the windings is 10 mm. The coil has 8 parallel windings. The frequency of the current is 400 kHz. The simulation results are verified by experimental investigations which are described in detail in [8].

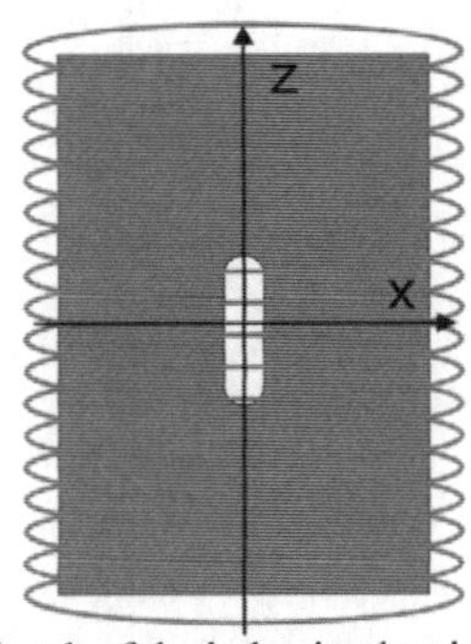

Fig. 4: Sketch of the induction heating system

Numerical investigation of the influence of holes

The following results show some main results of the study carried out in the research project. Detailed results are described in [6-8]. In this

paper only the principal effect of a hole on the electromagnetic field, the Joule heat distribution and the temperature distribution during longitudinal flux heating is described.
In the simulation the blank is located in the coil during the heating process (batch heating). The current in the induction coil is applied with a frequency of 400 kHz and an amplitude of 960 A. The heating process takes 20 s and ends with a temperature of around 950 °C.

The position of the hole is normal to the induction coil. The center of the hole is located at the x-coordinate 0 mm and the z-coordinate 0 mm. The hole has a width of 14 mm and a length of 96 mm. The radius of the corner is 3.5 mm.

Fig. 5 shows the Joule heat distribution in the blank close to the hole. The strip pattern is caused by the coil. Because of higher magnetic field intensity under the windings a higher current is induced and hence higher Joule heat is generated.

At the long sides of the hole less Joule heat is generated. This is understandable with a look at Fig. 6. The current doesn't flow parallel to the edge of the hole rather it turns around and flows back on the backside of the blank. The generated Joule heat results in a temperature distribution close to the hole (Fig. 7). The strips produced by the windings are noticeable and the attracted magnetic field causes a little higher temperature at the small endings. Due to this main results, the influence of holes and cut-outs are much less than expected before. But these influences are only low if the used frequency is high enough in order to generate a pronounced skin effect over the blank thickness which is necessary for pure longitudinal flux heating. If the skin effect is less pronounced currents will close in other ways, than only from top to down side of the blank. In this case local overheating close to the hole can occur.

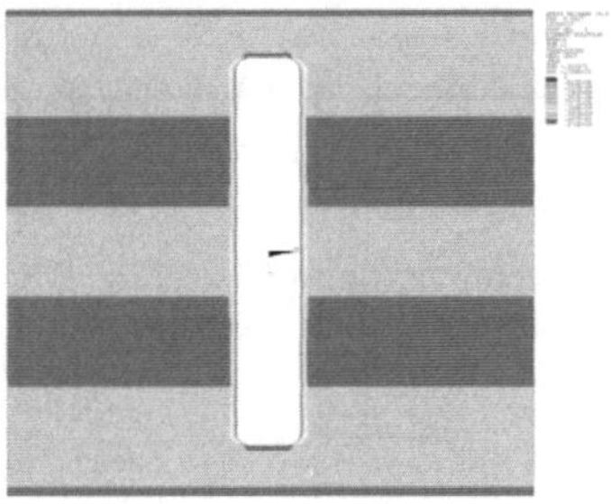

Fig. 5: Joule heat distribution close to the hole

Fig. 6: Current flow at the hole

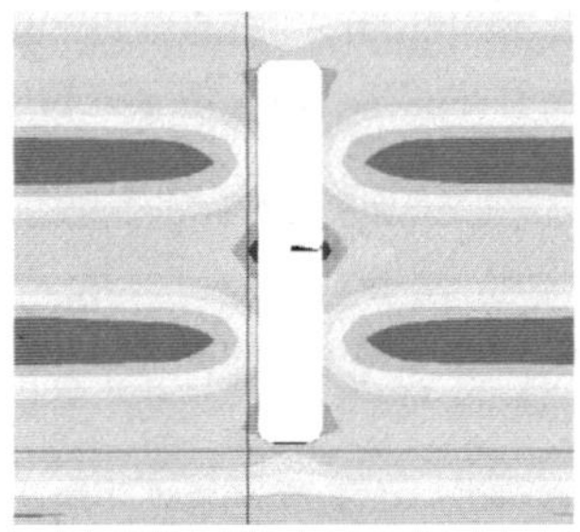

Fig. 7: Temperature distribution close to the hole

Conclusions

Induction heating is an effective heating method which allows a high energy efficient heating in many applications. The biggest advantage is the high heating speed and the high thermal efficiency if a fast heating is required. Induction heating is less suitable for holding processes. Therefore, in the case of necessary holding processes hybrid heating of a combined induction and conventional furnace heating can offer many advantages regarding energy efficiency and product quality.

For hot metal forming of car body parts, a comprehensive study for a single-stage induction heating process has been carried out. The influence of discontinuities such as holes, cut-outs and changes in cross-section has been investigated in detail by numerical simulations and verification by experiments.

One important result of the investigation is that in the case of longitudinal flux heating typically at the edge of holes a lower temperature exists in contrast to the general estimation, that holes generates temperature overheating at their edges. The lower temperature at the edges of holes is caused by the fact that the current close to a larger hole cannot cross the hole and tends to flow back on the down side of the blank. This is due to the general principle of longitudinal flux heating that currents are closing in the cross-section of a blank or strip.

Acknowledgment

The research project "Series-production of inductively heated blanks for hot sheet metal forming" (IFG-No. 18738 N) from the Research Association for steel Application (FOSTA) and the Research Association of Automotive Technology (FAT) is supported by the Federal Ministry of Economic Affairs and Energy through the German Federation of Industrial Research Associations (AiF) as part of the programme for promoting industrial cooperative research (IGF) on the basis of a decision by the German Bundestag.

References

1 Opitz, T.; Vibrans, T., Induction fast heating of formed blanks for hot metal forming in car body production, HTM Journal of Heat Treatment and Materials: Vol. 70, No. 6, pp. 261-266, 2015
2 Vibrans, T., Induction heating of formed blanks for hot metal forming, PhD thesis, Technical University Chemnitz, 2016

3 Tröster, T., Niewel, J.; Inductive heating of blanks and determination of corresponding process windows for press hardening, FOSTA Research Association for Steel Application, Vol. 805, 2014
4 Todzy, T., Process window for press hardening at fast heating of steel plates with AlSi coating, PhD thesis, University of Siegen, 2015
5 Schülbe, H.; Jestremski, M.; Nacke, B.: Induction heat treatment for press hardening process. 3rd Int. Conf. on Hot Sheet Metal Forming High-Performance Steel, Kassel, Germany, June 13-17, 2011
6 Dietrich, A., Nacke, B., Pfeifer, F., Marten, T., Tröster, T., Effects of holes in blanks for press hardening process due to induction heating. XVIII International UIE-Congress on Electrotechnologies for Material Processing, Hannover (Germany), June 6-9, 2017, Vulkan-Verlag Essen, ISDN 978-3-8027-3095-5, pp. 327-332
7 Dietrich, A., Nacke, B., Pfeifer, F., Marten, T., Tröster, T., Investigation of geometrical discontinuities in blanks for hot sheet metal forming process under the influence of induction heating. 5th International Conference on Steels in Cars and Trucks, Noordwijkerhout/Amsterdam, 2017
8 Dietrich, A., Nacke, B., Numerical investigation of effects on blanks for press hardening process during longitudinal flux heating. Proceedings of the VIII International Scientific Colloquium Modelling for Material Processing, Riga (Latvia), September 21-22, 2017, pp. 179-184
9 B. Nacke, E. Baake, Induction heating, Vulkan Verlag, 2016

IOP Conf. Series: Materials Science and Engineering **424** (2018) 012059 doi:10.1088/1757-899X/424/1/012059

Induction Preheating for the Submerged Arc Welded Steel Tube Production

C. Cincunegui[1] and P. Marino[1]

[1] Furnaces and Heat Transfer Department, Tenaris R&D center, Campana, Argentina

Corresponding author: ccincunegui@tenaris.com

Abstract
Tenaris is a leading supplier of tubes and related services for the world's energy industry that manufactures a complete range of seamless and welded steel tubular products. The welding facilities are located in North and South America, and perform three types of welding processes, namely Electric Resistance Welding (ERW), Spiral Submerged Arc Welding, and Longitudinal Submerged Arc Welding (LSAW). Regarding the ERW process, in recent years we studied both the welding using Comsol [1,2], and the ulterior seam heat treatment using an in-house developed FEM model [3,4]. During the LSAW process, after the plate is conformed through the UOE forming process, the butt joint of the pipe is welded in three phases, first of which aims at presenting the edges and joining them but without filling in the bevels, the second one is to fill in the seam region at the inside of the pipe and the third one to do this operation on the outside. The welds are made by melting with an electric arc the bare metal electrodes. Pressure is not applied on the pipe during welding. Depending on the final product properties sought, a preheating near the edges with a relatively narrow temperature range is required for these three welding stages. In this article we focus on the inductive preheating for the tack welding stage of a continuous production line; it involves an open tubular geometry whose edges are facing and separated approximately 10 mm. There is no need to reheat the full tube, but only a region of ±75 mm from the edges approximately. Given the geometry and the required localized preheating, usual solution would be bar type inductors as those used typically for seam annealing. Alternatively, cylindrical coils can be used instead of inductive bars. 2D and 3D numerical models of different inductor bars and cylindrical coils are used in order to assess both kinds of preheaters and to understand the advantages and disadvantages of each one.

Key words: induction heating – steel tubes – submerged arc welding

Introduction
The LSAW process that is studied in this work involves four main stages: first, water remaining after cleaning the already-formed steel sheet is evaporated in a combustion dryer, then the edges are joined in the tack welding process, subsequently the ID welder fills in the internal seam, and finally the OD welder does the same with the outer one, as shown in Fig. 1 (see reference [5]).

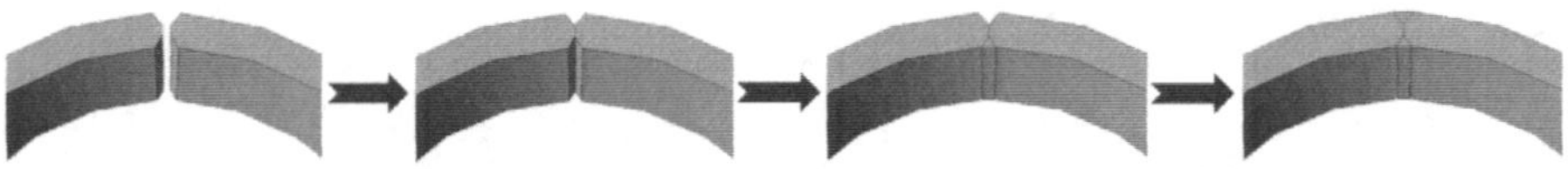

Fig. 1: Main processes in a LSAW line (after forming).

These three welding processes require tight temperature ranges in the zone to be welded. Besides, there is a maximum spot temperature that can be reached to avoid oxide formation. As the region of interest is about ± 75 mm surrounding the edges, for large OD products this represents a small percentage of the tube mass, and therefore heat conduction to the colder tube body cools down fast the preheated zone. On the other hand, if a wider preheated zone is generated, the process efficiency diminishes and the time to cool down such a large mass of steel is high and strongly depends on the mill conditions. Even more, the cooling time can become a bottleneck for the OD welder, given the fact that it is a process with a maximum acceptable temperature not very high.
Therefore, the preheating processes and the transference times from the furnaces to the welders are critical to meet the specifications, and the balance between preheating temperature (and distribution) and time is very delicate.
This study focuses on the tack welding process, defined as "a weld made to hold the parts of a weldment in proper alignment until the final welds are made" [6]. Although preheat is not required in this case according to this standard, there was a need to adapt the whole line for an out of range product that required a significant productivity improvement. The combustion chamber showed not to be sufficient for the previous necessary drying, and therefore the need of an extra preheating stage came up. Due to the reduced space available in the existing line, new induction heating equipment was

considered.
Given the fact that after the tack welding the strip edges are in contact and induced currents are able to close circumferentially, in what follows attention is focused on the tack welding preheater that requires an "open tube" geometry –a non-standard configuration in the industry–.

Initial proposals for the preheater
Induction bars were the first option that was considered because of the similarity with the standard seam annealing process. Two possibilities were analyzed:

- An annealer-like inductor, i.e., a central conductor with two returning branches symmetrically placed, parallel to the seam direction, as those used for the seam heat treatment in the Electric Resistance Welding (ERW) lines.
- A U-shape inductor, namely, two parallel conductors with opposite currents, or eventually same direction current, and a distant returning conductor. This kind of bar inductors are not very usual, given the fact that at the middle of both bars the magnetic flux density is null (first case) or lower than near one of the conductors (second case). As in the present case exists a separation between both zones to be heated (the edges), this issue could, in principle, be benefic for the considered geometry.

Both cases were simulated using COMSOL Multiphysics® software, in a 2D approach that disregards the heat diffusion in the axial direction and the effects of the finite length of the inductor bars and the formed steel. The model solves first the electromagnetic problem, and then uses the computed power density to feed the thermal problem, with boundary conditions trying to reflect the mill environment. Some results are shown in Fig. 2.

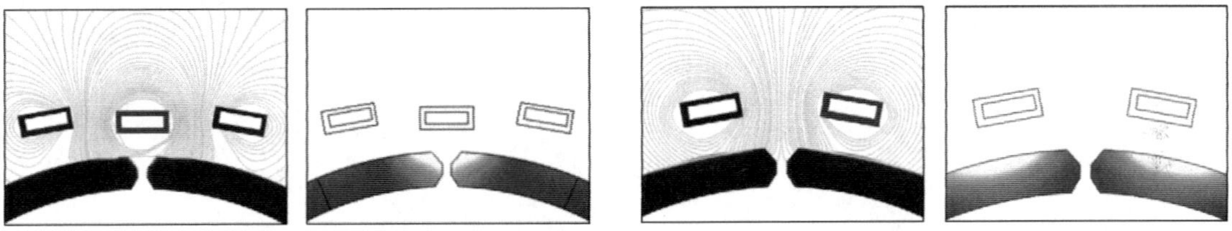

Fig. 2: The two figures on the left, an annealer-like inductor. First, the electromagnetic problem: a central conductor with free current circulating in one direction (red rectangle) and returning by the lateral conductors (blue ones). The color map in the formed steel sheet represents the power density distribution, while the magnetic flux density streamlines are depicted in grey. Second, temperature distribution at the preheater outlet. The travel time up to the tack welder inlet produces the needed homogenization in the region of interest. The two figures on the right, the same graphs but for a U-shape inductor.

The temperature distributions obtained from these simulations reach in some cases the desired conditions, depending on one hand on the conductors' transversal geometry and the gap with the steel, and on the other on the free current amplitude, the bars length, and location relative to the tack welder inlet. However, concerns regarding the tube bending and the consequent effect on the gap between the inductor bars and the steel surface, that highly affects the heating homogeneity, prevented the implementation of these options.

Final proposal for the preheater
In order to get a temperature distribution not so dependent of the gap between the inductor and the steel surface, interest was focused on a full body coil, that generates a more uniform inner magnetic flux density distribution. This kind of inductors have a circumferential free current that induces in a tube a current in the opposite direction and have a typical efficiency of 0.80 for reasonably good-tuned industrial equipment (as the one that was available in the plant for this operation). However, in this case the tube is not yet closed, and there is scarce literature of using a full body coil to heat an "open tube", that prevents the possibility of circumferential induced currents. The calculation of these currents requires to pose a 3D problem, for which the geometry and the mesh of the formed steel sheet plus the coil are shown in Fig. 3.

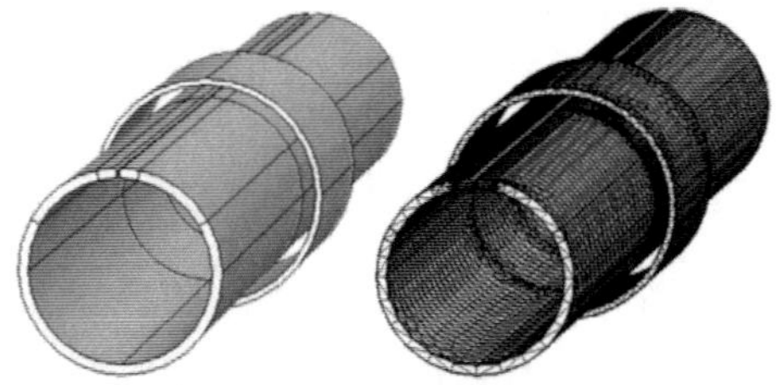

Fig. 3. Geometry and mesh for the formed steel sheet plus the full body coil.

Fig. 4 and Fig. 5 show the results obtained using color maps for the circumferential and the axial induced currents respectively. As schematized in Fig. 6, the current circulates along the outer surface circumferentially, and in the opposite direction respect to the free current of the coil. When it comes to one edge, most of the current surrounds the bevel and close the circuit returning along the inner tube surface. The rest of the current proceeds in a direction parallel to the strip edges, moving away from the coil center, and closing the circumferential path in a distributed way along tube sections that are out of the coil. As the length of steel that is under the coil is much shorter than the tube length, the current density magnitude is much smaller outside the coil, and the heating is therefore concentrated inside the coil, particularly near the bevels. Moreover, as the current surrounds the bevels, the power density in the zone of interest is distributed at the surface, both outer and inner. This fact improved the homogeneity of the heating through the thickness, as shown in Fig. 7, to be compared with Fig. 2 of the bar inductors mentioned before.

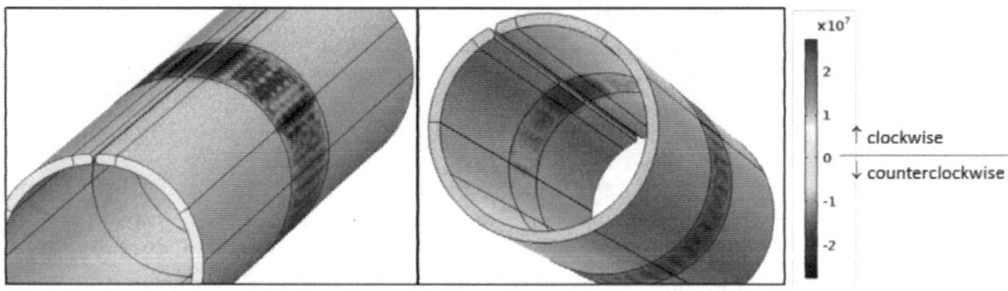

Fig. 4. Color map of the circumferential induced current density.

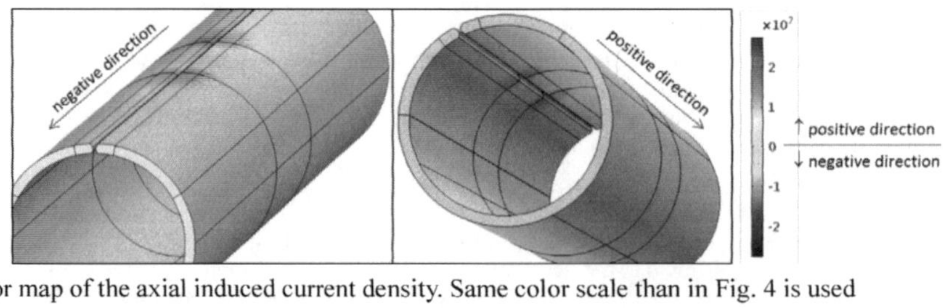

Fig. 5. Color map of the axial induced current density. Same color scale than in Fig. 4 is used

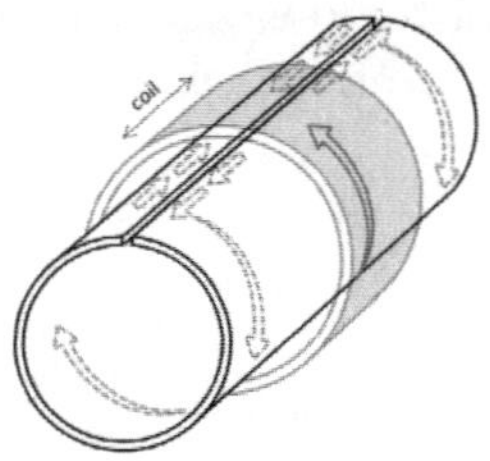

Fig. 6. Schema of the current circulation in the steel piece. The arrows in magenta represent the currents circulating on the outer surface, while the arrows in green account for those on the inner surface.

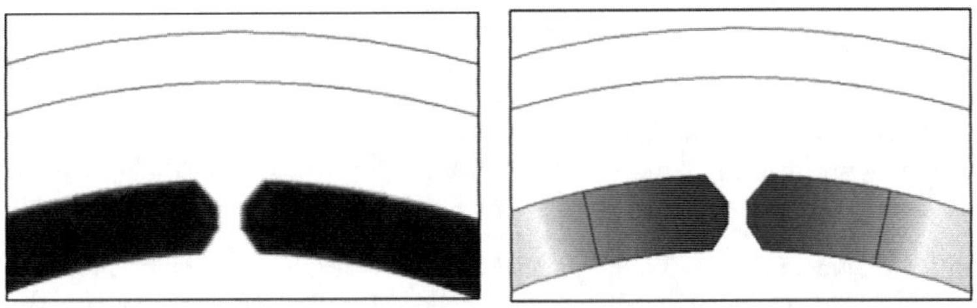

Fig. 7. Same as Fig. 2, but for a full-body coil. Power density at the middle of the coil length, on the left, and temperature distribution when the desired value is reached (at this plane), on the right.

Closing comments

Using numerical simulations it was verified that the preheating with a typical full body coil of a formed tubular steel sheet, not yet closed by the LSAW process, is feasible for the line conditions.

Moreover, this type of coil has two main advantages as compared with bar-type inductors. In first place, as the magnetic flux density is more uniform in the center of the coil, the distance of the bevels to the inductor is not as important as for a bar inductor, in which the field falls dawn quickly with the bar-tube gap. This issue becomes very important given that the localized heating could produce a considerable bending. In second place, the induced power is distributed both in the inner and outer surfaces of the zone of interest, not only at the outer side as in the bar case. The temperature distribution is therefore more homogeneous and improves the process.

Lastly, even though not shown here for lack of space, a static trial was performed that allowed to estimate the efficiency of the process in 0.30. Although obviously lower than the efficiency of heating a tube with a full-body coil, it has to be compared with the annealer-like inductor, for which the efficiency is estimated in 0.55 (below Curie temperature) in the case of a seam heat treatment.

References

1. C. Cincunegui and P. Marino, "Electric welding process in the tube manufacturing: comparison between a numerical model and a set of plant trials", Comsol Conference, Boston, MA, USA, 2014.
2. C. Cincunegui, P. Marino, D. Tieppo and E. Martinez, "Parametric analysis of the electric resistance heating for the welded tube manufacturing process", 20th IAS Steel Conference, Rosario, Santa Fe, Argentina, 2014.
3. C. Cincunegui and P. Marino, "Modeling of the seam annealing in welded tubes," 18th IAS Rolling Conference, Rosario, Santa Fe, Argentina, 2010.
4. C. Cincunegui, P. Marino, A. Sesa, R. Charreau, S. Nakoneczny, D. Johnson, D. Tieppo and E. Martinez, "Validation of a FEM model for the seam annealing", 19th IAS Rolling Conference, Rosario, Santa Fe, Argentina, 2012
5. M. Afzaal Malik, M. Ejaz Qureshi and N. Ullah Dar, "Numerical simulation of arc welding investigation of various process and heat source parameters", Failure of Engineering Materials & Structures, 30 (2007) 127–142
6. American Welding Society, "AWS A3.0: Standard Welding Terms and Definitions," 2010.

Simulating the magnetic field/transfer phenomenon of the tundish with channel type inducting heating

B. Yang[1,2], A. Y. Deng[1,2*] and E. G. Wang[1,2]

[1] Key Laboratory of Electromagnetic Processing of Materials (Ministry of Education), Northeastern University, NO. 3-11, Wenhua Road, Heping District, Shenyang, China
[2] School of Metallurgy, Northeastern University, NO. 3-11, Wenhua Road, Heping District, Shenyang, China

Correspomding author: dengay@epm.neu.edu.cn

Abstract
The tundish with channel type inducting heating is usually used to stabilize the superheat of molten steel, and remove the non-metallic inclusions. In order to research the transfer characteristics comprehensively, A mathematical model including electromagnetic field, flow and heat transfer is established to investigate the distribution of magnetic flux density, electromagnetic force (EMF), joule heat, flow and temperature field. The results indicate that the effective area of electromagnetic field mainly contains the channels, moreover, the region near the side walls contacting with channels also belongs to it. Therefore, the EMF generated in this area affects the flow pattern in pouring chamber and distribution chamber, and the pinch force produced in the channels also makes the molten steel rotate strongly. A comparison of flow field between the tundishs with and without induction heating implies that compared with the EMF, the thermal buoyancy has little influence on the flow of molten steel. The joule heat mainly distributes in the channels and accounts for 91.68%. Besides, after the induction heating being applied, the thermal drop can be compensated, and the temperature distribution gets more homogeneous.

Key words : numerical simulation, induction heating, tundish

1. Introduction

The inclusions removal and constant temperature casting with low temperature are two essential and tough issues in continuous casting[1]. The channel-type induction heating tundish inspired by the Marty's research and the application of induction furnace was developed to deal with the problems.[2] Compared with the traditional tundish, the channel-type induction heating tundish can cleanly and effectively compensate the temperature loss of the molten steel, besides, it is confirmed that is can promote the inclusions removal.

The electromagnetic characteristics of a induction heating furnace or tundish was always obtained by solving the Maxwell's equations using finite element method.[3, 4] It was found that the inductive current formed loops in the channels and chambers, and the magnetic field is rotational, on cross section, the EMF points to the interior of the channels. These basic phenomenon is also observed in the Vives's experienment.[5] The flow and temperature field was investigated by solving the Navier-Stokes equations using finite volume method. Jamil I. researched the effect of standard k-ε model, standard k-ω model, SST model and Zero Equation model on the flow pattern in induction heating furnace, and the results showed that there was no obvious difference between them.[6] In the channel type induction heating tundish, the inclusions removal was supposed by the absorption of the channel wall or floating to free surface, and the absorption behavior was forced by the electromagnetic pressure force.[7] However, it is reported that the movement of inclusions is most affected by the flow, namely, the influence of EMF on the movement of inclusions is very limited.[8] Besides,The previous works mostly concerned what happened in the channels, the phenomenon occurred in the chambers was ignored. Therefore, in order to promote the inclusions removal, it is essential to figure out the electromagnetic characteristics, flow and temperature distribution in the whole tundish, which have a non-negligible effect on the movement of inclusions.

In present work, based on the commercial software of ANSYS APDL and CFX, a mathematical model containing magnetic field, flow and heat transfer is established to gain a deep insight into the channel type induction heating tundish. The electromagnetic field distribution features, the flow and heat transfer behaviors in the tundish are qualitatively and quantitatively presented.

2. Mathematical model

The following assumptions are made to simplify the numerical model: (1) The molten steel is an incompressible continuous Newton fluid, and the physical properties are constant with temperature; (2) The slag layer is not modeled, thus, the effect of slag on the molten steel is neglected; (3) The displacement current is negligible. The electromagnetic mathematical model is based on the Maxwell's equations and the constitutive equations. [9, 10] The Navier-Stokes equations and standard k-ε model are employed to obtain the flow and temperature field:[11, 12]

Single-phase alternating current of 50Hz is applied on the induction coil, and the magnetic field intensity is regarded as zero at infinity when simulating the magnetic field. In the fluid and heat transfer model, uniform velocity calculated by the steel flux balance is applied at the inlet of the tundish, the temperature here is supposed as constant at 1748K. The surfaces of molten steel and the walls are set as free slip wall and no-slip wall, respectively. A pressure outlet is applied. According to the Sahai's research[13], the heat flux at the free surface, the bottom, the longitudinal walls, the transversal walls, and the channel walls are 15 kW/m^2, 1.8 kW/m^2, 4.6 kW/m^2, 4.0 kW/m^2, and 0.2 kW/m^2, respectively. Other simulation parameters are listed in Table 1.

Table 1 The simulation parameters

Parameter	Value	Parameter	Value
Conductivity of molten steel (S m^{-1})	7.14×10^5	Thermal expansivity (K^{-1})	1×10^{-4}
Relative permeability of molten steel	1	Length of tundish (m)	5
Density of molten steel (kg m^{-3})	7100	Width of tundish (m)	1.5
Viscosity of molten steel (Pa s)	6.2×10^{-3}	Depth of melt (m)	1
Specific heat of molten steel (J kg^{-1} K^{-1})	680	Diameter of the channel (m)	0.115
Thermal conductivity (W m^{-1} K^{-1})	30	Outlet diameter (m)	0.07

The electromagnetic characteristics and hydrodynamic problem are individually calculated by using the software ANSYS APDL and CFX. The coordinate system and geometric model are illustrated in Fig. 1, and the parameters of the tundish are also listed in Table 1.

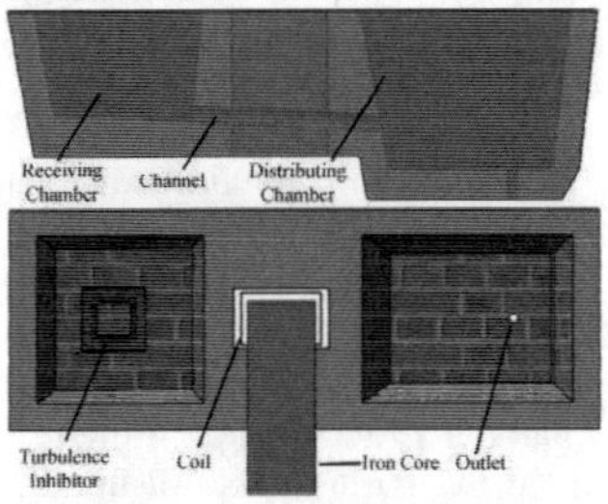

Fig. 1 The geometric model of the tundish

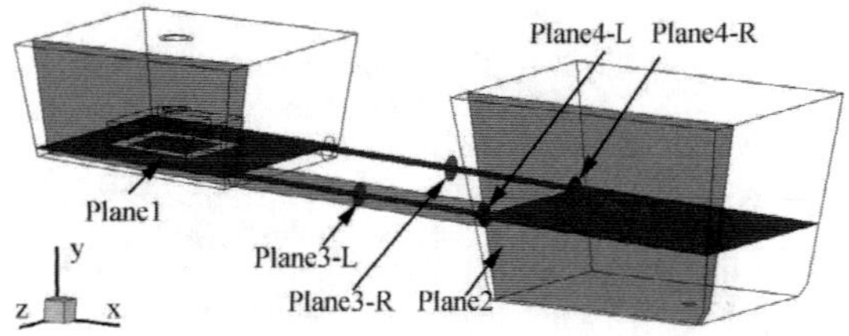

Fig. 2 The key planes in the tundish

3. Results and discussion

In this section, the distribution of magnetic induction intensity, EMF, joule heat, flow and temperature field is depicted in details at the heating power of 800 kW.The results are mainly revealed on some specific planes, for showing results in 3-D is too complicated to be understood clearly. Fig. 2 displays the planes where we will show the results. Plane 1 lies in the plane where the central axes of the two channels lie. Plane 2 is located on the center vertical section of the left channel. Plane 3 and Plane 4 are located on the center cross section and outlet of the channels, respectively.

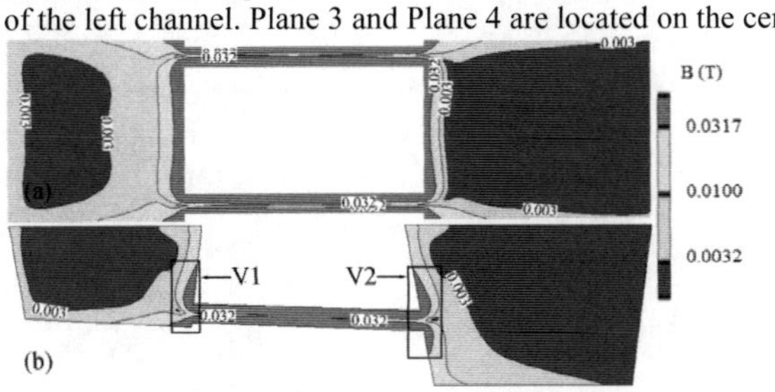

Fig. 3 The distribution of magnetic field

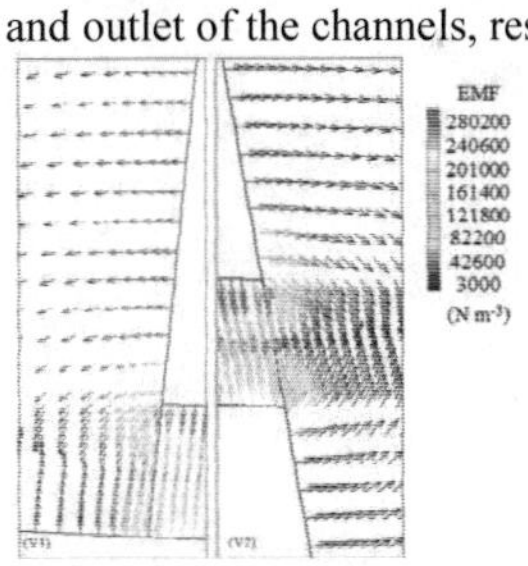

Fig. 4 The EMF vector around the throats of channels

The distribution of magnetic induction intensity is shown in Fig. 3. It can be seen that the magnetic induction intensity mainly distributes in the channels, in the two chambers, its value near the wall connected with the channels is significantly larger than that within other region. We also obtain the effective region of the magnetic field through two dimensionless numbers *Ha* and *N*, which can reflect the influence of EMF on the flow of molten steel with respect to shear and inertia force, respectively. In general, when the values of the two dimensionless numbers are smaller than 0.01, the influence of EMF can be omitted. When they reach the value of 1, the driving effect of EMF will be equivalent to that of the other two forces, amd the EMF will significantly affect the molten steel on condition that the dimensionless numbers reach the value of 0.1. Consequently, compared with the sheer force, the EMF has a significant impact on the flow over the entire tundish; compared with the inertial force, the EMF will show notably effect on the molten steel as long as the electromagnetic induction intensity reaches the value of 0.01T. Therefore, it can be seen in Fig. 3 that the effective area of magnetic field consists of the channels and the area near the walls connected with the channels.

The spatial EMF distribution should be consistent with that of the magnetic field, for it is generated by the interaction of the induced current and the magnetic field. Now, we focus on the EMF distribution within the marked regions in Fig. 3. It can be seen in Fig.4 that compared with the force near the joints of the channels and chambers, the force near the wall is smaller and points to the interior of the chambers, moreover, its direction is almost perpendicular to the Y-Z plane, which will push the molten steel there away from the walls. The force around the inlet and outlet of the channels is almost equivalent to that in the channels, and it also points to the inside of the chambers from the mouth of the channels, moreover, there is also some pinch effect to some extent. This complicated situation will strongly change the flow pattern and we will reveal it in later segments. It also can be seen that the EMF next to the upper edge is larger than that close by the lower edge. This phenomenon can be illustrated by Fig. 5 clearly. Compared with the EMF distribution on the center cross section (see Fig.5(a)), it is no longer symmetrical with regard to the Z direction at the exit of the channels (see Fig. 5(b)), and the force within the upper region is significantly larger, which will force the molten steel to flow downwards.

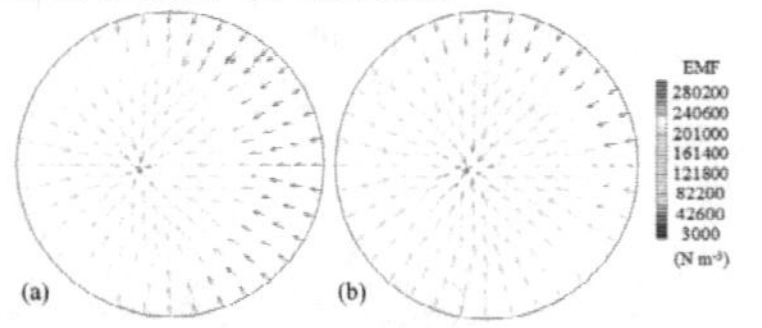

Fig. 5 The EMF distribution on different planes
(a) Plane 3-L (b) Plane 4-L

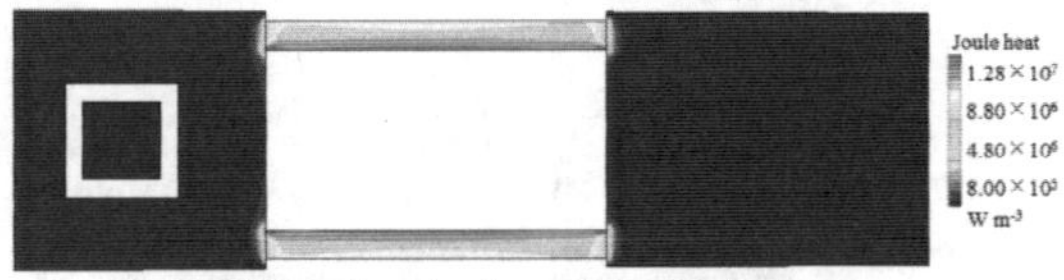

Fig. 6 The distribution of the joule heat

As can be seen in Fig. 6, the joule heat mainly distributes in the channels, and the proportion in the receiving chamber, channels, and distributing chamber is 4.47%, 91.68%, and 3.85%, respectively. The joule heat is evenly distributed along the axis of the channel, and it decreases gradually from the side near by the coil to the other side, which may give out implications for optimizing the channel structure to increase heating efficiency.

The temperature and flow variation against time in the tundish without the use of induction heating is shown in the Fig. 7 and Fig. 8 It can be seen that the temperature field in the receiving chamber is steady, and it gradually drops in the distributing chamber. The temperature drops to 1742K-1745K, and part of the molten steel rushing out from the channels directly flows to the outlet of the tundish at the time of 150s and 300s. The reason for this phenomenon is that the thermal buoyancy force associated with the temperature gradient is too small to conquer the effect of inertia force on the flow. From the time of 300s to 1500s, the temperature in the distributing chamber declines to 1739K-1744K, and the short-circuit flow is gently replaced by an upstream flow because of the improvement of the temperature difference.

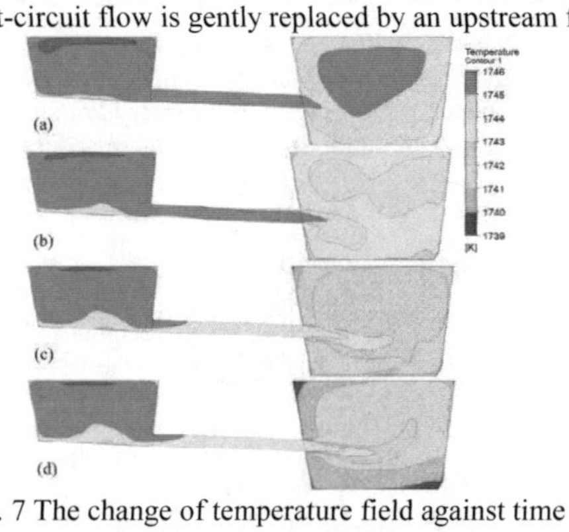

Fig. 7 The change of temperature field against time in the tundish without induction heating
(a) 150s (b) 300s (c) 780s (d) 1500s

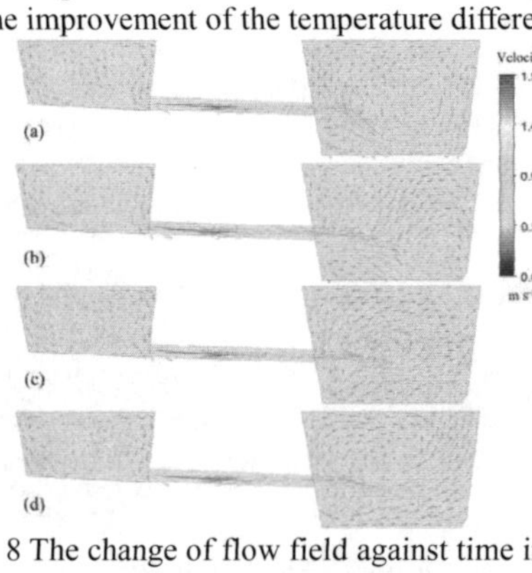

Fig. 8 The change of flow field against time in the tundish without induction heating
(a) 150s (b) 300s (c) 780s (d) 1500s

The Fig. 9 and Fig. 10 shows the temperature and the flow variation against time in the tundish with the use of induction heating. It can be clearly seen that the molten steel is continuously heated from 1749K to 1781K in the channels at the power of 800kW, and an adequate mixture between the heated molten steel and that in the distributing chamber promotes the temperature there from 1752K to 1777K. The flow field in the distributing chamber consists of a huge swirl and a small one located below the exit of the channels, and it does not change with time. Besides, several changes are notable. The molten steel forms dramatically secondary flow near the inlet of the channels in the receiving chamber, which means that part of the molten steel near the inlet of the channels is pushed away from there and generates two huge vortices on both sides of the receiving chamber. The molten steel can not timely ascend in the distributing chamber and it appears a downward trend compared with the situation that the induction heating is not applied. This is an undesirable situation for boosting the removal of the inclusions in general. There are two reasons account for this phenomenon. One of them is that the velocity of the molten steel is accelerated when it flows through the channels; another one is that the EMF pointing downwards near the exit of the channels is larger than that pointing upwards, which forces the molten steel to flow downwards at the outlet of the channels. This circumstance implies that the EMF and inertia force dominate the flow of the molten steel. Stable flow and homogeneous temperature field in the distributing chamber are the key element to maintain the quality of steel. After the induction heating is applied, the flow field is significantly stabilized and the temperature difference in the distributing chamber is reduced from 5K to 1K.

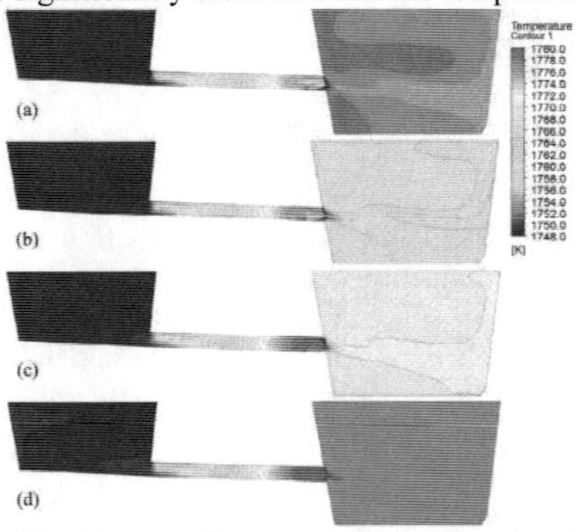

Fig. 9 The change of temperature field against time in the tundish with induction heating
(a) 150s (b) 300s (c) 780s (d) 1500s

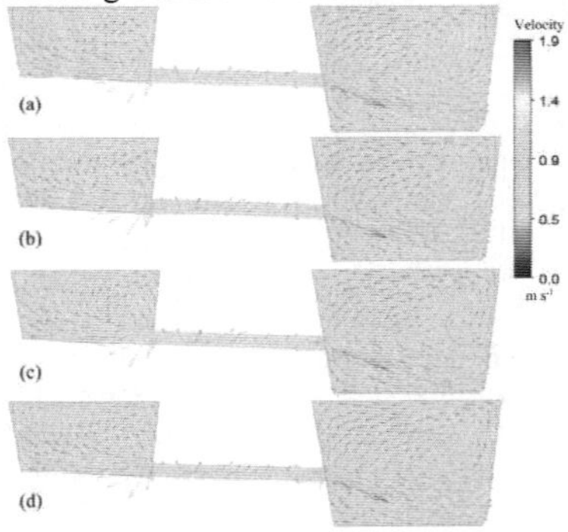

Fig. 10 The change of flow field against time in the tundish without induction heating
(a) 150s (b) 300s (c) 780s (d) 1500s

4. Conclusions

This work lays a foundation for researching the inclusions removal and optimizing the structure of channel type induction heating tundish in future by investigating the electromagnetic features, flow and temperature field in details. The important conclusions are summarized as follows: The magnetic field mainly distributes in the channels and the areas near the walls connected with the channels. The EMF located in the chambers is directed from the wall connected with the channels to the interior, which stabilizes the flow in the two chambers as a huge swirl flow. Compared with the EMF and inertia force, the thermal buoyancy force has a limited effect on the flow.

Acknowledgment

This work was supported by National Natural Science Foundation of China (51474065, 51574083), the Doctoral Scientific Research Foundation of Liaoning Province of China (20141008), and the Program of Introducing Talents of Discipline to Universities of China (B07015).

References

1. Y. Sahai, Metallurgical and Materials Transactions B. 47 (2016), 2095-2106.
2. P. Marty and A. Alemany, Metallurgical Applications of Magnetohydrodynamics. (1982), 245-259.
3. A.A. Alferenok and A.B. Kuvaldin, Russian Metallurgy (Metally). (2009), 741-747.
4. Q. Wang, B.K. Li and F. Tsukihashi, ISIJ International. 54 (2014), 311-320.
5. C. Vives and R. Ricou, Metallurgical and Materials Transactions B. 22 (1991), 193-209.
6. J.I. Ghojel, Progress in Computational Fluid Dynamics, an International Journal. 6 (2006), 435-445.
7. Q. Wang, F.S. Qi, B.K. Li and F. Tsukihashi, ISIJ International. 54 (2014), 2796-2805.
8. S. Pavlovs, A. Jakovics, E. Baake and B. Nacke, Proceeding of international conference on clean steel, City, 2012.
9. Y. Li, A.Y. Deng, H. Li, B. Yang and E.G. Wang, Metals. 8 (2018), 76-95.
10. F. Felten, Y. Fautrelle, Y. Du Terrail and O. Metais, Applied Mathematic Modelling. 28 (2004), 15-27.
11. B. E. Launder and D. B. Spalding, Computer Methods in Applied Mechanics and Engineering. 3 (1974), 269-289.
12. K. Hanjalić and S. Kenjereš, Flow, turbulence and combustion. 66 (2001), 427-451.
13. S. Chakraborty and Y. Sahai, ISIJ International. 31 (1991), 960-967.

IOP Conf. Series: Materials Science and Engineering **424** (2018) 012061 doi:10.1088/1757-899X/424/1/012061

Recent progress of induction heating technology in Baosteel

C.Y.Wu, X.L.Jin and Y.M.Zhou

Research Institute, Baoshan Iron&Steel Co. Ltd., Fujin Road, Baoshan District, Shanghai,China

Corresponding author: wucunyou@baosteel.com

Abstract
During researching on induction heating technology, attentions are focused on improvement of heating efficiency and heating effect. This paper briefly introduces recent research progress of induction heating technology in Baosteel, that is, development of full coil static induction heating technology of backup roller, developing of induction hardening technology of steel pipe, and research on sparking phenomenon during edge induction heating. A computer simulation method has been established and used in these research subjects, which can precisely calculate the whole heating processes and the system parameters. Based on experimental test and verification, the simulation model has been greatly improved and the corresponding computed results can be directly used as guidance for production process, which makes these research works more efficient and accurate.
Key words: induction heating, computer simulation, backup roller

Introduction

Efficiency, quality and cost are the three primary concerns of steel companies, while EPM technologies mostly deal with matters related to product quality. Regarding the induction heating technology, to maintain the output of a power source in a stable state as well as to obtain a desired temperature value are the most two important objects. Especially, the heating process and uniformity of temperature distribution are always the most concerned objectives in the application of induction heating technology, which mainly depend on reasonable designing of induction coil structure and precise controlling of heating process parameters. With the development of computer simulation technology, the accurate calculation of induction heating process becomes possible, which makes the application of this technology faster and more efficient. In addition, it is easy not only to obtain the heating temperature but also the induction heating system parameters by computer simulation[1-3].

Recently, the EPM team in Baosteel Research Institute has successfully carried out a full coil static induction heating technology for large forged back-up roller, a inner scanning induction heating technology for large steel pipe, as well as the edge heating technology for hot rolled steel plate. This paper gives a brief introduction to the development process and main difficulties of the above mentioned induction heating technologies. Three research topics during induction heating process, which are the end-effect and its countermeasures, the precise control of heating temperature as well as the electromagnetic field analysis under complicate working conditions, have been explored and will be mainly discussed.

Contents

1 Full coil static induction heating technology of backup roller

A full coil static induction heating technology is developed for induction hardening of backup roller, where a solenoid coil with multiple turns covers all the working surface of the heated backup roller during heating[4]. The roller rotates to improve circumferential temperature uniformity, while it is kept relative static to the induction coil in the axial direction. Compare to conventional heat treatment technologies, the full coil static induction heating technology has advantages of much deeper hardened layer and larger hardness value when used for heat treating of backup roller. By using of this technology, wear resistance and service life of the treated roller can be improved so as to reduce production and maintenance costs. Since a backup roller usually has a diameter larger than 1.5 meter and a height higher than 2 meter, it inevitably takes much longer heat treating time than that of conventional induction heating. Moreover, difference of temperature distribution along the whole roller surface is restricted within ±10 ℃ . These requirements make the industry application of this technology hard to realize, unless we can precisely calculate the induction heating system parameters and control the heating process[5-7].

In this study, the backup roller is heated from 450℃ to above 900℃， which has a great change of physical properties during the heating process and leads to a remarkable fluctuation of the heating system load. This phenomenon makes it difficult to precisely control the heating process. As shown in Fig.1, the equivalent inductance firstly drops with increasing of temperature and then finally reaches a stable value. Furthermore, the initial inductance is almost twice as much as that of high temperature, which causes the heating frequency increases almost 50% during the whole heating process under the condition that the load capacitance remains constant. This phenomenon not only greatly influences the stable running of the induction heating system, but also has an important effect on temperature distribution. Since the thickness of the induction heated layer is required to be more than 100mm, the heating frequency is usually under 100Hz. On the other hand, the parameters of the induction system changes with the size of the backup roller and its corresponding heating coil, which makes designing of the induction system parameters and control of the heating process more complicated.

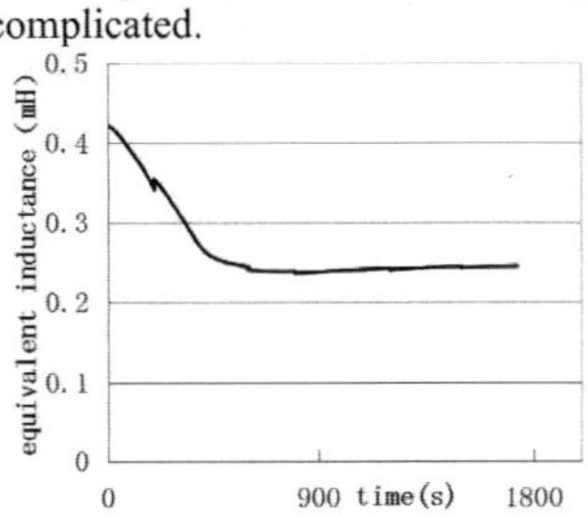

Fig. 1: Calculated equivalent inductance curve over time

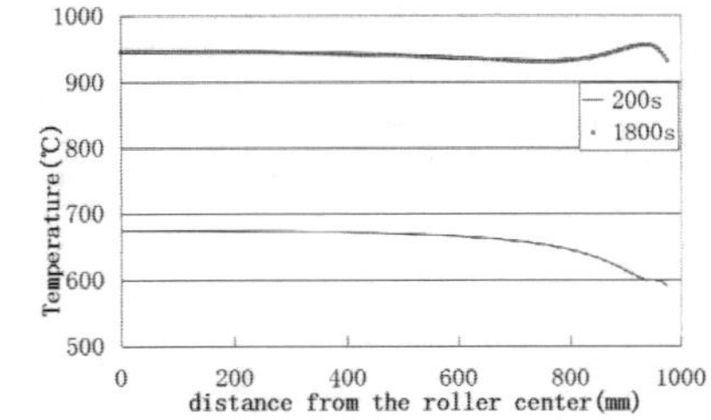

Fig. 2: Calculated temperature curves along the axial direction of roller surface at different temperature

Regarding the temperature distribution, it is relatively easy to obtain a desirable temperature distribution at the radius direction. However, the End Effect inevitably occurs when using a solenoid-type induction coil during the heating process, which makes the roller surface temperature difficult to control[8]. As can be seen in Fig.2, the roller surface temperature along the axial direction presents different features of nonlinear distribution over time, which mainly depends on the coil structure, roller height and coil to roller space. When the roll surface temperature is lower than the Curie temperature, the temperatures at the end part of the roller is much lower than that of the center part. However, when the roll surface temperature exceeds the Curie temperature during induction heating, the temperature at the end part of the roller will rise especially rapid and exceed that of other parts. This means simply changing the induction heating parameters can not improve the uniformity of temperature distribution. To solve this problem, an effective way has been proposed and conducted, which adjusts the heating coils currents separately based on the surface temperatures of the corresponding heated roller parts.

Regarding the hurdles mentioned above, a computer simulation was carried out to accurately simulate the induction heating process and temperature distribution. The simulation model was firstly verified an optimized by experiment results. And the final calculated temperature distribution of the roller along the longitudinal direction is shown in Fig.3, where it can be seen that the temperature uniformity of the roller surface layer is well controlled, while the inner part of the roller remains at relative low temperature. Results of an infrared temperature measurement show that the temperature difference of the entire roll surface is less than ±5 ℃, as shown in Fig.4., where S1~S5 are the temperature curves measured by infrared radiation thermometer at different locations of the roller surface during the final stage of the induction heating process. The five curves all but overlap with each other, which means a very uniform temperature distribution at the entire roller surface has been obtained.

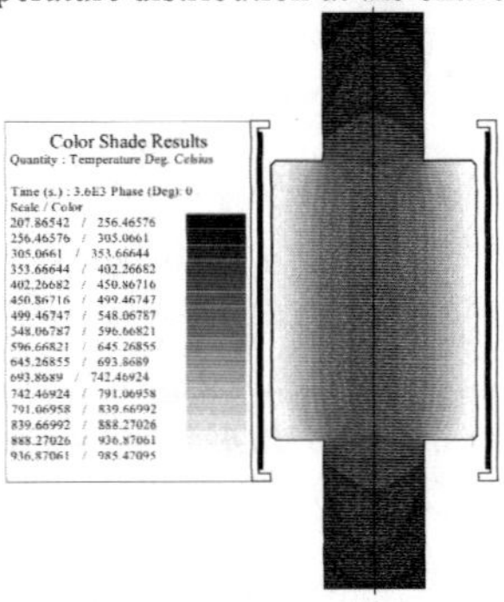

Fig. 3: The calculated temperature distribution after all induction heating stages.

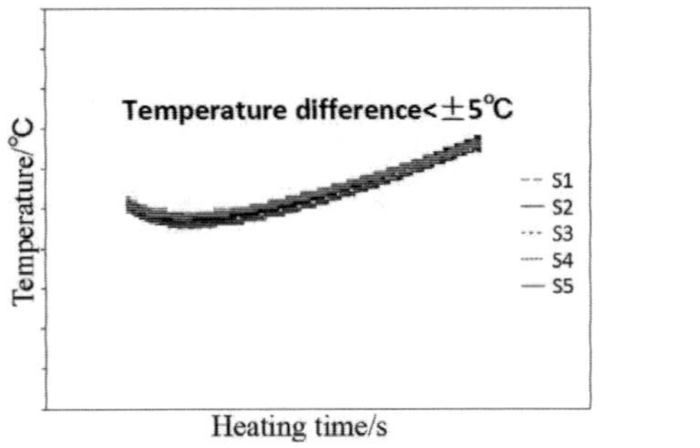

Fig. 4: Temperatures measured at different locations of the roller surface by infrared radiation thermometer.

2. Induction hardening technology of steel pipe

Wear resistance is one of the key performance indexes of steel pipe when it is used for material transportation, which directly affects its service life and cost performance. In the process of steel pipe production, the use of heat treatment technology to improve its hardness and wear resistance is an effective method, and has a wide range of applications in industry lines. However, for large diameter steel pipes that used for transportation, there are many difficulties in using conventional heat treatment methods. In this study, a scanning induction hardening technology was applied to the inner surface of the steel pipe, so as to improve its inner hardness and wear resistance. By carrying out a series of research works of composition study, computer simulation, physical experiment, temperature measurement, hardness testing and micro-structure analysis etc., a steel pipe with gradient strength distribution along the wall thickness direction has been obtained.

To improve its service life, the steel pipe is preferred to be induction hardened with more than 50% of its wall thickness. Since over-heating will make grain coarsening, it is important to control the maximum temperature below a certain temperature during the heating process. This peak temperature can be controlled by heating power, heating frequency and moving speed of the steel pipe. In order to precisely control the whole heating process, a computer simulation has been carried out. As we know that at a low moving speed, heat conduction becomes dominant, which makes temperature gradient hard to be obtained. As shown in Fig.5, there is a peak value of the inner surface temperature during heating, while the temperature difference along the radius direction soon becomes gentle out of the heating coil, where the temperature difference between the inner and outer surfaces is only within 20 degrees at the starting quenching position. Therefore, in order to increase the temperature gradient, moving speed of the steel pipe should be increased.

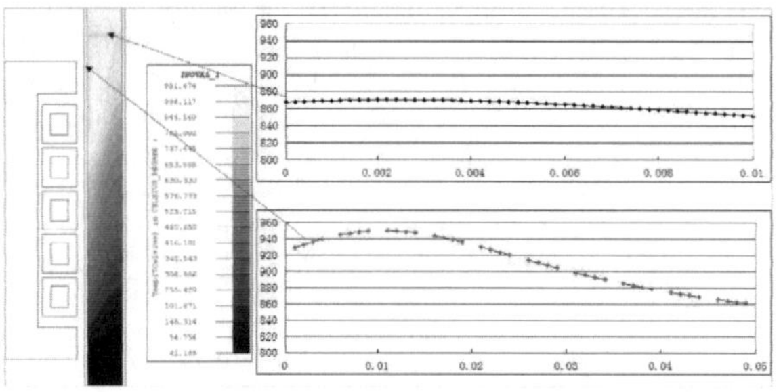

Fig. 5: Computer simulation model and calculated temperature distribution curves.

Fig. 6: Experimental equipment and temperature distribution of the induct heated steel pipe.

An induction hardening experiment has been carried out to study the real heating process and the mechanical properties of the treated steel pipe. The temperatures measured by thermocouples at different location of the steel pipe during the induction hardening process are in well accordance with the calculated results. As can be seen from Fig.6, a high temperature ring with uniform temperature distribution in the circumferential direction of the steel pipe can be seen during the experiment. This ring moves with constant speed and stable temperature distribution during the whole induction hardening process, which guarantees a 1400MPa strength of 10mm thickness of the steel pipe under a suitable steel composition.

3. Research on sparking phenomenon during edge induction heating

An edge heater having a "C" shape connecting core is used for induction heating of hot rolled steel plate. During application of edge heater, a spark phenomenon often occurs, which is induced by discharge arcing between the conveyor roller and the heated steel plate. In order to avoid the occurrence of spark, some preventive measures such as keeping conveyor rollers insulated to the ground, manufacturing the roller with unequal diameter, inputting the nearby coil currents with opposite direction and so on. Because of the bad conditions at the production line, iron scurf and water mist can soon destroy the insulation grade of the roller. It is very difficult to keep the high value of insulation for a long time, which brings great challenge to stable production and maintenance.

Analysis show that the main factors that causing spark phenomenon can be surmised as: (1) surface roughness of the roller; (2) the insulation value of the conveyor rollers; (3) magnitude of the induced eddy current in the contact positions between the conveyor roller and steel plate. The surface roughness and the insulation value of the roller basically depend on field maintenance of the conveyer rollers, while the induced eddy current is influenced by many factors, such as the deviation of the steel plate during moving, the width of the steel plate, the relative position of the coil, the heating power and so on. It is obvious that if the induced eddy current at the positions that conveyor roller contacting with steel plate is small, sparking between the steel plate and the roller is less likely to occur. In this study, the above mentioned factors that affect induced eddy current are analyzed by numerical simulation. The numerical simulation model was verified and optimized by experiment results to improve its accuracy.

It is found that there is a significant difference for different steel plate width in their current density distribution along the roller to plate contacting line. The maximum eddy current density at the contact line of narrow steel plate is bigger than that of wide steel plate. This result reveals that it is more likely to cause spark when heating a narrow steel plate. On the other hand, there are usually four coils connected in parallel on both sides of the steel plates. During the heating process, steel plates are easily shifted to one side, which cause the load match unbalanced for the induction heating power system, as shown in Fig.8. It is found that under the same input current, the induced current on more coil-covered side of the steel plate is larger than that on the other side. This result indicates that the probability of spark is greatly increased when the steel plate deviates to one side of the induction coils. Moreover, the above mentioned load match unbalance further strengthens the uneven distribution of input current in the coil, resulting in a great bias on the induced current in the steel plate. Thus, the steel plate will be heated to different temperatures at each side of the steel plate, resulting in uneven physical properties along the steel plate width. Therefore, it is strongly recommended that effective measures should be took to avoid bias of the steel plate so as to decrease spark phenomenon.

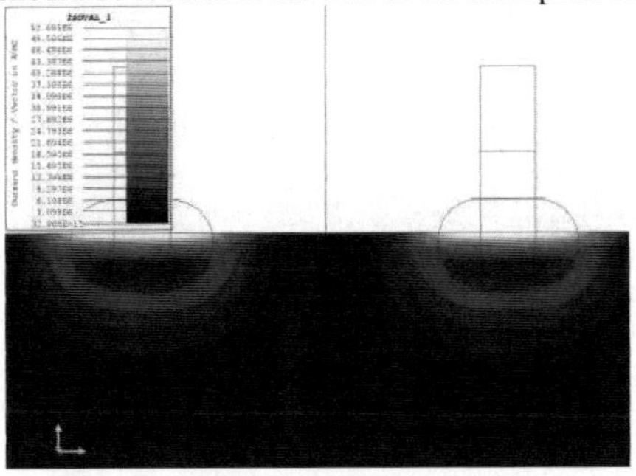

Fig. 7: Current density distribution along the steel plate surface.

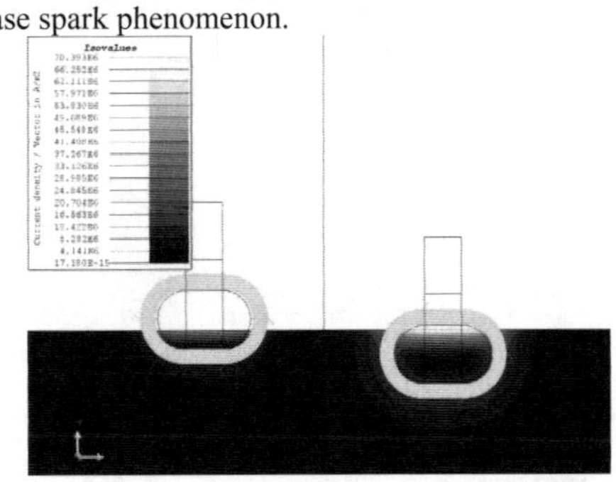

Fig. 8: Current density distribution along the steel plate surface when the steel plate is off-tracking.

Conclusions

This paper mainly focuses on recent researches about applications of induction heating technology in Baosteel, where three research topics related to roller, plate and pipe has been studied. Although shapes of the treated work pieces are very simple in these cases, special requirements make the industry application of induction heating technically difficult. Through the above mentioned researches, following conclusions can be obtained:

1) Computation simulation is an indispensable way in application of modern induction heating technology, which can accurately guide industry process;
2) Sufficient attentions should be paid to non-electrical factors that greatly affect induction heating process, such as moving speed, surrounding condition and so on.

References

1.Yu. A. Samoilovich, Steel in Translation, 45(2015), 73-79
2.S.Nobuo,M.Shinji,E.Shigeru, JFE technical report, 2008, 11,1-6
3.M.Behulova,B.Masek,L.W. Meyer, Materialpruefung, 2006, 48,217-224
4.Union Electric Steel, Forged Roll Heat Treatment[OL], http://www.uniones.com/the-ues-difference/forged-roll-technology/forged-roll-heat-treatment.
5.L.Zhigang,J.Haichao,Z;Feng,et al., Heat Treatment of Metals, 36(2011):115-118
6.C.Tianxiang,W.Zhanjun, Heat Treatment of Metals ,41(2016),149-153
7.W.Cunyou,J.Xiaoli,Z.Yueming,Heat Treatment of Metals, 43(2018),216-221
8.R.E. Haimbaugh,Practical induction heat treating[M],ASM international,2001,50-51

***** Formatting Issue - Best Available Paper/Graphic *****

IOP Conf. Series: Materials Science and Engineering **424** (2018) 012062 doi:10.1088/1757-899X/424/1/012062

$$T\big|_{t=t_{MF}+t_{b1}+t_{HF}} \geq Ac_m(v_{ih}) \quad (2)$$

the uniform austenite microstructure is obtained. If we have smaller final temperature after heating

$$Ac_{3m}(v_{ih}) \geq T\big|_{t=t_{MF}+t_{b1}+t_{HF}} \geq Ac_m(v_{ih}) \quad (3)$$

austenite microstructure could not be fully uniform. All three critical temperatures depend on velocity of induction heating v_{ih}. Dependences of the critical temperatures on heating rate v_{ih} determined from the Time-Temperature-Austenitization (TTA) diagram for the investigated steel are presented in Fig.1.

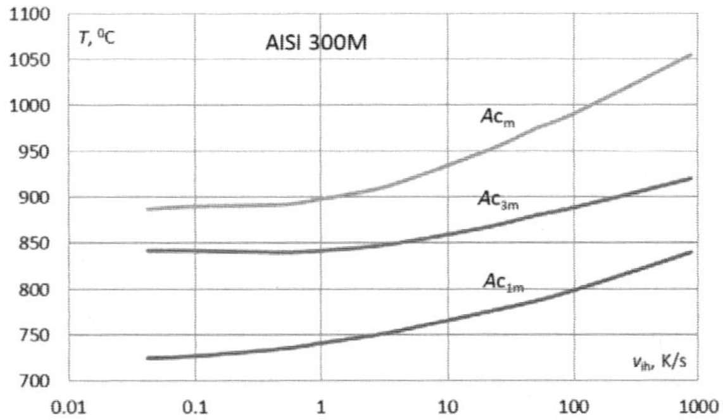

Fig. 1: Dependence of critical temperatures Ac_m, Ac_{3m}, Ac_{1m} on velocity of induction heating.

For the low heating rate $v_{ih} = 0.1$ K/s the modified critical temperature $Ac_m' = 882.5^0$C and it is almost the same value as for the conventional heating in furnaces. If the heating rate is distinctly bigger ($v_{ih} = 900$ K/s) the modified critical temperature $Ac_m'' = 1045^0$C and it is 162.5 K higher than Ac_m'. In the same conditions as previously the modified upper critical temperature $Ac_{3m}' = 842^0$C and $Ac_{3m}'' = 920^0$C and it is 78 K of difference between them. For the modified lower critical temperature $Ac_1' = 725.5^0$C and $Ac_{1m2} = 830^0$C and it is 104.5 K of difference between them. The CDFIH process is terminated with the intensive cooling to the temperature guaranteed termination of martensite transformation (Ms_f – martensite finish temperature). Exemplary temperature dependence of gear wheel in the surface zone on time for the CDFIH process is presented in Fig. 2.

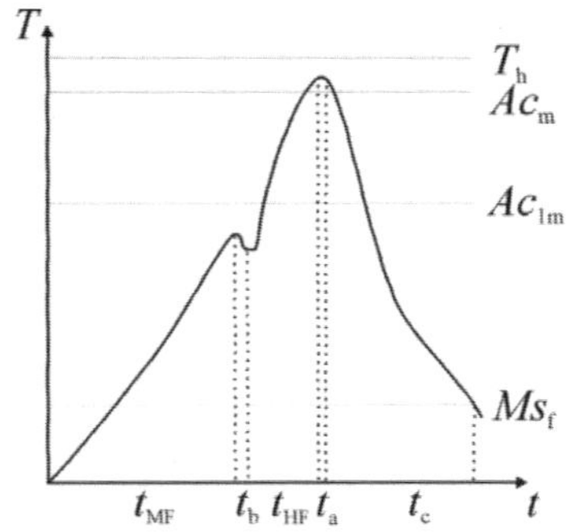

Fig. 2: Exemplary temperature dependence on time in surface zone of the body for the CDFIH process. (Notations in text)

The hardening temperature T_h could be slightly bigger than the modified critical temperature Ac_m.

$$T_h = T\big|_{t=t_{MF}+t_{b1}+t_{HF}} + T \quad (4)$$

where $\Delta T = 20....40^0$C.

IOP Conf. Series: Materials Science and Engineering **424** (2018) 012062 doi:10.1088/1757-899X/424/1/012062

For the hardening process important influence have several parameters which should recognized by numerical simulation and experiments. The first of them is influence of the prior microstructure [1]. The second one is the magnetic permeability of steel which changes rapidly because of magnetic transformation at the temperature of Curie point Ac_2 [4]. The third one is inaccuracy of temperature characteristics of material properties and heat transfer coefficients [5, 6]. An incorrect value of such properties like electric conductivity, density, thermal conductivity and specific heat could cause differences of calculated final temperature of about 100 K [5]. And finally a way of cooling and the cooling rate. The uniform martensitic microstructure at the surface zone is obtained if the cooling rate is big enough and the final temperature T_f in the whole hardened zone is smaller than the Ms_f temperature [1, 7].

Illustrative Example

Let us consider an example of the CDFIH process provided for the small gear wheel made of steel AISI 300M. Main parameters and dimensions of the gear are as follow: teeth number n = 16, width of the tooth ring b = 6 mm, root diameter d_e = 35.6 mm, tip diameter d_i = 26.9 mm, hole diameter d_h = 0.016 m. The arrangement of the CDFIH system, location of sensor points on working surface and view of prior microstructure are presented in Fig.3.

a) b) c)

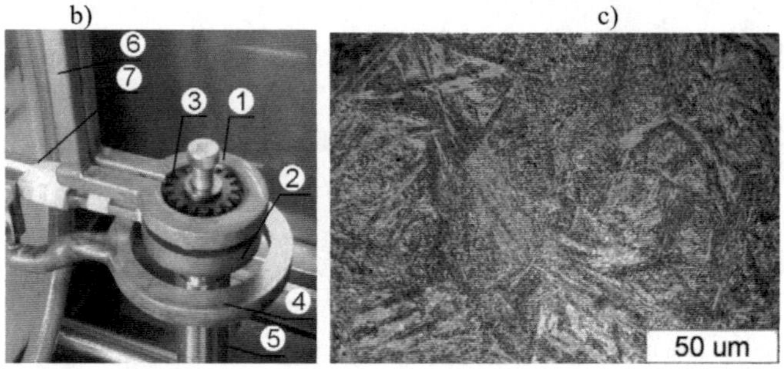

Fig. 3: Arrangement of CDFIH system (a) 1 – MF inductor, 2 – HF inductor with flux concentrator, 3 – gear wheel, 4 – sprayer, 5 – cylinder, 6 – MF bus-bars, 7 – HF bus-bars. Location of points A...G on the working surface (b). Prior microstructure of steel AISI 300M. Tempered martensite with some tempered bainite. Mag. x1000.

At beginning the gear wheel 3 is located inside the MF inductor 1 and heated to the temperature of about the lower critical temperature Ac_{1m}. Then it is quickly removed to the next position inside the HF inductor 2. When the temperature exceeds the modified critical temperature Ac_m gear is removed to the final position inside sprayer 4. As a quenchant the polymer solution is applied. In order to obtain expected hardness distribution along external diameter of the element the gear wheel mounted on cylinder 5 rotates with velocity of about 2...5 r/s. In order to minimize electrical losses lengths and distances between MF bus-bars 6 and HF bus-bars 7 are as minimal as possible. Computations are provided by means of the Flux 3D software for coupled electromagnetic and temperature field during induction heating and cooling and QT steel software supported by several own procedures for hardness and microstructure fields. Let us choose points A.....G located at the working surface of the tooth (Fig.3b). Prior microstructure of the material is presented in Fig. 3c. Several computations make possible to recognize correct parameters and shorten a time necessary to find optimal parameters during experiments are provided on the specialized laboratory stand located in the Silesian University of Technology in Katowice [3]. After computations following parameters of the process are selected: MF induction heating: current I_{MF} = 1450 A, time of heating t_{MF} = 4 s, its frequency f = 36 kHz, austenitization t_a = 0.1 s, HF induction heating: current I_{HF} = 520 A, time t_{HF} = 0.4 s, frequency f = 242 kHz. Modified critical temperatures and hardening temperature: lower critical temperature Ac_{1m} = 700^{0}C, upper critical temperature Ac_{3m} = 700^{0}C, critical temperature Ac_m = 700^{0}C, hardening temperature T_h = 1030^{0}C, rotation velocity: 5 r/s, quenchant: Aqua Quench 140, its concentration – 10 %, Tempering: time t = 3600 s, temperature T = 160^{0}C. Results of computations and measurements as well as their comparison are presented in Figs. 4-5. Oscillograms of MF and HF inductor currents are presented in Fig. 4a. Distribution of hardness along the line A...G located at the working surface is presented in Fig. 4b. It is not fully uniform, but differences between points A located at the root of the tooth and G placed in the top is equal to 20 HV and it matches general requirements. Hardness along the line AG is not fully uniform, but difference between points A and G is equal to 20 HV and it matches requirements. In order to control a shape of contour pattern the Surface Depth Hardening (SDH) coefficient is calculated along the line A...G. Let us calculate the SDH coefficient defined as the distance perpendicular to the surface point to a point where the hardness decreases to the 80 % of its maximal value at the surface (Fig.5a). Corresponding microstructures are presented in Figs. 5b, c.

*** Formatting Issue - Best Available Paper/Graphic ***

a) b)

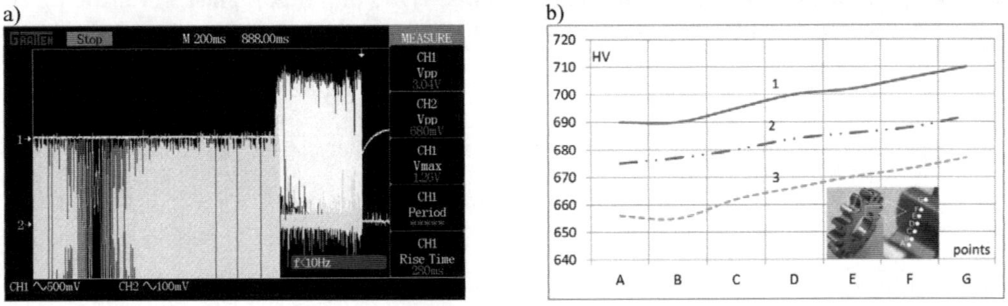

Fig. 4: Oscillograms of inductor currents MF current in blue, HF current in yellow (a). Hardness distribution along the surface line A.....G. (b) 1 – induction hardening (computations), 2 – induction hardening (measurements), 3 – induction hardening and tempering (measurements).

a) b) c)

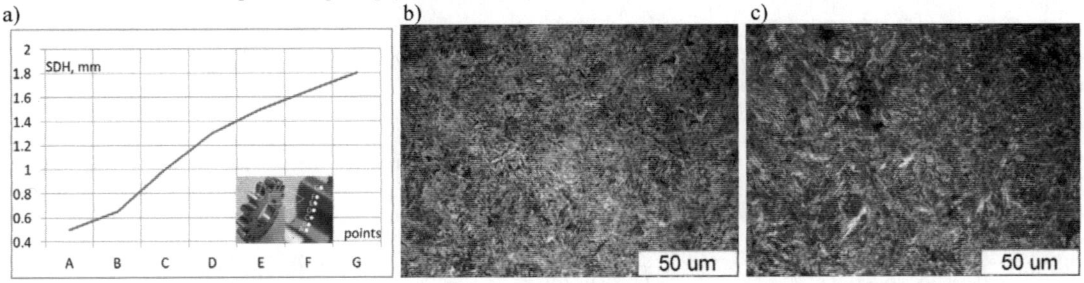

Fig. 5: Surface Depth Hardening (SDH) distribution along the surface line A.....G (a), Microstructure of steel AISI 300M. Contour zone with martensitic microstructure Mag. x1000 (b) Transient zone Mixed microstructure with martensite, bainite and ferrite mag. x1000 (c).

Summary
The induction contour hardening (ICH) process for gear wheels made of steel AISI 300 M is analyzed in the paper. The CDFIH method is applied. Prior microstructure of the material consists of the tempered martensite. Numerical model of the process dealing with coupled electromagnetic, temperature and hardness fields is elaborated. Illustrative example is presented. Calculated and measured hardness distribution are compared and acceptable accordance is achieved. A shape of contour zone is not uniform and the SDH coefficient changes between 0.5 mm at the root of the gear to 1.8 mm at the top of the tooth. Next research activities should be aimed at obtaining of more uniform shape of the contour zone.

Acknowledgment
The paper was prepared within the NCBiR project PBS2/A5/41/2014

References

1. V. Rudnev, G. Totten, Induction Heating and Heat Treatment, *ASM International*, 4C (2014).
2. J. Barglik, Mathematical modelling of induction surface hardening, *COMPEL* 35 (2016), pp. 1403 - 1417.
3. J. Barglik, A. Smalcerz, A. Smagór, G. Kopeć, Experimental Stand for Investigation of Induction Hardening of Steel Elements. Metalurgija 57 (2018) Vol. 4 pp. 341 - 344.
4. J. Barglik, A. Smalcerz, Influence of the magnetic permeability on modelling of induction surface hardening. *COMPEL* 36, (2017), 555 - 564
5. J. Barglik, A. Smalcerz, A. Smagór, Induction hardening of gear wheels of steel 41Cr4, International Journal for Applied Electromagnetics and Mechanics, 57, (2018), S3 - S12.
6. S. Schubotz, B. Nacke, Modeling and verification of convective heat transfer coefficient for induction applications, International Journal for Electromagnetics and Mechanics, 53, (2017), 79 - 88.
7. S. Lupi, Fundamentals of Electroheat, Electrical Technologies for Process Heating, Springer, (2017).

IOP Conf. Series: Materials Science and Engineering **424** (2018) 012063 doi:10.1088/1757-899X/424/1/012063

Nuclear waste treatment by induction heating and stirring of a metal/glass bath: the PIVIC process

P. Charvin[1], F. Lemont[1], A. Russello[1]

[1] CEA, DEN,DE2D,SEVT,LPTI, 30207 Bagnols sur Cèze, France

Corresponding author: patrice.charvin@cea.fr

Abstract
A fusion process based on low-frequency induction heating has been developed to treat mixed radioactive waste and produce biphasic glass/metal canisters compatible with a storage in a geological repository. The direct transfer of inductive power to the metal load ensures that fusion occurs directly in the final container and avoids a pouring stage. A thermal equilibrium develops during fusion between the power transmitted by induction and the power evacuated in the furnace's cooling circuits. The generator's electrical parameters are used to determine the amount of metal present in the canister.

Keywords: nuclear waste, induction heating, electromagnetic stirring

Introduction

Several facilities are required to separate and recycle the valuable components of spent nuclear fuel [1]. These facilities, and notably those used to prepare new nuclear fuel, produce contaminated secondary waste. These surface-contaminated wastes come in various forms, a large proportion of which are organic (gloves and glovebox equipment, cable jackets, waste bags...), with also metals (hydraulic cylinders, tools, transfer boxes...) and ceramics (laboratory glassware, HEPA filter media).
Low-level wastes with activities less than 370 Bq/g of alpha radiation are sorted out by kind and treated at CENTRACO by incineration or fusion [2]. Weakly contaminated metals are treated by fusion in several countries (France, Sweden, Germany) [3] because this process offers several advantages:

- A reduction in waste volume (greater than can be achieved by compaction)
- A reduction in the surface/volume ratio (reduced corrosion)
- A partial decontamination of the metals, as the radionuclides migrate in the slag
- An homogenization of the remaining radionuclides.

Intermediate-level wastes are stored in 120 L metal drums. The treatment process being developed by the CEA (French Alternative Energies and Atomic Energy Commission) for these types of waste involves incinerating the organic fraction before fusing the residues (ash and metals) [4] into biphasic glass/metal waste packages that meet the requirements for a future geological storage. The work described in this article covers the design and development of the fusion module and the fusion tests carried out with simulated residues.

Description of the fusion process

Induction heating

Previous studies have shown that induction heating is well-suited for the fusion of ferrous metals with a supernatant oxide slag [5]. Indeed, electromagnetic stirring increases the exchanges between the metallic and the slag phase. A 500 kW Inductotherm VIP generator delivers low-frequency (50 Hz) alternating current (AC) to an induction coil 680 mm in diameter and 350 mm in height with 19 turns. At this frequency, the electromagnetic skin depth δ_B (Eq. 1) is about 80 mm in stainless steel. The very high value (several meters) obtained for liquid glass confirms its transparency to the magnetic field.

$$\delta_B = \sqrt{2/(\mu_0 \sigma \omega)} \qquad (1)$$

With μ_0 the permeability of free space, σ the electrical conductivity of the material, and ω the angular frequency.

In-can Setup

The fusion process is carried out in-can, meaning that the packet is prepared directly in the final canister. The canister is attached to the side of the furnace, forming part of the latter. The metal and glass phases are fused in the canister by power transfer into the metallic component. The multilayer design ensures that metal pieces are fused without degrading the metal canister itself. The inner walls of the metal canister are cladded with an insulator and the waste residues are placed in a smaller-diameter ceramic crucible. This crucible is heated to high temperatures by contact with the molten material. The canister is placed on a water-cooled plate inside a divided crucible, which is also cooled by circulating water. This setup allows the narrow external metallic shell to be cooled. A thermal gradient thus forms in the isolating cladding.

The in-can principle imposes batch production with a variable filling level. However, it avoids any pouring of molten metal and glass. Furthermore, the canister forms a barrier between the molten metal and the cooling water, thereby limiting the risk of water/metal interactions compared with cold crucibles. The insulator reduces the power required to achieve fusion in spite of the screening effect from the external metal shell.

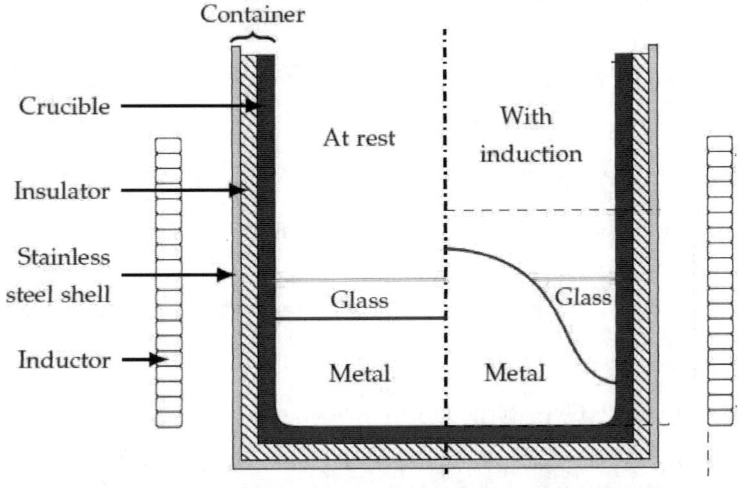

Fig. 1: Schematic cross-sectional view of the melting module.

Arrangement of the liquid phases

The different densities and chemical nature of the glass and metal phases ensure that they are well separated in the liquid state (Fig. 2), and after cooling (Fig. 3), with the vitrified phase on top. Nonetheless, the electromagnetic field from the inductor generates electromagnetic pressure that deforms the liquid metallic phase. This forms a dome [6,7] that can emerge in the center of the surface (Fig. 2) depending on the amount of glass present and the power used [8]. After the generator is turned off, the metal migrates to the bottom of the crucible and the glass forms a layer of even thickness on top. A relatively low cooling rate is used that avoid freezing the dome and let the liquid phases reorganize under the effect of gravity.

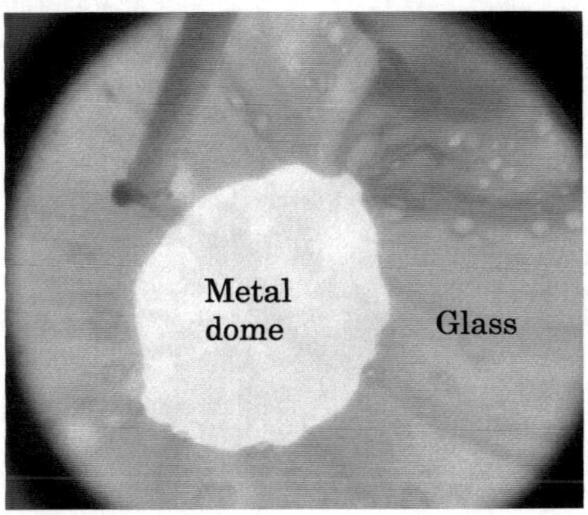

Fig. 2: Surface of the molten metal/glass bath.

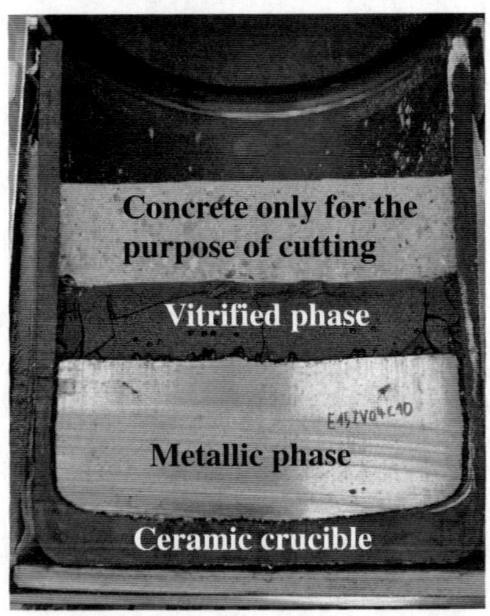

Fig. 3: Cross-section of a canister at the end of a test

Thermal transfers

Power delivered by the inductor

The inductor transfers power to all metallic elements inside the coil. The divided crucible and the divided plate absorb only a very small amount of this power. The magnetic field then reaches the metal canister. Despite the thinness of the outer walls, this shields the magnetic field and receives a substantial proportion of the transferred power. It contributes to the inductive load inside the coil.

The residual magnetic field that passes through the metal canister transfers power to the massive metal pieces in the ceramic crucible, which along with the insulating cladding are transparent to the electromagnetic field at this frequency. The more metal is present in the crucible, the higher the transferred power is.

Power evacuated in the cooling circuits

The power delivered to the metal canister and the metallic contents of the ceramic crucible heat these two elements. A set of springs keeps the metal canister in contact with the water-cooled divided crucible. Power is dissipated via the thermal gradient between the two according to Equation (2).

$$P_{released} = h . S . \Delta T = h . S . (T_{canister} - T_{water}) \tag{2}$$

Where h is the heat-transfer coefficient between the canister and the divided crucible and S is the exchange surface area.

When the temperature difference (ΔT) is large enough for all the induced power ($P_{transfered}$) to be transferred by conduction ($P_{released}$) according to Equation (3), the metal canister ceases to heat up and its temperature stabilizes.

$$P_{released} = P_{transfered} \tag{3}$$

The same thermal transfer occurs between the bottom of the canister and the cooled plate.

The thermal power transferred to the metallic pieces in the ceramic crucible leads to the same heating phenomenon. The metallic pieces need to be massive given the electromagnetic skin depth at this frequency. However, an insulator has been placed around the ceramic crucible to avoid thermal losses. The glass frit placed on top of the metal load acts similarly as an insulator for the upper surface. The temperature of the metal in the ceramic crucible thus increases until the melting point of the metal is reached. The dome that forms once the metal becomes liquid pushes the glass frit to the sides. This drastically increases the thermal losses from the surface because of the hot surface of the metal dome. In contact with the metallic phase, the glass frit also heats up until it melts, but at a lower temperature than the metal does.

Thermal equilibrium

The surface of the liquid baths is much hotter than the rest of the furnace, which leads to substantial heat loss. The temperature of the liquid baths increases until the thermal equilibrium defined by Equation (3) is reached. The released power is the sum of the power lost through all the edges of the bath (sides, bottom and top). A substantial portion of the power lost through the upper surface is radiative because of the surface's temperature. The T^4 dependence of the Stephan-Boltzmann law (Equation 4) means that the power lost through the surface increases rapidly with the temperature.

$$P_{radiation} = \sigma . \varepsilon . S . (T^4_{surface} - T^4_{water}) \tag{4}$$

Where σ is the Stefan-Boltzmann constant, ε, the emissivity of the material, and S, the surface area.

Thermal equilibrium thus occurs at a temperature close to the melting point of the metal. If the power transferred is too low, more power is lost as soon as the metal dome forms. Surface cooling freezes the dome to maintain thermal equilibrium between the received and emitted power. To avoid this, the canister has to be loaded with at least 70-80 kg of metal to ensure that the amount of power it receives is sufficient to maintain fusion after the dome forms.

Electrical parameters

The charge in the induction coil can be monitored using the generator's electrical parameters. However, no significant variation is observed during the fusion of the metal pieces.

Metal pieces are added to the crucible during the waste treatment process. These increase the height of the metallic phase. The charge in the induction coil therefore increases, which affects the generator's electrical parameters. The current intensity in the coil decreases each time metal is added (by 2 A/kg, Fig. 4). These changes in the electrical parameters are used to determine the amount of metal in the ceramic crucible (Fig. 5). In parallel, power transfer into the metal phase is improved, which increases the temperature at the surface before thermal equilibrium is reestablished.

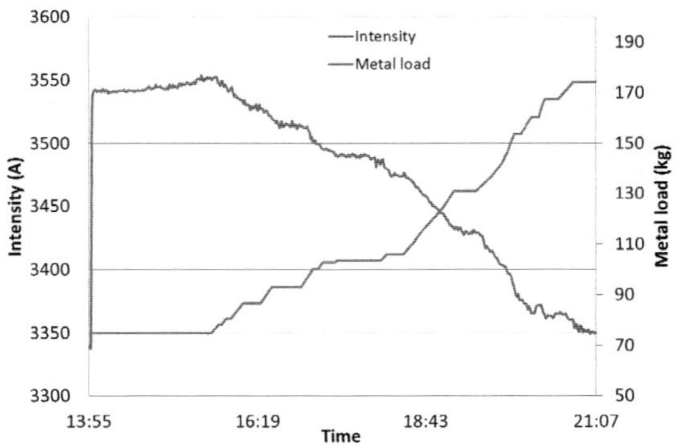

Fig. 4: Intensity variations during the feeding step

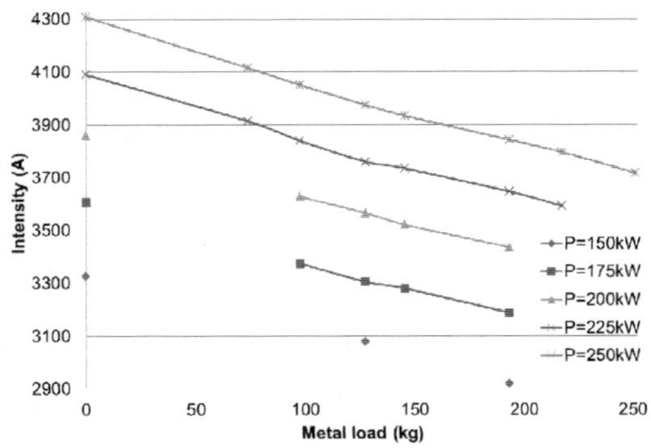

Fig. 5: Intensity vs metal load at several powers

Conclusion

A fusion process based on low-frequency induction heating has been developed to treat mixed radioactive waste and produce biphasic glass/metal canisters compatible with a storage in a geological repository. Preliminary tests have validated the method of direct induction heating inside the final container (in-can principle). The two phases are heated and stirred by the electromagnetic field, directly for the metallic phase and indirectly for the glass phase via exchanges at the glass-metal interface. The transfers of heat and momentum through this interface are sufficient to keep the glass molten. An analysis of the heat transfers shows that an equilibrium is established close to the metal's melting temperature between the power transferred by induction and the power evacuated by the cooling circuits. The generator's electrical parameters are used to determine the amount of metal present in the canister and thus monitor the filling process.

Acknowledgment

The PIVIC project is a partnership between CEA, ORANO and ANDRA. It is supported by the French government program "Programme d'Investissements d'Avenir" whose management has been entrusted to ANDRA.

References

1. P. Bernard, Progress in Nuclear Energy, 49 (2007), 583-588
2. J.M. Ducoulombier, B. Lantes, M. Bordier, Nuclear Engineering and Design, 176, (1997), 27-34
3. H-W. Seo, D-H. Lee, D.S.Kessel, C-L. Kim, Annals of Nuclear Energy, 110 (2017), 633-647
4. R. Boen, P. Charvin, F. Lemont, A. Russello, Patent EP 3031054, (2016)
5. B. Heshmatpour, G.L. Copeland, Nuclear and chemical waste management, 2 (1981) 25-31
6. K. Pericleous, V. Bojarevics, G. Djambazov, R.A. Harding, M. Wickins, Applied Mathematical modelling, 30 (2006), 1262-1280
7. G. Sugilal, J.J. Shashikumar, M.H. Rao, K. Banerjee, G.K. Dey, Materials Today : Proceedings, 3 (2016), 2942-2950
8. R. Bourrou, A. Gagnoud, O. Budenkova, P. Charvin, C. Lafon, F. Lemont, EPM 2018 Conference

Induction heat treatment of large diameter pipes and coils

V. Demidovich[1], V. Andrushkevich[2], Yu. Perevalov[2]

[1]Department of Electrotechnology, Faculty of Electrical Engineering and Automation, St.Petersburg Electrotechnical University (LETI), prof. Popov 5, St.Petersburg, Russia
[2] Russian Technologies of Induction Heating (RTIH) Ltd., prof. Popov 23, St.Petersburg, Russia

Corresponding author V. Demidovich: vbdemidovich@mail.ru.

Abstract
New approach of designing and control of induction heat treatment installations with utilizing of numerical models is developed. Models include two-dimensional simulation of coupling electromagnetic and temperature fields in cylindrical systems for processing of pipes and rolls. Also, thermal and structural tension during heating and cooling of pipes and rolls are simulated. These data allow defining structure, hardness, the size of grain and other properties of the pipes and rolls. They allow optimizing design and a choice of equipment, a heat treatment mode for achievement of the maximum quality and minimization of energy consumption. The developed models were used not only for the design of induction heat treatment systems of pipes and rolls, but also for a digital control of these complexes.

Keywords: Coupled Electromagnetic and Temperature Fields, Electromagnetic processing of metals, Induction heating, Multiphysics Problems

Introduction
Induction heating technology is widely used in steel industry. Induction heating installations in the metallurgical industry have a power level that may exceed several hundreds of kilowatts or even tens of megawatts. The frequency range of used power sources is from tens Hz to MHz. In spite of the fact that induction hardening has successfully been applied in industry for many years [1], there is a growing demand in industry for a more precise process control. Also, it is necessary to develop new technology of induction heating of pipes and rolls for rolling mills with big diameter. Heat treatment of pipes and rolls is important stage in manufacturing of high quality steel products like lengthy pipes and rolls with big diameters for rolling mills. Hardening and annealing are main types of heat treatment of pipes and rolls. Modern heat treatment means strict observance of temperature evaluation during heating, controlled cooling and, probably, repeated heating. Using of computer simulation is a powerful and necessary approach for solving these problems.

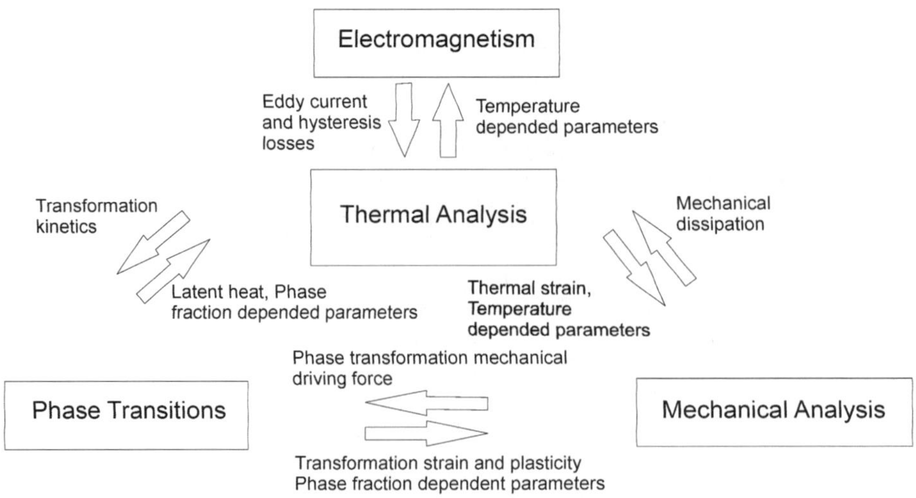

Fig. 1. Heat treatment physical phenomena

IOP Conf. Series: Materials Science and Engineering **424** (2018) 012064 doi:10.1088/1757-899X/424/1/012064

Mathematical models

From the physical point of view the induction hardening is a very complicated process (Fig. 1). We should include all the significant phenomena to the mathematical description to achieve an acceptable calculation result. Moreover, it is very important to understand that all phenomena are coupled with each other.

It is obvious that induction hardening mathematical description should include the electromagnetic, thermal and phase transformations phenomena. But the resulting workpiece properties are strongly depend on stress-strain state (residual stress) due to summation stress effect with the external load. Moreover, stress is an additional driving force for the phase transformation (Fig. 1). That's why it is desirable to describe this phenomenon too.

Effective numerical models of technological process of heat treatment are developed. They include two-dimensional modeling of electromagnetic and temperature fields in system [2]. Also, thermal tension during heating and cooling of pipes and rolls are simulated. These data allow defining structure, hardness, the size of grain and other properties of metal of the pipes and rolls.

The developed two-dimensional models were extremely effective not only at design of induction heat treatment systems of pipes and rolls, but also for a digital control these complexes. Using modular structure of induction heaters gives significant advantages in induction heating technology, since it allows optimizing the operation modes of equipment in relation to: energy consumption; scale formation; the ability to maintain a constant exit temperature.

Induction heat treatment of pipes

The heat treatment of pipes is important stage in production of high quality pipes with big diameters. Main types of heat treatment of pipes are annealing, normalization, aging, hardening etc. Dispersive hardening for two-phase (duplex) steel significantly improves properties of pipes.

Heat treatments of pipes usually carry out in gas-fired furnaces or electric furnaces. Gas-fired furnaces are not flexible due to big inertia. Besides combustion products contain gases which interact with metal that leads to deterioration of its quality and reduction of service life. Advantages of induction furnaces, such as absence of thermal inertia and high energy efficiency, are especially shown in the conditions of flexible change of temperature - time profile of the heat treatment. For pipes of small diameter, the greatest distribution was gained by induction installations of continuous action when the heating - cooling mode by direct placement of a zone of cooling behind the inductor is easily realized.

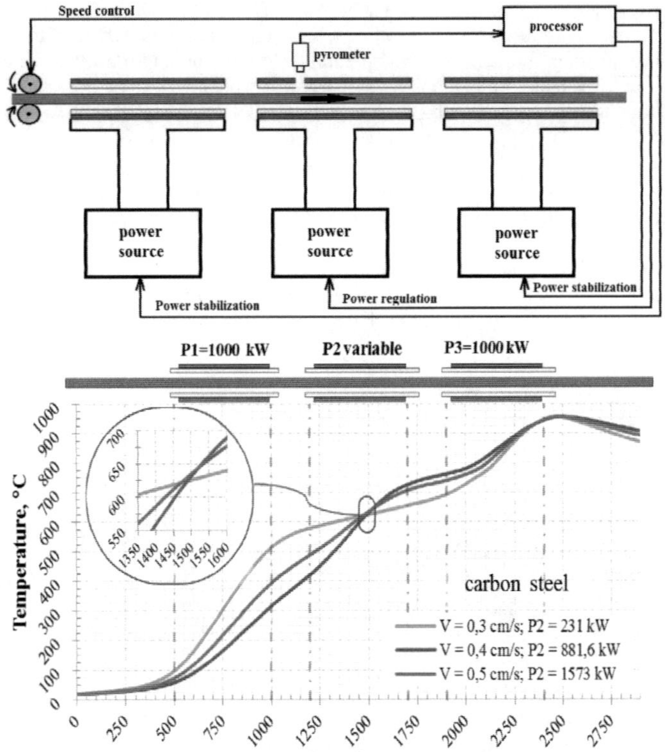

Fig. 2. Temperature control for a modular induction heater with a single control point

Advanced digital control allows effective processing of possible disturbances, such as: changes in the speed, diameter and initial temperature of the workpiece, and therefore to minimize the amount of rejects during heating [2].

To fully realize the advantages of modular induction units, an advanced digital control system should include control algorithms for stationary and non-stationary operation modes of the heater that occur during the initial start-up of a heater or when the technological lines stop and restart after a stop, with a change in the productivity of processing lines. Non-stationary modes can be characterized by a strong change in the temperature field of the load and the electrical parameters of the heater, and potentially can lead to a large percentage of rejects.

Investigation of the regulation of non-stationary operation modes of modular induction heaters was performed to determine the possibilities of reducing the changes in the temperature field of the load at the exit of the induction heater by means of an automatic control system with one temperature control point within the heater when controlling heating during the transient process, i.e. only one induction module (active zone).

The results of numerical modeling of the process of heating cylindrical billets using several modules with independent frequency converters have shown a high efficiency of induction heaters with this configuration. They enable the creation of a highly efficient and reliable control system for an induction heater based on three or more modules. In the installation of the three modules, the heating control of the workpiece occurs only on the second module (the "active zone") along the pyrometer, the sighting point of which is located at a certain point inside the second inductor, while the power in the first and third inductor is maintained at a constant level (Fig.2).

Temperature curves for this type of control (Fig. 2) obtained for different velocity of pipes intersect at one point inside the second induction module and have the same temperature at the exit from the last module of the induction heater. For heated thin bodies with linear properties, the existence of this point was analytically proved [3].

For cylindrical nonlinear bodies, modeling showed that, while maintaining a constant temperature at the point of intersection of temperature curves, by creating a system of automatic power control of the generator with feedback on the temperature fixed by an optical pyrometer at the control point, this method of regulation allows almost completely to remove fluctuations in the temperature at the exit of an induction heating installation during the temperature transient processes.

However, realization of a necessary mode of heat treatment by space distribution of heating and cooling zones at a constant speed of movement of pipes in many cases is difficult to pipes of big diameter and demands unfairly big power of heating. And in this case, it is preferable to realize a stage induction heating of lengthy pipes that provides flexibility at realization of heat treatment. In this way heat treatment line consists of two separate zones: induction oscillating heating and quenching. After heating to the required temperature, the pipe quickly moves to the cooling zone.

Induction heating of pipes of big diameter possesses several features which need to be considered at equipment design.

1. For each diameter and thickness of pipes wall there is an optimum frequency of current with strong expressing of the maximum of electric efficiency. Therefore, a proper frequency choice is important at designing of energy saving induction heating systems.

2. In practice there is wide nomenclature of pipes diameters and thickness of a pipes wall. Procedure of a choice of number of replaceable inductors is developed for a guarantee of heating of any pipe with electric efficiency not lower than 10-15% from the maximum.

3. Special attention is paid to a lining choice at high-temperature heat treatment because of a big surface of thermal radiation. As temperature differences on thickness of a wall of pipes usually don't limit power, the choice of power of inductors is defined not only by productivity, but also achievement of high thermal efficiency.

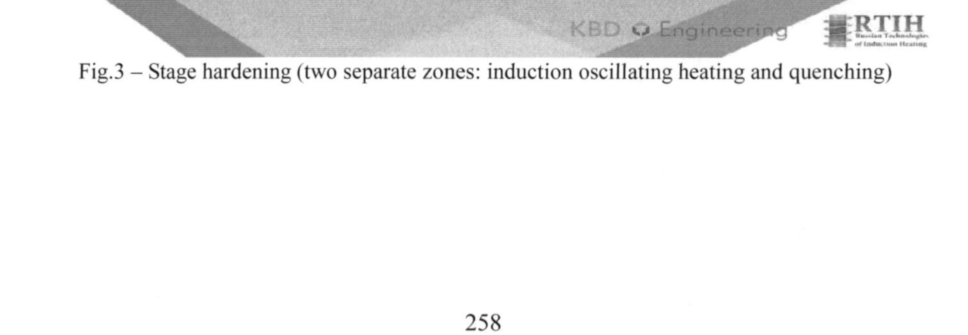

Fig.3 – Stage hardening (two separate zones: induction oscillating heating and quenching)

IOP Conf. Series: Materials Science and Engineering **424** (2018) 012064 doi:10.1088/1757-899X/424/1/012064

Induction heat treatment of large diameter rolls

The first use of induction heating for surface hardening of rolls was carried out in the early 1920s at the company Midvale Steel Company (USA) [1].

Rolls of rolling mills are made of steels with high hardenability. They must have a high strength of the surface layers and a uniform distribution of hardness along the length of the roll. At the same time, the rolls must resist strong shock loads and prevent the formation of cracks, chips, peeling both during the heat treatment and during the rolling process.

In order to meet these requirements, induction surface hardening is ideally suited for heat treatment of rolls. Along with the high quality parameters of the rolls, induction surface hardening allows to provide automation and repeatability of quenching, absence of deformations and high energy efficiency of heat treatment of rolls.

Due to the large dimensions of the rolls (length is up to several meters) simultaneous hardening is impossible due to the required considerable power and difficulty with water cooling. Therefore, continuous-sequential quenching with a vertical roll arrangement is used. Typical design of the roll and the machine for hardening are shown in Fig. 4.

To obtain the required distribution of the hardness and structure of the metal immediately before intense water cooling, the necessary temperature distribution over the section in the working zone of the roll must be in the interval:

$$1000\ ^{\circ}C > T > 750\ ^{\circ}C$$

The distribution of the temperature over the roll section greatly affects the thermal stresses both during the heating before quenching and on the residual thermal stresses after quenching. It is also necessary to consider the time and intensity of cooling in order to avoid self-release.

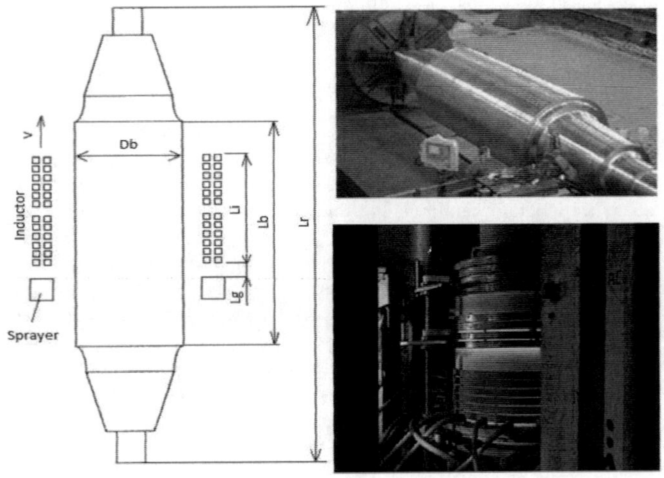

Fig. 4 – Induction hardening of large rolls

Conclusions

Induction heating technology is widely used in steel industry. In spite of the fact that induction hardening has successfully been applied in industry for many years, there is a growing demand in industry for a more precise control and quality assurance of the processed products.

New technologies for induction heat treatment of large diameter pipes (40"– 62") and rolls of rolling mills with a barrel diameter of up to 65" are developed. Developed computer control systems use specialized numerical models that provide high quality heat treatment and minimize residual stresses.

Acknowledgment

Financial support from Ministry of Education and Science of Russia was carried out in the framework of the state work "Organization of Scientific Research" (assignment No.8.6037.2017 / VU) is gratefully acknowledged.

References

1. A. Muehlbauer: History of Induction Heating and Melting. (2008), Vulkan Verlag.
2. V. Demidovich: Computer simulation and optimal designing of energy-saving technologies of the induction heating of metals, Thermal Engineering, (2012), V. 59, Issue 14, 1023—1034.
3. V.Demidovitch, J.Nelson, E.Blake: New generation of galvanizing lines utilizing induction heating technology, The 3rd International Symposium on Electromagnetic Processing of Materials, (2000), Nagoya, Japan, 224-229

Design of an one-sided transverse flux induction coil by using a numerical optimization algorithm

M. Schulze[1], A. Nikanorov[1,2], B. Nacke[1,2]

[1]Institute of Electrotechnology, Leibniz University Hannover, Wilhelm-Busch-Str. 4, 30167 Hannover, Germany
[2]Department of Electrotechnology and Converter Engineering, St. Petersburg Electrotechnical University, Prof. Popov Str. 5, St. Petersburg, Russia

Corresponding author: schulze@etp.uni-hannover.de

Abstract
The transverse flux heating (TFH) concept offers very high electrical efficiency in combination with unique technological flexibility. Numerous advantages make this method beyond competition to be applied in e.g. continuous strip production and processing lines. However, all potential advantages of TFH can be realized in practice only by optimal design of the inductor shape and optimal control of the heaters, using numerical modelling and optimization techniques.
The paper describes a successive approach to optimal design of a single layer induction coil, which will be used for one-sided TFH of moving thin steel strip with constant width. The peculiarity of the design refers to the very slow movement speed of the strip and the target temperature, which is set below Curie.

Key words: induction heat treatment, one-sided transverse flux heating, strip heating, numerical optimization

Introduction

The continuous heating of thin metal strip is an important process step in many industrial production lines. The areas of application of strip heating ranges from heat treatment, coating and drying of sheets and strips to the use in slab and casting plants for intermediate heating before milling. The thicknesses of the strip could vary between 0.1 mm and 60 mm at widths up to 2000 mm. Due to the different fields of application of strip heating for a wide variety of materials, the strip temperatures to be achieved are in the range of a few 10 °C for drying and up to 1200 °C for hot-forming. The manufacture of semi-finished products of different composition and qualities often requires a heat treatment, such as tempering, annealing or galvannealing in order to achieve desired material properties. Due to targeted heating, different degrees of hardness, grain sizes or surface textures can be set. The use of sheet and strip as semi-finished products is particularly important in the metalworking industry, e.g. the automotive or electrical industry. [1]
Commonly used materials are non-magnetic metals such as copper, brass and aluminum, precious metals such as gold and silver, and various steels and alloys. [2]
Today, conventional heaters such as gas or oil fired furnaces are still widely spread. All of these installations are working by the principle of heat transfer to the workpiece through the surface. Based on the laws of thermodynamics, the atmosphere in the oven always has to be warmer than the goal temperature of the workpiece. Although this technique ensures homogeneous and through heating of the material, there are a variety of disadvantages and restrictions like the large size of the installations, long heating time, high thermal losses, limited power density and speed of motion, intensive scale formation on the materials surface and environmental pollution.
Considering the aspect of time, on the one hand for the heating of the material and on the other hand the warm-up and cooling time of the furnace, a high waste of energy and inflexibility for the integration in modern production lines is deduced.
Induction heating offers solutions to the above mentioned disadvantages and restrictions of conventional heaters. Based on the direct and contactless heating, the heat is generated directly inside the workpiece. The advantages are amongst others theoretically unlimited power density, very short heating times, low floor space, easy and fast process control.
For the heating of continuous moving metal strip, two processes have become established in industry: the longitudinal flux heating (LFH) and the TFH. Both methods are based on the same heating method, but differ by the direction of the magnetic flux, which is parallel to the workpiece plane for LFH and normal for TFH. Due to the different flux orientation, the eddy-currents are also induced in a different way. For LFH the eddy-currents are flowing in the cross-section of the metal strip. Especially for thin metal strip the LFH is limited by the penetration depth and furthermore by the frequency. If the penetration depth is not small enough, the induced eddy-currents may cancel out each other. Since the eddy-currents in the TFH flow in the surface of the metal strip, there is no limit to a decrease of the strip thickness and is thus particularly suitable for the heating of thin metal strip. In addition to this the induced current flows as an image of the inductor shape, which gives the opportunity to control the Joule heat distribution inside the metal strip and thus the temperature distribution.

Characterization of a transverse flux heating system

The typical setup of a TFH system consists of a symmetrical arrangement of two induction coils and if necessary of two adapted flux guiding elements, one above and the other below the metal strip. Neglecting connections to the generator, Fig. 1 shows a principle sketch of a TFH heater.

The characteristic dimensions of a TFH system are given by the pole pitch *t*, coupling gap *h*, length of the induction coil b_i, width of the induction conductor a_i in the regular zone and a_{je} in the coil head zone (can also exist of two or more windings if the gap between each conductor is much smaller than the width of the conductor). The influence of these geometrical parameters have been examined in [1-3].

The following shows how to design successively a one-sided TFH coil by using two-dimensional and three-dimensional numerical simulations and optimization.

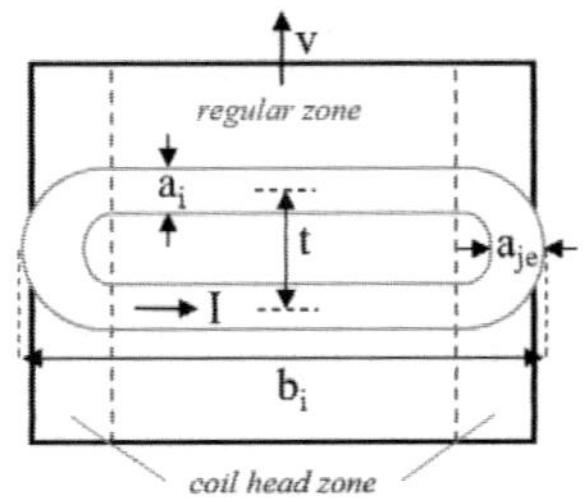

Fig. 1: Typical setup of a TFH coil.

Defining the goal and analytical preliminary investigation

The main goal of the inductor design is to achieve a homogenous temperature distribution with a specific setpoint temperature T_{sp} in a defined distance d_{sp} after passing through the inductor. The width and the thickness of the strip are fixed. A second goal is the demand for a robust and constructible design. In order to determine a suitable starting range for the geometrical dimensions of the inductor, expert values or the available space for the integration of the inductor into the system can be used. The latter is a compromise between a filigree and a maximum possible dimension of the inductor. Important geometrical parameters that have to be dimensioned are the conductor profile and the pole pitch *t*. From these dimensions all further distances can be derived based on the investigations of [1] and [3].

2D numerical modelling of the regular zone

The next step in inductor design involves two-dimensional numerical investigations on the infinitely wide strip. As a result, an infinitely wide regular zone of the TFH inductor is mapped. These calculations are made with ANSYS Mechanical additionally using the electromagnetic package. Fig. 2 shows a close-up of the regular zone of the two-dimensional numerical model based on the predefined inductor concept. The electromagnetic Modell (EM) can be used to calculate (harmonic solution) the optimum frequency for the best electromagnetic coupling between the inductor and the workpiece and thus the maximum electrical efficiency. Furthermore, sensitivity studies on the influence of e.g. the coupling gap, the strip thickness or the influence of the magnetic permeability (μ_r) of the strip can be executed with little computational effort. Fig. 3 shows the results of an exemplary frequency analyses. It is typical for TFH, that there is a wide range of frequency wherein the electrical efficiency is changing less. For higher frequencies, the penetration depth is decreasing and the heating behavior is changing from TFH to LFH [1-4]. Within these frequencies, a special feature of TFH is that it is not sensitive to the magnetic permeability of the strip material. This is shown in Fig. 3 and is especially of interest if the magnetic data of the steel strip is not exactly known and the maximum temperature does not exceed Curie-temperature.

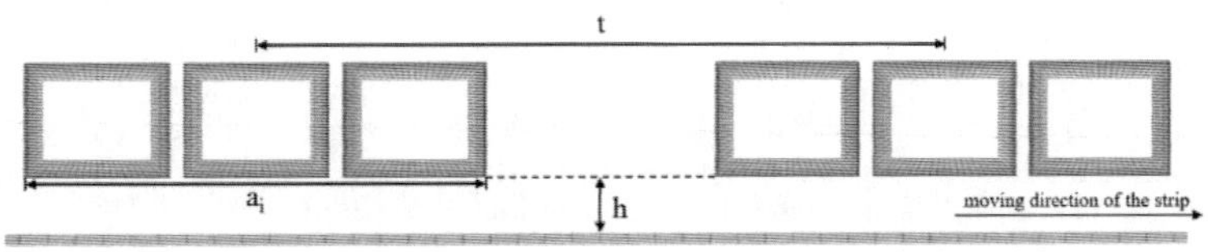

Fig. 2 Sketch of the two-dimensional electromagnetic numerical model.

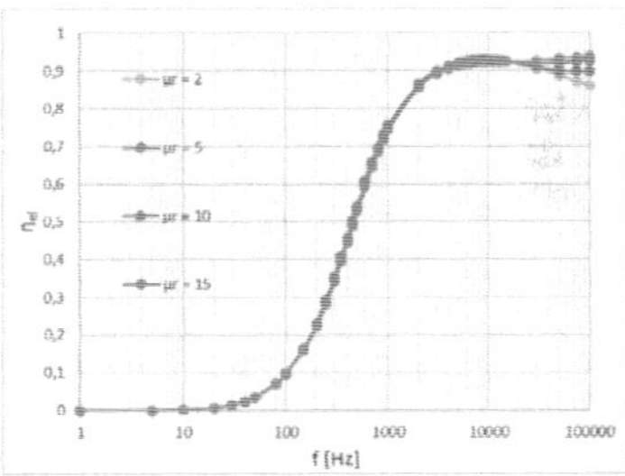

Fig. 3: Sensitivity studies of the electrical efficiency for different magnetic permeability of the strip.

Subsequently, investigations are carried out on the moving strip by means of a transient thermal analysis. The necessary thermal model (TM) consists only of the strip and is coupled with the Joule heat sources of the EM. Convective heat losses and radiation losses on the bottom and top of the strip are taken into account. The strip movement is described by means of a procedure which is performed after each transient calculation step. Each nodal temperature value is shifted by a distance in the direction of movement, which results from the velocity *v* and the time step *t*. If the shifting distance is smaller or bigger than the distance between the nodes, the value gets linearly interpolated. For this procedure a mapped meshing of the strip is required. Now, the inductor current must be set, to reach the desired target temperature T_{sp} at the point with the distance d_{sp} behind the inductor. For this a manual parameter study or an automated evaluation can be done. For the latter, e.g. a loop can be used that calculates multiple current flows. Based on the results, the needed current can be interpolated. A typical TFH surface temperature distribution along the strip moving at speed v is shown in Fig. 4.

This concludes the preliminary two-dimensional investigations and initial estimates for the expected power range and costs for peripheral elements such as generator and resonant circuit can be done. Since an idealized system was calculated until now and losses due to the coil head and connections must be added to the manufactured TFH system, the assumed power should be increased by some percent to gain a representative value.

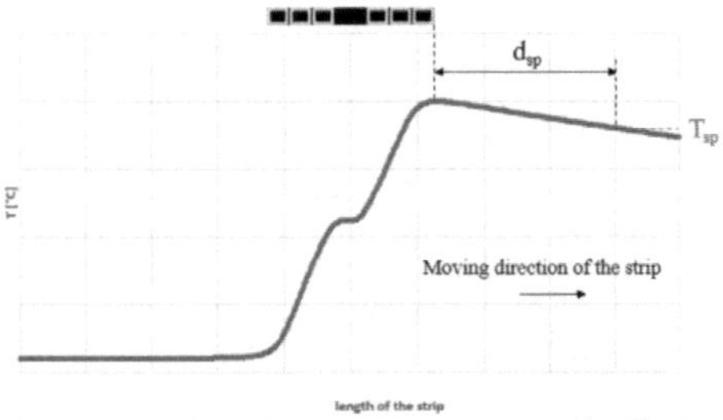

Fig. 4: Quasi-stationary temperature distribution along the surface of a moving strip.

Three-dimensional electromagnetic and thermal modelling

The next step in coil design involves numerical studies on the three-dimensional EM (harmonic solution) and TM (transient solution). This step is necessary to map the influence of the coil head on the strip edge effects. Due to the symmetrical structure of the TFH system, it is possible to examine a quarter- or a half model. This saves computing time and enables a fast and in case of the half model a yet physically correct simulation of the heating process. Fig. 5 shows examples of both model types for a round coil head. Other possible head shapes are e.g. rectangular or lamp shape. To ensure that the frequency analysis from the two-dimensional EM model has yielded a comparable result for the quarter or half model, in Fig. 6 the frequency analysis for the three-dimensional half-model with the two-dimensional model for an exemplary calculation with the same material properties is compared. The figure shows that the electrical efficiency in the three-dimensional model is lower. This is due to the coil head losses, which are not shown in the two-dimensional model. However, the optimal frequency is in the same range.The aim of this investigation step is to design the coil head in such a way that the temperature distribution with the target temperature T_{sp} in a distance d_{sp} after passing through the inductor is as homogeneous as possible. To achieve this, an automated optimization algorithm will be used.

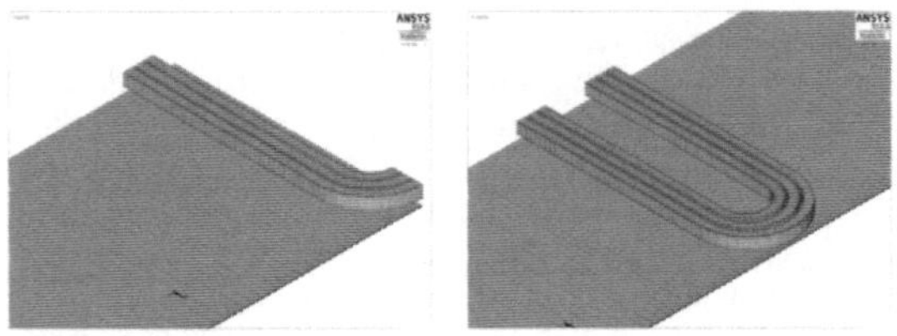

Fig. 5: Example of a quarter - (left) and a half model of a TFH coil (right).

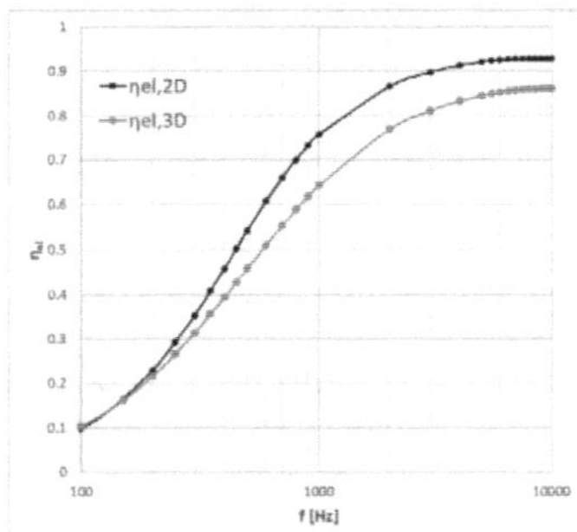

Fig. 6: Comparison between two- and three-dimensional frequency analysis.

Numerical calculation of the coil head using a genetic optimization algorithm

The used optimization algorithm is a genetic algorithm, which is characterized by the fact that the global minimum is found within a given range for three different parameters. The exact mode of operation is described inter alia [5,6] and the optimization structure is shown in Fig. 7.

Before the optimization procedure is started, the three-dimensional EM and TM must be parameterized. Based on the results of the two-dimensional examination, initially parameters are selected that have an influence on the coil head area. A good start is the inductor length b_i, the inductor cross-section a_{je} and the current. This choice ensures that the regular area remains unchanged, thus remaining comparability between the models. The next step before starting the algorithm is to set the range for each parameter. A good estimate to use for the current is the calculated value from the two-dimensional calculation as a lower limit and the same value plus 20 % as the upper limit. The current is needed to reach the target temperature T_{sp}. A possible choice for a_{je} is to narrow and enlarge the actual value each by 20 %. For the parameter range of b_i, the values chosen are those which, at the lower limit, ensure that the coil head is completely inside the strip and with the inner winding outside the edge of the strip as an upper limit. Fig 8 shows the successive approach to the optimal coil design step by step.

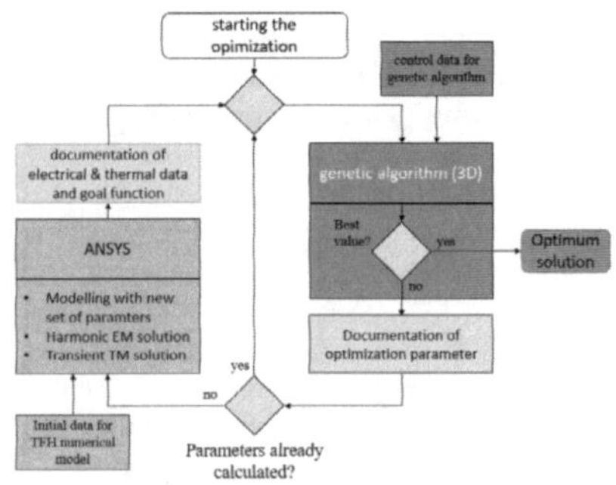

Fig. 7: Structure of optimization procedure.

IOP Conf. Series: Materials Science and Engineering **424** (2018) 012065 doi:10.1088/1757-899X/424/1/012065

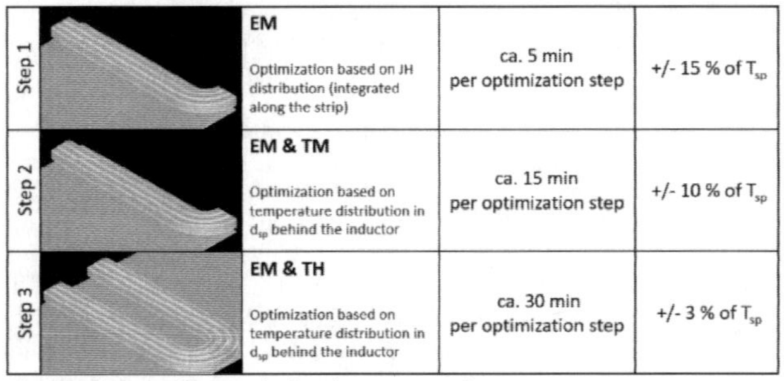

Fig. 8: Successive optimization approach.

"Step 1" includes an optimization based on the integral Joule heat source distribution (JH) using a quarter EM. The JH is normalized to the maximum value in the regular zone in order to obtain a qualitative course of the heat source distribution and thus to ensure a good comparability of the goal function. With the simulation computer used a duration of about five minutes per each optimization step in test examinations on a EM with 180,000 elements was reached. The temperature distribution was then validated on a half model and a maximum temperature deviation of 15 % could be determined. When analyzing the optimization results for each parameter, it should be considered that these are in the middle of their interval. If one or more parameters tend to a limit, the affected interval must be increased in that direction of range violation and the opposite limit can be decreased. Then the optimization has to be restarted. In order to a further reduction of the deviation from the set temperature T_{sp}, the EM is coupled with TM in "Step 2". The goal function of the optimization algorithm is given by the maximum deviation from T_{sp}. Due to the transient solution of the TM, the computing time increases to 15 min per step. However, it is possible to reduce the parameter range based on the optimization results from "Step 1". This makes it possible to find the optimal solution faster. In this step, a minimum deviation from the target temperature of 10 % could be validated on the equivalent half-model. Further minimization is not possible on the quarter model. This is due to the fact that the heat sources on the returning inductor side are not taken into account and thus do not flow into the temperature calculation. For very fast strip movements (e.g. v > 10 m/min), sufficient accuracy can be achieved, as described in [2]. For slow strip movements, integral effects become more influential and analysis on the half model is necessary to further minimize the temperature deviation as it is shown in "Step 3". In this optimization step, the EM and the TM of the half inductor arrangement are mapped. As a result, the number of elements to be calculated and, therefore, the computing time doubled to 30 min per step. However, the calculation of the goal function has the highest accuracy and also offers the possibility of electrical data e.g. the inductors voltage and inductance. The least deviation that could be achieved with these optimization parameters is ± 3 %. Further investigation with other optimization parameters combinations may provide further improvement in maximum deviation from T_{sp}.

Conclusions

In this paper, a successive approach to optimal design of an inductive transverse flux heating system using numerical simulations and a genetic optimization algorithm is shown. The optimization takes place with respect to a goal function which describes a homogeneous temperature distribution at a defined distance after passing through the inductor. The hierarchical structure that consists of several steps is described. First, a coil concept was set by e.g. expert values. Next, preliminary calculations or sensitivity studies have been carried out based on a two-dimensional numerical model which maps the cross section through the regular zone of the inductor. Amongst others, it has been shown that the transverse flux heating is not sensitive to the magnetic permeability of the strip. In the next step, a three-dimensional numerical model was created, parameterized and coupled with a genetic algorithm. Depending on the temperature homogeneity to be achieved, step by step the complexity of the model and thus the associated computational effort can be increased. Starting with a quarter electromagnetic model, where the goal function is the normalized Joule Heat distribution, this can be extended by an additional thermal model and the goal function can be adapted to the temperature distribution. In a final step with maximum accuracy, the optimization can be carried out on an electromagnetic coupled with a thermal model mapping the half inductor setup. Thus, a successive improvement in accuracy from ± 15 % to ± 3 % could be achieved.

References

1. G. Nauvertat, Fortschritt-Berichte VDI, Reihe 19, Nr. 129, 2000
2. A. Aliferov et al., Induction heating technologies. Intensive Course Specific 1, St. Petersburg, Publishing House of ETU, 2013
3. A. Ruhnke, Fortschritt-Berichte VDI, Reihe Reihe 19, Nr. 119, 2000
4. B. Nacke, E. Baake, Induction heating, Vulkan Verlag, 2016
5. S. Galunin et al., Proceedings of the International Seminar on Heating by Internal Sources, Padua, September 12-14, 2001
6. B. Nacke, International Symposium on Heating by Electromagnetic Sources, Padua, June 19-22, 2007

IOP Conf. Series: Materials Science and Engineering **424** (2018) 012066 doi:10.1088/1757-899X/424/1/012066

Potential of open source simulation tools for induction heating

V. Geza[1], M. Scepanskis[2], R. Vilums[2] and A. Eimuss[2]

[1] University of Latvia, Faculty of Physics and Mathematics, LV-1002, Riga, Latvia

[2] CENOS LLC, Kalnciema iela 106A-2A, Rīga, Latvia, LV-1046

Corresponding author : vadims.geza@lu.lv

Abstract

The paper describes simulation of induction heating process of ferromagnetic steel, when a full non-linear model is used: a temperature dependent B-H curve is taken into account. The full B(H,T) model is essential to obtain accurate results, which are confirmed by experimental validation. The paper also points out sensitivity of the model to slight variation of magnetic properties of material like Curie temperature, which might vary for steel of different grades. All results in the paper are obtained using open source simulation tools that demonstrates high accuracy compared to experimental results and benchmarked commercial software ANSYS Classic.

Keywords: induction heating, simulation, open source

Introduction

Numerical simulation procedure is mentioned frequently as a major factor for the successful design of complex induction heating and hardening coils [1]. While manufacturing of physical prototypes cost significant price in terms of materials, machining expenses, labor costs, engineering and, finally, time, practice shows that simulation might be an efficient way to shorten a design workflow down to a single iteration if numerical methods are used wisely prior to physical prototyping. This paper aims to examine necessity to implement a full non-linear magnetic model for induction heating of ferromagnetic materials like carbon steel. Such model simultaneously includes a B-H curve and its quantitative change with temperature raise. If taken into account, a sophisticated B(H,T) model may increase accuracy of numerical prediction of heating process, thus, adding value to a simulation software. However, such non-linear modelling cannot be performed analytically or using simple calculation sheets, FEM software is necessary.

Regarding FEM software, traditionally, only enterprise level companies could afford numerical simulation tools, since not only high costs of software itself, but also long learning period and the highest qualification of engineers needed in order to run simulations in time-efficient way. This paper has the aim to demonstrate that such times will likely pass away very soon! Open source tools, which are not only free, but also provide access to their source code, nowadays, are able to perform very accurate and time-efficient calculations. Since full access to the source code is provided, such software are very cooperative to slightly changes of models, including such enhancement as B(H,T) non-linearity.

Nevertheless, beside accuracy, the biggest concern for open source simulation software is the significant time needed in order to set a case in an open source tools. Since they usually provide low quality user experience and, frequently, have no user interface at all, open source tools require up to thousand additional hours per year if compared with user-friendly commercial software.

Because of all mentioned factor, only enterprise level companies with significant R&D departments are able to run simulations in time-efficient way and, therefore, are able to produce efficient coils for heating and hardening of complex parts. However, CENOS platform developed the unique technology, which is able to connect the best of various open source tools in simulation platform, which is simple in use and focused on induction heating.

The present paper will demonstrate simulation cases of induction heating of a billet, which are performed using GetDP open source software, coupled with pre-processing tool Salome and post-processing tool ParaView, powered in time- and cost-efficient way by CENOS platform. The results of the simulation are compared with experimental works 1) by Scurtu & Turewicz at Leibniz University of Hanover [2] and 2) Di Luozzo et al. at University of Buenos Aires [3].

Mathematical model

Magnetic vector potential A and electric scalar potential V formulation is used (AV-formulation):

$$\nabla \times (\nu \nabla \times \vec{A}) - \nabla(\nu \nabla \vec{A}) + \sigma \left(\frac{\partial \vec{A}}{\partial t} + \nabla V\right) = 0, \ in \ \Omega_1; \tag{1}$$

$$\nabla \times (\nu \nabla \times \vec{A}) - \nabla(\nu \nabla \vec{A}) = J_s, \ \ in \ \Omega_2; \tag{2}$$

$$-\nabla \left(\sigma \frac{\partial \vec{A}}{\partial t} + \sigma \nabla V\right) = 0, \ \ in \ \Omega_1 \tag{3}$$

Here, $\nu=1/\mu$ - reluctivity, μ - permeability, σ – electrical conductivity, J_s – source current density. Ω_1 stands for a electrically conducting eddy current domain (a workpiece to be heated), Ω_2 stands for a non-conducting domain (air) and a domain with a current source (a coil).

Temperature field is determined by solving the heat transfer equation:

$$\rho(T)c_p(T)\frac{\partial T}{\partial t} = \nabla(\lambda(T)\nabla T) + Q. \quad (4)$$

Here, ρ – density, c_p – specific heat capacity, λ – thermal conductivity, Q – Joule heat source. On the outer surface of the tube, both convective and radiation heat losses are set:

$$q = h(T - T_{amb}) + \sigma_B \varepsilon (T^4 - T_{amb}^4). \quad (5)$$

Here, σ_B – Stefan Boltzmann constant, h – heat transfer coefficient, ε – emissivity.
To account for nonlinear phenomena in time harmonic simulation, fictitious time independent material is introduced. Magnetic field intensity change over period is not sinusoidal in ferromagnetic material, but linearization approach finds effective B_{eff} that leads to same integral energy as it would be in transient simulation:

$$B_{eff}(H) = \frac{2}{H}\int_0^H B(H')dH'. \quad (6)$$

The mathematical model is solved with the Finite Element Method (FEM), coded in open source tool GetDP. For time-efficiency, CENOS platform was used to set material and numerical parameters of the model in graphic mode as well as to combine GetDP computational algorithms with pre-processing tool Salome and post-processing tool ParaView.

Validation cases

Validation of numerical models were performed using 3 different cases, each of the case stands for the respective experimental results (the references are provided):

- **Case 1.1 – linear model** for aluminum, ref. [2];
- **Case 1.2 – μ(T) model** for carbon steel, ref. [2];
- **Case 2.0 – B(H,T) model** for carbon steel, ref. [3].

In the Case 2.0, the full non-linear magnetic model is represented by magnetic permeability as a function of both magnetic field intensity and temperature – μ(H,T):

$$\mu(H,T) = \left(\frac{B(H)}{H} - \mu_0\right)\left(1 - \left(\frac{T}{T_C}\right)^{\alpha}\right) + \mu_0. \quad (7)$$

where T_C is the Curie temperature, α is characteristic exponent of permeability temperature dependence and μ_0 is vacuum magnetic permeability. In the Case 2.0, α = 6; the Curie temperature T_C=735 °C.

Case 1.1: linear model for aluminum

Case 1.1 represents the simple linear model for induction heating of the aluminum billet. Fig.3 demonstrates that the results, obtained using the open source software GetDP, perfectly match both results of benchmark simulation by ANSYS and the experimental results [2]. The results represents temperature in the middle of the billet, at the symmetry axis, over heating time.

*Figure 1 – **Case 1.1 & 1.2**: 3D rendering of the parts (right) and the scheme of 2D axially symmetric system (left). On the scheme: 1 – the billet, 2 – the coil.*

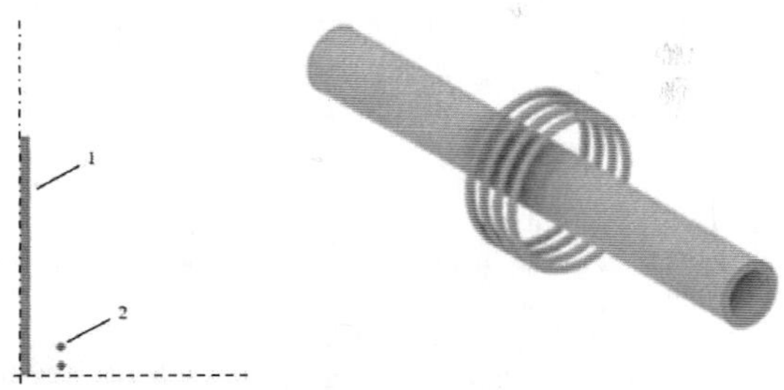

*Figure 2 – **Case 2.0**: 3D rendering of the tube and the coil (right) and the scheme of 2D axially symmetric system (left). On the scheme: 1 – the tube, 2 – the coil.*

Case 1.2: μ(T) model for steel

It is obvious, that induction heating simulation of steel requires to consider non-linear properties of ferromagnetic materials. Frequently, simulation is limited to temperature dependence of magnetic properties μ(T) and does not take into account a B-H curve. Such approach is demonstrated also in [2], where experimental results of induction heating of a steel billet are obtained by Scurtu & Turewicz at Leibniz University of Hanover. Beside the experimental results, the authors published numerical results by ANSYS Classic, which is well-known accurate benchmark software (see Fig.4). While the simulation by Scurtu & Turewicz is performed taking into account only temperature-dependence of magnetic permeability μ(T), the simulated temperature perfectly matched the asymptotic (steady state) value, however, significantly

IOP Conf. Series: Materials Science and Engineering **424** (2018) 012066 doi:10.1088/1757-899X/424/1/012066

underestimates temperature during transient heating. E.g., at 10th second, the benchmark model of ANSYS predicts the surface temperature of the steel billet 100 °C less than measured during the experiment.
The simulation by open source tool GetDP coincides with the benchmark simulation, even more, slightly better predicts transient temperature during heating (see Fig. 4). The last fact is just because of adaptive time step, which allowed more accurately resolve the temperature raise. Nevertheless, the figure clearly demonstrates inability of the simple μ(T) model to predict temperature at the surface of the steel billet during the heating process.
Does a B-H curve help predicting the temperature accurately? While there is no exact information available neither regarding the grade of the steel used in the experiment, nor B-H properties of it, the calibration of analytical B-H model demonstrate improvement of simulation result on Fig. 4 (orange curve). However, one can recognize overestimated temperature after 10th second. Since B-H curve is calibrated and does not ground in material properties of the steel, we decided to double check necessity of the B(H,T) model in the Case 2.0, which provides accurate experimental data.

Table 1 – Case 1: Parameters of the validation cases

	Case 1.1	Case 1.2	Case 2.0
	Geometric parameters		
Radius of the billet	30 mm		Inner: 16.5 mm/ Outer: 24.5
Length of the billet	150 mm		150 mm
Inner radius of the coil	70 mm		35 mm
Length of the coil	400 mm		50 mm
Number of windings	29		4
	Operational parameters		
Current	1.3 kA	1.3 kA	725 A
Frequency	1.9 kHz	1.9 kHz	15 kHz
	Material properties		
Material	aluminum	steel	
Thermal conductivity, Heat capacity, Electric conductivity	$\lambda(T)$, $c_p(T)$, $\sigma(T)$ [2]	$\lambda(T)$, $c_p(T)$, $\sigma(T)$ [3]	
Density	2.45 g/cm^3	7.87 g/cm^3	
	Magnetic properties		
Over temperature	**constant μ = 1**	**μ(T)** [2]	**B(H,T)**, see Eq.(7)
B-H curve	-	**B(H,T)** [3]	**B(H,T)** [3]

Case 2.0: B(H,T) model validation
The Case 2.0 demonstrates accurate validation of B(H,T) model in respect to experimental data obtained by Di Luozzo et al. at the University of Buenos Aires [3]. The non-linear model is represented by Eq.(7).
Fig.5 presents the essential results, comparing temperature over time at the surface of the billet. Temperature maxima is located in the coil region and it falls rapidly outside of coil where no heating source is present. The curves represent the points at different distance from the plane or mirror-symmetry of the system, and demonstrate good match of simulation results with the accurate experiment.

Discussion of the results
In general, good agreement between numerical and experimental results is achieved. Main discrepancies are as follows:

- change in heating rate at the symmetry plane appears 2 seconds earlier in numerical results rather than according to the experimental data. Constant voltage regime was carried out at the experiment, while constant current regime is used in 2D simulation. So, while the current in the simulation model was constant during all simulation time, it was, obviously, increasing during the initial short time moment in the experiment;
- change in the heating rate at the symmetry plane appears at lower temperatures ($\Delta T \approx 15$ K) in simulation results than in experimental results. We would like to argue here that the precise Curie point for the steel used is not known. Simulations at different Curie temperatures show that this change always appears slightly above (10-15 K) Curie point.

Change in heating rate appear at 747 °C, which is 10 degrees higher than Curie point. The same character is observed in simulations for variation of Curie temperature. Furthermore, increased heating rate appears immediately after Curie point for short period. This might be explained with electromagnetic effect of joined materials [5], which leads to local Joule heat maxima in non-magnetic part of steel above Curie point.

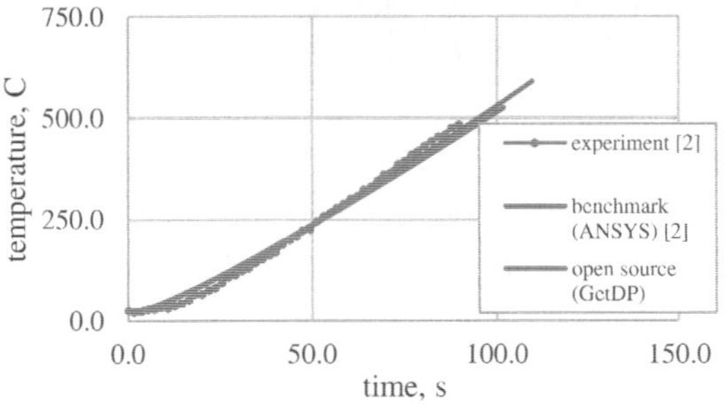

*Figure 3 – **Case 1.1**: Temperature at the middle of the aluminum billet over heating time. Experimental results and benchmark calculation ref. [2].*

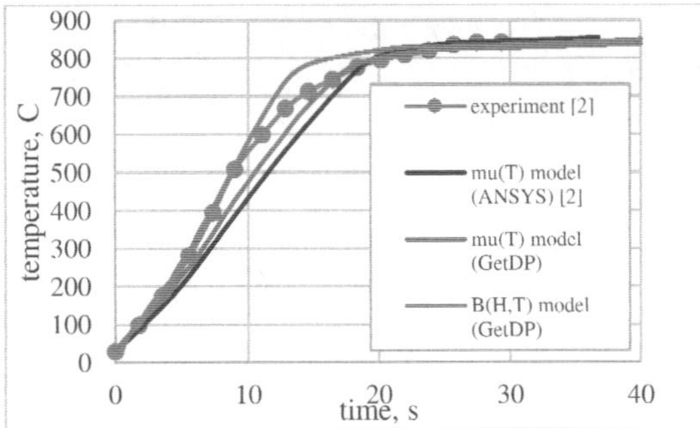

*Figure 4 – **Case 1.2**: Temperature at the surface of the steel billet over heating time. Experimental results and benchmark calculation of **μ(T)** and **B(H, T) model** [2].*

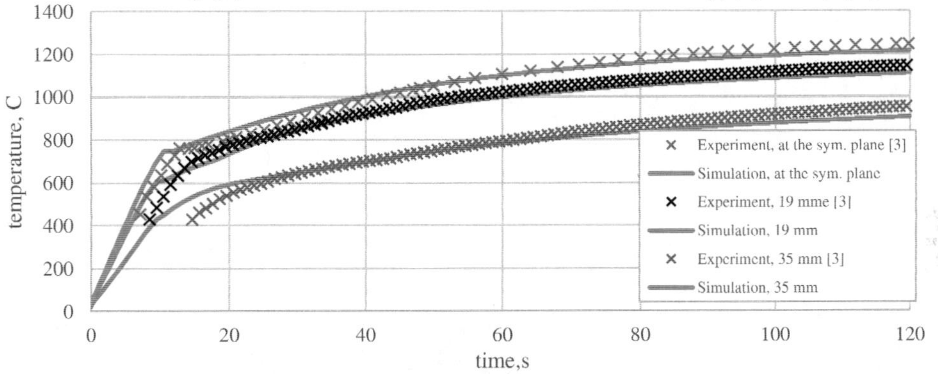

Figure 5 – Case 2.0: Simulation and experimental results (ref. [3]) of temperature at the surface of the workpiece. The curves are measured at different locations, the distance from the symmetry plane is specified in the legend.

Conclusions

Open source software GetDP has proven to be very reliable for modeling induction heating applications, especially due to its capability of capturing temperature dependent material properties and non-linear magnetic effects. Results obtained with GetDP are in good agreement both with ANSYS commercial software and experimental results.

Non-magnetic material heating has best agreement with experiment due to simpler mathematical model which leads to weaker coupling between thermal and electromagnetic problems.

Cases with ferromagnetic materials have good agreement but some details of heating curve are not captured. First reason for that are assumptions made, e.g. linearization of ferromagnetic material for time harmonic simulation. Second reason is insufficient knowledge of exact material properties, which turns out to be essential input data required for simulations. Even slight variation in Curie point had significant influence on temperature evolution

Cenos platform has proven its value for users of open source software as a great tool for simpler, faster and more efficient work.

References

1. ASM Handbook, Volume 4C: Induction Heating and Heat Treatment, ed. By Rudnev, V. & Totten, G. E., ASM International (Materials Park, OH, 2014).
2. Scurtu, G. L. & Turewicz, P., "Numerical Modeling of Static Induction Heating," *Proc Engineering Numerical Modeling and Simulation*, Sinaia, Romania, September 2012.
3. Di Luozzo, N., Fontana, M., Arcondo, B., "Modelling of Induction Heating of Carbon Steel Tubes: Mathematical Analysis, Numerical Simulation and Validation," *J Alloys Compd*, Vol. 536S (2012), pp. S564-S568.
4. Paoli, Gerhard & Biro, O & Buchgraber, Gerhard. (1998). Complex representation in nonlinear time harmonic eddy current problems. Magnetics, IEEE Transactions on. 34. 2625 - 2628. 10.1109/20.717607.
5. Rudnev, V., Loveless, D., Cook, R. (2018). Handbook of Induction Heating, Second Edition. Boca Raton: CRC Press.

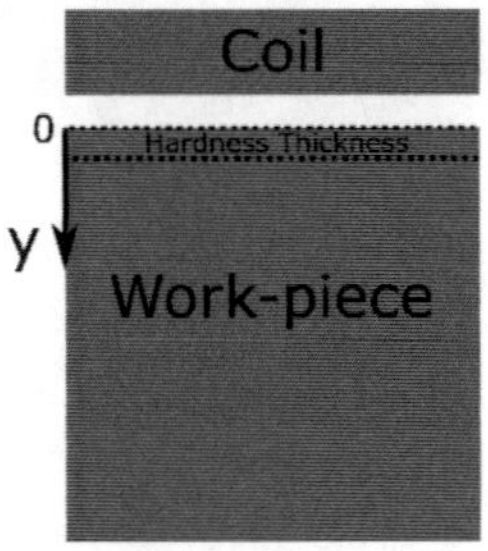

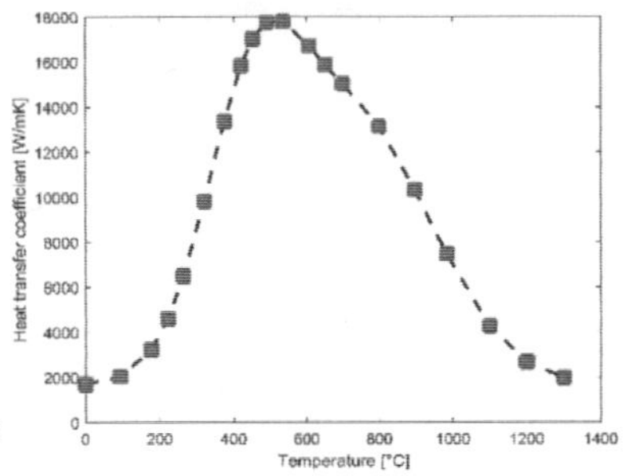

distribution is easily obtainable from the microstructure calculatioı

Fig. 1: Geometry of the 1D model. The coil is a massive piece of copper.

Fig. 2: Dependency of the heat transfer coefficient on temperature for simulating the quenching phase.

Multi-objective optimization using the 1D model

A great benefit of the one-dimensional model is its non-dependence on the work-piece geometry. Once the 1D model has been coupled with an optimization algorithm, it establishes a tool able to estimate the optimal heating time and frequency. The current is a parameter of optimization too. However, the optimal current value cannot be directly used in the next hierarchical steps because it must be adapted to the surface power that is needed in order to heat the work-piece. At the same time, results of 1D model enable us to estimate the necessary power density and therefore the installation power needed. The parameters that affect results of optimization are the steel grade, the cooling medium, the cooling time and rate, plus, of course, the hardening depth. The carbon steel C56 is considered in this study. The cooling medium is water with an estimated quenching effect visible in Fig. 2. Results of optimization are thus a function only of the hardening depth.

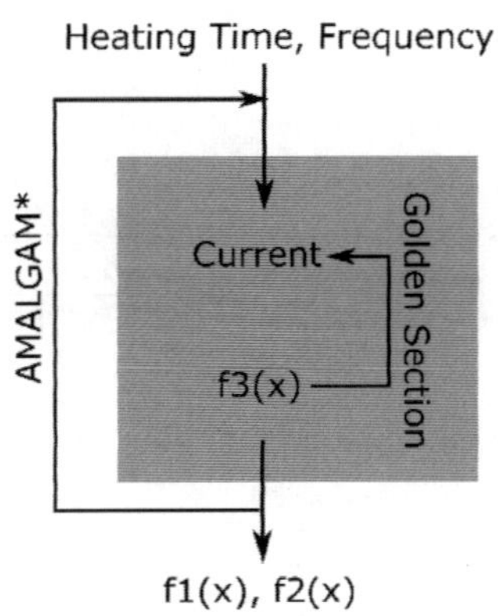

Fig. 3: Double-loop algorithm.

For the first investigations, we decided to simplify the numerical model and to consider only the heating phase. In doing so, we identify the hardening depth with the part of the work-piece that has been heated up over the AC_3 temperature, i.e. over 850°C. In other words, for example in the case of 3 mm hardening depth, a temperature of 850 °C has to be achieved 3 mm distant from the surface. The multi-objective optimization algorithm AMALGAM* [13] has been employed for the analysis. A three dimensional vector x of investigated variables (heating time, frequency and current) has been defined. Our first attempt focused on the trade-off between three conflicting functions, i.e. surface temperature $f_1(x)$, energy consumption $f_2(x)$ and heating accuracy $f_3(x)$. Heating accuracy in this case means how the evaluated temperature in the hardening depth differs from AC_3. It turned out that such approach is ineffective in our case, since we are interested only in the part of the Pareto front in which $f_3(x)$ trends to zero.

The next step we took was to modify the optimization algorithm. Similar to [14], a double loop was introduced (Fig. 3). AMALGAM* was used in the "external loop" where the vector variable is now two dimensional (heating time and frequency) and two objective functions (f_1 and f_2) are considered. The "internal loop" is simply intended to find the right current that zeroes f_3 and it is based on a golden section search and a parabolic interpolation. In fact, given a frequency and a heating time, there is only one value of current that able to reach the AC_3 temperature in the hardening depth. As a result, each point of the obtained Pareto front is a potential solution. Despite the considerable computational time, this approach is preferable.

The following presents the results obtained with 1 and 3 mm as hardening depth. Table 1 summarizes the search ranges of the variables.

	1 mm Hardening depth	**3 mm Hardening depth**
Heating time [s]	3 - 20	3 -20
Frequency [kHz]	10-100	1-20

Table 1: Variable ranges

Fig. 4 shows respective 1 and 3 mm Pareto fronts after 15 generations using a population of 25 individuals. Figs. 5a and 5b indicate instead the "image" of the Pareto front, i.e. the values that the variables assume, for the 1 mm case. As expected, the point with minimum surface temperature is characterized by the maximum time and the minimum frequency. But

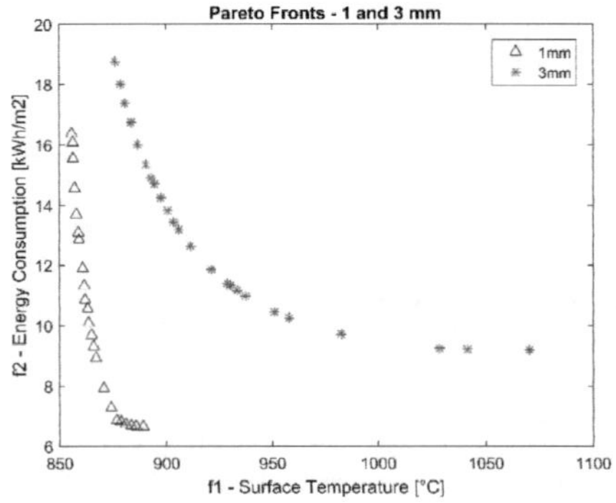
Pareto Fronts - 1 and 3 mm
1mm
3mm
f2 - Energy Consumption [kWh/m2]
f1 - Surface Temperature [°C]

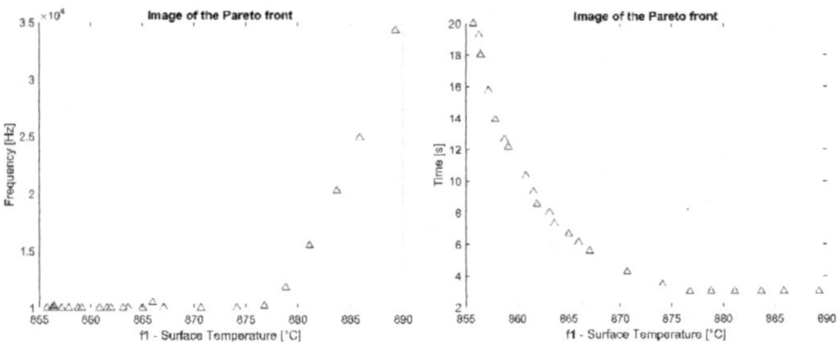
Image of the Pareto front
Frequency [Hz]
f1 - Surface Temperature [°C]
Image of the Pareto front
Time [s]
f1 - Surface Temperature [°C]

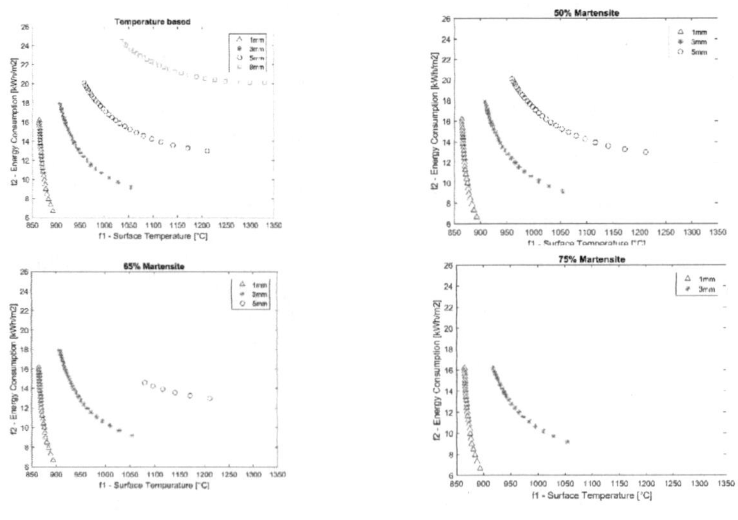
Temperature based
50% Martensite
65% Martensite
75% Martensite
f2 - Energy Consumption [kWh/m2]
f1 - Surface Temperature [°C]

IOP Conf. Series: Materials Science and Engineering **424** (2018) 012068 doi:10.1088/1757-899X/424/1/012068

Three-dimensional numerical computation of gas-liquid two-phase flow under pseudo microgravity environment using a superconducting electromagnet

S. Kikuchi[1], T. Tagawa[2]

[1] Graduate School of Aerospace Engineering, Tokyo Metropolitan University, Hino, Tokyo, 191-0065, Japan
[2] Department of Aeronautics and Astronautics, Tokyo Metropolitan University , Hino, Tokyo, 191-0065, Japan

Corresponding author: kikuchi-shinji@ed.tmu.ac.jp

Abstract
By using the strong magnetic field generated by a superconducting electromagnet, theoretically it is possible to produce the pseudo microgravity environment on the ground. The final goal of this research is to obtain knowledge of fluid control under a strong magnetic field using a superconducting electromagnet. In this study, in order to examine a method for accurate CFD computation, we carried out computation for a gas-liquid two-phase flow under a strong magnetic field. A milk crown phenomenon inside the bore of superconducting electromagnet is studied. The liquid is assumed to be an aqueous gadolinium nitrate solution (15%). The CLSVOF method is used for the computation of the two-phase flow and the HSMAC method is used for the pressure correction. We evaluate how much the interior of the superconducting electromagnet is in a pseudo microgravity environment by using the equation expressing the gravity level. In addition, in order to speed up the calculation, GPU parallel computation is performed, realizing about 22 times faster than CPU single computation. As a result, we succeeded in investigating a method for computation a gas-liquid two-phase flow in a strong magnetic field.

Keywords : Superconducting electromagnet, Two-phase flow, Computation, GPU

1. Introduction

In the microgravity environment, even substances with different specific gravities are uniformly dispersed, and convection caused by the density difference does not occur. Moreover, since impurities do not mix in such an environment, development of new materials including nano-skeletons has been conducted. Under the high magnetic field produced by a superconducting magnet, the magnetic body force acting in ordinary fluids such as air or water is almost the same order as the gravitational body force. Therefore, we are convinced that the pseudo microgravity environment on the ground can theoretically be reproducible, and the above-mentioned development can be done on the ground. In order to acquire the control techniques for such two-phase flows by the use of high magnetic field, we compute the behavior of gas-liquid two-phase flow inside the superconducting electromagnet.

2. Computational model

The computational model used in this study is shown in **Fig.1**. A cuboid container was installed inside the bore of the superconducting electromagnet. The interior is assumed to be gas-liquid two-phase flow (A gadolinium nitrate aqueous solution and air). The concentration of gadolinium nitrate in the aqueous solution is 15% in weight. For the superconducting electromagnet, a helium-free superconducting magnet manufactured by Sumitomo Heavy Industries, Ltd. is modeled.

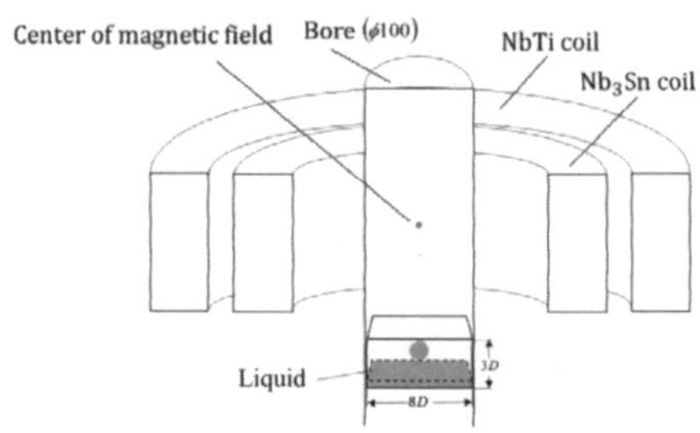

Fig. 1: Computational model in this study.

3. Computation method

3.1 Handling of phase[1) 2)]

In this study, CLSVOF method is used as an interface capturing method along with the two-phase flow computation. The CLSVOF method is a method of solving the advection equation of the interface using the smoothed Heaviside function (H_α) calculated from the Level Set function (ϕ) with the property of distance function. In a multiphase flow such as a gas-liquid two-phase flow, physical properties differ in each phase. Therefore, the physical property values are set using the smoothed Heaviside function.

3.2 Governing equations[3) 4)]

The dimensionless governing equations used in this study are shown below.

$$\vec{\nabla}\cdot\vec{U}=0 \tag{1}$$

$$\frac{\partial\vec{U}}{\partial\tau}+\left(\vec{U}\cdot\vec{\nabla}\right)\vec{U}=-\frac{1}{\rho_c}\vec{\nabla}P+\frac{1}{\rho_c}\vec{\nabla}\cdot\left(2\mu_c\boldsymbol{E}\right)+\frac{\Gamma}{\rho_c}\vec{F}_{SF}+M\vec{F}_M-G\vec{e}_z \tag{2}$$

$$\frac{\partial C}{\partial\tau}+\vec{\nabla}\cdot\left(C\vec{U}\right)-C\vec{\nabla}\cdot\vec{U}=0 \tag{3}$$

$$\vec{B}=-\frac{1}{4\pi}\iiint_V\frac{\vec{R}\times d\vec{s}}{R^3}dV \tag{4}$$

Here, R in equation (4) is the distance between the point on the coil and the computation point, $\vec{R}$ is the vector from the point on the coil to the computation point, and V is the volume of the coil. The fluid handled in this study assumed incompressible Newtonian fluid. For the discretization of the space, an equally spaced staggered grid system is used. As a discretization method, the Euler explicit method is used as the time derivative term, the third order accuracy upstream difference method is used as the advection term, and the other terms such as the diffusion term and the interfacial force term are approximated with the second order central difference method. The boundary conditions in this study are slip-free condition, no inflow / outflow condition. $\boldsymbol{E}$ (Strain rate tensor), $\vec{F}_{SF}$ (interfacial force term) and $\vec{F}_M$ (magnetization force term) appearing in equation (2) are shown below.

$$\boldsymbol{E}=\frac{1}{2}\left(\vec{\nabla}\otimes\vec{U}+{}^t\left(\vec{\nabla}\otimes\vec{U}\right)\right) \tag{5}$$

$$\vec{F}_{SF}=\kappa\vec{\nabla}H_\alpha=-\vec{\nabla}\cdot\left(\frac{\vec{\nabla}\phi}{\left|\vec{\nabla}\phi\right|}\right)\vec{\nabla}H_\alpha \tag{6}$$

$$\vec{F}_M=\chi_c\left(\vec{B}\cdot\vec{\nabla}\right)\vec{B} \tag{7}$$

The gravitational level inside the superconducting electromagnet X_G was obtained by using the following evaluation formula at the central axis of the bore.

$$X_G=M_GB_z\frac{\partial B_z}{\partial Z}-1 \tag{8}$$

The dimensionless variables and numbers were set as follows.

$$\vec{X}=\frac{\vec{x}}{D},\ \vec{U}=\frac{\vec{u}}{\mu_G/(\rho_GD)},\ \vec{B}=\frac{\vec{b}}{i\mu_m/D},\ \tau=\frac{t}{\rho_GD^2/\mu_G},\ P=\frac{p}{\mu_G^2/(\rho_GD^2)},\ \rho_c=\frac{\rho}{\rho_G},\ \mu_c=\frac{\mu}{\mu_G},$$

$$\chi_c=\frac{\chi}{\chi_G},\ \bar{\rho}=\frac{\rho_L}{\rho_G},\ \bar{\mu}=\frac{\mu_L}{\mu_G},\ \bar{\chi}=\frac{\chi_L}{\chi_G},\ G=\frac{g\rho_G^2D^3}{\mu_G^2},\ \Gamma=\frac{\gamma\rho_GD}{\mu_G^2},\ M=\frac{\chi_G\rho_G^2\mu_mi^2}{\mu_G^2},\ M_G=\frac{\chi_Li^2\mu_m}{gD^3}$$

Here, μ_m is the magnetic permeability in vacuum and i the electric current of the coil of the superconducting magnet and sets 164A. The subscripts G and L of each physical property value represent gas and liquid respectively.

The computational conditions were set as follows as shown in **Table 1**.

Table 1 Computational condition.

Parameter	Value	Parameter	Value
Number of grids (X, Y, Z)	384, 384, 144	Representative length (D)	0.005 (m)
Computation time(τ)	8.0	Gravitational acceleration (g)	9.8 (m/s^2)
Time step ($\Delta\tau$)	5.0×10^{-4}	Surface tension (γ)	0.053 (N/m)
Reynolds number (Re)	815.97	Impact velocity (u_0)	2.35 (m/s)
Weber number (We)	0.64	Gas density (ρ_G)	1.25 (kg/m^3)
Froude number (Fr)	10.62	Gas viscosity (μ_G)	1.8×10^{-5} (Pa · s)
Dimensionless number about magnetic field (M)	7.34×10^{-5}	Gas susceptibility (χ_G)	3.0×10^{-7} (m^3/kg)
		Liquid density (ρ_L)	1032 (kg/m^3)
Dimensionless number about gravity level(M_G)	3.12×10^{-3}	Liquid viscosity (μ_L)	1.7×10^{-3} (Pa · s)
Density ratio ($\bar{\rho}$)	825	Liquid susceptibility (χ_L)	1.13×10^{-7} (m^3/kg)
Viscosity ratio ($\bar{\mu}$)	93.6	Current value (i)	164 (A)
Magnetic susceptibility ratio ($\bar{\chi}$)	0.38	Permeability in vacuum (μ_m)	$4\pi \times 10^{-7}$ (H/m)

The concentration of the gadolinium nitrate aqueous solution is 15%. It is expressed by the following equation showing concentration dependency by Matsushima et al.[5] n represents the concentration (kg-solute/kg-solution) of the aqueous solution.

$$\chi_L = (8.377 \times 10^{-7})n - 8.460 \times 10^{-9} \tag{9}$$

3.3 Parallel computation on GPU[4]

In this study, by a parallel computation on GPU, high speed computation has been realized. However, it is impossible to correct the pressure field repetitively by using the HSMAC method in usual parallel computation. Because the pressure correction has a dependant relationship with adjacent points, so it is impossible to simply compute each grid by parallelization. Therefore, when correcting the pressure field by parallel computation, the Red and Black method is used. In this model, the computation of time has been improved about 22 times faster by the parallel calculation on GPU.

4. Result

Computational result of gravity level inside the bore of superconducting electromagnet at the central axis when the central magnetic field is 9.8 Tesla is shown in **Fig. 2**. The horizontal axis represents magnetic force and the vertical axis represent bore height. The container is installed so that zero gravity point accord with the interface as shown in **Fig.3**. The gravity level in the bore was upward maximum of about 2.8 G, downward maximum of about 4.8 G, and there was a large gradient.

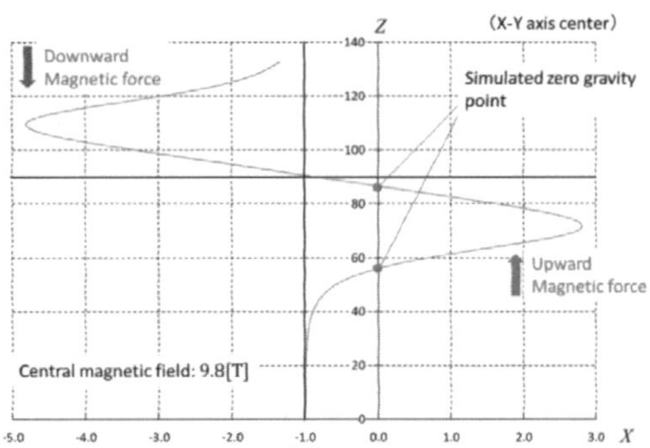

Fig. 2: Gravity level inside the bore of superconducting electromagnet at the central axis when the central magnetic field is 9.8 Tesla.

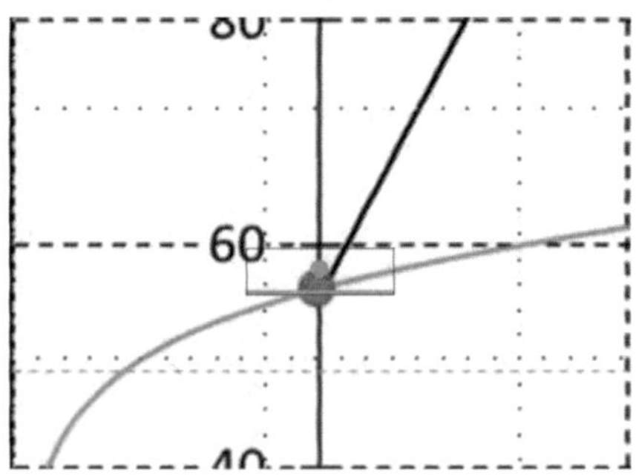

Fig. 3: Placement of the container. (Real scale)

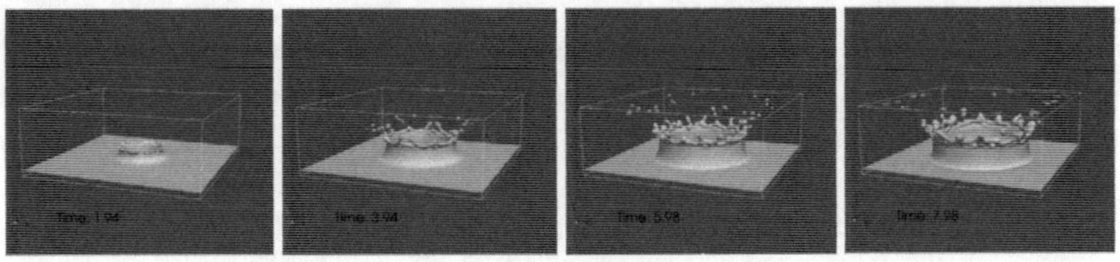

(a) No magnetic field

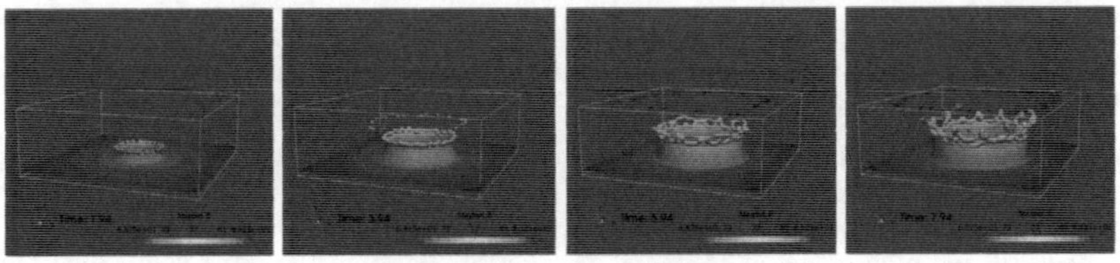

(b) With magnetic field

Fig. 4: Computational result.

The computational results are shown in **Fig. 4**. The color of (b) in **Fig. 4** represents the magnitude of the upward magnetic force. The central magnetic field is 9.8 Tesla. In the two results, the volume error during implementing computation was 0.02% or less, which resulted from taking advantage of the characteristics of the CLSVOF method. In the case with magnetic force, it can be seen that the droplet is separated due to the influence of the upward magnetic force and that it reaches the upper wall more quickly than the case without magnetic field. This is because the gradient of the magnetic field is large in the container and the magnetic force exceeds the gravity considerably in most part of the container.

5. Summary

We succeeded in real scale computation of the gas-liquid two-phase fluid inside the superconducting electromagnet and speeding up the computation. However, the region of the pseudo microgravity environment with the superconducting electromagnet modeled this time was very narrow. There was the incline of the big magnetic field in the computation domain, and the liquid greatly caught the influence of the magnetic power at the upper part of the computation domain. We would like to compute different models including the effect of wettability and conduct computation close to the result of real phenomena.

6. References

1. K. Yokoi, "Efficient implementation of THINC scheme: A simple and practical smoothed VOF algorithm," J. Comput. Phys. vol.232, (2007), pp. 1985-2002.
2. Y. Iwamoto, "Numerical computation on water drop magnetic floating phenomena," Kyushu University, master's thesis at Department of Energy and Environmental Engineering, (2006) (In Japanese).
3. M.M Francois, S.J. Cummins, E.D. Dendy, D.B. Kothe, J.M. Sicilian and M.W. Williams, "A balanced-force algorithm for continuous and sharp interface tension models within a volume tracking framework," J. Comput. Phys., vol.213, (2006), pp. 141-173.
4. Y. Anzai, "Development of multiphase flow numerical analysis method on GPU", Tokyo Metropolitan University, master's thesis at Department of Aerospace Engineering, (2015) (In Japanese).
5. K. Matsushima, M. Kaneda and K. Suga, "Effect of a Strong Magnetic Field on the Natural Convection off a Droplet on Lyophobic Surface", Osaka Prefecture University, The Japan Association of Fluid Mechanics Annual Meeting, (2009).

Numerical investigation of a double frequency approach for longitudinal HF welding of cladded pipes

W.Ebel[1], M.Kroll[2], E.Baake[1], A. Nikanorov[1]

[1] Institute of Electrotechnology, Leibniz Universität Hannover, Wilhelm-Busch-Str. 4, D-30167 Hannover, Germany

[2] Institute for Machine Tools and Production Processes, Technische Universität Chemnitz Reichenhainer Str. 70, D-09126 Chemnitz, Germany

Corresponding author: baake@etp.uni-hannover.de

Abstract

This article contains findings of simulation research on longitudinal induction welding of cladded pipes with use of simultaneous double frequency. Solutions are proposed to reach the required temperature distribution at the welding edge for the cladding composite of S355 and Alloy 625 with single and simultaneous double frequency. An advanced consideration of magnetic and other material properties was performed to simulate the dominating physical effects of high frequency (HF) welding. The background of use and advantages of simultaneous double frequency are presented. In the context of the research, a correlation for welding speed, frequency and temperature distribution with industrial relevance was found for the cladded pipe welding.

Key words: HFI-welding, longitudinal welding, pipes, cladded pipes, induction welding

Introduction

Today, cladded pipes are used in niche applications in natural gas transportation systems, power plants and the chemical industry. The combination of two different metals within one pipe offers high corrosion and abrasion resistance in an economical way. Until today the cladded pipes are manufactured by discontinuous arc welding, weld cladding or mechanical bracing with comparatively low production rates. Bended cladded sheet material, which is metallurgically bonded by explosive cladding, cold roll or hot roll bonding, serves as the feedstock for the welding process. Induction heating is a state of the art process in longitudinal seam welding of single material pipes. It is widely used in the almost all industries. Despite its very high productivity (feed speed of up to 200 m/min [2]) and energy efficiency, inductive longitudinal seam welding is not applied in the manufacturing of cladded pipes. With less than 5 m/min [3], [4] arc welding as the conventional joining process is far less productive.

In this paper, a 2D model of pipe welding is used to examine the dominating process parameters for a common material combination of the fine-grained ferritic carbon steel S355 and the nickel base Alloy 625. S355 exhibits a good weldability and is applied as an outer carrier material. The inner layer of Alloy 625 is characterised by a great resistivity against all kinds of corrosion.

Solution Approach

In this work, a 2D dynamic model is created to determine the influence of the main pipe welding parameters, such as feed rate, Vee angle, distance of inductor to the welding point, geometry of cross sections and the electrical parameters. The dynamic model is able to represent the 3D induction welding as shown in [5]. This can be done, as due to proximity effect, where the inductor have only a localized effect on current distribution, but a weak effect on the welding edge current.

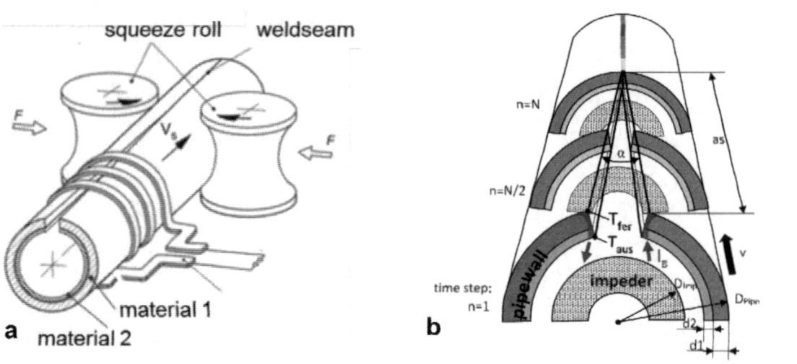

c

Par	*Description*
$d1$	thickness of carrier material
$d2$	thickness of cladded material
D_{pipe}	pipe diameter
D_{imp}	impeder diameter
as	distance btw. current feeding point and welding point
α	welding angle
v	welding rate
I_B	welding current
f_C	welding frequency
$mat1$	carrier material
$mat2$	cladding material

Fig. 1. Overview of inductive pipe welding process (a), 2D model steps along the pipe edge move (b) model parameters considered by the simulation (c)

The basic geometries of the inductive welding is shown in Fig. 3 a, the model geometry and considered parameters for the numerical research in Fig. 1 b, c. The solution was calculated by a weak electromagnetic – thermal coupling algorithm in the commercial FEM software ANSYS® Mechanical as shown in [6]. After each time step the results were transferred between the electromagnetic and thermal solutions. The electromagnetic solution was calculated in several iterations to adapt the magnetic permeability according to magnetic field strength.

Material Properties and Influence

Electro- and thermo-physical properties of the cladded pipe materials are pivotal to reproduce the physical effects in the simulation. Due to the non-linearity of the material parameters for the alloys used, the process behaviour is highly sensitive to the change of any input properties. To obtain precise results, physical properties of used material need to be considered and determined from ambient temperature to the melting (or liquefaction) point. In this context, the specific electrical resistance and the thermal conductivity are considered as shown in Fig. 2 a, b.

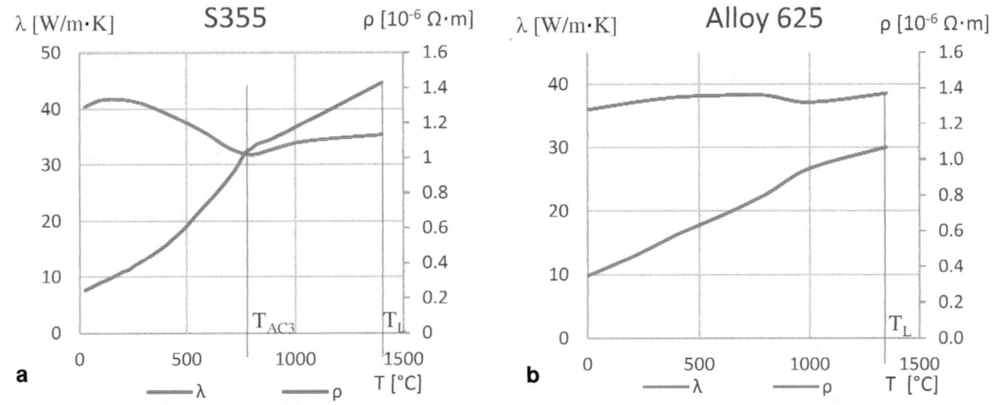

Fig. 2. Specific electrical resistance ρ and thermal conductivity λ of S355 [9], [13] (a) and Alloy 625 [10] (extrapolated to the melting points) (b)

The liquefaction temperatures of the alloys used are given as $T_L(\text{S355}) = 1471°\text{C}$ [7] and $T_L(\text{Alloy 625}) = 1340°\text{C}$ [8]. The function for the temperature dependent specific heat capacity for each alloy [9], [10] is complemented by the enthalpy of the fusion distribution function $h(T)$, which integral equalizes the enthalpy of fusion value L_f. The heat of fusion is calculated by the pure element enthalpy of fusion and the mass proportion of each element in the alloy. For the S355, the heat of fusion is determined as $L_{f,S355} = 266\ kJ/kg$ which is close to reference value $L_{f,S355,lit} = 268\ kJ/kg$ [11]. For the nickel-based Alloy 625, the enthalpy of fusion is calculated as $L_{f,A.625} = 338.4\ kJ/kg$.
In the context of material precision, a particular look on the relative magnetic permeability of the backing steel S355 and the ferrite material of the impeder is made. Except the consideration of temperature depending change up to the Curie temperature, magnetic field dependence must also be taken into account. Equation (**1**) is an approximation function for the real part of magnetic permeability μ_r', introduced by Fisher [12, pp. 286–299] and transferred from the magnetic flux B(H) to the relative magnetic permeability $\mu_r(H)$ relation. By choosing $a_F = 390$ and $b_F = 0.697$ [13, p. 67] as coefficients for the soft magnetic material S355 [14], the magnetic field dependency of material S355 is taken into account. For the impeder an adapted magnetic saturation function, with a starting pitch is used as shown in Fig. 3 b.

$$\mu_r'(H) = \frac{1}{\mu_0(a_F + b_F * |H|)} \quad (1)$$

For the temperature-field dependence of the magnetic permeability, equation (2) is used, with a function φ(T) representing the temperature dependence of the permeability. This function is taken from [15, p. 76] and redrawn in (Fig. 3 c).

$$\mu_r(H, T) = \varphi(T) * (\mu_r(H) - 1) + 1 \quad (2)$$

The saturation effect is correlated with the welding current and can be indirectly used for the process control by two input parameters. The change of current frequency influences the electromagnetic skin depth and therefore the required welding current. The second influence on the saturation is the welding rate, which is also positively correlated to the required current. Due to the complexity of the process, both input parameters have side effects. For example, a lower feed rate increases the effect of the divergent heat conductivity of materials by extending the process time.

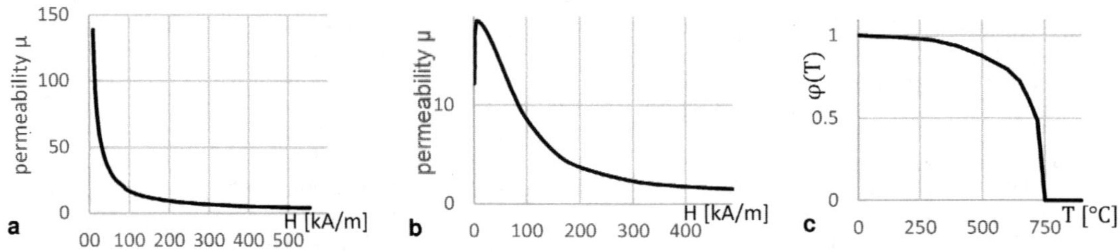

Fig. 3. Magnetic field dependent permeability of S355 at 20°C (a), permeability of the ferrite impeder (b) and temperature dependent function of permeability φ(T) referring to [15, p. 76] (c)

The 2D dynamic model has some limitations in the representation of the process: For example the longitudinal heat transfer and effect of adverse current circuit on the pipe inside cannot be displayed. Despite this the model represents the main physical effects of welding in the cross section and can be used for the task of finding matching process parameters.

Single Frequency Welding

The correct simulation of the welding of cladded pipes strongly relates to the material properties and their proportion to each other. The temperatures T_{fer} for the S355 and T_{aus} (Fig. 1 b) for the Alloy 625 at both edges of the joint serve as the target variable (Fig. 1 b). The aim is to reach the liquefaction temperature of both alloys when the joining surfaces get in contact. Welding with single frequency (SF) is possible by adapting the frequency and the feed rate to the geometrical parameters. Therefore, a feed rate – frequency curve can be found for each material combination and geometrical setup. For a given setup (Fig. 4 a) a corresponding correlation of feed rate – welding velocity is presented in Fig. 4 b.

Parameter	*Value*
$d1$	$2\ mm$
$d2$	$1\ mm$
D_{pipe}	$60\ mm$
D_{imp}	$42\ mm$
as	$90\ mm$
α	$2°$
$mat1$	$S355$
$mat2$	$Alloy\ 625$

a

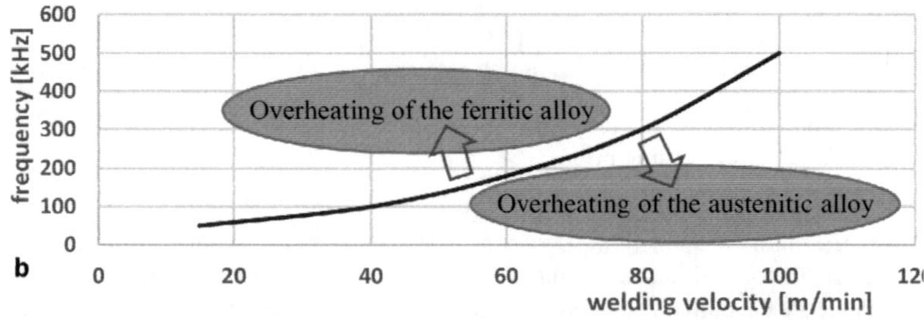

Fig. 4. Geometrical and material parameter set for single frequency simulation (a), frequency – welding velocity curve for convenient welding temperature at the join edge

As shown in Fig. 4 b, the frequency should match the chosen welding speed to melt both alloys at the desired position. Industrial welding generators usually do not offer the high enough flexibility to adapt the frequency in a wide range to the demanded feed rate in practical use. Therefore, the use of double frequency for the longitudinal welding of cladded pipe was investigated.

Simultaneous Double Frequency Welding

Welding with two simultaneous frequencies permits a higher flexibility for the operative adaptation of the welding process to industrial scale variations in pipe geometry and material. For a fixed geometrical and material setup, the usable range of feed speeds depends on the spread of the upper and lower frequency of the energy source.

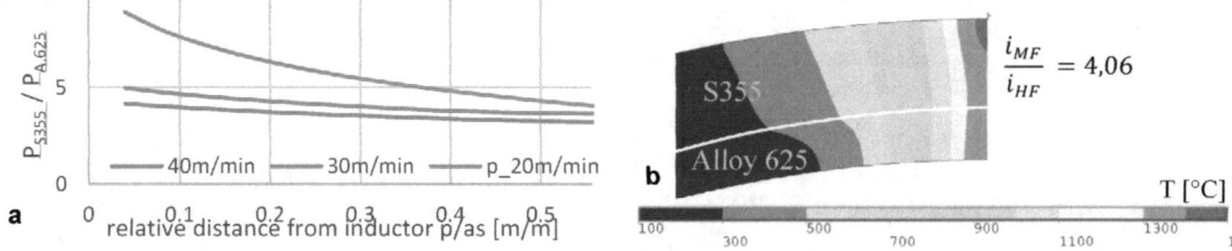

Fig. 5. Total power proportion between the layers for different welding velocities (a), resulting temperature distribution for the welding velocity of $v = 40\ m/min$

To lower the welding speed, it is necessary to increase the relative absorbed power in the ferritic steel layer in proportion to the austenitic alloy. This is necessary as the ferritic S355 offers a higher heat conductivity (Fig. 5 a). The examined frequencies for the research are $f_{HF} = 220\ kHz$ for upper and $f_{MF} = 10\ kHz$ for the lower frequency. From Fig. 4 it can be taken, that with the chosen frequencies the minimum and maximum welding velocities are $v_{min} = 10\ m/min$ and $v_{max} = 68\ m/min$. For the welding velocity of $v = 40\ m/min$, the desired temperature distribution can be reached by setting a MF to HF current ratio of $i_{MF}/i_{HF} = 4{,}06$ (Fig. 5). Further suitable material combinations, such as S235 for the carrier material and X5CrNi18 or X2CrNiMo17 as cladding materials, can be found for the presented longitudinal seam welding approach.

Conclusion

By the use of FEM analysis, the theoretical capability of an inductive longitudinal seam welding process for cladded pipes can be demonstrated. Despite some simplification, the main dominating effects of the longitudinal seam welding were considered by the developed model. Industrial scale process windows for both, single frequency and simultaneous double frequency are presented in this work. The verification was performed on a cladding material combination of S355 and Alloy 625 with strongly divergent material properties, where several effects were used to reach the melting temperatures in the joint edges. The developed approach was also applied to other material combinations, which are not presented in this paper. Potentially, a wide range of metal combinations comprising ferritic alloys as carrier and austenitic alloys as cladding material can be welded by the presented approach. Therefore, a precise material properties determination needs to be existent. Future investigations will highlight the influence of inductor and the full 3D model. Also, thicker-walled cladded pipes are about to be examined for the inductive HF welding.

Acknowledgment

The IGF-project 19041 BG / FOSTA No. P 1183 of the Research Association for Steel Application (FOSTA) was supported by German Federation of Industrial Research Associations (AiF) in the frame of the Industrial Collective Research program (IGF) from Federal Ministry for Economic Affairs and Energy.

References

1. B. Nacke and E. Baake, *Induktives Erwärmen*. Essen: Vulkan Verlag GmbH, 2014.
2. B. Grande and O. Waerstad, "Consistent quality in high-frequency tube and pipe welding," 2011.
3. X. Meng, G. Qin, Y. Zhang, B. Fu, and Z. Zou, "High speed TIG-MAG hybrid arc welding of mild steel plate," *J Mater Process Technol*, vol. 214, no. 11, pp. 2417–2424, 2014.
4. J. Sabbaghzadeh, M. Azizi, and M. J. Torkamany, "Numerical and experimental investigation of seam welding with a pulsed laser," *Opt Laser Technol*, vol. 40, no. 2, pp. 289–296, 2008.
5. A. Nikanorov, E. Baake, H. Brauer, and C. Weil, "Approaches for Numerical Simulation of High Frequency Tube Welding Process," in *Modelling for Electromagnetic Processing*, 2014, no. 1, pp. 445–450.
6. W. Ebel, M. Kroll, A. Nikanorov, and E. Baake, "Nummerische Simulation des HF-Längsnahtschweißens," *Prozesswärme*, vol. 2, 2018.
7. E. Le Guen, M. Carin, R. Fabbro, F. Coste, and P. Le Masson, "3D heat transfer model of hybrid laser Nd:Yag-MAG welding of S355 steel and experimental validation," *Int J Heat Mass Transf*, vol. 54, no. 7–8, pp. 1313–1322, 2011.
8. Corrosion Materials, "Alloy 625," 2008. [Online]. Available: www.corrosionmaterials.com.
9. U. Peil and M. Wichers, "Schweißen unter Betriebsbeanspruchung – Werkstoffkennwerte zur Temperaturfeldberechnung für einen S 355 J2 G3," *Stahlbau*, vol. 74, no. 11. pp. 249–257, 2005.
10. M. Woite GmbH, "2.4856," 2017. [Online]. Available: http://woite-edelstahl.info/24856de.html.
11. J. Winczek and E. Gawrońska, "The modeling of heat affected zone (HAZ) in submerged arc welding (SAW) surfacing steel element," *Metalurgija*, 2016.
12. J. Fischer and H. Moser, "Die nachbildung von Magnetisierungskurven durch einfache algebraische oder transzendente Funktionen," *Arch für Elektrotechnik*, vol. 42, no. 5, pp. 286–299, 1956.
13. C. Nörenberg, "Bestimmung elektromagnetischer Materialeigenschaften für Rohrstähle mit Hilfe experimenteller Untersuchungen und numerischer Simulation.," Leibniz Universität Hannover, 2017.
14. K. Lim and M. Hammond, "Universal Loss Chart for the Calculation of Eddy Current Losses in Thick Steel Plates," *Proc Inst Electr Eng*, vol. 117 (4), pp. 857–864, 1970.
15. B. Nacke, "Ein Verfahren zur numerischen Simulation induktiver Erwärmungsprozesse und dessen technische Anwendung.. Universität Hannover," Leibniz Universität Hannover, 1987.

IOP Conf. Series: Materials Science and Engineering **424** (2018) 012070 doi:10.1088/1757-899X/424/1/012070

Effect of Electromagnetic Field on Microstructure and Properties of Cu-Cr-Co-Si Alloy

Xinglong Sun [1], Jinchuan Jie [1*], Tingju Li [2]

[1] Key Laboratory of Solidification Control and Digital Preparation Technology (Liaoning Province), School of Materials Science and Engineering, Dalian University of Technology, Dalian 116024, China
[2] Laboratory of Special Processing of Raw Materials, School of Materials Science and Engineering, Dalian University of Technology, Dalian 116024, China

Corresponding author: jiejc@dlut.edu.cn

Abstract

The Cu-Cr-Co-Si alloy is a newly designed copper alloy with high mechanical properties and a proper electrical conductivity. The intermediate frequency electromagnetic field was applied during the solidification process. The results showed that the grain size remarkably refined with the application of electromagnetic field, the structure of the as-cast Cu-1Cr-1Co-0.2Si alloy transformed from coarse grain to homogeneous equiaxed grain due to the electromagnetic force. The microhardness and tensile strength of the alloy with the electromagnetic field intensity of 10 mT increased by 27.4% and 14.9% respectively compared with as-cast condition after solid solution and aging treatment. While, the electrical conductivity of the ingots with electromagnetic field is 53.1% IACS, which is slightly lower than that of the as-cast condition.

Keywords: Cu-Cr-Co-Si alloy, Electromagnetic field, Microstructure, Hardness, Tensile strength, Electrical conductivity.

Introduction

Cu-Cr alloys are the most potential alloys that meet comprehensive performance of high strength and high electrical conductivity, which have been widely used in industrial applications such as lead frames, railway contact wires, connectors, heat transfer elements and resistance welding electrode materials[1-3]. Alloying has been proved to be an effectively method to improve the mechanical properties with a slightly decrease of electrical conductivity[4-6]. Hence, a new type Cu-Cr alloy with Co and Si additions was designed. The high strength is attribute to the precipitation and particle dispersion strengthening mechanisms, while the electrical conductivity is decreased due to solute atom that dissolve into the Cu matrix. The main structural requirement of the Cu-Cr-Co-Si alloy is the homogeneous distribution of refined precipitation. Hence, a long time of homogenizing annealing is needed to eliminate the segregation of solutes ahead of solution treatment[7,8]. However, being in high temperature for a long time can result in a severe grain coarsening[9,10]. Thus, the use of magnetic field is a promising topic in the fabrication of materials.

Magnetic field has been successfully applied into the improvement of the solidification structure[11]. Extensive researches in the past demonstrated that the melt acted by appropriate magnetic field has the advantages of refining solidification structure and reducing segregation[12,13]. Therefore, electromagnetic field is considered to apply into the solidification of Cu-Cr-Co-Si alloy to improve the structure and to reduce the segregation of solutes, which helps to shorten the homogenizing annealing. This study investigated the influence of magnetic field on the microstructure, precipitate distribution and properties of Cu-Cr-Co-Si alloy.

Experimental Procedures

The Cu-1Cr-1Co-0.2Si alloy was prepared in a 5 kg intermediate frequency induction furnace with electrolytic copper (99.97 wt.%), pure chromium (99.8 wt.%), pure cobalt (99.9 wt.%) and pure silicon (99.9 wt.%). At first, the electrolytic copper, cobalt, and silicon were melted in the furnace. Pure chromium was added to the melt as the temperature reached 1300 ^{o}C. Then the temperature was held at 1300 ^{o}C for 10 min to ensure the adequately melting of

Cr, Co and Si elements. Finally, the melt was poured into the mold with a diameter of 60 mm, the magnetic field was imposed during the solidification process. Fig. 1 is the schematics of the magnetic field experiment instruments. The magnetic flux density, *B*, used in this experiment was 10 mT. The Cu-1Cr-1Co-0.2Si alloy solidified without magnetic field was prepared as a comparison. The ingots were solution treated at 990 °C

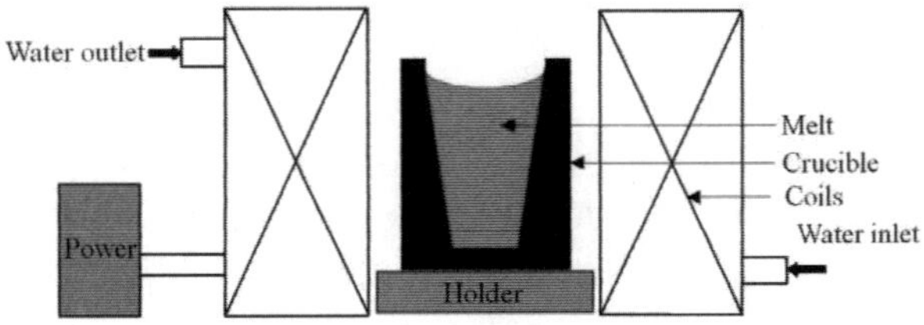

Fig. 1 The schematics of the electromagnetic field experiment instrument.

The microstructures were observed by optical microscope (OM, MEF4, Leica company, Solms, Germany) and field emission scanning electron microscope (FESEM, SUPRA55, Carl Zeiss company, Oberkochen, Germany). Mechanical properties were tested by a universal electronic tensile-testing machine. At least three standard tensile bars of φ10mm were tested for each sample. All tensile tests were performed at room temperature with a strain rate of 1 mm/min. The microhardness was measured using a Vickers hardness tester under 300 g loads maintaining 15s. The electrical conductivity was measured using a D60K conductivity measuring instrument, which is characterized by International Annealed Copper Standard (% IACS).

Results

Macrostructures and microstructures.

Fig. 2 shows the macrostructures and SEM images of Cu-Cr-Co-Si alloy with and without magnetic field. It can be seen that the macrostructure of as-cast Cu-Cr-Co-Si alloy without magnetic field are mainly coarse columnar grains. With the application of the magnetic field, the solidification structure tends to homogeneous equiaxed grains and the average grain size refined from 3000 μm to 750 μm (Fig. 2a-b). Two kind of newly formed phases are formed during the melting and solidification (Fig. 2c-d). The EDS analysis confirms the phase with gray color and a large size is $Cr_{15}Co_9Si_6$, while the other phase with white and a small size is Co_2Si. Meanwhile, the number of the particles distributes more uniform with the application of magnetic field.

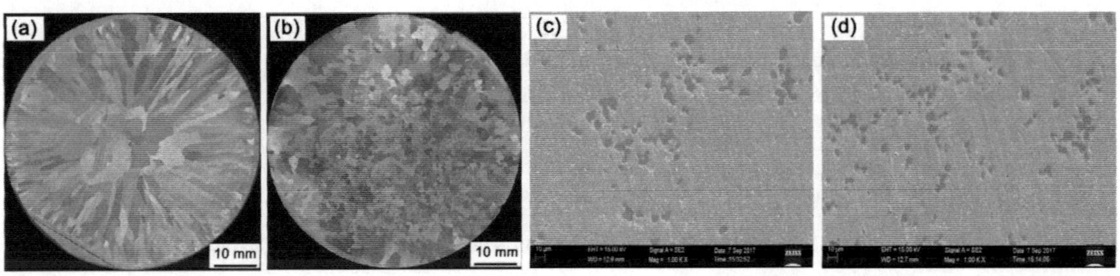

Fig.2 Macrostructures and SEM images of as-cast Cu-1Cr-1Co-0.2Si alloy with and without electromagnetic field: (a) and (c) *B*=0; (b) and (d) *B*=10 mT

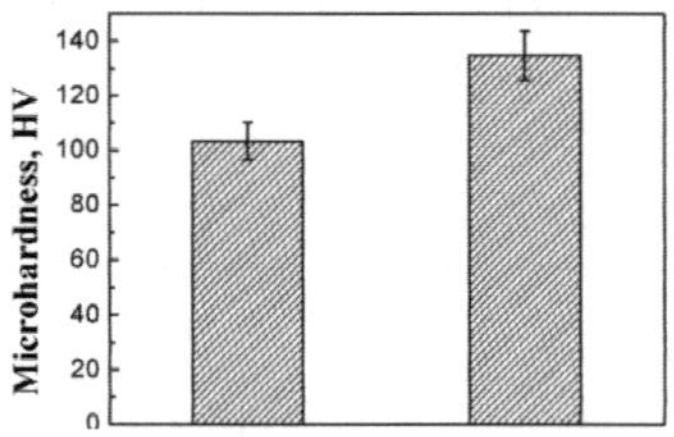

Fig. 3 Microhardness comparsion of the Cu-1Cr-1Co-0.2Si alloy with and without electromagnetic field.

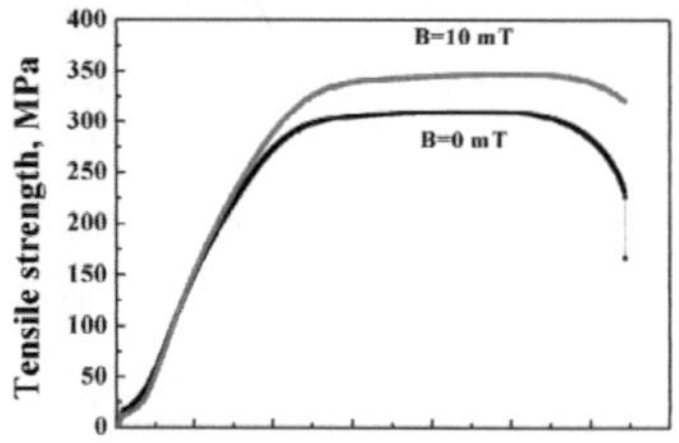

Fig. 4 Stress-strain curves of the Cu-1Cr-1Co-0.2Si alloy with and without electromagnetic field.

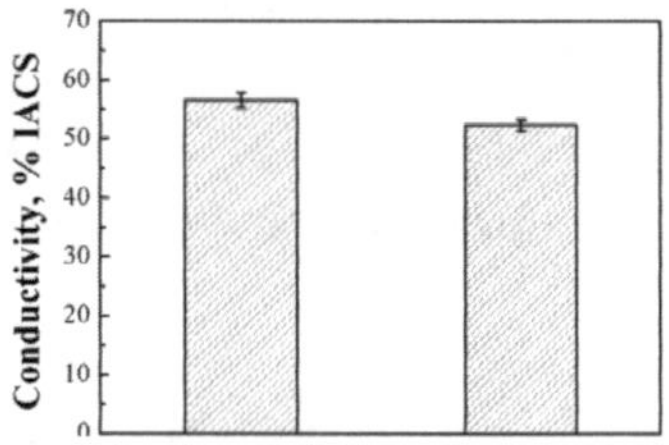

Fig. 5 Electrical conductivity comparsion of the Cu-1Cr-1Co-0.2Si alloy with and without electromagnetic field.

Properties

Fig. 3 presents the microhardness comparsion of the Cu-1Cr-1Co-0.2Si alloy with and without magnetic field. It can be seen that the average values of microhardness under magnetic field B=0 and B=10 mT are 103.3 HV and 131.6 HV, respectively. With the application of magnetic field, the microhardness increases by 27.4 %. Fig. 4 exhibits the stress-strain curves of the Cu-1Cr-1Co-0.2Si alloy with and without magnetic field. The tensile strength of Cu-1Cr-1Co-0.2Si alloy solidified under magnetic field B=0 and B=10 mT are 306.4 MPa and 352.1 MPa, respectively. With the application of magnetic field, there is an evident increment in tensile strength and elongation. The increment of tensile strength is in correspondence with the particle distribution are shown in Fig. 2c-d. The slight improvement of elongation under electromagnetic field is due to the grain refinement as shown in Fig. 2a-b. Since the macrostructure of Cu-1Cr-1Co-0.2Si alloy transforms to refined equiaxed grains under the effect of magnetic field, the difference between deformation within the grain and grain boundaries grows less. Thus total deformation becomes more uniform under the action of external force. The electrical conductivity comparsion of Cu-1Cr-1Co-0.2Si alloy with and without electromagnetic field is shown in Fig. 5. The average values of electrical conductivity under electromagnetic field B=0 and B=10 mT are 57.6 %IACS and 53.1 %IACS, respectively. With the application of electromagnetic field, there is a light decrease in electrical conductivity, which is due to the increase of interface scattering.

Discussion

The effect of magnetic field on grain refinement.

The coils create a rotating magnetic field when alternating current is imposed. To the melt, under the effect of magnetic field, generates an induced eddy current in turn. As a result, the melt is subjected to a Lorentz force under the common action of induced eddy current and rotating magnetic field, which can be expressed as follows [14]:

$$F = J \times B \tag{1}$$

where J is the induced eddy current, and B is the magnetic flux density.

The low temperature melt near the edge of graphite is taken to the center by the electromagnetic body force, and the temperature gradient in the melt reduces. The compulsive convection caused by magnetic field generates shearing force, and plenty of dendrites are broken off [15, 16]. The flow transports these fragments from the interdendritic spacing to the region ahead of the solidification front. In this area, the fragments offer cores for nucleation. Since the nucleation rate increases and temperature field becomes homogeneous, the solidification structure transfers from big columnar dendrites to refined equiaxed grains over the cross–section.

The effect of magnetic field on the properties.

Particle strengthening occurs due to the Orowan dislocation bypass mechanism. This mechanism comprises a dislocation bow–cut and bypassing between un-deformable particles, leaving a dislocation loop around the particles. According to the Orowan strengthening model, the strength increment can be calculated as follows[17,18]:

$$\Delta\sigma_{Or} = 0.84 \frac{MGb}{2\pi r\sqrt{1-\upsilon}\left(\sqrt{\frac{3\pi}{2f}}-\pi/4\right)} \ln\left(\frac{\pi r}{4b}\right) \quad (2)$$

where M is the Taylor factor for the fcc copper matrix, b is the Burgers vector, G is the shear modulus of the matrix, υ is the Poisson's ratio of the copper matrix, r and f represent the average radius and volume fraction of the particle precipitates, respectively. With the application of magnetic field, the precipitation particles are refined and their amount increases, resulting in an improvement of tensile strength and microhardness.

The resistivity of copper matrix can be partitioned into the contribution of four main scattering mechanisms [19]:

$$1/\sigma_{alloy} = \rho_{alloy} = \rho_{imp} + \rho_{def} + \rho_{int} + \rho_{pho} \quad (3)$$

where ρ_{pho} is the resistivity contribution from phonon scattering, ρ_{def} is the dislocation scattering, ρ_{int} is the interface scattering, and ρ_{imp} is the impurity scattering. The Alloy with and without magnetic field have similar values of ρ_{imp}, ρ_{def} and ρ_{pho}. Since the amount of precipitate increases with the application of magnetic field, the more interface produced between the Cu matrix and the particles increases the ρ_{int}. Thus, with the application of magnetic field, the total resistivity of aged Cu-1Cr-1Co-0.2Si alloy is increased and the electrical conductivity is reduced.

Summary

(1) The macrostructure of Cu-1Cr-1Co-0.2Si alloy transform from coarse columnar grains to refined equiaxed grains with the application of magnetic field.

(2) The microhardness and tensile strength of aged Cu-1Cr-1Co-0.2Si alloy increases with increasing the magnetic field.

(3) The electrical conductivity of aged Cu-1Cr-1Co-0.2Si alloy exhibits a slight decrease with increasing magnetic field.

Acknowledgement

The authors gratefully acknowledge the supports of National Key Research and Development Program of China (Nos. 2016YFB0301303, 2017YFA0403800).

References

1. Q.J. Wang, Z.Z. Du, L. Luo, W. Wang: J. Alloy Compound. Vol.526 (2012), 39–44.
2. C.D. Xia, W. Zhang, Z.Y. Kang: Mater. Sci. Eng. A.Vol.538 (2012), 295–301.
3. M. Bizjak, B. Karpe, J. Kovac: Appl. Surf. Sci Vol.277 (2013), 83–87.
4. J.H. Su, Q.M. Dong, P. Liu, H.J. Li, and B.X. Kang: Mater. Sci. Eng. A. Vol.392(2005), 422.
5. S.J. Sun, S. Sakai, H.G. Suzuki. Mater. Sci. Eng. A. Vol.303 (2000), 187–196.
6. J.E. Bailey and P.B. Hirsch: Philos. Mag. Vol.5 (1960), 485.
7. S.G. Jia, X.M. Ning, P. Liu, M.S. Zheng, G.S. Zhou, Met. Mater. Int. 15 (2009) 555–558.
8. S.G. Jia, M.S. Zheng, P. Liu, F.Z. Ren, B.H. Tian, G.S. Zhou and H.F. Lou: Mater. Sci. Eng. A Vol. 419 (2006), p. 8
9. T.H. Kim and J.K. Park: Mater. Sci. Technol. Vol. 29 (2013), p. 1414
10. M. Akiyama, Y. Neishi and Y. Adachi: Eng. Computation Vol. 20 (2003), p. 499
11. X. Zhang, D.L. Wang, S.X. Zhang, Y.W. Ma, W.S. Yang, Y. Wang, S: J. Magn. Magn. Mater. Vol. 322 (2010), p. 302
12. T.J. Li, Z.Q. Cao, J.Z. Jin, and Z.F. Zhang: Mater. Trans. JIM Vol. 42 (2001), p. 281
13. W.Z. Jin, F.D. Bai, T.J. Li, and G.M. Yin: Mater. Lett. Vol. 62 (2008), p. 1585
14. C. Vives: Metall. Mater. Trans. B. Vol.20 (1989), p. 623
15. S. Eckert, P. A. Nikrityuk, B. Willers, D. Rabiger, N: Eur. Phys. J. Special Topics Vol. 220 (2013), p. 123
16. J. Stiller, K. Koal, W.E. Nagel and J. Pal, A. Cramer: Eur. Phys. J. Special Topics Vol. 220(2013), p. 111
17. J.L. Xu and S.X. Jin: Plasma Physics, Atom Energy Press, Beijing (1998).
18. J. Li, T. Wang, J. Xu, Z. Yan, J. Sun, J. Xu, Z. Cao and T. Li: Mater. Sci. Technol. Vol. 27 (2011), p. 676
19. L. Qu, E.G. Wang, K. Han, X.W. Zuo: J. Appl. Phys. Vol.113(2013), 173708.

IOP Conf. Series: Materials Science and Engineering **424** (2018) 012071 doi:10.1088/1757-899X/424/1/012071

Design improvements for increasing lifetime of single-shot coils applied at rotating workpieces

S. Schubotz[1], B. Nacke[2]

[1] EFD Induction GmbH, Lehener Str. 91, 79106 Freiburg im Breisgau, Germany
[2] Institute of Electrotechnology, Leibniz Universität Hannover, Wilhelm-Busch-Str. 4, 30167 Hannover, Germany

Corresponding author: Stefan.Schubotz@efd-induction.com

Abstract
In the industry, induction hardening of rotationally symmetrical workpieces by a single-shot process is a widespread method. Due to only partial superimposition of the workpiece areas to be heated by the coil, high power densities are often needed there. These lead to local hot spots, amounting to an intensive material stress and often result in a short lifetime of the inductor. In this paper, some numerically investigated models will be presented, revealing approaches, how to reduce mechanical stress in the single-shot coil and thus, enabling an increase of service life.

Key words : induction hardening, single-shot-coil, numerical simulation, thermal fatigue

Introduction

Induction hardening of rotationally symmetrical workpieces is usually carried out step by step with a ring coil in a scanning process or at once by a single-shot coil. In the second case, the inductor consists of two long conductors (main leads) and two bridges (Fig. 1). Since the inductor covers only a part of the workpiece in the axial direction (Fig. 2), the entire workpiece zone to be hardened is heated due to rotation almost simultaneously. For this purpose, high power densities are required in the inductor which lead to an inhomogeneous heat source distribution due to electromagnetic effects. After consideration of fluid dynamic behaviour, the resulting temperature field in the coil shows a comparably uneven distribution. At locations of high temperatures, the copper tries to expand but is constrained by colder material. In this way, large tensile and compressive stresses occur in the copper due to high temperature differences. In the usual case of cyclic usage of the hardening inductor, the stresses promote the formation of cracks and these can lead, in long term, to material fatigue and thus to coil failure. In particular, cracks often occur at the workpiece-facing sides of the connection between main conductor and bridge (see Fig. 3). As there is a graphical correlation between lifetime and stress amplitude by Wöhler curves, some approaches for changing the design with the aim to reduce the maximum load and thus, to increase coil lifetime are presented in this paper.

Approach

After determining the heat sources based on the electromagnetic field distribution in the inductor, a determination of the temperature field has been accomplished, taking into account the turbulent flow in the coil. Here, the LES (Large Eddy Simulation) model has proven to be very suitable in experimental comparison. Afterwards, the mechanical stresses in the coil are calculated based on the temperature distribution. All simulations are done with the commercial tool ANSYS. For calculating the equivalent stress σ_E in each point of the model, the stress intensity is determined according to the shear stress hypothesis of Tresca, which is depending on the three principal stresses σ_1, σ_2 und σ_3 as follows:

$$\sigma_E = \max(|\sigma_1 - \sigma_2|, |\sigma_2 - \sigma_3|, |\sigma_3 - \sigma_1|) \quad (1)$$

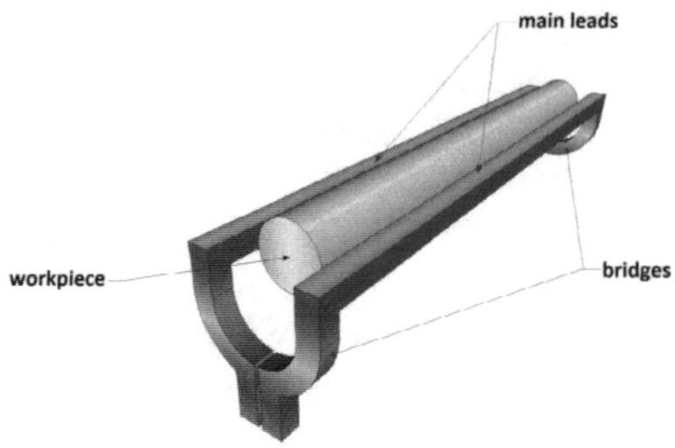

Fig. 1: Single-shot coil with main leads and bridges

Fig. 2: Cross-section of single-shot coil and workpiece. Red lines expose geometrical superimposing zone

IOP Conf. Series: Materials Science and Engineering **424** (2018) 012071 doi:10.1088/1757-899X/424/1/012071

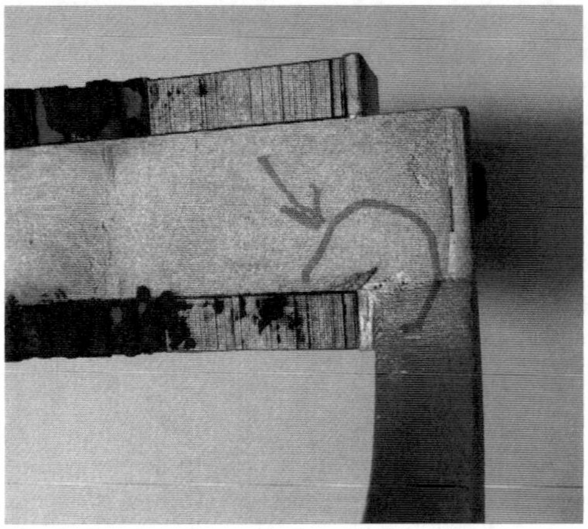

Fig. 3: Crack at the workpiece-facing corner

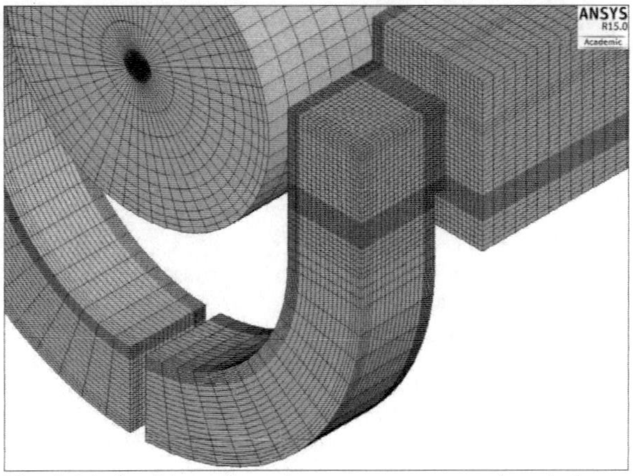

Fig. 4: Mesh of the coil, coolant, concentrators and workpiece

The equivalent stress as a reference value provides information about the load on the inductor and should reveal a conclusion about the success of the coil adjustment.

Reference-Model

In this investigation, a single-shot coil with a length of 350 mm is considered for workpieces with a diameter of 35 mm and a coupling distance (space between workpiece-facing sides of main leads and outer diameter of workpiece) of 2.5 mm. A square copper profile is used with an edge length of 10 mm and a wall thickness of 1.5 mm. To improve the efficiency, single-shot coils for hardening applications are usually equipped with field guiding elements (so-called concentrators). Therefore, these components are also taken into account in this simulation, whereas the magneto-dielectric material Fluxtrol100 [1] is placed along the main leads, leaving a small distance to the bridges of 1.5 mm (→ concentrator length: 327 mm). Usually, single-shot inductors are fixed at the sides of the main leads (often by screws or bolts) for stabilization. As a result, the coil can expand in length and height but almost no width change is possible, so the coupling distance remains approximately constant. As follows, the mechanical boundary conditions of this model are defined as the widths of the two outer sides of the Fluxtrol are set as fixed. The model with the associated mesh is shown in Fig. 4. As it can be seen, particularly the areas are finely meshed where high power densities have been simulated previously and where also often cracks occurred in practice that have been responsible for coil failure.

In this model, one heating step with a duration of one second, a cooling water throughput of 8 l/min, an electrical current with a frequency of 8.5 kHz and an effective value of about 3200 A is simulated. The power converted into the workpiece is around 105 kW and heats up the part to approximately 255 °C. This is far beyond hardening temperature, but previous results have revealed that the stress in the coil does not change much at further heating steps. The temperature distribution in the coil is shown in Fig. 5. As it can be seen, the highest temperature (about 153 °C) prevails near the corner on the workpiece-facing side between the main conductor and the bridge. Since the water temperatures rises in the coil, the

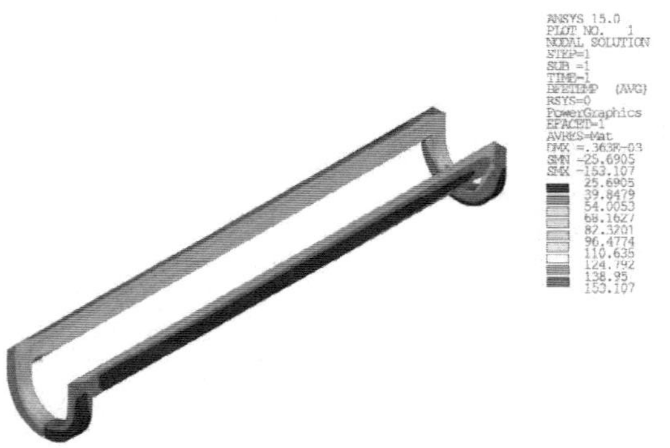

Fig. 5: Temperature distribution in the coil

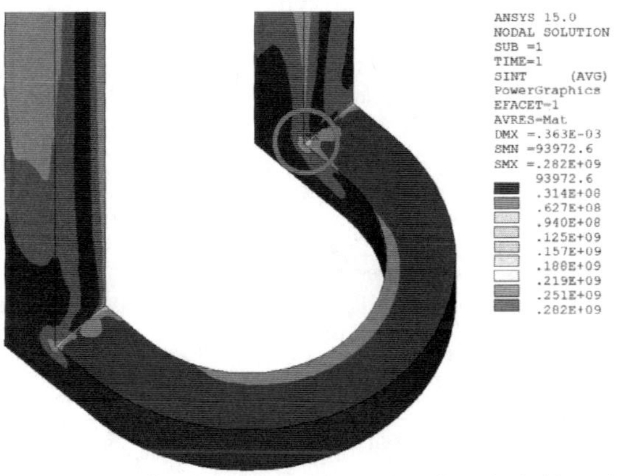

Fig. 6: Stress distribution; peak stress location circled in red

IOP Conf. Series: Materials Science and Engineering **424** (2018) 012071 doi:10.1088/1757-899X/424/1/012071

highest temperature is found in the corner before the outlet. The equivalent stress shown in Fig. 6 displays a maximum in a different corner but also at the workpiece facing side between a main lead and a bridge. Generally, the simulation reveals that the equivalent stress is similar in all four corners. Nevertheless, the supplying leads are not fixed in this simulation whereas this also approximates the reality since there is oftentimes no additional mounting. In most cases, only some tape is used to keep the insulation between the supplying leads in place which leaves some space for the copper to thermally expand. In contrast, a lower degree of mobility at the corners near the closed bridge can result in higher stress, although the highest temperature is at a different (last) corner.

Optimization of coil design

Now, some constructive approaches shall be tested in context of a sensitivity analysis with the aim to reduce the maximum mechanical stress. As part of the geometry modification of the various models, different power densities and temperature fields are generated in the workpiece to be heated. Therefore, the current in the inductor is adjusted in each simulation so that the power converted into the workpiece remains constant. The volume flow is also defined as a constant reference variable, meaning if the coil cross-section changes, an adaptation of the water velocity takes place in each case accordingly. First, it will be examined whether an adjustment of the width or height of the used copper profile leads to an improvement. In addition, due to frequent occurrence of the highest power density in corner regions, using of copper profiles with rounded corners will be tested, whereas it is expected that the power density and thus also the temperature maximum and equivalent stress amplitude at the problematic location reduces. Due to the so-called slot effect, the use of field-guiding elements leads to a displacement of the current to the workpiece-facing side of the main conductors, emerging the generation of hot spots. Therefore, the distance between the lengths of the concentrator-equipped main leads and the bridges of the inductor shall be varied. Since a reduction of the concentrator length would reduce the heating of the workpiece in the end region correspondingly, the inductor length will be increased instead. Finally, another approach is followed with the aim to reduce the current demand greatly. For this purpose, an inductor with four instead of two main conductors is constructed. The current reduction should generate significantly lower power densities in the inductor, whereas a lower maximum stress amplitude is expected.

Results

In order to establish a relationship between the determined equivalent stress and the lifetime of the coil, the available Wöhler curve from the data sheet of the applied copper (CU-HCP) is used, displaying the fatigue strength for the range of 10^5 - 10^8 cycles [2]. For this purpose, the approach of [3] is utilized, where it is assumed that the microstructure of the originally soft annealed copper deforms during cyclic use, resulting in a time-averaged stress of zero. Thus, the magnitude of stress in the heating on (+) and off (-) state is the same and the amplitude corresponds to the half of the simulated amplitude. Therefore, at an equivalent stress of 141 MPa (= 282 MPa / 2), the coil withstands approximately 620.000 alternating loads, leading to an estimated inductor life of about 310.000 cycles.

By reducing the profile height and width, stress can be decreased to 136 MPa and 121 MPA respectively. Basically, shortening the profile height reduces the superimposed area of the workpiece by the coil (red marked zone in Fig. 2), thus, leading to a higher demanded power density in the main conductors. This negative effect is reduced because the coupling distance between the straight profile and the round workpiece is highest at the edges. Although the current is smaller due to a higher inductor impedance, the power density in the critical spot has raised which leads to an increase of the peak temperature by about 9 °C. However, [3] has found out that the difference between peak and average temperature in a 2D coil cross-section has a vital impact on coil-lifetime. A similar circumstance is also observed in this 3D-simulation,

Fig. 7: Stress at a coil with rounded square profile. Peak stress location is circled in red

Fig. 8: Coil with four main leads (meander shape) Peak stress location is circled in red

	Modification	Fatigue strength [N/mm²]	Estimated lifetime (cycles)
1	(Reference model)	141	310.000
2	Profile height: 8 mm	136	390.000
3	Profile width: 8 mm	121	840.000
4	Coil length: 356 mm	123,5	730.000
5	Corner Radius: 3,5 mm	74	> 10^8
6	Meander shape	108,5	1.620.000

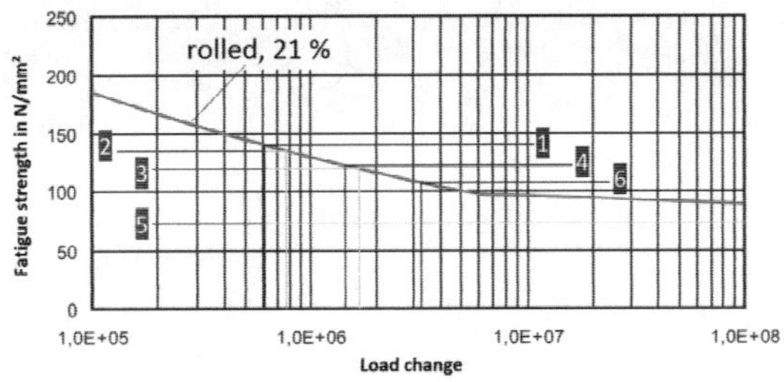

Fig. 9: Simulated fatigue strengths of coil adjustments (left) and correlation to expected load changes (right, [2])

where the temperature of the area around the critical spot is much higher when the profile width is 8 mm, leading to a smaller difference between peak and average temperature. Besides that, assuming a constant volumetric flow, the smaller coil cross-section results in a rising water velocity, thereby improving the heat transfer coefficient which also reduces this temperature difference. The sum of these different effects finally results in a reduction of the stress.
Raising the water speed meaning an increase of the heat transfer coefficient is also a positive consequence when the profile width is reduced. In this case, the peak temperature in the corner decreases by over 5 °C in comparison to the reference model. Another advantage when lowering the width of the cross-section is that due to the narrower copper profile, the magnetic field lines close over a shorter distance and thus, slightly reduce the power loss and finally the power consumption. Therefore, the disadvantage of an increased resistance due to a reduction of the copper cross-section only plays a minor role because of current displacement effects.
The simple approach of increasing the coil length and thus, the distance between the field-guiding elements and the bridge, also has a beneficial impact on the mechanical stresses (141 → 123.5 MPa). Raising the gap by only 3 mm from 1.5 to 4.5 mm has already decreased the load by more than 12 %. An additional increase can further reduce the stresses significantly, but by extending the length of the inductor, the hardening zone is also affected. Therefore, depending on the workpiece, a redesign of the end connections might be necessary.
Applying a copper profile which has the shape of a rounded square with 3.5 mm corner radius leads to a large reduction of the mechanical stresses in the coil (see Fig. 7). As a result, it is possible to influence the problematic zone specifically, since the current distributes across a bigger area, thus reducing the power density in the corner. In this case, the increased inductor impedance and the smaller flow cross-section lead to additional advantages due to reduced power requirement and a higher fluid velocity.
The constructively most complicated approach, using four instead of two main conductors, also has a positive effect on coil lifetime (see Fig. 8). In this case, doubling of the superimposed area between inductor and workpiece at a high coil impedance results in a significantly lower current demand which reveals a great reduction of the power density.
All calculated nominal stress amplitudes of the different simulations are summarized in a table and they are exposed with the associated Wöhler curve in Fig. 9.

Conclusion
In this paper, the issue of a short lifetime of single-shot coils for rotationally symmetrical workpieces has been presented. Some investigated, constructive approaches have shown that with little effort, thermally induced, mechanical stresses can be reduced significantly. Especially by using of copper profiles with rounded corners, service life can be improved greatly. In comparison, the meander shaped coil does not bring such a big advantage but it creates a lot more manufacturing labour. Besides that, the efficiency of this coil is about 4 % lower, compared to the reference model. Nevertheless, this variant is still an interesting approach if workpieces with a large diameter have to be hardened, resulting in a big necessary surface power density demand in the coil. In this case, if a coil with a rounded profile does not give sufficient service life, one could think of using a meander-shaped coil, maybe also with six or even eight main leads.
In the further course, the exact influences of the design adjustments have to be analysed in the context of parameter studies. Additionally, since the connection between the main conductor and the bridge is oftentimes soldered, there is usually a small radius in the corner between these two components, which influence on coil lifetime shall also be considered. Besides that, electromagnetic forces also put stress on the coil and are therefore to be investigated. Finally designing guidelines have to be defined, allowing the coil manufacturer to build the single-shot inductor according to, whereas this shall result in a long coil life time, even at high power levels.

References
1. Fluxtrol Inc.: Fluxtrol 100, https://fluxtrol.com/inc/pdf/Fluxtrol-100-Specs.pdf
2. Deutsches Kupferinstitut, Cu-HCP-Datenblatt
3. K. Svendsen, Thermal Fatigue in High Current Density Induction Coils, Dissertation (2012)

Investigation of refining mechanism in pure Al under pulsed magnetic field

S.P. Yue, Z.L. Zheng, J.C. Jie *, Z.K. Guo, T.J. Li

Key Laboratory of solidification Control and Digital Preparation Technology (Liaoning Province), School of Materials Science and Engineering, Dalian University of Technology, Dalian 116024, China
Corresponding author: Tel. / Fax: +86-411-8470-6220.

**Corresponding author: jiejc@dlut.edu.cn*

Abstract
The application of pulsed magnetic field can refine the macrostructure of metal effectively in solidification process. This paper presents experimental and numerical investigations on the refining mechanism of PMF on the solidification structure of pure aluminum. Forced convection induced by PMF was applied to the pure Al to investigate its effects on the grain refinement. Experiment results indicated that the forced convection promoted a thermal homogenization in the melt which is beneficial to the survival and growth of the nucleus and resulted in a significant refinement in the pure Al.

Keywords: Macrostructure; Force convection; Grain refinement; Aluminum

Introduction

Grain size is one of the significant factors that affect the processing and mechanical properties of material [1, 2]. Many methods can be used to refine the grain of alloy. Apart from addition of particles which can act as substrates for heterogeneous nucleation [3, 4], applying ultrasonic wave or electrical current, or magnetic field during solidification process to enhance grain nucleation and control grain growth [5]. A number of mechanisms for the grain refinement or CET under electromagnetic field have been proposed, such the fragmentation of dendrites induced due to the circulation of solute rich inter-dendritic liquid [6], the separate of dendrite fragments from the wall and mushy zone [7]. However, the influence of impurity element on the grain refinement has not clarity up to now.

In the research, the solidification structure evolution of commercial-purity aluminum and high-purity aluminum with a forced convection induced by PMF during the unidirectional solidification process was investigated for an in-depth understanding of the influence both the forced convection and the impurity elements, and the corresponding mechanism was discussed.

Experimental

Fig. 1 shows a schematic view of the experimental setup which consists of a straight solenoid connected with a self-prepared power which includes a capacitor bank of 1000 μF, temperature measurement system, cylindrical graphite mold. All experiments used a pulse frequency of 10 Hz, a half sinusoidal waveform with a peak voltage range between 0 V and 400 V and the internal was 100 V. Commercial-purity (CP) aluminum (99.7 wt. %) and high-purity aluminum (99.997 wt. %) were charged and held in a resistance furnace for 30 min at temperature 100 K higher than the melting point. Furthermore, the pouring temperature for each experiment was fixed 60 K higher than the melting point and the mold temperature was maintained at 973 K before casting. The solidification process was rendered a near isothermal condition by use of thermal insulating blanket on the lateral and bottom surfaces of the mold. The liquid melt was poured into graphite mold with an inner diameter of 45 mm and height of 100 mm which was placed on a water-cooled copper mold for 30 s, and then transferred into the PMF immediately. The K-type thermocouples were arranged vertically along the axis of the specimen to monitor the distribution of temperature during the solidification process.

The specimens were sectioned longitudinally along the mid-plane and then grounded on SiC papers and polished to remove any scratches for metallographic analysis. The etching reagent used to reveal the macrostructure was a solution of 75 ml HCl, 25 ml HNO3 and 5 ml HF. A CanoScan (LiDE120) was employed to obtain the photographs of the macrostructure. Quantitative measurements of the grain size were determined by the mean linear intercept method using Nano Measurer software package.

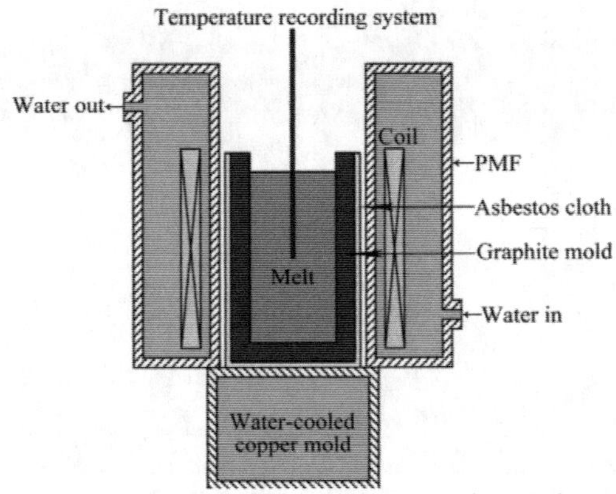

Fig. 1 The schematic view of the experimental setup

Result and Discussion

Fig. 2 is an image sequence map to show the macrostructure of the commercial-purity aluminum solidified on the water-cooled copper mold for 30 s before the PMF treatment with different excitation voltages. It can be seen that compared with the solidification structure in the absence of PMF at the bottom of the specimen, a columnar-to-equiaxed transition occurs in the upper portion of the specimen and the growth of the columnar dendrites is suppressed. By combination of Fig. 2 (a)-(d), it should be mentioned that the application of PMF during the solidification process can generate the isotropic fine equaxied grains and the average grains size decreases with increasing the excitation voltages.

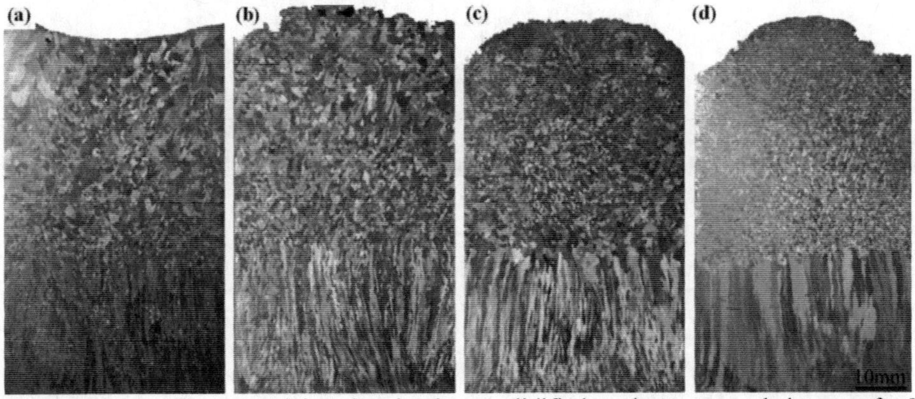

Fig. 2 The macrostructure of commercial-purity aluminum solidified on the water-cooled copper for 30 s before PMF treatment with different excitation voltages: (a) 100 V. (b) 200 V. (c) 300 V. (d) 400 V

The representative solidification macrostructure with same processing conditions are shown in Fig. 3. As can be seen from Fig. 3, a significant discrepancy exists in the solidification structure between the commercial-purity aluminum and high-purity aluminum. Fig.3 a shows a dominance of columnar grains growing opposite to the direction of heat flow and there are no fine equiaxed grains along the longitudinal section of the specimen under the PMF which the excitation voltage is 100V. With increasing the excitation voltage, fine equiaxed begin to appear in the upper portion of the specimen but its grains size is larger than the commercial-purity aluminum at the same solidification conditions.

In the convention casting process for the case in the absence of the PMF, the heat dissipation of the liquid melt near the wall of the graphite mold is fast during the solidification. Upon initiation of cooling, a certain under-cooling is establishes in the liquid, and the grains begin to nucleate first at the wall where the temperature is the lowest. The already nucleated grains continue grow opposite the direction of heat flow. However, when the PMF is imposed to the liquid, a Lorentz force will be generated by the coupling between the induction current and magnetic field which caused a forced convection in the liquid. Hence, a large number already nucleated grains originate from fragmentation both at the walls and the front of the columnar crystal will advected into the bulk melt. The schematic sketch of refinement mechanism under PMF can be illustrated as shown in Fig. 4.

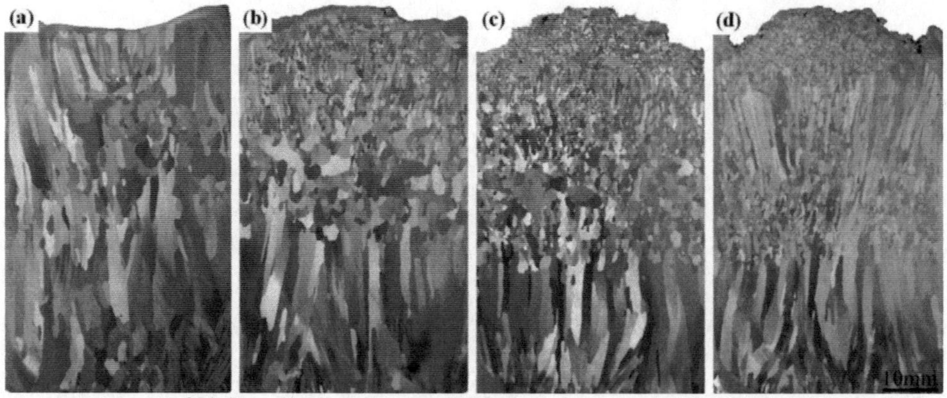

Fig. 3 The macrostructure of high-purity aluminum solidified on the water-cooled copper for 30 s before PMF treatment with different excitation voltages: (a) 100 V. (b) 200 V. (c) 300 V. (d) 400 V

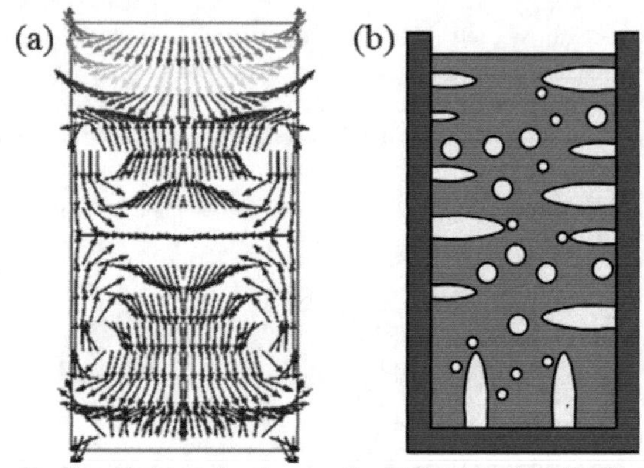

Fig .4 The schematic sketch of refinement mechanism under PMF: (a) The distribution of flow field under PMF (b) The schematic sketch of nucleus multiplication

In order to exact illustrate the refinement mechanism of pure aluminum under PMF, the distribution of two major impurity elements Fe and Si between the commercial-purity and the high-purity was experimentally determined by electron probe microanalysis (EPMA).

As can be seen in Fig. 5, compared with high-purity aluminum, the distribution of Fe element on the grain boundary is successive. According to the Al-Fe binary alloy diagrams, the intermetallic compound Al_3Fe formed at the temperature of 928 K. Due to the magnetic oscillation caused by the PMF, the Al_3Fe with low melting point will be deformed and bent, which will destroy the solidified shell. The fragmentation of solidified shell at the wall are advected into bulk melt by the convection flow.

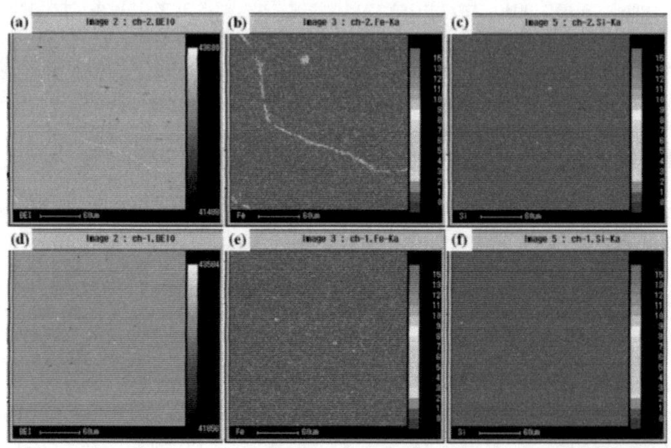

Fig. 5 The distribution of impurity elements in commercial-purity aluminum and high-purity aluminum: (a)-(c) Commercial-purity aluminum. (d)-(f) High-purity aluminum.

Submission

In the present study, the refinement mechanism of both commercial-purity aluminum and high-purity aluminum was investigated. The results show that for the case where the PMF was applied, a columnar-to-equiaxed transition occurs in the upper of the melt. Furthermore, compared with the high-purity aluminum, due to the distribution of Fe element on the grain boundary, the solidification structure of commercial-purity is smaller. It is indicated the impurity elements play an important role in the refinement of pure aluminum.

Acknowledgements

The authors gratefully acknowledge the supports of National Key Research and Development Program of China (Nos. 2017YFA0403800), the National Natural Science Foundation of China (Nos. 51771040, 51501028, 51690 163, 51525401), and Fundamental Research Funds for the Central Universities of China (No. DUT17JC44).

References

1. M.A. Easton, M. Qian, A. Prasad, D.H. Stjohn, Current Opinion in Solid State & Materials Science, 20 (2016) 13-24.
2. N. Wang, Z. Wang, K.T. Aust, U. Erb, Acta Metallurgica Et Materialia, 43 (1995) 519-528.
3. Y. Xu, D. Casari, Q. Du, R.H. Mathiesen, L. Arnberg, Y. Li, Acta Materialia, 140 (2017).
4. Y. Xu, D. Casari, R.H. Mathiesen, Y. Li, Acta Materialia, 149 (2018) 312-325.
5. D. Räbiger, Y. Zhang, V. Galindo, S. Franke, B. Willers, S. Eckert, Acta Materialia, 79 (2014) 327-338.
6. E. Liotti, A. Lui, R. Vincent, S. Kumar, Z. Guo, T. Connolley, I.P. Dolbnya, M. Hart, L. Arnberg, R.H. Mathiesen, Acta Materialia, 70 (2014) 228-239.
7. X. Liao, Q. Zhai, J. Luo, W. Chen, Y. Gong, Acta Materialia, 55 (2007) 3103-3109.

Microstructure Evolution of Cu-15Ni-8Sn Alloy Prepared by Vertical Semi-continuous Casting with EMS

Zhongkai Guo[1], Jinchuan Jie[1]*, Shipeng Yue[1], Tingju Li[1], Qingtao Guo[2]

[1]Key Laboratory of Solidification Control and Digital Preparation Technology(Liaoning Province), School of Material Science and Engineering, Dalian University of Technology, Dalian 116024, Liaoning, China
[2]State Key Laboratory of Metal Material for Marine Equipment and Application, Anshan, 114000, China

*Corresponding author: jiejc@dlut.edu.cn

Abstract
In the present study, vertical semi-continuous casting equipped with the electromagnetic stirring (EMS) was applied in order to control the solidification process and obtain the anticipated homogeneous microstructure for Cu-15Ni-8Sn alloy. The experimental results indicate that the microstructure of as-cast Cu-15Ni-8Sn (wt.%) alloy prepared by conventional casting consists of Sn-depleted α phase (α Cu(Ni) solid solution), Sn-rich γ phase ($(Cu_xNi_{1-x})_3Sn$), and intermediate transition region (α+γ), and most of the Sn element mainly segregates at dendritic grain boundary. Discriminatively, some equiaxed grains come into being in as-cast Cu-15Ni-8Sn alloy prepared by vertical semi-continuous casting, and Sn-rich γ phases distribute on the matrix and grain boundary uniformly. Furthermore, the size of equiaxed grains becomes smaller after applying EMS, which reduces the segregation of Sn element to a large degree. In this study, microstructure evolution mechanism of Cu-15Ni-8Sn alloy prepared by conventional casting and by vertical semi-continuous casting without and with EMS is analyzed.

Keywords: Cu-15Ni-8Sn, Vertical semi-continuous casting, Electromagnetic stirring, Microstructure evolution

1. Introduction
A commercial Cu-Ni-Sn alloy, strengthened by spinodal decomposition [1], is widely used in electronic industry, aerospace, mechanical systems and so on due to its high strength, high elastic, good mechanical properties, and electrical conductivity, etc [2-6]. However, some previous studies [3, 4] indicate that a serious segregation phenomenon of Sn element exists in the conventional casting process for Cu-Ni-Sn alloy as a result of the low-melting-point of Sn compared to Cu and Ni, which has a negative influence on the subsequent processing and mechanical properties. Therefore, in recent decades, great efforts have been carried out to explore for new preparation technology such as powder metallurgy [7], rapid solidification [8, 9], spray deposition [10] and so on to reduce the segregation of Sn. Although Cu-Ni-Sn alloy has been successfully prepared by such above-mentioned methods, it should be noted that the high cost and small size limitation of samples via such complex preparation process have become to the great obstacles for industrial production and application.

Therefore, the industrial production and application of Cu-Ni-Sn alloy become increasingly significant. Grain refinement is one of the alternative means to reduce the segregation. Grain refinement can be achieved by several methods, i.e., (a) the addition of grain refiner in aluminum alloys [11]; (b) the stirring of melt by mechanical equipment or electromagnetic field [12-14]. In the present study, vertical semi-continuous casting equipped with electromagnetic stirring (EMS) is applied for the industrial production of Cu-15Ni-8Sn alloy to reduce the segregation of Sn. The microstructure evolution mechanism of Cu-15Ni-8Sn alloy prepared by conventional casting and by vertical semi-continuous casting without and with EMS is analyzed. The results provide a fundamental understanding for the industrial preparation of Cu-Ni-Sn alloy, which has tremendous scientific and technological relevance.

2. Experimental
2.1 Preparation of Cu-15Ni-8Sn alloy
Cu-15Ni-8Sn (wt%) alloy billet (the cross section is 100 mm × 100 mm, the length is 500 mm) was prepared by vertical semi-continuous casting with EMS using an intermediate frequency induction furnace. The raw materials are electrolytic copper plate (99.95 wt.%), electrolytic nickel plate (99.95 wt.%) and industrial pure tin (99.95 wt.%). Industrial pure tin was added into the furnace after the electrolytic copper and electrolytic nickel were completely melting, then the temperature was held at 1350°C for 10 minutes to guarantee that the mixing melt was homogeneous. In order to avoid the oxidation of the melt surface, the carbon powder was used to cover the surface of the alloy melt. Magnesium-copper master alloy was used for degassing in the casting process. Then, the melt was poured into a crystallizer with a vibration equipment and an intermediate frequency induction coil. The coil was superimposed on a certain height of the crystallizer where the solid-liquid interface was located exactly at the height. The schematic diagram of experiment device is shown in Fig.1. The power of the intermediate frequency magnetic field are 0 KW, 5 KW, 10 KW, respectively. In addition, as a comparison experiment, the Cu-15Ni-8Sn

alloy ingot (the diameter is 40mm, the length is 200mm) was prepared by using an intermediate frequency induction furnace under argon atmosphere.

2.2 Microstructure observation and phase identification

In order to analyze the solidification behavior of Cu-15Ni-8Sn alloy, the as-obtained samples were polished and etched by a solution consisting of 95 ml C_2H_5OH + 5 ml HCl + 5 g $FeCl_3$ for morphology observation and phase identification. Microstructure was observed using an Olympus GX51 optical microscope and a scanning electron microscope (SEM, Zeiss supra 55) equipped with an energy dispersive spectrometer (EDS) operated in secondary electron mode and with an accelerating voltage of 15 kV.

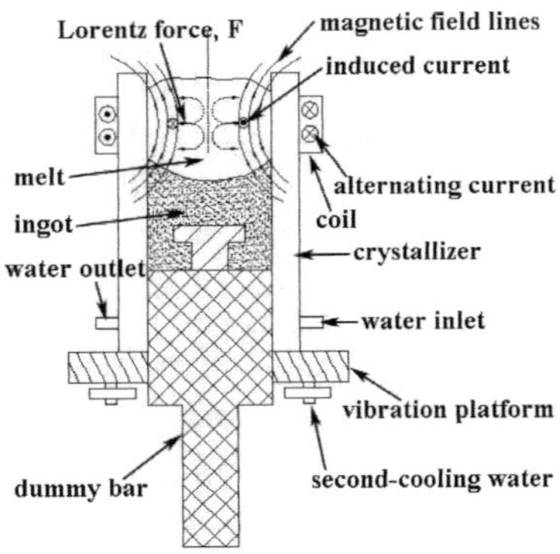

Fig. 1 The schematic diagram of experiment device

3. Results and discussion

3.1 Microstructure of as-cast Cu-15Ni-8Sn alloy prepared by conventional casting

Fig. 2 shows the microstructure of as-cast Cu-15Ni-8Sn alloy prepared by conventional casting. It can be clearly seen that the developed dendritic microstructure which exhibits various growth orientations exists in as-cast Cu-15Ni-8Sn alloy as shown in Fig. 2(a). The dark grey phase relates to matrix and the light grey phase corresponds to dendritic grain boundary. Most of Sn element segregates at the dendritic grain boundary which presents a width of hundreds of micrometers. The dendritic matrix consists of Sn-depleted α matrix, nubby Sn-rich γ phase, and slight intermediate transition region as indicated by Fig. 2(b). There are some nubby Sn-rich γ phases located at dendritic grain boundary as shown in Fig.2(c), which can be identified as $(Cu_xNi_{1-x})_3Sn$ according to the EDS result in Fig. 2(d). In addition, lamellar (α+γ) phases can also be clearly found at grain boundary, which is similar to the pearlite in steel.

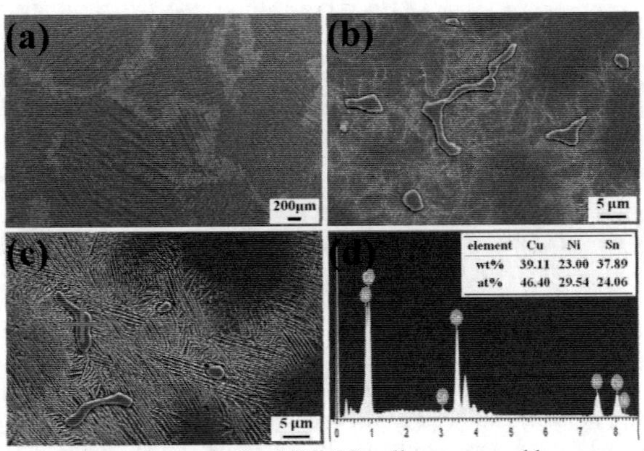

Fig. 2 (a), (b), (c) Microstructure of as-cast Cu-15Ni-8Sn alloy prepared by conventional casting; (d) the corresponding EDS result of the point in (c).

As is well known, phase diagram is an important tool for understanding and ascertaining the solidification behavior of an alloy. Fig. 3(a) shows the quasi binary phase diagram of Cu-15Ni-XSn (wt.%) calculated by CALPHAD. According to the phase diagram, the melting point of Cu-15Ni-8Sn alloy is about 1110°C, which is in agreement with that calculated by Jyrki Miettinen [15]. In addition, the temperature range of liquid-solid transformation is about 190°C (from about 1110°C to 920°C), which can be considered to be the main reason for the Sn segregation. As for equilibrium simulation, the solid fraction versus temperature curve gently falls. There is only a single α Cu(Ni) phase after the solidification is completed at about 920°C. However, the small platform on Scheil simulation curve can be ascertained as the formation temperature of γ phase, which is about 830 °C. During practical solidification process for Cu-15Ni-8Sn alloy, the firstly formed phase possesses a composition of kC_0, which is smaller than the average composition of the alloy. Consequently, the low-melting-point Sn-rich melt are pushed to dendritic boundaries, causing the segregation of Sn along grain boundaries. Most of the Sn-rich melt can solidify to supersaturated α phase, and only a few fraction of Sn-rich melt forms nubby γ phases via liquid-solid transformation. Subsequently lamellar γ phase would precipitate from supersaturated α solid solution and exhibit lamellar (α+γ) phases as indicated in Fig. 2(c). In addition, there are also some Sn-rich droplets existed in dendritic matrix due to insufficient diffusion for Sn atoms. These Sn-rich droplets would solidify to γ phases at about 830 ℃ just as demonstrated in Fig. 2(b).

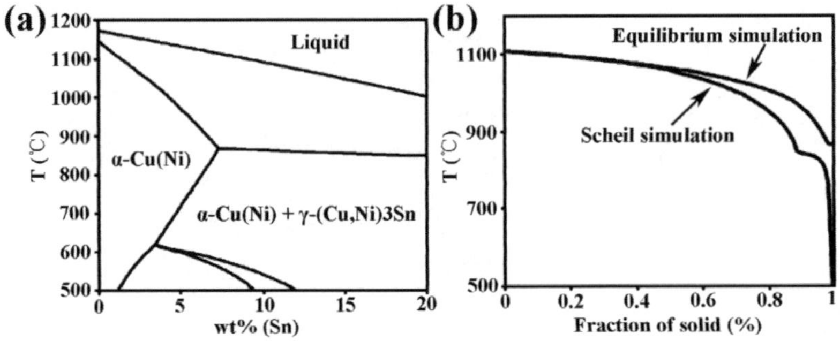

Fig. 3 (a) Quasi binary phase diagram of Cu15Ni-XSn alloy, (b) Sheil and equilibrium simulations of Cu-Ni-Sn solidification.

3.2 Microstructure of as-cast Cu-15Ni-8Sn alloy prepared by vertical semi-continuous casting without and with EMS

Fig. 4 shows microstructure of the as-cast Cu-15Ni-8Sn alloy prepared by vertical semi-continuous casting under different powers of the intermediate frequency magnetic field. Some coarse equiaxed grains come into being as shown in Fig4. (a) when the vibration caused by mechanical equipment is introduced during the casting process. The Sn-rich γ phases distribute on the matrix and grain boundary equally. Furthermore, the size of equiaxed grains becomes smaller with the increasing power of the intermediate frequency magnetic field after the EMS is applied. Especially, the size of equiaxed grains can be reduced to about 100 μm as demonstrated in Fig4. (c) when the power of the intermediate frequency magnetic field is 10 KW, which reduce the segregation of Sn to a large degree.

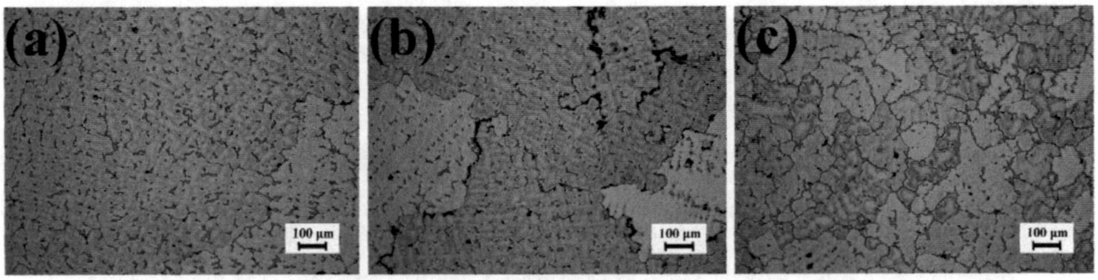

Fig. 4 Microstructures of the as-cast Cu15Ni-8Sn alloy under different powers of the intermediate frequency magnetic field: (a) 0 KW; (b) 5 KW; (c) 10 KW.

On the one hand, the vibration device of crystallizer is introduced in order to reduce the friction between the ingot surface and the wall of the crystallizer. Consequently, the ingot can be successfully prepared and the quality of the ingot surface has been improved greatly. On the other hand, the vibration of crystallizer can break up the dendrite arms and promote the formation of equiaxed grains. Moreover, the refining effect on the grain becomes more obvious after the EMS is applied. The fundamental principle of EMS is to create a Lorentz force, F, acting on the alloy melt. The alternating electromagnetic field, B, would be born when the coil is connected to an alternating current. Subsequently, the induced current, J, can be generated in the alloy melt. Lorentz force, F, can be caused by the interaction of induced current and alternating electromagnetic field. It can be given by

$$F = J \times B = \nabla\left(\frac{\mu B^2}{2}\right)\left(\frac{1}{\mu}\right) + (B \bullet \nabla)B \quad (1)$$

where the μ is the permeability. The first item on the right of the formula is the radial partial force which is vertical to the side of the melt and points to the center of the melt. The contact pressure between the melt and the wall of crystallizer can be reduced by the radial partial force, thus achieving the purpose of soft-contact electromagnetic semi-continuous casting. The second item is the cornering force caused by the edging-effect which can stir the melt by generating a vertical downward whirl. During the solidification of Cu-15Ni-8Sn alloy, dendrite arms can be fragmented by the Lorentz force and transformed through the mushy zone by forced convection. Some fragments could survive in the melt and become the cores of heterogeneous nucleation, which lead to the formation of the refined equiaxed grains. In addition, the temperature gradient could be lowered under EMS, which extends the undercooling region and contributes to the dendritic-to-equiaxed transition.

4. Conclusions:

Cu-15Ni-8Sn alloy was prepared by conventional casting and by vertical semi-continuous casting without and with EMS. The microstructure evolution mechanism is investigated in the present study. The microstructure of as-cast Cu-15Ni-8Sn alloy prepared by conventional casting consists of Sn-depleted α phase (α Cu(Ni) solid solution), Sn-rich γ phase ($(Cu_xNi_{1-x})_3Sn$), and intermediate transition region (α+γ), and most of the Sn element mainly segregates at dendritic grain boundary. Whereas, some equiaxed grains come into being in as-cast Cu-15Ni-8Sn alloy prepared by vertical semi-continuous casting, and Sn-rich γ phases distribute on the matrix and grain boundary uniformly. Furthermore, the size of equiaxed grains becomes smaller with the increasing power of the intermediate frequency magnetic field after EMS is applied, which reduces the segregation of Sn element to a large degree. The grain refinement can be ascribed to the double effects of the vibration of crystallizer and EMS. During solidification, dendrite arms can be fragmented and become the cores of heterogeneous nucleation, consequently, forming refined equiaxed grains after solidification.

Acknowledgment

The authors gratefully acknowledge the supports of National Key Research and Development Program of China (Nos. 2017YFA0403800), the National Natural Science Foundation of China (Nos. 51771040, 51690163, 51525401), State Key Laboratory of Metal Material for Marine Equipment and Application (No. SKLMEA-K201701)

References

1. L.H. Schwartz, S. Mahajan, J.T. Plewes, Acta Metallurgica, 22 (1974) 601-609.
2. W.R. Cribb, M.J. Gedeon, F.C. Grensing, Advanced Materials & Processes, 171 (2013) 20-25.
3. W.R. Cribb, F.C. Grensing, (2009).
4. W.R. Cribb, F.C. Grensing, Canadian Metallurgical Quarterly, 50 (2013) 232-239.
5. C.R. Scorey, S. Chin, M.J. White, R.J. Livak, JOM, 36 (1984) 52-54.
6. J.C. Rhu, S.K. Sang, C.J. Yun, S.Z. Han, J.K. Chang, Metallurgical & Materials Transactions A, 30 (1999) 2649-2657.
7. P. Goudeau, A. Naudon, J.M. Welter, Scripta Metallurgica, 22 (1988) 1019-1022.
8. L.E. Collins, J.R. Barry, Materials Science & Engineering, 98 (1988) 335-338.
9. L. Deyong, R. Tremblay, R. Angers, Materials Science & Engineering A, 124 (1990) 223-231.
10. R.H. Cookey, J.V. Wood, Powder Metallurgy, 33 (2013) 335-338.
11. A.L. Greer, A.M. Bunn, A. Tronche, P.V. Evans, D.J. Bristow, Acta Materialia, 48 (2000) 2823-2835.
12. T. Campanella, C. Charbon, M. Rappaz, Metallurgical & Materials Transactions A, 35 (2004) 3201-3210.
13. W.D. Griffiths, D.G. Mccartney, Materials Science & Engineering A, 216 (1996) 47-60.
14. D. Lu, Y. Jiang, G. Guan, R. Zhou, Z. Li, R. Zhou, Journal of Materials Processing Tech, 189 (2007) 13-18.
15. J. Miettinen, Calphad-computer Coupling of Phase Diagrams & Thermochemistry, 27 (2003) 309-318.

IOP Conf. Series: Materials Science and Engineering **424** (2018) 012074 doi:10.1088/1757-899X/424/1/012074

High-Density β-FeSi2 crystals with 3D alignment fabricated by an oscillating magnetic field

Keizo Ono[1], Ayumu Hashimoto[1], Keiji Kurokawa[1], Noriaki Nakatsuka[1], Kohei Morishita[1], Hideyuki Yasuda[1*]

Department of Materials Science and Engineering, Kyoto University, Kyoto 606-8501, Japan

**Corresponding author: yasuda.hideyuki.6s@kyoto-u.ac.jp.ac.jp*

Abstract
This paper demonstrates three-dimensional alignment of β-$FeSi_2$ in sintered bulk specimens. The three-dimensional alignment was achieved by using an oscillating magnetic field during slip-casting under a magnetic field of 6T. Sintering using direct electric current heating was performed in various conditions. The degree of alignment did not significantly degrade during sintering process. Apparent density became as high as 97%. The degree of alignment tended to increase as grain size increased.

Keywords: Magnetic orientation, crystallographic alignment, magnetic anisotropy, anisotropic magnetic field

Introduction
Single crystals are desired to measure physical properties and to improve materials performance. Not a few compounds are produced through the solid-solid reaction, of which rate is extremely small, comparing to that in the melt growth. Thus, it is often difficult or nearly impossible to fabricate single crystals for the compounds by melt growth processes and solidification processes. One of the possible solutions is to fabricate materials with crystallographically aligned texture by using magnetic fields.

Crystals with lower symmetry lattice structure (*i.e.* hexagonal, orthorhombic symmetry) exhibit magnetic anisotropy. If a magnetic field with sufficient intensity is applied, a certain crystallographic orientation tends to align to reduce the magnetic energy. For example, the uni-axial alignment has been realized by imposing static magnetic fields for various kinds of materials such as metallic, inorganic and organic substances [1-11]. The aligned structure with micro-meter scale was also fabricated by irradiating laser on the surface of non-equilibrium solid solution of Bi-Mn alloys [12].

The three-dimensional alignment has paid much attention, because the three-dimensional alignment [13] leads to fabrication of pseudo single crystals. The magnetic susceptibilities of a-, b- and c-axes are different each other for a material with low symmetry like the orthorhombic structure. An elliptic magnetic field, in which the direction of the magnetic field rotates with the modulated angular velocity, is used for the three-dimensional alignment. When crystals suspended in a viscous fluid cannot follow the magnetic field rotation, they are likely to rotate in accordance with the time-averaged magnetic field. If the conditions (magnetic field intensity, rotation speed, viscosity, size and density of crystalline particle and magnitude of magnetic anisotropy) satisfies the requirement, the three-dimensional alignment is achieved in accordance with the magnetic anisotropy. For example, the three-dimensional alignment has been obtained for L-alanine [13, 14] and β-$FeSi_2$ (the orthorhombic structure) [15, 16]. However, the aligned crystals suspended in a resin or textured compacts with lower density were obtained. From a viewpoint of application, it is strongly desired to fabricate t bulk materials, which consist of a single phase, with the three-dimensional alignment. Thus, it is a critical issue to fabricate the three-dimensionally aligned bulk materials.

Intermetallic compound β-FeSi2, which is produced through the peritectoid reaction (solid-solid reaction) as shown in Fig.1, has received considerable attention because of its high potential as optoelectronic and thermoelectric materials. The single crystals are

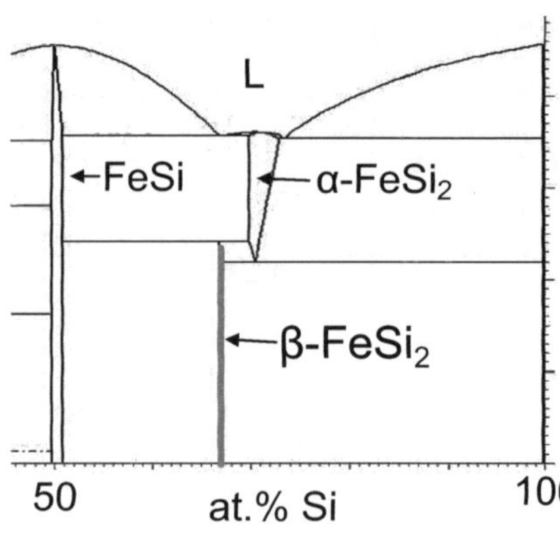

Figure 1 Phase diagram of Fe-Si system.

desired for measuring intrinsic physical properties and for improving materials performance. A motivation of this study is to fabricate the three-dimensionally aligned $FeSi_2$ with higher density. We will demonstrate processing of bulk β-$FeSi_2$ with higher density.

Experiments

Mother alloy with composition of 33.3at%Fe-66.7at%Si was made by conventional arc melting. The ingot was annealed at 1173K for 24h to proceed the solid-solid reaction. After annealing, the ingot was mechanically crashed into the particles (diameter: 2.2 μm, ranging from sub-μm to 20μm). Since grain size of β-$FeSi_2$ in the ingot was sufficiently larger than the crashed particle diameter, most of particles consist of a single grain of β-$FeSi_2$ phase.

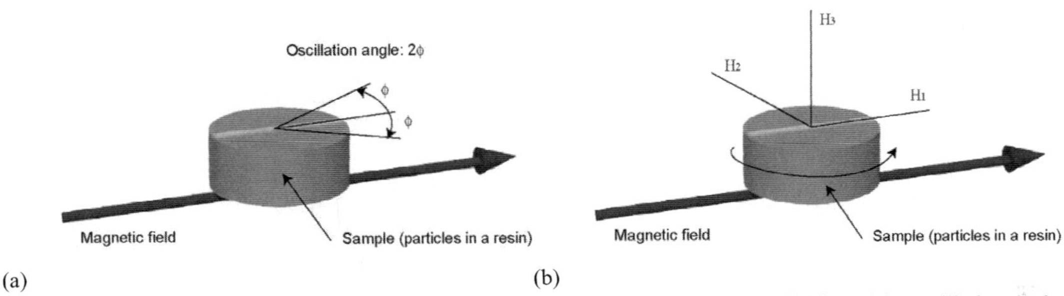

Fig. 2. Oscillating magnetic field by changing the rotation direction periodically. (a) Definition of the oscillating angle and (b) definition of axes in accordance with the oscillating magnetic field.

Figure 2 shows a schematic illustration of oscillating magnetic field (referred to as the oscillating magnetic field). The rotating direction was periodically changed. In this study, the oscillation angle 2φ and rotation speed were 135 degrees and 9 rpm, respectively. The procedures for the alignment is essentially the same as those in the previous works [14,15].

The β-$FeSi_2$ particles (10 vol%) was suspended in ethylene glycol (viscosity is 20 times larger than that of water). Ultrasonic vibration was imposed for 15min for homogenization. The slurry was cast into a mold of which bottom was made of plaster (inner diameter: 5mm). A time-dependent magnetic field (6T in H_1 direction) was imposed during slip-casting. It took 12-24 hours to remove the fluid (ethylene glycol) from the bottom plaster plate. Sedimented particles (refer to as a precursor) was obtained. Dimension of precursor was 5mm in diameter and 1-2 mm in thickness.

Molds made of graphite or WC were used for sintering. The precursor was once sintered at 1073K for 10min -30min to produce a green compact. The green compact was sintered at 1073 K to increase apparent density. of β-$FeSi_2$. Pressure of sintering ranged from 50 to 200 MPa.

Since magnetic susceptibilities of β-$FeSi_2$ are $\chi_c > \chi_b > \chi_a$ [15], the following relationship is expected for the three-dimensional alignment; H_1 // c-axis, H_2 // b-axis and H_3 // a-axis.

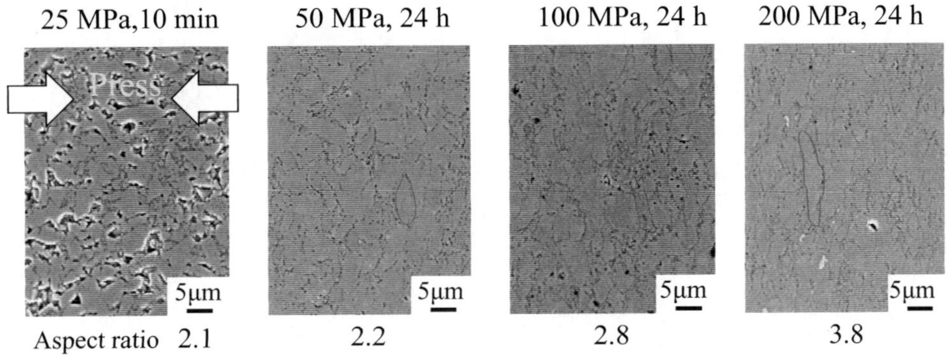

Figure 3 Microstructure of sintered specimens. Press axis is indicated by arrows on the photograph. Average aspect ratio of grains is also shown below the microstructure.

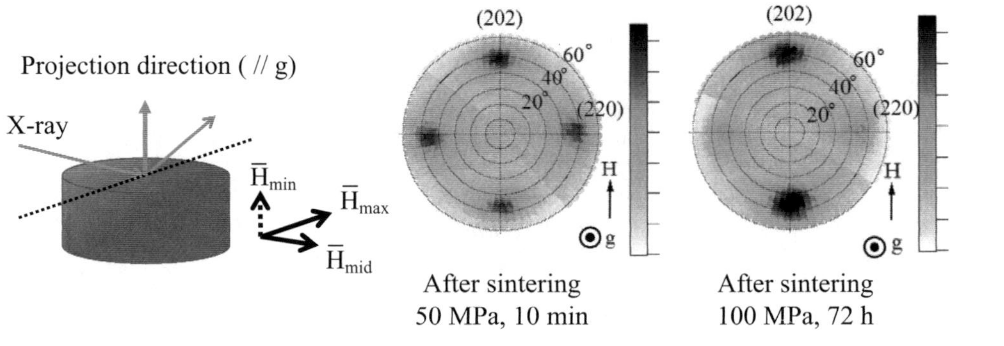

Figure 4 (a) Configuration of measurement of crystallographic orientation and (a) distribution of normal vectors of (220) and (202) planes. Four peaks on the projection indicate the three-dimensional alignment.

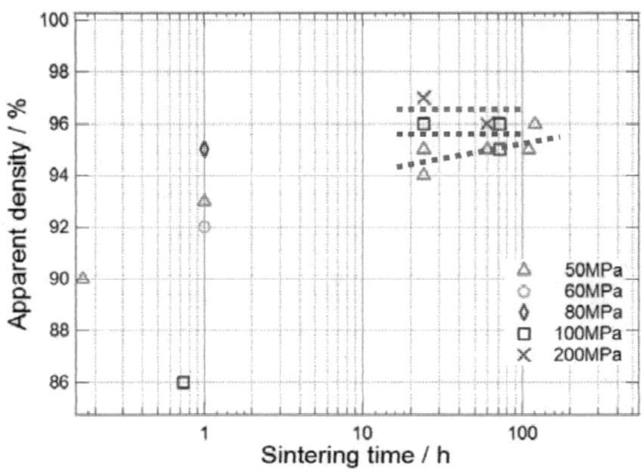

Figure 5 Apparent density change as a function of sintering pressure and duration.

Results and discussion

Figure 3 shows microstructure of sintered specimens. Pressure and sintering duration are listed in the figure. Although porosity was clearly observed at 25MPa, dense specimens were produced at a pressure more than 50MPa. In addition, aspect ratio of grains increased with increasing pressure of sintering. The change suggested that brittle deformation of β-$FeSi_2$ grains occurred during sintering and consequently contributed to decrease in porosity. Crystallographic alignment was measured by X-ray diffraction, as shown in Fig. 4. Diffraction of (220) and (202) peaks are plotted in Fig. 4(b). When the three-dimensional alignment is achieved, four peaks appears on the stereo projection. As shown in Fig. 4(b), the four peaks were clearly remained after sintering even for 72h under a pressure of 100MPa. Thus, β-$FeSi_2$ compacts with the three-dimensional alignment was fabricated by the proposed processes.

Figure 5 shows apparent density, which should be unity for β-$FeSi_2$ single crystal, as a function of sintering time. Apparent density increased with increasing sintering duration and pressure. Apparent density was less than 96% for 1hour-sintering at any pressure.

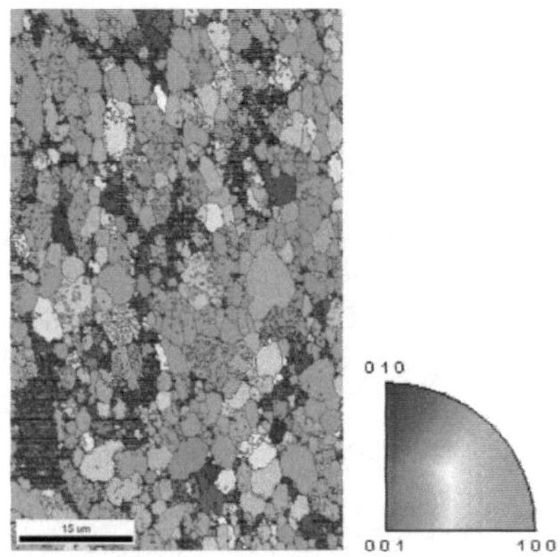

Figure 6 Inverse pole figure of the compacts sintered for 72 h, 100 MPa. The cross-section is perpendicular to the rotation axis (H_3 in Fig. 1)

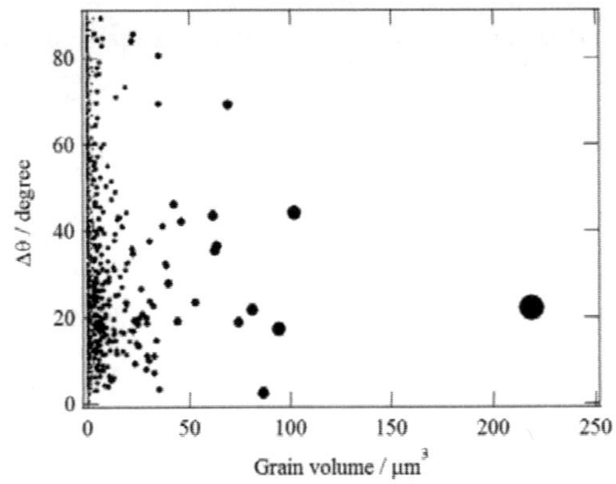

(a)

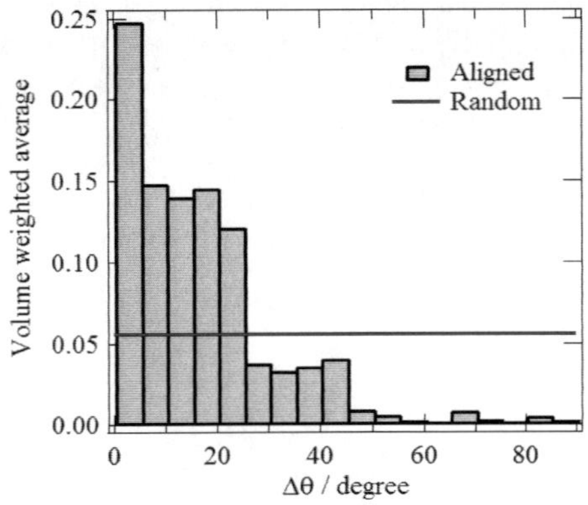

(b)

Figure 7 (a) Relationship between the misalignment (Δθ) and grain volume estimated by grain diameter and (b) volume-weighted histogram. The misalignment Δθ is the angle between [100] and H_3 defined in Fig. 1. Size of marks in (a) is proportional to grain volume.

Appropriate sintering condition (duration > 24hours, pressure > 100MPa) achieved dense compact with the three-dimensional alignment.

Figure 6 shows inverse pole figure of a compact sintered for 72 hours under 100 MPa. Green color indicates that normal direction of (100) plane is perpendicular to the cross-section (H_1-H_2 plane). The crystallographic orientation in the sintered compact agreed with the expected one. However, there are some grains with different colors such as blue, violet and pink. A misalignment (Δθ) was defined as angle difference between <100> direction and H_3 direction. Figure 7(a) shows the relationship between the misalignment (Δθ) and grain volume (estimated from grain diameters on the cross-section). Mark size is proportional to grain volume. The misalignment was likely to decrease with increasing grain volume. The tendency suggested that larger particles tend to have higher alignment in the precursor and was still observed in the sintered compacts. Although some grains have relatively high misalignment, misalignment for most of grains was less than 20 deg and clear alignment was observed in grain-scale aspect, as shown in Fig. 7(b).

Conclusions

We tried to fabricate bulk β-$FeSi_2$ compacts with the three-dimensional alignment. The alignment was initially obtained by slip-casting under the oscillating magnetic field (maximum 6T). The crystallographic alignment remained even after sintering. This study showed the sintering condition (temperature: 1073 K, pressure: 50-200 MPa and duration: 24 hours or more) for the three-dimensional alignment of β-$FeSi_2$ compacts. The apparent density was as high as 97%.

Acknowledgement

This work was partially supported by Grant-in-Aid for Scientific Research (S) , provided by JSPS, Japan.

References

1. D.E. Farrell, B.S. Chandrasekhar, M.R. DeGuire, M.M. Fang, V.G. Kogan, R.J. Klem, D.K. Finnemore, Phys. Rev. B 36 (1987) 4025
2. A. Lusnikov, L.L. Miller, R.W. McCallum, S. Mitra, W.C. Lee, D.C. Johnson, J. Appl. Phys. 65 (1989) 3136
3. P.de Rango, M. Lees, P. Lejay, A. Sulpice, R. Tournier, M. Ingold, P. Germi, M. Pernet, Nature 349 (1991) 770
4. S. Sarkar, P.S. Nicholson, Appl. Phys. Lett. 61 (1992) 494
5. R.H. Arendt, M.F. Garbauskas, K.W. Lay, J.E. Tkaczyk, Physca C 176 (1991) 131
6. A. Holloway, R.W. McCallum, S.R. Arrasmith, J. Mater. Res., 8 (1993) 727
7. S. Stassen, A. Rulmont, Ph Vanderbemden, A. Vanderschueren, Z. Gabelica, R. Cloots, M. Ausloos, J. Appl. Phys. 79 (1996) 553
8. R.W. Roberts, Phys. Rev., 104 (1956) 607
9. H. Yasuda, I. Ohnaka, O. Kawakami, K. Ueno, K. Kishio, ISIJ International, 43 (2003) 942

IOP Conf. Series: Materials Science and Engineering **424** (2018) 012074 doi:10.1088/1757-899X/424/1/012074

10. T.S. Suzuki, Y. Sakka, Jpn. J. Appl. Phys. 41 (2002) L1272
11. A. Nakahira, S. Konishi, Y. Honda, H. Yasuda, I. Ohnaka, Trnas. Mater. Res. Soc. Jpn. 28 (2003) 287
12. H. Yasuda, I. Ohnaka, Y. Yamamoto, A.S. Wismogroho, T. Takezawa, K. Kishio, Materials Transaction 43 (2003) 2550
13. T. Kimura, M. Yoshino, Langmuir 21 (2005) 4805
14. T. Kimura , F. Kimura, M. Yoshino, Langmuir 22 (2006) 3464
15. N. Nakatsuka, H. Yasuda, T. Nagira, M. Yoshiya, J. Physics: Conf. Ser., 165 (2009) 012021
16. N. Nakatsuka, K. Kurokawa, T. Nagira, M. Yoshiya, H. Yasuda, J. Iron Steel Res. Int., 19 (Suppl 1-1) (2012) 183-186

Mono-dispersed droplets formation from capillary jet of liquid metal by applying an electric field

Y. Hamaguchi[1a], K. Matsumoto[1, 2], S. Shimasaki [3b] and S. Taniguchi [4]

[1] Bachelor student of National Institute of Technology, Kagawa College, Takamatsu, Japan
[2] Now master student of Tsukuba University, Tsukuba, Japan
[3] National Institute of Technology, Kagawa College, Takamatsu, Japan
[4] Emeritus Professor of Tohoku University, Sendai, Japan

Corresponding author: [a] hamaguchi.416@gmail.com, [b] shimasaki@t.kagawa-nct.ac.jp

Abstract
Purpose of our research is to develop a manufacturing process of mono-dispersed metal balls for spherical silicon solar cell. The spherical solar silicon cell is a solar cell that uses silicon spheres of 1 mm in diameter and has drawn great attention due to its high productivity. Silicon spheres are manufactured by a jet dropping method. In this method, silicon droplets are generated from breakup of a capillary jet of molten silicon. The droplets are solidified while failing in a cooling tower, and then spherical silicon balls are obtained. However, it is found to occur that the silicon balls are not mono-dispersed due to coalescence of the droplets during falling. In order to solve this problem, we proposed the droplet electrification method by applying high voltage electric field. A repulsive force will act between each charged droplets and prevent them from coalescence. In our experiments, capillary jets of water, glycerol and molten gallium were used instead of molten silicon to investigate the effect of the electrification. The experimental results are found that coalescence frequency is depending on a dimensionless parameter, which represents a ratio of electrostatic potential over kinetic energy of droplets.

Key words: spherical silicon solar cell; droplet; applying an electric field

Introduction
Solar power generation attracts attention as an environmentally friendly power generation method. Among them, a spherical silicon solar cell as shown in Fig. 1 was developed. The spherical silicon solar cell is a solar cell that uses silicon spheres of 1 mm in diameter and has several advantages as follows: (1) Productivity is high because the spheres are directory produced from molten silicon. (2) Material yield of silicon is high because there is no cutting process of silicon ingot. (3) There is little directivity of light. The spherical silicon solar cell can accept wide angle of light. (4) A module of the spherical silicon solar cell has high flexibility. It can be fit to a curved surface.
However, the spherical silicon solar cell has still a problem of production process. A mass-production method of silicon spheres at low cost has not been established yet. In the dropping method, silicon droplets are formed by breakup of a capillary jet of molten silicon. Then the molten droplets are allowed to solidify in free fall through a cooling tower and the solidified particles are collected at a bottom of the tower. The problem of the process is that the size of the particles is not uniform, because the jet breakup is not uniform and the droplets coalesce due to velocity difference between droplets in free fall. As a result, the productivity of the spheres is decreased.

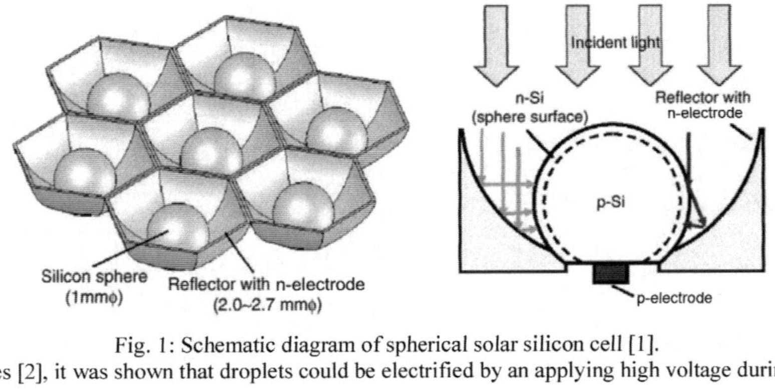

Fig. 1: Schematic diagram of spherical solar silicon cell [1].

In previous studies [2], it was shown that droplets could be electrified by an applying high voltage during disrupting of a capillary jet. Then, repulsive forces acting on electrified droplets prevented them from coalescence. In order to design manufacturing equipment, it is necessary to know a charge amount per droplet to prevent coalescence. The purpose of this research is to construct a prediction model of a charge amount per droplet to prevent coalescence. Therefore, we investigate the relationship between a charge amount per droplet and coalescence frequency.

Principle of droplet electrification

Coalescence of droplets can be prevented by electrification of droplets. The principle of droplet electrification is shown in Fig. 2. A capillary jet is surrounded by a grounded cylindrical electrode. The jet has a high-electrostatic potential, which is maintained by a high-voltage power supply of several thousand volts. The capillary jet have an equipotential surface charge, and the charge will be trapped by droplets detaching from the tip of the jet. The charged droplets have a repulsive force preventing them from coalescence.

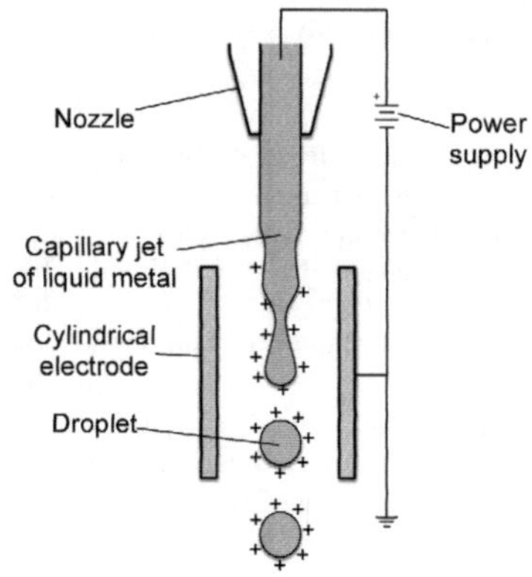

Fig. 2: Prevention of droplets coalescence by electrification [2].

Experimental method

The experimental equipment is shown in Fig. 3. The liquid, which was eclectically connected to the high-voltage power supply, was issued from the nozzle and the capillary jet was formed. The cylindrical electrode was arranged near the breakup point of the liquid jet, and the electrode was connected to ground. In this experiment, a charge amount per droplet was measured by the electrometer and motion of the droplet during dropping was observed by the high-speed camera. A coalescence frequency was measured from the image of the high-speed camera.
Experimental conditions are shown in Table 1. Cold model experiments were carried out. Water, glycerol and gallium were used as jet liquid. Effects of the experimental conditions of liquid medium, nozzle diameter, inner diameter of cylindrical electrode and applied voltage on coalescence prevention of droplets were examined.

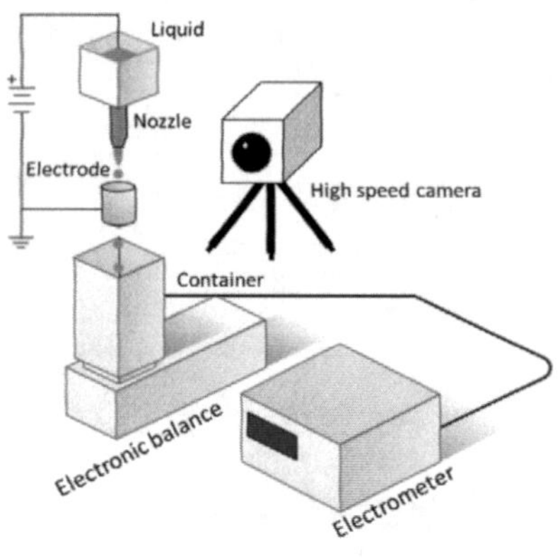

Fig. 3: Experimental equipment.

Table 1: Experimental conditions.

liquid	water, glycerol, gallium
voltage, φ / kV	0–3.0
Inner diameter of cylindrical electrode, d_e / mm	11.6, 17.5
nozzle diameter, d_n / mm	1.58, 2.14
photographing speed, F / s^{-1}	8000

IOP Conf. Series: Materials Science and Engineering **424** (2018) 012075 doi:10.1088/1757-899X/424/1/012075

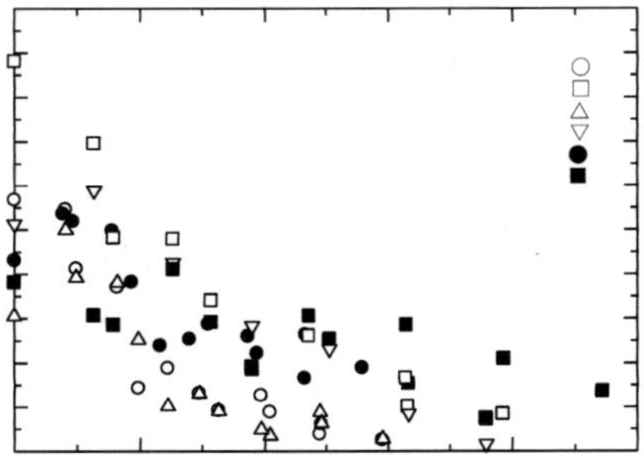

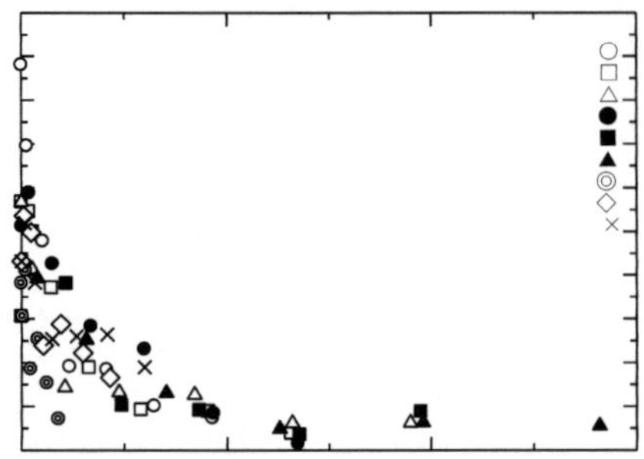

Structure Formation of Magnetic Particles under Magnetic Fields toward Anisotropic Materials

T. Ando[1], D. Katayama[1], N. Hirota[2], O. Koike[3], R. Tatsumi[4], M. Yamato[5]

[1] College of Industrial Technology, Nihon University, Narashino 275-8575, Japan
[2] Fine Particle Engineering Group, National Institute for Materials Science, Tsukuba 304-0047, Japan
[3] Products Innovation Association, Tokyo 113-8656, Japan
[4] Environmental Science Center, The University of Tokyo, Tokyo 113-8656, Japan
[5] Graduate School of Urban Environmental Sciences, Tokyo Metropolitan University, Hachioji 192-0397, Japan

Corresponding author : ando.tsutomu@nihon-u.ac.jp

Abstract
We performed numerical simulation of structure formation of magnetic particles under magnetic field toward anisotropic material development. In this study, we examined the unsteady process of the structure formation and its feature using magnetic particles with diameter of micrometers. This study evaluated the process of structure forming with various particle diameters and particle volume concentrations using the Non-dimensional Boundary Area (NBA). The structure formed by the magnetic particles does not depend on the particle diameter but depends on the particle volume concentration. In the case of particle volume concentration ϕ = 5 vol%, the bundle of chain-like cluster is not made and each chain-like cluster by magnetic particles is almost single chain. However, in the case of more than ϕ = 10 vol%, the bundle structure formed by contacts of multiple chain-like clusters is proceeding. On the process of structure formation by magnetic particles, firstly, the single particle rotates by itself with orienting the magnetic moment to the direction of the applied magnetic field. After then, particles connect with each other and the chain-like cluster is formed.

Keywords : Magnetic particles, Chain-like cluster, Anisotropic Material, Discrete element method, Simulation

Introduction

In recent material processing, the use of magnetic fields that can control the orientation of texture and structure of material without contact has received attention[1-3]. There are two methods in order to improve the mechanical properties of the material using magnetic fields. One is a method of orienting the magnetic anisotropic substances to the same direction in the material. Another is a method of controlling the direction of the structure formed by the aggregated particles in the medium. For the latter, when a magnetic field is applied to the magnetic particles dispersed in the medium, the magnetic moments of all the magnetic particles are aligned in parallel to the magnetic field, and the particles make chain-like cluster with connection in a parallel direction to the magnetic field by the magnetic dipole interaction. So, the magnetic field is useful tool on development of anisotropic materials.

In this time, we performed numerical simulations of the structure formation of magnetic particles with diameter of micrometer toward anisotropic material development under magnetic fields. We examined the unsteady process of the formation and its feature using DEM (Discrete Element Method)[4] including the magnetic effect. A lot of simulation researches for magnetic particles has been already published by A. Satoh[5]. But he deals with magnetic particles in magnetic fluids and particle size is on the order of tens of nanometers. Therefore, he calculates using Brownian dynamics method in which particles do not collide directly with each other. He also discusses mainly the rheology of magnetic fluid due to the structure of magnetic particle, and not structure formation in unsteady process. On the other hand, Kawamoto *et. al* dealt magnetic particles of 35 to 107 μm[6]. They measured the length of the chain-cluster formed by magnetic particles in air and oil under various magnetic fields by experiment and numerical simulation. However, they also didn't discuss the unsteady process of structure formation and the structure of chain-like cluster on various particle volume concentrations.

Simulation model

Figure 1 shows the schematic model of numerical simulation. The cube of simulation region is a part of the inside of the container and the periodic boundary condition for a three-dimensional system is used. Magnetic particles are randomly dispersed in the container at the initial condition. Here, it was assumed that the magnetic field gradient is negligible in the calculation region. In addition, it is assumed that the particles have a single magnetic domain with saturation magnetization. In this study, we ignored the influence of gravity. In the initial state, the position of particles and the direction of magnetic moments are given randomly. After applying the magnetic field, the direction of the magnetic moment of the particle is aligned parallel to the direction of the magnetic field by the magnetic torque as shown in Fig. 2.

In the figure, $\boldsymbol{B}$ is the direction of the magnetic flux density, and the magnetic moment $\boldsymbol{m}$ of the particle is expressed using the meridian.

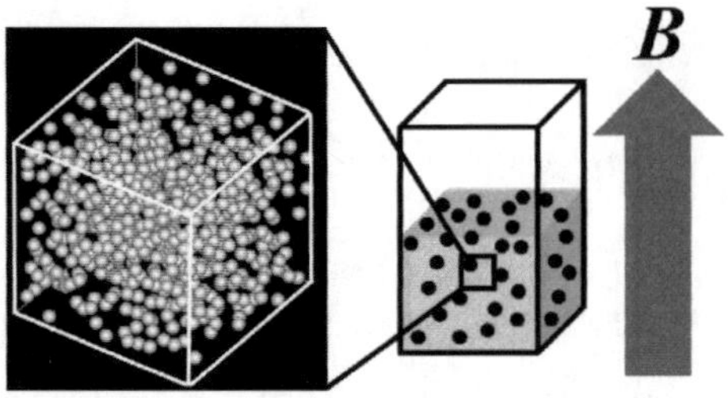

Fig. 1: Simulation model and initial state of particles.

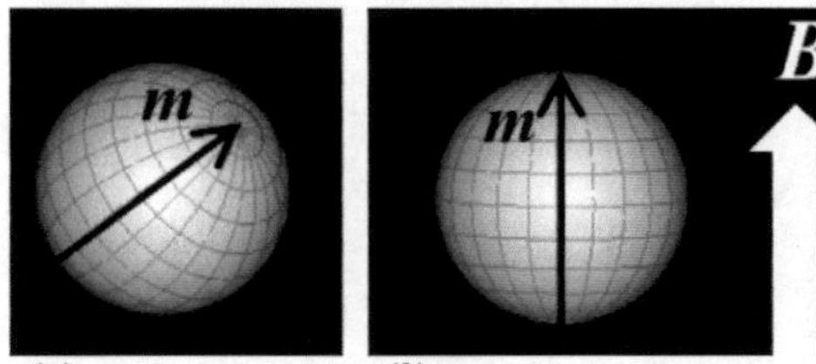

(*a*) (*b*)

Fig. 2: Direction of magnetic moment;
(*a*) Without magnetic field, (*b*) With magnetic field.

The translational motion of particles governed by Newton's equation and the rotational motion of particles are expressed by the following equations (1) and (2), respectively.

$$m_p \frac{d\boldsymbol{V}}{dt} = \boldsymbol{F}^m + \boldsymbol{F}^c + \boldsymbol{F}^w - \xi_{tra}\boldsymbol{V} + \boldsymbol{R}_{tra} \, , \tag{1}$$

$$I \frac{d\boldsymbol{\omega}}{dt} = \boldsymbol{T}^H + \boldsymbol{T}^m + \boldsymbol{T}^c - \xi_{rot}\boldsymbol{\omega} + \boldsymbol{R}_{rot} \, , \tag{2}$$

Where m_p is the mass of the particle, V is the translational velocity of the particle, t is time, $\boldsymbol{F}^m$ is the magnetic dipole interaction force, $\boldsymbol{F}^c$ is the contact force, $\boldsymbol{F}^w$ is the van der Waals force, $\boldsymbol{R}_{tra}$ is the random force by thermal fluctuation of medium, I is the moment of inertia of the particle, ω is the angular velocity of the particle, $\boldsymbol{T}^H$ is the magnetic torque due to the applied magnetic field, $\boldsymbol{T}^m$ is the magnetic dipole interaction torque, $\boldsymbol{T}^c$ is the contact torque, and $\boldsymbol{R}_{rot}$ is the random torque by thermal fluctuation of medium. The coefficients of fluid resistance for the translational and rotational motions, ξ_{tra} and ξ_{rot}, are expressed by the following equations, $\xi_{tra} = 6\pi\eta r$ and $\xi_{rot} = 8\pi\eta r^3$, respectively. Here r is the particle radius and η is the viscous coefficient of the solvent. Note that all the variables in bold represent vectors.

The interactions $\boldsymbol{F}^m$ and $\boldsymbol{T}^m$ acting between magnetic dipoles, which form a coupling structure of particles, are expressed by Eqs. (3) and (4)[3)]. The magnetic field Torque $\boldsymbol{T}^H$, which align the direction of magnetic moment of particles, is expressed by Eq. (5)[3)].

$$\boldsymbol{F}^m = \sum_{i \neq j}^{N} \left[-\frac{3\mu_0 |\boldsymbol{m}_i|^2}{4\pi l_{ij}{}^4} [-(\boldsymbol{n}_i \cdot \boldsymbol{n}_j)\boldsymbol{t}_{ij} + 5(\boldsymbol{n}_i \cdot \boldsymbol{t}_{ij})(\boldsymbol{n}_j \cdot \boldsymbol{t}_{ij})\boldsymbol{t}_{ij} - \{(\boldsymbol{n}_j \cdot \boldsymbol{t}_{ij})\boldsymbol{n}_i + (\boldsymbol{n}_i \cdot \boldsymbol{t}_{ij})\boldsymbol{n}_j\}] \right], \tag{3}$$

$$\boldsymbol{T}^m = \sum_{i \neq j}^{N} \left[-\frac{\mu_0 |\boldsymbol{m}_i|^2}{4\pi l_{ij}{}^3} \{\boldsymbol{n}_i \times \boldsymbol{n}_j - 3(\boldsymbol{n}_i \cdot \boldsymbol{t}_{ij})\boldsymbol{n}_i \times \boldsymbol{t}_{ij}\} \right], \tag{4}$$

$$\boldsymbol{T}^H = \mu_0 \boldsymbol{m} \times \boldsymbol{H} \, , \tag{5}$$

Where N is the number of particles to be calculated, μ_0 is the space permeability, $\boldsymbol{m}_i$ is the magnetic moment of the particle, $\boldsymbol{n}_i$, $\boldsymbol{n}_j$ are the unit vector ($\boldsymbol{n} = \boldsymbol{m} / |\boldsymbol{m}|$), l_{ij}, $\boldsymbol{t}_{ij}$ are the interparticle distance and the unit vector in that direction, and $\boldsymbol{H}$ is the applied magnetic field. Here, subscripts i and j are used to distinguish particles. The magnetic moment of the particle is expressed as $\boldsymbol{m} = \boldsymbol{M}V_p$, where the magnetization vector of the particle is $\boldsymbol{M}$ and the particle volume is V_p. Therefore, the force and torque by the magnetic effect depend on the particle diameter.

Simulation condition and evaluation method

Table 1 shows physical properties of particle and solvent. The particle is a spherical body with physical properties identical to those of a nickel particle. The solvent is assumed to be an ultraviolet curable resin. Table 2 shows the simulation condition. The computational domain is $(x, y, z) = (30d_p, 30\ d_p, 30\ d_p)$ and the periodic boundary condition is employed. It was assumed that the nickel particles had the saturation magnetization under this applied magnetic field $|\boldsymbol{B}| = 1$ T. In this study, the particle diameter d_p and the particle volume concentration ϕ are variable numbers.

Table 1 Physical properties.

Nomenclature		Value
Particle (Nickel)	Density ρ_p	8.9×10^3 [kg/m^3]
	Saturation magnetization $\|\boldsymbol{M}\|$	490.1×10^3 [A/m]
	Diameter d_p	2.5, 10 [μm]
Solvent	Viscosity η	20 [mPa·s]

Table 2 Simulation conditions.

Nomenclature	Value
Computational domain	$30d_p\times30d_p\times30d_p$
Cutoff distance R_{coff}	$10d_p$
Applied magnetic flux density $\|\boldsymbol{B}\|$	1 [T]
Temperature T	293 [K]
Particle volume concentration ϕ	5, 10, 15 [vol%]

As the evaluation method of the structure formed by the magnetic particles, we introduce the Non-dimensional Boundary Area (NBA) defined by Eq. (6).

$$\mathrm{NBA} = \frac{1}{N_{all}}\left\{\frac{1}{c_{max}}\sum_{c=0}^{c_{max}}(c_{max}-c)n(c)\right\}, \tag{6}$$

Where N_{all} is the total number of particles, and c is the coordination number. The coordination number is maximum value $c_{max} = 12$ when all the particles are perfect spheres and the all particle is same size. $n(c)$ denotes the number of particle at the coordination number c. The NBA becomes close to 1 when the particles are dispersed in the solvent, and the NBA becomes close to 0 when the particles are aggregated. If the particles form a complete one-dimensional structure, that is, the contact number of all the particles is 2, the value of NBA is 0.8333. Therefore, it means that structure of bundle formed by contacts of multiple chain-like clusters is proceeding when the value of NBA becomes 0.8333 or below.

We define the degree of orientation of the particle by the following Eq. (7).

$$\overline{\cos\theta} = \frac{1}{N_{all}}\sum_{i=1}^{N_{all}}[\boldsymbol{h}\cdot\boldsymbol{n}_i], \tag{7}$$

Where $\boldsymbol{h}$ is a unit vector in the direction of the applied magnetic field ($\boldsymbol{h} = \boldsymbol{H} / |\boldsymbol{H}|$). In the case of $\overline{\cos\theta} = 1$, it means that the directions of magnetic moment of all the particles are aligned with that of the applied magnetic field.

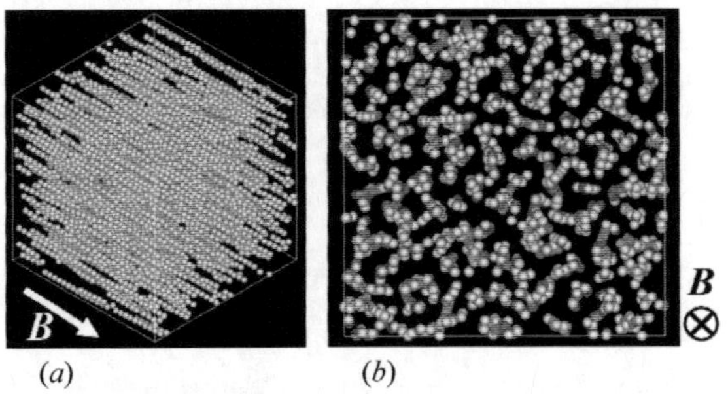

Fig. 3: Simulation result of magnetic particles (d_p =2.5 μm, ϕ = 10 vol%); (*a*) Isometric view, (*b*) Cross-sectional view perpendicular to the magnetic field.

Result and discussion

As an example of these simulation results, Fig. 3 shows the structure formed by magnetic particles with a particle diameter d_p = 2.5 μm and a particle volume concentration ϕ=10 vol% under the applied magnetic filed $|\boldsymbol{B}|$ = 1 T. Figure 3 (*a*) shows that the particles form a coupling structure in the direction parallel to the applied magnetic filed, and Fig. 3 (*b*) shows that the respective coupling structures are hold with some intervals.

Figures 4 and 5 show the time courses of NBA for the particle diameters of d_p = 2.5 μm and 10 μm, respectively. The representative time t_0 is derived from the magnetic dipole interaction force F_0 and the mass of the particle m_p ($t_0 = (m_p d_p/|\boldsymbol{F}_0^m|)^{1/2}$). Here, $F_0(=|\boldsymbol{F}_0^m|)$ is the repulsion force if the magnetic dipole moments of two particles are same direction and the distance between the surfaces of these two particles is 0. From Figs. 4 and 5, it is seen that both values of NBA for the particle volume concentration ϕ = 5 vol% are around 0.833 and that each chain-like cluster by magnetic particles is almost single chain. However, in the case of more than ϕ = 10 vol%, the values of NBA are significantly lower than 0.833, and the bundle structure formed by contacts of multiple chain-like clusters is proceeding. The structures formed by the magnetic particles with diameters d_p = 2.5 μm and 10 μm are almost same. From these results, it was confirmed that the structure depends only on the particle volume concentration in the medium.

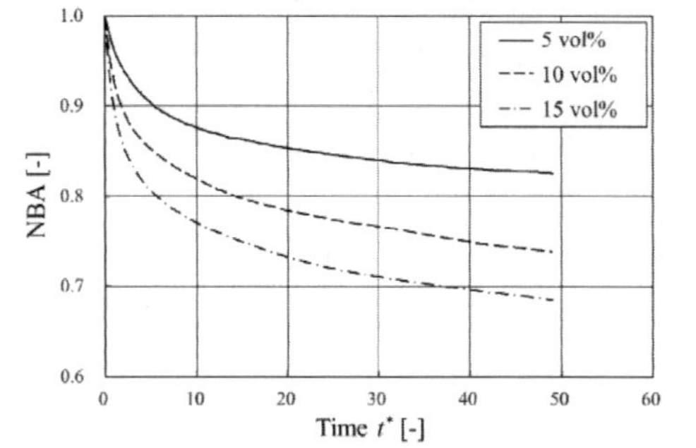

Fig. 4: Time course of NBA for particle diameter of d_p = 2.5 μm.

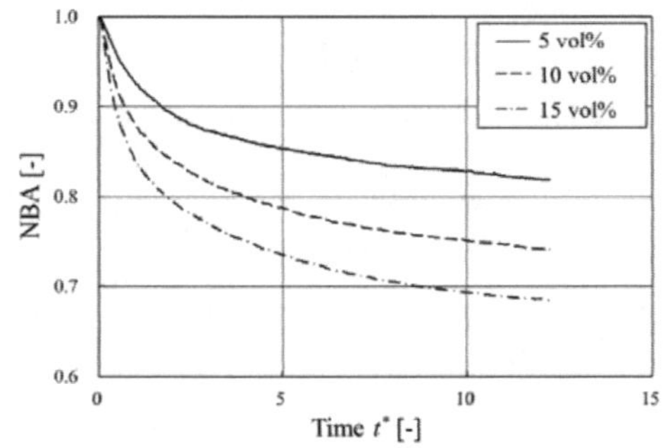

Fig. 5: Time course of NBA for particle diameter of d_p = 10 μm.

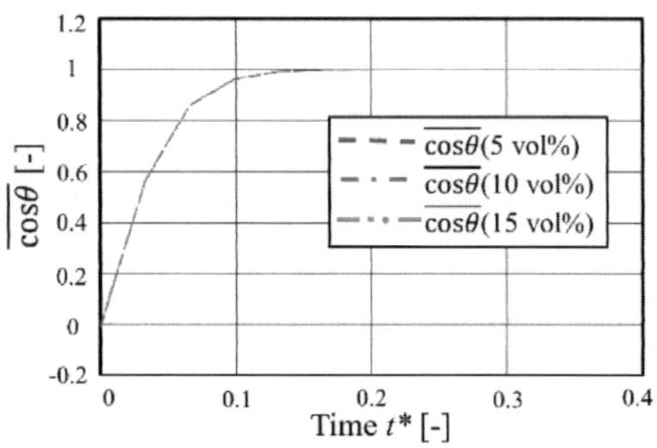

Fig. 6: Time course of particle orientation, $\overline{\cos\theta}$, for d_p=2.5 μm .

Figure 6 shows the time course of the particle orientation, $\overline{\cos\theta}$, for particle diameter d_p=2.5 μm. This result shows that three lines for different particle volume concentration are overlapped. Therefore, it is seen that the particle orientation does not depend on particle volume concentration. Compared with curves of NBA in Fig. 4, curves of $\overline{\cos\theta}$ in Fig. 6 are saturated quickly. Therefore, we can see that there are two steps when magnetic particles form a structure under the applied magnetic field. Firstly, the single magnetic particle rotates by itself with orienting the magnetic moment to the direction of the applied magnetic field. After then, particles connect with each other and the chain-like cluster is formed. In addition, in the case of the particle volume concentration more than ϕ=10 vol%, the bundle formation by contacts of multiple chain-like clusters occurs after the single chain-like cluster structure, or occurs at the same time as the structure formation of the chain-like cluster.

Conclusion

We conducted a numerical simulation of structure formation by magnetic particles under magnetic fields toward anisotropic material development. The three-dimensional simulation with various particle diameters and particle volume concentrations was carried out using DEM. The following results were confirmed in the present study under the condition that the gravity force was ignored.

1. The structure formed by particles, that is the non-dimensional Boundary Area (NBA), does not depend on the particle diameter but depends on the particle volume concentration.
2. In the case of particle volume concentration ϕ = 5 vol%, the bundle formation of chain-like cluster is not occurred and each chain-like cluster by magnetic particles is almost single chain. However, in the case of more than ϕ = 10 vol%, the bundle formations are proceeding.
3. On the process of structure formation by magnetic particles, firstly, the single magnetic particle rotates by itself with orienting the magnetic moment to the direction of the applied magnetic field. After then, particles connect with each other and the chain-like cluster is formed.

References

1. T. Kimura, Polymer Journal, 35 (2003), 823-843.
2. M. Yamato, *et. al*, Polymer, 55 (2014), 6546-6551.
3. S. Tanaka, T. Takahashi, K. Uematsu, Journal of the European Ceramic Society, 34 (2014), 3723-3728.
4. M. Fujita and Y. Yamaguchi, Journal of Chemical Engineering of Japan, 39 (2006), 83-89.
5. A. Satoh, Introduction to Molecular-Microsimulation of Colloidal Dispersions, Elsevier, (2003).
6. H. Kawamoto, *et. al*, Journal of Advanced Mechanical Design, Systems and Manufacturing, 1 (2007), 669-680.

Structural Transformation of Bilayer Ferro-foams Caused by Homogeneous Static Magnetic Field

Wang Chen[1], Hua-Qiao Dong[1], Song-Bao Wang[1], Zuo-Sheng Lei[1*], Jia-Hong Guo[2]

[1] State Key Laboratory of Advanced Special Steel & Shanghai Key Laboratory of Advanced Ferro-metallurgy & School of Materials Science and Engineering, Shanghai University, Shanghai 200072, China
[2] Shanghai Institute of Applied Mathematics and Mechanics, Shanghai University, 20072, Shanghai

* *Corresponding author: lei_zsh@staff.shu.edu.cn*

Abstract
Structural transformation of honeycomb structure bilayer ferro-foams caused by homogeneous magnetic field is investigated in this paper. Previous researchers have proved that monodisperse bilayer foams with honeycomb structure can transform reversibly by changing foam liquid fraction. However, we have observed that, when applying a homogeneous magnetic field, bilayer foams can also perform this transformation at a constant liquid fraction. This self-organizational behavior of foam structure is believed to be based on energy minimization principle, the magnetic interaction of Ferro-foams Plateau borders change the total energy of the whole system, which trigger the structure transformation process.

Keywords: Honeycomb Structure; Magnetic interactive potential energy; Ferro-Foams.

Introduction

Monodisperse bilayer foams are composed of series of identical units. There are two kinds of these units. One is honeycomb unit, which is the basic cell constituting honeycombs in nature, and it is once believed to have minimal surface area with given volume [1, 2]. The other is discovered by L.Fejes Tóth[3] in 1964. It is proved to be 0.35% less in surface area than that of honeycomb unit, and now named Tóth unit, as shown in Fig. 1（We call the monodisperse bilayer foams structure composed of honeycomb units honeycomb structure and Tóth units Tóth structure below. But no counterpart in nature can be found of the Tóth structure. Until 1994, Weaire and Phelan[4] find Tóth structure in dry bilayer soap foams for the first time, and also realize that Tóth structure transform to honeycomb structure by adding liquid to the dry foams. Researchers believe this structural transformation is a conduct of foam self-organization for minimizing system energy. Afterwards, Reinhard Höhler et al[5] analyze the energy of ordered monodisperse foam systems, and conclude an energy(**E**)-liquid fraction(**Φ**) relation. It shows that, foams prefer Kelvin structure with low liquid fraction and fcc structure with high liquid fraction. There exit a critical transformation liquid fraction point between these two structures. It is believed that foam structures are chose by minimizing their energy. So, is it possible to realize structure transforms by exerting a kind of energy into the foam system while keep the liquid fraction constant?

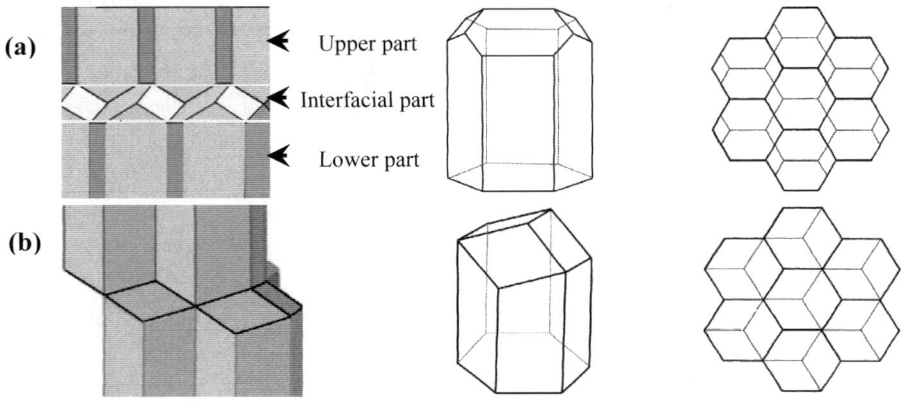

Fig. 1: (a) Tóth unit, vertical distribution and a view of such units projected on a plane; (b) honeycomb unit, Vertical distribution and a view of such units projected on a plane.

In this paper, we prepare monodisperse bilayer ferro-fluid foams at certain specific liquid fractions in Hele-Shaw cells. Then a vertical homogeneous magnetic field is imposed to observe the foam structural transformation behavior. Results show that foam unit cell changes from Tóth unit to honeycomb unit at a specific liquid fraction. And when removing the field, foam restores to its initial Tóth unit. A possible explanation based on foams system energy analysis was given. This study opens possible applications in the field of metal foam preparation, but also as a model experimental system to study the structrue contolling in the foam physics.

Method

The homogeneous magnetic field is generated by Helmholtz double coils[6], as shown in Fig. 2. The liquid is ferro-fluid and the parameters concerned are shown in Table.1 below. Surfactant (Sodium dodecyl benzene sulfonate, SDBS) oncentration is set at 0.4g/L to maintain foams stabilization.

Table.1 Physical parameters of magnetic fluid

Carrier fluid	Color	Saturation magnetic susceptibility(A/m)	Density (kg/m^3)	Initial density magnetic susceptibility(A/m)	Surface tension (N/cm)
water	black	15915	1180	0.6	2.6×10^{-4}

The bilayer foams are prepared in a Hele-Shaw cell (30mm wide × 60mm high × 3mm thick). As illustrated in Fig. 2, a flat syringe needle is used to control bubble sizes. And it is connected to a gas flowmeter. In this experiment, inner diameter of the needle is 0.41mm, and gas flow is set at 5ml/min. In this case, the generated monodisperse bubbles[7] will fill the Hele-Shaw cell in two layers. Then, the cell is placed horizontally on the symmetry plane of Helmholtz double coils which provides a homogeneous magnetic field. Experimental phenomena are recorded from the top by a digital camera.

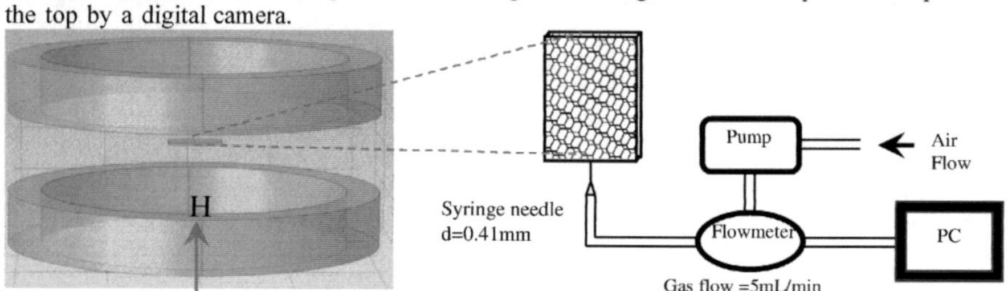

Fig. 2: Helmholtz double coils and foam preparation device system.

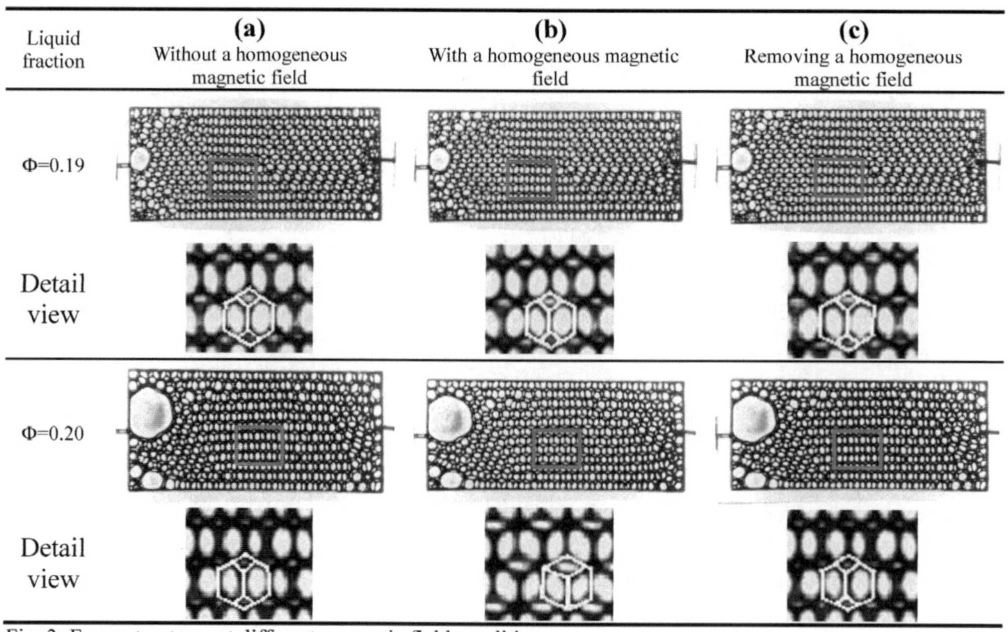

Fig. 3: Foam structures at different magnetic field conditions.

Results & Discussions

In this experiment, bilayer foams at different liquid fractions are prepared by foam preparation device. Foam transformation critical value is found at about 0.28. The value of the liquid fraction is obtained from measurements of the volume of Hele-Shaw cell and the volume of containing ferro-fluid. This estimate may not be entirely accurate, but can at least serve as a nominal value for future research. Afterwards, we have observed foam structures at certain specific liquid fractions before or after applying a homogeneous magnetic field. And the results show that, when liquid fraction is below 0.19(including 0.19), bilayer foams structure take the form of Tóth unit whether applying a homogeneous magnetic field or not. When liquid fraction comes up to 0.20, foams still remain as Tóth structures before applying a magnetic field. But, after applying the field, foams transform from Tóth structures to honeycomb structures. When removing the field, foams restore to their initial Tóth structures, which are shown in Fig. 3.

The total free energy is supposed to determine foam structures. So when applying homogeneous magnetic field to foam systems its effect should be taken into considerations on the transformation process.

When magnetic substances are put in a magnetic field, they will be magnetized to generate induced magnetic fields [8-10]. After that, these induced fields will interact leading to system energy change, which can be considered as magnetic moment interaction. In our experiment, the magnetic substance is ferro-fluid, which is mainly existed in plateau borders. So, plateau borders can be viewed to be composed of dozens of moment units.

As to bilayer ferro-foams, they can be divided into three layers in vertical direction, which are upper plateau borders, lower plateau borders and middle interfacial plateau borders in Fig. 1. We just take the upper and lower parts into consideration when figuring out magnetic interactive potential energy, because the middle interfacial plateau borders occupy a low percentage in height. The total magnetic interactive potential energy consists of interactive energy of plateau borders in same layers and different layers. As to plateau borders in same layer, the interactive energy of these two types of foams are equal. However, when considering the energy generated by plateau borders in different layers, border spacing of Tóth foams is different from that of honeycomb, as shown in Fig. 4, which leads to the difference of magnetic interactive potential energy. And this is the basic reason that causes the difference of foam system energy.

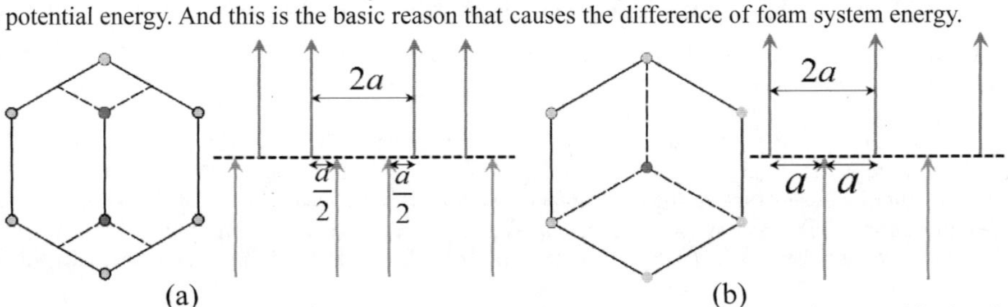

Fig. 4: (a) top view of a Tóth unit and its longitudinal section; (b) top view of a honeycomb unit and its longitudinal section

Given to small interfacial part, we views it as a plane when analyzing average plateau border spacing in the longitudinal section of bilayer ferro-foams, as shown in Fig. 4. Average plateau border spacing is defined as the average distance d of two adjacent plateau borders in a regular hexagon prism bottom S, which can be written as:

$$n\pi \cdot \left(\frac{d}{2}\right)^2 \approx S \tag{1}$$

Sides of foam units are obtained by image analysis. In our experiment, the side of a regular hexagon prism bottom is 0.5mm. Then, the average plateau border spacing d_{TH} of Tóth foams is 0.45mm, and d_{HC} is 0.53mm.

For any magnetic mass point dL_2 in plateau border L_2 and dL_1 in plateau border L_1. Supposing that, there is a magnetic moment $\vec{m}_2$ at dL_2 and another magnetic moment $\vec{m}_1$ at dL_1. These two magnetic moments will interact to generate magnetic interactive potential energy. Then the total magnetic interactive potential energy $W_{L_1-L_2}$ of plateau border L_1 and L_2 can be solved. For Tóth structure foams, the energy between two plateau borders is

$$W_{L_1-L_2} = -\frac{\mu_0 m_1 m_2}{4\pi} \times 1.07 \tag{2}$$

And for honeycomb structure foams, the energy between two plateau borders is

$$W_{L_1-L_2} = -\frac{\mu_0 m_1 m_2}{4\pi} \times 0.81 \tag{3}$$

Let us note that among equation (2) and (3), the negative sign represents attractive energy. Obviously, the absorbed energy of Tóth structure foams is bigger than that of honeycomb structure foams in two plateau borders model. This is the same in the whole bilayer foam systems for plateau borders of different layers. Since the interactive energy of these two types of foams are equal in same layers, we can conclude Tóth structure foams absorb more magnetic interactive potential

energy than honeycomb structure foams in whole bilayer foam systems, denoted $W_{TH} > W_{HC}$.
To stack bilayer foams in form of Tóth structure or honeycomb structures a question of energy. As shown in Fig. 5, the black curves are energy trends of these two structures as foam liquid fraction changes without magnetic field[5]. Since

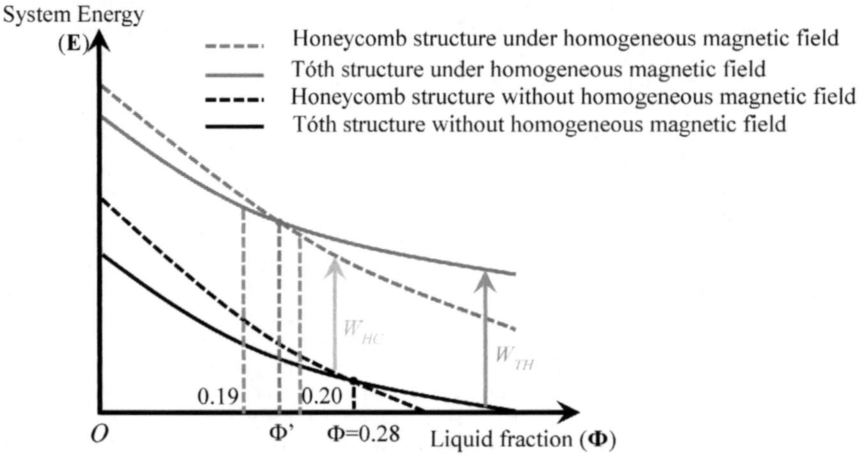

Fig. 5: A predicted relation between system energy and bilayer foams liquid fraction after applying a homogeneous magnetic field (red curves) and a known relation between them without magnetic field (black curves).

foam system inclines to stay in a minimal energy , bilayer foams present different structures at different liquid fraction range. According to the black curves, bilayer foams will change their structures when foam liquid fraction above 0.28. However, when applying a homogeneous magnetic field, the foam at liquid fraction 0.20. Which should take the form of Tóth structure without magnetic field, transform into honeycomb structure. This phenomenon is novel but can be explained by foam system energy absorption. As introduced above, after applying a homogeneous magnetic field, the absorbed energy of bilayer Tóth structure foams is bigger than that of honeycomb structure foams. As shown in Fig. 5, the red curves present the foam system energy after applying a homogeneous magnetic field. Due to$W_{TH} > W_{HC}$, the energy curve of Tóth structure foams lifts higher than that of honeycomb structure foams, which results that the critical transformation liquid fraction shift left, as Φ' shown in Fig. 5. In our experiment, after applying a homogeneous magnetic field, bilayer foams at liquid fraction 0.19 do not change their structures, while bilayer foams at liquid fraction 0.20 change. We can conclude that, after applying a homogeneous magnetic field, the critical transformation liquid fraction is in the range of 0.19 to 0.20. Comparing to the critical transformation liquid fraction (Φ=0.28) without the field, the transformation liquid fraction is 8-9% ahead.

Conclusion

Introducing a homogeneous magnetic field will lead to a structural transformation at a certain specific liquid fraction below the critical points, and this transformation can be explained by magnetic interactive potential energy generated by magnetic moments.

Acknowledgment

This project is financially supported by National Science Foundation of China (NO.51274137).

References

1. Mackenzie D. P, Science, 1999, 285(5432): 1338-1339.
2. Weaire D and Aste T, CRC Press, 2008.
3. Tóth L.F, Bull Am Math Soc, 1964, 70(4): 468-481.
4. Weaire D and Phelan R, Nature, 1994, 367: 123.
5. Höhler R, Yip Cheung Sang Y, Lorenceau E et al, Langmuir, 2008, 24(2): 418-425.
6. Boyer F and Falcon E, Physical review letters, 2008, 101(24): 244502.
7. Drenckhan W, Elias F, Hutzler S et al, Journal of applied physics, 2003, 93(12): 10078-10083.
8. Takayama T, Ikezoe Y, Uetake H et al, Physica B: Condensed Matter, 2004, 346: 272-276.
9. Hirota N, Takayama T, Beaugnon E et al, Journal of magnetism and magnetic materials, 2005, 293(1): 87-92.
10. Ando T, Hirota N, Satoh A et al, Journal of magnetism and magnetic materials, 2006, 303(1): 39-48.

Multi-frequency inductive system for magnesium level detection in a titanium reduction reactor

N. Krauter[1], F. Stefani[1], T. Gundrum[1], T. Wondrak[1], P. Frick[2] and R. Khalilov[2]

[1] Helmholtz Zentrum Dresden Rossendorf (Department of Magnetohydrodynamics, Dresden, Germany)
[2] Institute of Continuous Media Mechanics (Russian Academy of Sciences Ural Branch, 614013 Perm, Russia)

Corresponding author: n.krauter@hzdr.de

Abstract
The inductive measurement of the magnesium level in a titanium reduction reactor is a challenging task because the state of the art measurement techniques are hampered by the formation of titanium sponge rings within the reactor. By distorting the magnetic field between excitation and sensor coils, which is used to determine the magnesium level, these rings lead to a considerable measurement error. We present a novel approach to magnesium level detection using the existing infrastructure of the titanium reduction reactor, while considering the unknown size, position and electrical conductivity of the titanium sponge ring [1]. Based on numerical simulations, we demonstrate how to solve the resulting inverse problem by applying a look-up-table method comprising tens of thousands of pre-calculated parameter combinations.

Keywords: Titanium, Kroll process, level detection, inductive measurements

Introduction

The Kroll process is commonly used to produce titanium by reducing titanium tetrachloride with magnesium inside a steel crucible (reduction reactor) at temperatures of about 850 °C [2]. The corresponding reaction equation is called Kroll reduction reaction [3]: $TiCl_4 + 2Mg \rightarrow 2MgCl_2 + Ti$. Induction coils, which are located close to the crucible, are used to heat the Mg, additional heat is generated by the exothermic reaction of $TiCl_4$ and Mg. The $TiCl_4$ has to be fed slowly onto the molten magnesium, therefore the whole process typically takes a few days. Solid titanium is formed at the bottom but also at the walls of the reduction reactor [4] and will be processed further after the Mg and $MgCl_2$ have been removed. In order to control the process, it is important to know certain parameters like the present temperature or the level of molten Mg within the reduction reactor. This article will focus on a new approach to reliably determine the position of the Mg surface via inductive methods. Because of the extreme conditions inside the crucible and also to avoid contamination of the Ti, the Mg level z_{MM} has to be measured from outside. For this, we propose the following measurement setup (Fig. 1, Fig. 2):

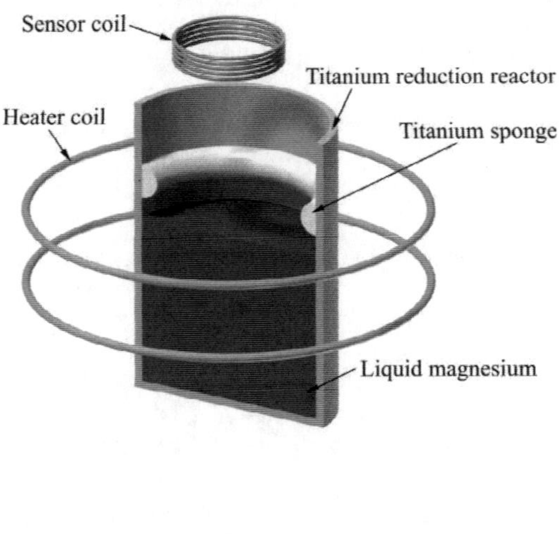

Fig. 1: Simplified illustration of the titanium reduction reactor and the corresponding measurement setup.

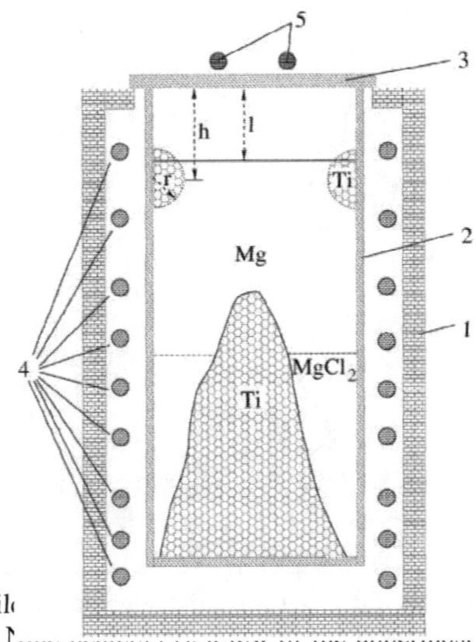

Fig. 2: Detail[illegible] covered by a lid (3). [illegible] ones are utilized as excitation coils. A sensitive receiver coil (5) is positioned above the retort.

A sensitive sensor coil is positioned above the reactor, while the heater coils are used as excitation coils with different frequencies. This setup has the advantage that parts of the already existing infrastructure can be used in order to perform the measurements. Previous inductive approaches for measuring z_{Mg} produce increasing measurement errors when the Ti sponge forms at the reactor walls [5]. This Ti ring strongly impacts the measurements since it has an electrical conductivity σ_{Ti} in the same order of magnitude as the liquid Mg. Since the measurement signals are not only influenced by z_{Mg} but also by σ_{Ti}, the sponge cross section radius r_{Ti} and vertical sponge position h_{Ti}, this constitutes a complicated inverse problem which requires significant computational power to solve. To save computational power during the measurements, we propose here a multi-coil multi-frequency approach with pre-calculated numerical simulations.

Numerical simulation

In order to determine z_{Mg} with sufficient accuracy, although the shape, position and electrical conductivity of the Ti sponge are unknown, a large number of numerical simulations for different parameter combinations of z_{Mg}, σ_{Ti}, r_{Ti} and h_{Ti} are pre-calculated for a certain number of frequencies n_f and excitation coils n_C. A two dimensional, axisymmetric simulation model has been designed in Comsol Multiphysics under the assumption of a few simplifications. Since the reactor has a total height of 4 m we concentrate only on the top 1 m of the reactor.

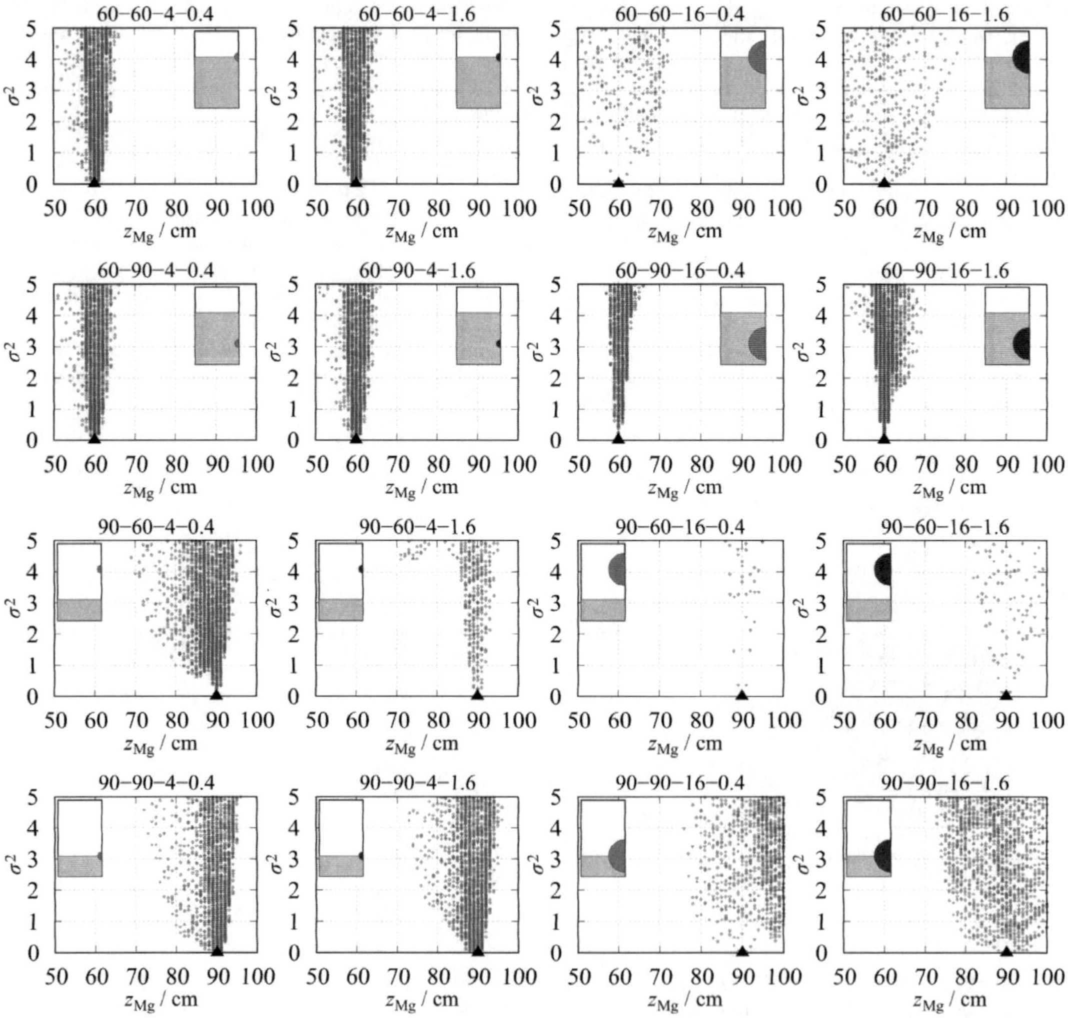

Fig. 3: Simulation results for the mean squared error of 16 representative parameter combinations. Each parameter combination is displayed above the respective chart by four numbers: z_{Mg}(cm) - h_{Ti} (cm) - r_{Ti} (cm) - σ_{Ti} (MS/m).

The Ti sponge ring at the walls is assumed to have a half-torus like shape (Fig. 1) with a homogeneously distributed although unknown σ_{Ti}. Four heater coils are used to induce eddy currents within the Mg and Ti; these metals distort the magnetic field which is detected by the sensor coil. By measuring the amplitude and phase of the induced voltage, z_{Mg} can ultimately be inferred. Only one heater coil is activated at a time, the coil closest to the top is named C_1, the lowermost coil C_4. The excitation current has an amplitude of 200 A. We use three different excitation frequencies (1, 10, 50 Hz) to "scan" the interior of the reactor.

Depending on which excitation coil is activated and which frequency is used to generate a signal, voltages with different amplitudes and phase shifts are induced in the sensor coil. After calculating these voltages for a large number of combinations of the unknowns z_{Mg}, σ_{Ti}, r_{Ti} and h_{Ti} (in this example we have a total number of 67,881), they are stored in a so called look-up table (LuT). Then it becomes possible to identify an unknown parameter combination by comparing the measured voltages to all of the pre-calculated values contained in the LuT and calculating the deviation between them. The parameter combination with the lowest deviation corresponds most likely to the unknown parameter combination. In order to quantify the deviation, we use the mean squared error (MSE) σ^2, which is calculated for each of the four amplitudes and phase shifts per frequency and subsequently summed up. Because the amplitudes decrease for an increasing distance between heater and sensor coil, weight factors are introduced for the amplitudes of the coils C_2, C_3 and C_4 to guarantee that each coil has a comparable impact on the MSE. The results for 16 representative parameter combinations are displayed in Fig. 3. There is a small panel within each chart which illustrates the respective parameter combination. The Ti sponge is represented by a half circle, its color for low σ_{Ti} is medium grey, for high σ_{Ti} dark grey. Additionally the exact values of each parameter are displayed directly above the chart. The magnesium level z_{Mg} is indicated by the light grey area.

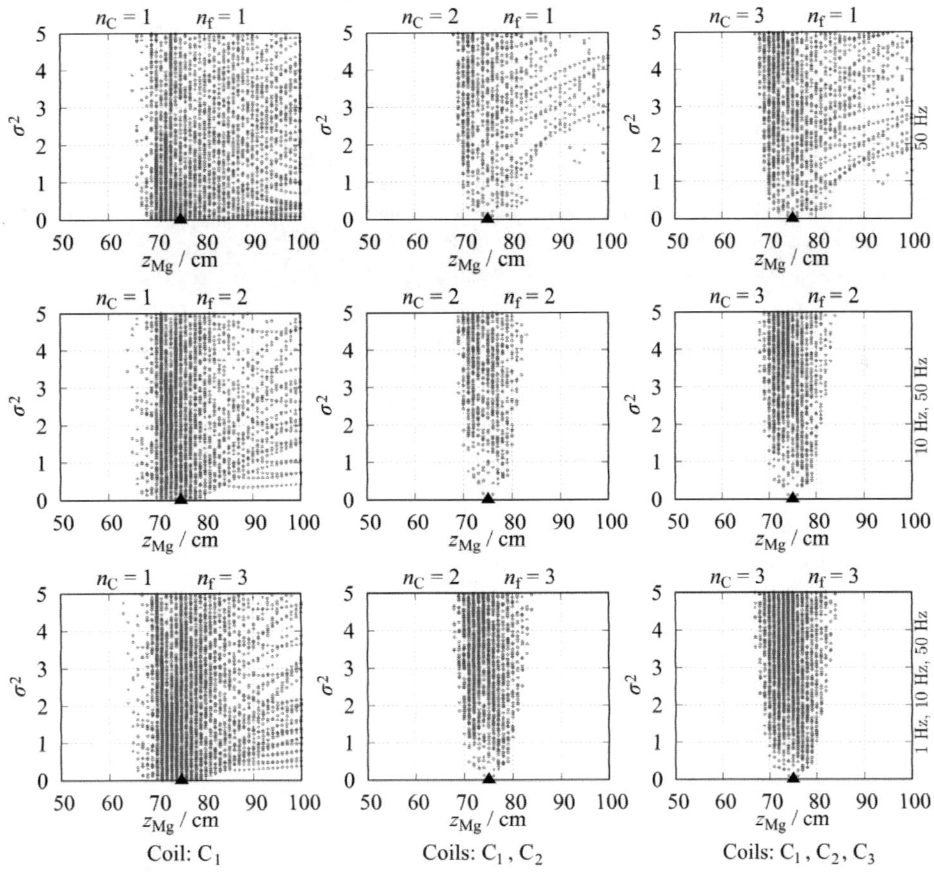

Fig. 4: MSE for a different number of coils n_C and frequencies n_f for the specific parameter combination 75-75-10-1.0.

The charts show the distribution of the parameter combinations that have a MSE of less than 5, the "measured" value is indicated by the black triangle ▲; (because there are no measurement errors in frame of the numerical simulation, its

MSE is always 0). In Fig. 3 it can be seen that the lowest corresponds indeed to the Magnesium level of the unknown parameter combination, therefore may be ascertained even if there is a measurement error. It can also be seen that for some parameter combinations (e.g. 60-90-16-0.4) there is a very sharp minimum of , therefore z_{Mg} can be determined with high accuracy, other parameter combinations display a shallow minimum (90-90-16-1.6) where it becomes more difficult to identify the actual . The influence of the Ti sponge is especially apparent for parameter combinations with a high where the problem is generally ill-conditioned, with the exception when the Ti sponge is submerged in Mg. It is also possible to determine the other parameters , r_{Ti} and h_{Ti} but in most cases they are conditioned even worse. That is because either an increase of or r_{Ti} result in stronger eddy currents, hence with the proposed measurement setup it becomes very difficult to separate the two effects.
Fig. 4 gives an overview over the effects of increasing the number of coils n_C and the number of frequencies n_f. The parameter combination (75-75-10-1.0) used in this example lies exactly in the middle of the investigated parameter space. When using only one coil (C_1) at one excitation frequency it becomes almost impossible to determine z because does not converge to a certain point. By increasing only n_C or only n_f the results become slightly better but there is still a high uncertainty regarding . Only by using multiple excitation (heater) coils with multiple excitation frequencies a clear minimum of can be observed and therefore the actual z_{Mg} can be determined with high accuracy.

Conclusions

We have demonstrated by means of numerical simulations that the Mg level in the Kroll process can be determined via inductive measurements from the outside of the titanium reduction reactor. The underlying inverse problem is solved by creating a look-up table with thousands of pre-calculated values for different parameter combinations. Previous attempts for the determination of z_{Mg} have been hampered by the presence of Ti sponge rings which form at the walls inside the reactor. Although performing the numerical simulations for the creation of the LuT takes a moderate amount of time, the corresponding parameter combination (including z_{Mg}) can be obtained within seconds during the actual measurements, by simply determining which solution has the lowest deviation from the measurement. Therefore the results can be displayed almost in real time, even with a standard PC, because only a few thousand comparison operations have to be performed to identify the most likely parameter combination that corresponds to the measurement. For this first investigation we have made some simplifications regarding the geometry of the Ti sponge ring which might be different in a real reactor. In future investigations, a full 3D simulation model with an asymmetric sponge ring will be developed. There are also small scale experiments planned to validate the simulation models. The general technique is not limited to the titanium production, but can be easily adjusted to other level detection problems.

Acknowledgement

This study was partially supported by the Government of Perm Krai, research Project No C-26/060359 11.03.2016.

References

1. N. Krauter et al. Metal. and Mat. Trans. B (2018). https://doi.org/10.1007/s11663-018-1291-y
2. Ch. R. V. S. Nagesh et al., Transactions of the Indian Institute of Metals (2008), vol. 61, issue 5, 341-348
3. W. Kroll, J. Electrochem. Soc. (1940) vol. 78, issue 1, 35-47
4. H. N. Bae et al., Materials Science Forum (2010), vols. 654-656, pp. 839-842,
5. W.A. Barannikov et al., Russian Patent RU 2334952 (2007)

MHD-equipment and technologies of semi-continuous billet casting of high-strength Al-alloys

M.A. Slazhniev[1], K.H. Kim[1], H.S. Sim[1], S.W. Kim[1], W.J. Kim[1]

[1]Dong San Tech. Co., ltd., 169 Hamansandanro, Hamangun, Gyeongnam, Korea

Corresponding author: slazhnevmykola@gmail.com

Abstract
In this study, there are presented design and conceptual approaches developments for 2 (two) kinds of the original complex technological foundry equipment for the melting, holding and managed electromagnetic magneto-hydrodynamic high-strength aluminum alloys pouring at continuous billet casting processes.
First kind performed as MHD-complex device created on base of the magneto-dynamic installation (MDI MDN-6A-0.63-WMD-M), that has applied as mixer-batcher with 630 kg capacity by Al-alloys with electromagnetic pump (up to 0.35 bar pressure). An experimental approbation for workability of the developed technologies for MHD-semi continuous billet casting, by means open stream melt feeding into 7 inch crystallizer and by progressive hot-top closed process, at the producing 1.0 meter of 7' billet from Al-Zn-Mg-Cu wrought alloy has been executed.
As second kind, there are developed conceptual design of 500 kg capacity by Al-Alloy pilot plant of the foundry equipment for the melting, holding and managed electromagnetic continuous billet casting, comprised upgraded tilting reverberator melting furnace combined with multifunctional horizontal magneto-hydrodynamic units (MHDU), as a W-shaped induction channel with 2 inductors (2x20kW) and external C-type electromagnet (0.01~0.3T). MHDU provide homogenization melt temperature (650~850°C) and it chemical compositions (at circulation melt intermixing 0.5~3.0kg/s) and at the switching electromagnet of electromagnetic melt pouring with managed flow rate 0.05kg/s ~ 15kg/s to tundish of the 3~10 inches billet continuous casting machine (CCM).

Keywords : High-strength aluminum alloys, continuous billet casting, electromagnetic stirring, electromagnetic pump.

Introduction

At the developing of rational equipment for the semi or continuous billet casting technologies from high-strength aluminum alloys 6xxx and 7xxx series, the main trend are accented on the creation multifunctional and multitask equipment, combined several metallurgical and casting operations in one electro-technological device.

These thing are concerned for the serious decreasing of the energy using, simplification exploitation modes and creation of new possibilities of high-effective melt processing methods for rising alloys properties, prevention casting defects due to the absence or insufficiency chemical, temperatures and structural inhomogeneity in the liquid state of alloy. Second reason in the decreasing of the quantity melting and foundry equipment's using are concluded, and accenting using only one electric energy for all operation from melting, holding, refining up to the pouring are consisted. Especially for the continuous castings technologies as usual applying induction crucible melting furnaces, induction channels furnaces or crucible electric resistance furnaces [1]. As more advantage oven for the melt melting can be suggested an reverberate melting furnace with upper location heaters, when the main steel case of furnace lined inside and have a widely by capacity of aluminum alloys range. However as a disadvantage have an insufficient melt stirring into main capacity and cannot guaranteed melt homogeneity and require application of the artificial melt stirring means. As a next melt processing stage continuous casting technologies require melt holding, degassing, refining at the providing melt temperature and chemical homogeneity. For these by the usual approaches using the mixer-batcher or intermediate heated vessel with maintaining melt temperature range in the holding mode.

As a most like for this case, have a perfect possibilities an induction channels furnaces [1, 2], that to perform artificial melt stirring and inductions contactless heating before pouring melt to the molds or crystallizers. As a finishing device at the continuous casting processes are used the rows means for the liquid alloys transportation with high accuracy operating parameters by speed, mass flow-rate, temperature and with prevention for the second oxidation or involving gases or hydrogen methods [2, 3]. Usual there are tilting mixer-batcher, pneumatic pressure dosing devises or electromagnetic magneto-dynamic pump (EMP) with contactless action on the melt. As well magneto dynamic pumps (MDP) as a most progressive equipment can be selected for the continuous casting technologies [2~4]. Assuming dedicated the usual equipment are consist as minimum 3 divided devices, that are required own electrical energy rates and as minimum two times melt overfilling from each devices as following step by step.

MHD-complex device for semi-continuous billet casting of the high strength Al-alloys (7xxx series)

For reasons creating a full-closed technological cycle of production of high-strength aluminum alloy at development for special technology of continuous casting has been accepted, where are the one foundry unit, for melting and further

controllable contactless electromagnetic pouring directly into the crystallizer of continuous casting machine are using. MHD-complex device and comprised methods for manufacturing, preparing, processing and controllable contactless electromagnetic pouring melt directly to the crystallizer by the MHD-tundish means can be succeed applied. As main unit in this case applied the magneto dynamic installation (MDI) as a mixer-batcher [2, 3~5], electromagnetic pump (EMP) and induction-melting furnace. The classic solution of MDI for aluminum alloys MDN-6A (630 kg capacity by Al-alloy), designed with dual combined W-shaped channel, with two independently controlled inductors for each channel, and additional electromagnet for creation of external alternating magnetic field (B) [1].
The main working principle MHD-equipment based on the interaction of AC-current (I) (5~20kA) with external AC-magnetic field (B) in the working area (WA) (fig. 1). As result by the Fleming's Left Hand Rule in WA are creating the unidirectional pulsating (120Hz) pulse electromagnetic force, - Fem (70~285N). It is produce pumping of molten metal from EMP WA through central channel (fig. 1), follow or into crucible by circulation mode or to casting molds/tundish, by EM-pump with managed pressure 7~35kPa and flow rate 0.1~5kg/sec by low electromagnetic pressure principle.
MDI allows providing the number technologies for electromagnetic stirring of liquid aluminum alloy in the crucible and induction channels with controlled melt flow rate 1~5 m/sec, using this electromagnetic pump. In addition, for melt processing the MDI can use the high-density of AC-electric current (up to 60 A/mm^2) and pulsing electromagnetic forces PEMF (120 Hz) for purpose homogenization, dispersion of micro-segregation in melt, dissolve ligatures and modifiers dispersing of intermetallic compounds [5]. On these means MHD-equipment for continuous billet casting, named as MHD-tundish (fig. 1, 2) has assembled and tested. It is consisted the MDI combined with special metal-duct/tundish for controlled electromagnetic pouring Al-alloy from mixer-batcher to the crystallizer of the CCM.

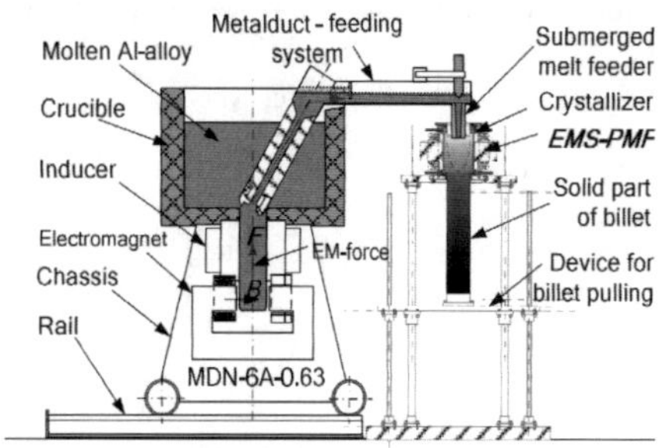

Fig. 1: Scheme of 7075-alloy casting from magneto-dynamic installation to the semi-continuous 3~10" billet casting machine

Fig. 2: Photos pilot-plant installation for high-strength alloy casting from magneto-dynamic installation semi-continuously billet casting and 7" billet casting process, (analog 7075 alloy)

For practical verification workability of MHD-technologies of continuous/semi-continuous billet casting, there are of 2 types for MHD-tundish, EMS, Cu-water-cooled crystallizer and billet puling stand has been developed and installed.
On first stage for feeding melt from MDI to 7" EMS-crystallizer has been used metal-duct/tundish with stream melt feeding system by opened meniscus mode, and further - original closed melt feeding process at the using hot-top system. For opened meniscus casting mode as tundish has been used modified metal-duct, consisted primary ceramic tube with graphite tip, installed to the central mouth of the W-shaped induction channel into crucible MDI, from where are the melt are pumping inside (fig. 2). On the upper end metal duct was installed metallic cap as an tundish, that has horizontal channel with height 250 mm and length 500 mm. On the opposite end of tundish was connected the cylindrical vessel with diameter 200 mm. In this bottom place was manufactured hole with diameter 70 mm, and outside in the vertical plane to this hole was attached removable crater with output hole with diameter 8 mm.
Selection gating hole for 8 mm diameter to provide create mass flow-rate aluminum alloys from tundish, at the conditions hydrostatic pressure 0.2~0.25 meter in ranges 0.05~0.1 kg/sec. Crater was need for the stabilization melt flows lines and focusing melt stream to the center of crystallizer. Height of crater selected as 100 mm, from viewpoints possibilities on 10~30mm submerging it end under melt level into crystallizer for the prevention occurs melt foam and it involving to the billet body or catching in mushy zone.
Casting process has realized by switching MDI to the electromagnetic pumping mode and starting controllable melt filling to the cavity of tundish with speed in range 0.5~0.6 kg/sec, 690~710°C. Visual slowly filling melt to the crystallizer cavity, for formation full filling it chamber up to the touching hot-top insertion during no less than 30~40 second. After filling crystallizer cavity – delaying no more than 10~20 second to switching billet pulling device for 60~100mm/min and starting semi-continuous billet casting process. At the maintaining electromagnetic melt feeding into tundish in range 0.5~0.8 kg/sec - continuing filling tundish cavity up to the level 150~160 mm from it bottom side and after reducing voltage on electromagnet for value, when melt level in tundish maintaining on set level. After reaching required lengths billet, MDI switching to holding mode.

Study of the MHD-Physical modification effect on microstructure refinement of the Al-Zn-Mg-Cu (Zr-Free) high-strength wrought aluminum alloy in 7" billet at semi-continuous casting

As an alloy for the continuous billet casting was selected experimental Al-5.5Zn-2.2Mg-0.6Cu-0.1Mn-0.23Cr-0.1Fe-0.1Si-0.1Ti wrought alloy composition without admissions Zr.

For manufacturing alloy used unique methods, when MDI furnace started by pure aluminum in quantity 50 kg, and all necessary component was added step by step into MDI crucible in the master alloy state.

The final capacity melted alloy reached for 300 kg as for test mode. By the original way, the argon gas degassing melt has performed during 20 minutes for 300 kg, at the Ar-gas rate 5l/min at using lateral static tuyere in the special lateral stirring mode [5]. Final content of hydrogen had no more than 0.1 ppm.

For homogenization alloy state the 2 main working modes used, as permanent holding mode at slowly (2~3 kg/sec) melt circular electromagnetic stirring and intake reverse mode intensively stirring by pulsing (100~120Hz) electromagnetic force with 15~20kPa pressure, (70~143.5N), during 10 minutes. After that has been performed continuous billet casting from the magneto-dynamic installation for two main conditions without EMS and with application electromagnetic melt stirring into Cu-water-cooling crystallizer for recognizing effect processing melt into MDI by the MHD means and combined effect MHD-melt processing into MDI and it EMS-processing into crystallized during billet solidification.

For compare efficiency applying MHD-processing 7xxx-series Al-alloy has produced 2 experimental casting at using melt from conventional melting furnace with/without EMS (fig. 3), and pouring from MDI, with/without EMS (fig. 4).

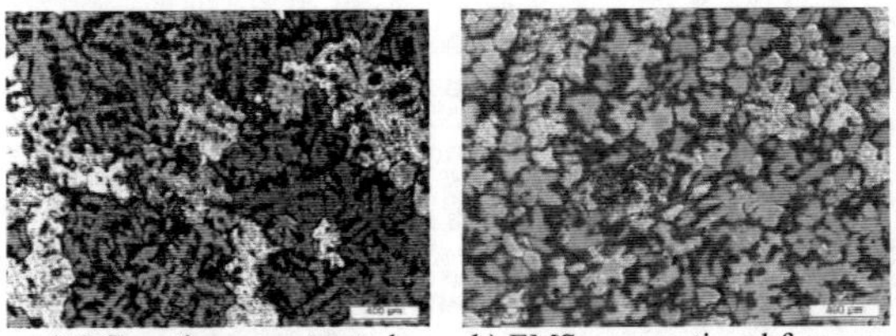

a) DC casting, non-treated b) EMS, conventinoal furnace

Fig. 3. Microstructure for central part 7" billet from Al-Zn-Mg-Cu alloy melted in conventional melting furnace: a) DC casting; b) EMS-processing into crystallizer.

a) DC casting, MHD-processed b) EMS, MHD-processed

Fig. 4. Microstructure for central part 7" billet from Al-Zn-Mg-Cu alloy melted and poured from magneto-dynamic installation:: a) DC casting from MDI; b) EMS-processing into crystallizer.

At MHD processing (fig. 4a), by the means of unidirectional 120Hz pulse electromagnetic force (70~143.5Newton), created in working area (WA) electromagnetic pump are provide the fine distribution intermetallic particles, as emulsion state in liquid melt. Indicated microstructure state reached during 10 minute intensively processing and keeping at holding mode permanently as ready to the cast mold pouring conditions.

It is allow forming advantageous conditions for further non-dendrite microstructure formation in EMS-mold (fig. 4b), with resulted grain size no more 50~80μm for quantity melt from 50~600 kg as for one capacity of MDI crucible.

Analysis of the 7" billet at casting to the water-cooled crystallizer of continuous casting machine shown that MHD-processed melt have not hot cracks (fig. 4), porosity or any tensions into billet. However non-treated alloy (Fig. 3a), poured from conventional melting furnace, have prone to the hot crack appearance in the central part of billet, and large dendrite microstructure with 450~650μm for DC casting state (fig. 3a), and 150~250μm dendrite's for EMS (fig. 3b).

As well, MHD-process can by presented as in-time melt processing into magneto-hydrodynamic tundish, for creation high-effective technology manufacturing 3"~10"billets with improved properties and solving problem hot-cracks occurs into wide billets at it manufacturing.

Development of MHD-technological foundry equipment for semi-continuous billet casting high strength Al-alloys

For creation specialized for continuous casting processes equipment there are focusing attention for concepts creation combined as one device the 3 main aggregates, - melting furnace, mixing-batching device and confirmed early means about electromagnetic pump, for managed melt pouring in the continuous casting mode (fig. 5, 6). At first developing device could be have a melt capacity no less than 500 kg or more (up to 10 tons), provide independent alloys melting from ingots state and overheating to required temperature. From this viewpoint and according to required semi-continuous working mode, as a main melting device was selected the reverberator-melting furnace with minimal capacity 500 kg. Existed reverberate furnace have a wide range by capacity from 500 kg up to 50 tons by aluminum alloys, using simple melt heating method, safety at the exploitation and not require expensive changeable ceramic crucible and can realize melting process without harmful effect for inductions channels as well.

The main bath of furnace have a steel string case (fig. 5), linen inside by the fireproof mass, have a tilting mechanism for pouring melt and use the upper radiation heating melt methods from the electric resistance heating elements. Case of furnace fully closed and heat insulated inside, have a long time servicing.

IOP Conf. Series: Materials Science and Engineering **424** (2018) 012079 doi:10.1088/1757-899X/424/1/012079

On the front side installed loading door, for removing slag, flux processing and loading ingots or melt scrap. On opposite side located the pouring nozzle, for outfling melt by the gravity methods at the furnace case tilting by the hydraulic cylinders. As well, this construction can be use also for continuous casting technologies, but take into account that pouring speed in this case controlling only by the tilting angle, and cannot provide the maintaining the melt flow rate in stable value ranges. Additionally by using the upper melt heating way also to difficult are provide the homogeneity melt temperature in the on surface and in the down part.
Especially appointed to the continuous casting technologies was decided improve reverberator melting furnace by the attaching multifunctional horizontal magneto-hydrodynamic units MHDU, as a W-shaped induction channel with 2 inductors and electromagnet fig. 4, 5.

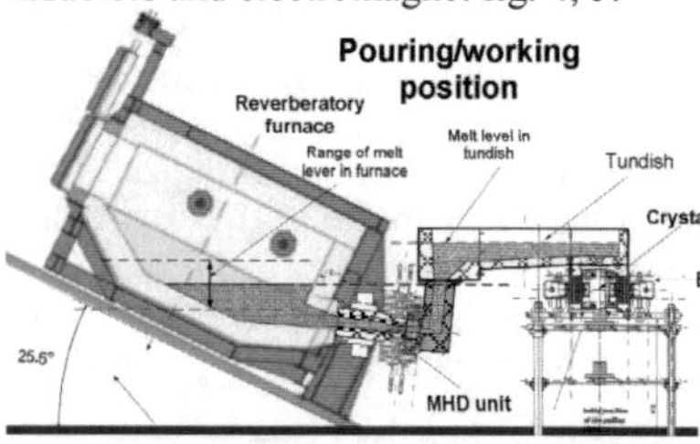

Fig. 5. Device and technological scheme for pilot plant reverberator-melting furnace with MHD unit and tundish for continuous billet casting.

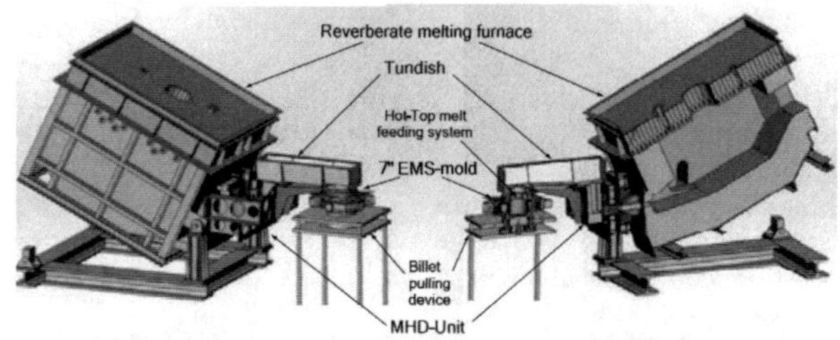

Fig. 6. 3D-model of the reverberator melting furnace 500 kg capacity by Al alloy, adopted for the assembling with MHD Unit.

In this case magneto hydro-dynamic unit (MDU) to provide the continuous contactless high effective induction heating with artificial continuous melt stirring into channels and into all volume of the reverberator melting furnace during melting, refining, holding and controlled electromagnetics pouring to crystallizer of continuous billet casting machine. The prototype of foundry melting, processing and pouring device presented on fig. 4, 5. Magneto-hydrodynamic unit are provide additional controllable induction melt heating with power 2~50kW, continuous melt stirring into channel and into furnace crucible by the electromagnetic force means – named as electromagnetic pump. There are provide the homogenization of melt temperature and chemical compositions and at switching electromagnet, also provide electromagnetic melt pouring from the pouring nozzle with managed flow rate 0.5 kg/s ~ 15 kg/s to the mold, tundish of the continuous casting machine (CCM), as presented on fig. 4 – working position.
The main construction of reverberator furnace, as was described early, executed as an base case, that have tilting possibilities in one direction to MHD-unit side on 24~26 degree. For tilting furnace the case installed on the frame, which have the 2 tilting support points knots, and on opposite directions knots for attaching of two hydraulic cylinders.

Conclusion
In this study presented design and conceptual approaches for 2 (two) kinds of the original technological foundry equipment for the melting hydrodynamic high-strength aluminum alloys pouring at continuous billet casting processes., holding and managed electromagnetic magneto-
First kind performed on base of the magneto-dynamic installation (MDI MDN-6A-0.63-WMD-M), that has applied as mixer-batcher with 630 kg capacity by Al-alloys with electromagnetic pump (up to 0.35 bar pressure). An experimental approbation developed technologies for MHD-semi continuous billet casting, by means open stream feeding into 7"crystallizer and by progressive hot-top closed process, at the producing 1.0 meter of 7" billet from Al-Zn-Mg-Cu wrought alloy has been execute. There are show efficiency of MHD melt processing – providing non-dendrite microstructure state and full elimination crack-ability as fore DC and for EMS billet casting modes.
As second, developed an conceptual design of 500 kg capacity by Al-alloy pilot-plant foundry equipment, comprised upgraded tilting reverberator melting furnace combined with multifunctional horizontal magneto-hydrodynamic units (MHDU), as a W-shaped induction channel with 2 inductors (2x20kW) and external C-type electromagnet (0.01~0.3T). MHDU provide homogenization melt temperature (650~850°C) and chemical compositions, at circulation melt intermixing 0.5~3.0kg/s, and electromagnetic melt pouring with managed flow rate 0.05kg/s ~ 15kg/s to tundish of the 3"~10" billet casting for continuous casting machine.

References
1. http://www.induga.com/pdf/vertical-casting.pdf
2. L. Puzhaylo, A. Seryi, A. Gordynya, S. Polivoda, (2017), Patent of Ukraine № UA 119406 U.
3. V. Fixsen, Dubodelov, M. Şlazhniev and et., (2005), Casting processes, 4, p. 5-10.
4. V. Polishchuk, M. Tsin, R. Horn & et. (1989), Kyiv, Naykova dumka, 256.
5. M. Slazhniev, K. Kim, H. Sim, S. Kim. (2015), EPM2015, France (Cannes).

IOP Conf. Series: Materials Science and Engineering **424** (2018) 012080 doi:10.1088/1757-899X/424/1/012080

Fundamental Studies on Induction Heating and Stirring of Non-Metallic Molten Fluid

N. Yoshikawa[1], K. Watanabe[2], T. Igarashi [2] and S.Komarov[1]

[1] Graduate School of Environmental Studies (GSES), Dept. of Materials Science and Engineering,Tohoku University 6-6-02, aza-Aoba, Aramaki, Aoba-ku, Sendai, Japan, 980-8579
[2] Graduate student at Tohoku University, GSES

Corresponding author: yoshin@tohoku.ac.jp

Abstract
Induction heating of non-metallic molten fluid has been investigated, mainly performed on melt glass stirring. Because silicate glass has relatively high viscosity and low electric conductivity for induction stirring, their experiments require very high temperature above 1500°C, and are difficult for observation and measurements. In this report, observation and simulation of heat generation and flow in molten salt are presented. And flow states of molten vanadium oxide glass is observed, both of them have low melting points. It is intended to investigate them from both experimental and theoretical directions, as a fundamental research. It is demonstrated that molten NaCl-KCl possessed the low impedance value in frequency range of hundred kHz to 1MHz, preferable for giving rise to induction current generation. Circulating flow of the fluids from the container wall to the center was observed. Temperature measurement in the melt was also conducted at various positions with respect to the coil configuration. The flow velocity and temperature distribution were analyzed by numerical simulation, and compared with the measurements, taking account of the transparent nature of the molten salts. They are shown to explain the experimental results, mostly. Molten vanadium oxide glass having the similar physical properties exhibited the similar flow characteristics.

Key words : non-metallic fluid, flow, observation, temperature distribution, electric conductivity, simulation,

Introduction

Induction heating and stirring of molten metal has been applied for melting and refining of metals, and for recycling of metal scraps etc. The flow and heat transfer analysis are one of the major subjects in this conference. Many researches have been performed so far, and contributed to promoting understanding of the processes. On the other hand, induction of heating and stirring of non-metallic fluid has been performed relatively less. One of its major purposes is a necessity of heating and stirring of glass for solidification stabilization of high-level radioactive wastes in confining in a canister, before burial in the deep undergrounds.

It is known that the silicate glass has relatively high viscosity and low electric conductivity for induction stirring even above 1500°C. Therefore, impeller stirring has been taken into consideration instead of induction stirring [1,2]. Because very high temperature experiments are required for reducing the viscosity and increasing the conductivity, relatively less experimental studies have been conducted for measurement of temperature and flow states than that by the theoretical analysis [3,4]. Besides, because the physical properties of molten glass have large temperature dependences, the flow simulation becomes more complicated, compared with that of molten metal cases.

In order to obtain a basic understanding of high frequency heating and stirring of non-metallic molten fluids, this study is intended to investigate them both from the experimental and the theoretical directions, as a fundamental study. The authors selected NaCl-KCl molten salt as an object for their analysis in the first step, because of much low viscosity and relatively high electric conductivity, comparing with silicate glasses as demonstrated in Fig. 1.

In this study, starting with impedance measurement of molten NaCl-KCl, temperature distribution and flow velocity are measured, then discussed by comparison with the simulation. In addition, it is intended to introduce our 'on going observation studies of flow states' by selecting a vanadium oxide glass, as the second step. Vanadium oxide is known to be a semi-conductor and is applied for some electric devices for special purposes, because of having relatively higher electric conductivity (Fig. 1). Vanadium oxide glass is known to possess mostly similar physical viscosity and electric conductance with that of NaCl-KCl [5], although these properties are known to be critically dependent on the valence of vanadium. A care has been undertaken by considering this stability of its physical properties, and it is to be incorporated into the simulation inputs in future studies.

Experimental

Specimens used in this study were commercially available salt (NaCl) and KCl, V_2O_5, and Na_2CO_3 powders of reagent grade, NaCl is manually mixed with KCl in a container made of silica. V_2O_5 and Na_2CO_3 were mixed, similarly. Because electric conductivity of these powder mixtures are too low to be inductively heated at room temperature,

SUS304 stainless steel bar is inductively heated (400kHz, max 3kW, RF Nestle Kanagawa, Japan), and the powder mixture were melted once in a container made of silica having inner diameter of 40mm and 50mm in depth. Then, the metal bar is removed and the melts were inductively heated.

Impedance measurement was conducted using an impedance analyzer in a frequency range between 4Hz and 5MHz.

Temperature measurement during the induction heating was conducted using a thin sheath K-type thermo couple (ϕ0.3mm) and a thermo-viewer (Nippon Avionics TH9100MR). In order to determine the heat transfer coefficient (h [W/m^2/K], cooling curves of various positions of the melts were obtained, then fitted by one dimensional heat transfer equation. This h value was input for the boundary conditions in a numerical simulation of the heat transfer. Flow and magnetic field distribution were calculated by using a commercial software COMSOL Multiphysics ver. 5.2a. The value of heat transfer coefficient was almost interpreted as that of radiation, although the detailed analysis is not described in this report.

Results and Discussion

1. Measurement of electric conductivity of molten NaCl-KCl

Platinum plate electrodes were immersed in NaCl-50mol%KCl melt and impedance of the melt was measured with variation of temperature and frequency. The result of the impedance measurement is plotted in Fig. 2. It was shown that the impedance decreases as an increase of temperature and that the impedance (resistivity) become lower above 10kHz, having a minimum around hundred kHz to MHz. This result convinced us a validity of selecting the induction frequency of 400kHz for the present study.

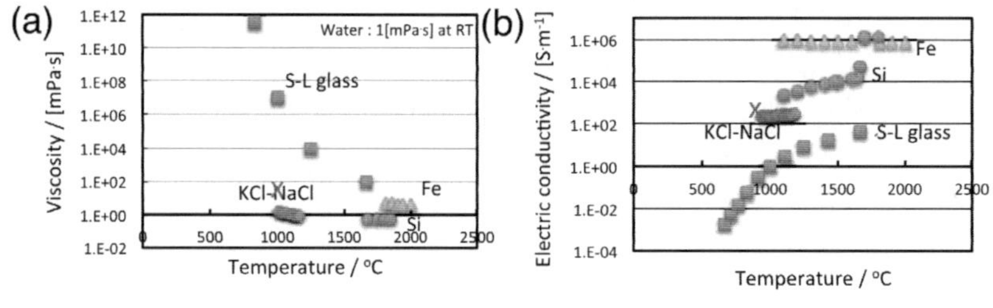

Fig. 1: Temperature dependence of (a) viscosity and (b) electric conductivity of various melts. Vanadium oxide glass values are indicated with keys of x.

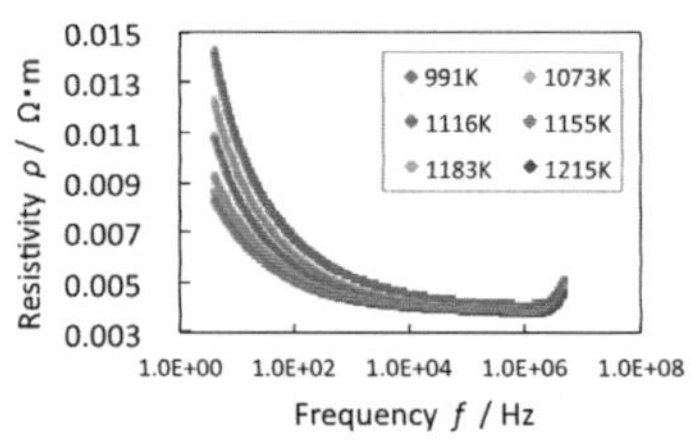

Fig. 2: Frequency dependence of NaCl-50%KCl melt impedance (resistivity), with variation of temperature.

2. Temperature measurement

After NaCl-KCl was completely melted by induction heating of stainless steel bar, (immersed in the powder mixture in advance). The melt was inductively heated, as the photos shown in Fig. 3, where states of heating change obviously, as the increase of induction voltage or the input power. The relationship between the input power and the temperature is plotted in Fig. 4. The melt container is placed in a water cooling coil, and the configuration is shown in Fig. 5.

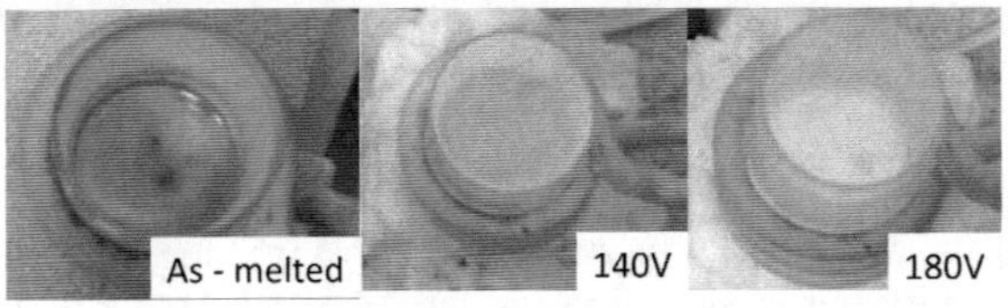

Fig. 3: Sketch of the casting flow problem with magnet pole faces and numerical grid for the simulations.

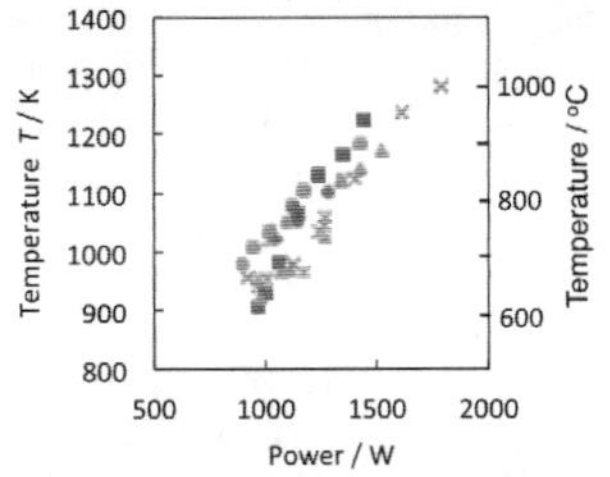

Fig. 4: Relationship between the induction power and the temperature of NaCl-KCl melt. Different marks correspond to the measurement trials.

Temperature distribution in the melt was measured with a thermo-couple and a thermo-viewer. They are demonstrated in Fig. 6. It was shown the bottom temperature is higher because of thermal insulation. Measurement of the temperature in the middle depth was not possible because of fluctuation, probably due to the flow. One of the finding is that the temperature value by thermo-viewer (*mark) is higher than that of bottom and the surface in the case of larger mass (having the large depth, (a2 in Fig. 5)). This discrepancy might be caused by an occurrence of highest temperature positioned between them. The measured value by thermo-viewer might be due to the wave length of light transmitted through the transparent melt. The radiation heat transfer in the transparent material is one of the problems to be taken into consideration for the analysis of temperature distributions in an inductively heated melts. Further study is being carried out.

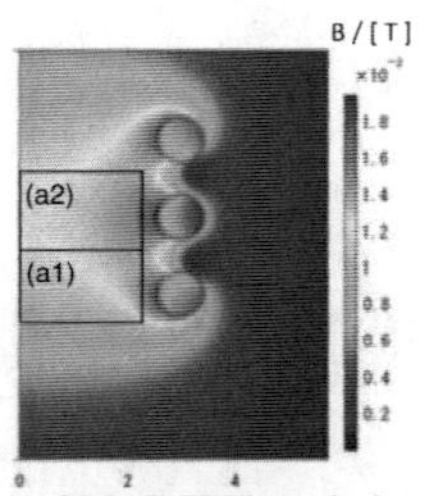

Fig. 5: Configuration of NaC-KCl melt in the induction coil (height of 35 and 70g specimen corresponds to rectangles (a1) and (a2)), and the calculated distribution of magnetic flux density (coil current 300A).

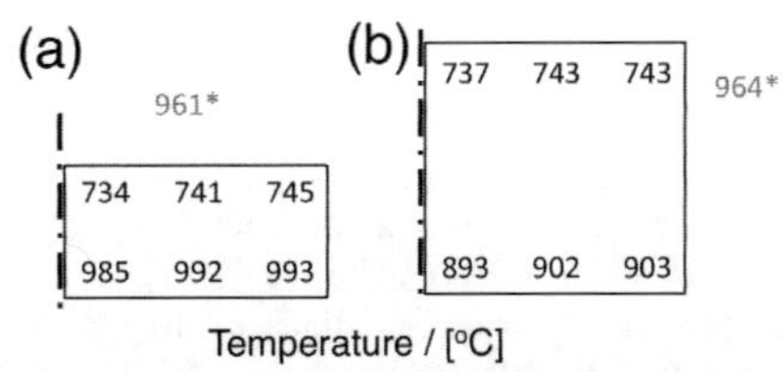

Fig. 6: Measured temperature values with a thermocouple and with a thermo-viewer, shown within and out of the rectangles (with an asterisk), corresponding to the quantity of the melts (a1, a2 in Fig. 5), respectively.

3. Induction stirring of molten NaCl-KCl and vanadium oxide glass

Flow of the NaCl-KCl melt driven by induction and its flow pattern was observed by tracer particles as a photograph shown in Fig. 7. It was demonstrated that circulating flow having down stream in the central position of the container existed. The flow state of NaCl-KCl is not clear in Fig. 7 (hence the flow directions are schematically indicated with arrows.). However, it was much easier in observation of flow pattern in vanadium oxide glass melt cases (without tracer particles), as shown in Fig. 8. The similar flow pattern is observed, because the physical properties of the melt are almost similar, as demonstrated in Fig. 1.

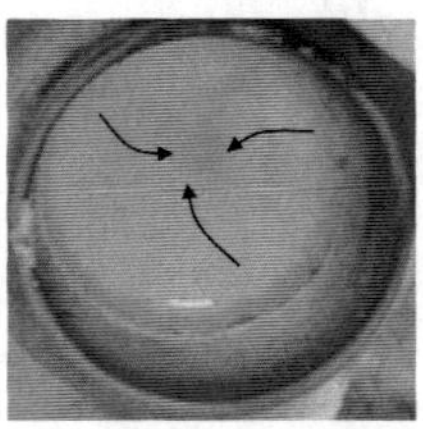

Fig. 7: Photograph of inductively stirred NaCl-KCl melt.

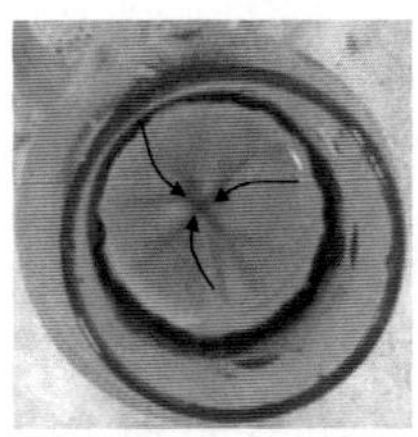

Fig. 8:Photograph of inductively stirred vanadium oxide melt.

IOP Conf. Series: Materials Science and Engineering **424** (2018) 012080 doi:10.1088/1757-899X/424/1/012080

4. Numerical simulation

In order to simulate the measured temperature and the observed flow velocity distributions in the NaCl-KCl melt, numerical simulation was performed. The software solved the equation for the magnetic field distribution (magnetic flux density $\boldsymbol{B}$ [T]) expressed by Eq. 1. The induction current ($\boldsymbol{J}$ [A/m^2]) and the Lorentz force ($\boldsymbol{f}$[N/m^3]) exerting to the melt is expressed by Eq. 2. The analysis of electromagnetic field is done in terms of vector potential ($\boldsymbol{A}$) in the software. The Lorentz force $\boldsymbol{f}$ is incorporated into the Navier-Stokes equation expressed in Eq. 3 as a force term. And the induction current $\boldsymbol{J}$ is also incorporated into the heat transfer equation expressed in Eq. 4 as a heat source term. The electromagnetic analysis and the flow/heat transfer analysis are assumed not mutually coupled. The output of the former is input into the latter calculation. But the flow and heat transfer equations (Eq. 3 and 4) are mutually coupled. The Parameters appeared in Eq. 1 are as follows: $\boldsymbol{B} = \mu\boldsymbol{H}$, $\boldsymbol{H}$ is a magnetic field [A/m], μ is a magnetic permeability [H/m], $\boldsymbol{E}$ is an electric field [V/m]. The other parameters t, $\boldsymbol{v}$, σ, p, ρ η and g are the time [s], velocity vector [m/s], electric conductivity [S/m], pressure [Pa], density [kg/m^3], viscosity [Pa·s] and gravity acceleration [m/s^2], respectively.

$$\frac{\partial \boldsymbol{B}}{\partial t} = \nabla \times (\boldsymbol{v} \times \boldsymbol{B}) + \frac{\nabla^2 \boldsymbol{B}}{\sigma\mu_m} \tag{1}$$

$$\boldsymbol{f} = \boldsymbol{J} \times \boldsymbol{B} \ , \quad \boldsymbol{J} = \sigma\left(\boldsymbol{E} + \boldsymbol{v} \times \boldsymbol{B}\right) \tag{2}$$

$$\rho\left(\frac{\partial \boldsymbol{v}}{\partial t} + (\boldsymbol{v} \cdot \nabla)\boldsymbol{v}\right) = \eta\nabla^2\boldsymbol{v} - \nabla p + \boldsymbol{f} + \rho g \tag{3}$$

$$\rho C_p\left(\frac{\partial T}{\partial t} + \boldsymbol{v} \cdot \nabla T\right) = \lambda\nabla^2 T + \frac{\boldsymbol{J}^2}{2\sigma} \tag{4}$$

Boundary conditions used for heat and flow simulations are shown in Fig. 9. The upper surface and right side surface (S1) are cooled (as the heat transfer coefficient (h) provided as the boundary condition) and thermal insulation condition imposed at the bottom. As can be seen in Fig. 5, a magnetic field (magnetic flux density B) is distributed around the coil, and B is penetrated deepest into the specimen at the height of the center coil. This distribution influenced the position of heating source and the temperature distribution in the melt.

Fig. 10 demonstrates the simulated temperature distribution, corresponding to the specimen configuration shown in Fig. 5. It can be seen that the highest temperature region appears in the upper internal area a bit from the right edge (S1), which is close to the center coil. As the distance (r) from the coil increases, the heat source intensity becomes small, thus the symmetrical axis (left side (S2) in left side, adiabatic boundary condition) area has low temperature. The calculated temperature distribution looks mostly agree with the tendency of measured temperature distribution. In Fig. 10, the calculated flow velocity is also indicated. The flow patterns are consistent with the observation, as shown in Fig. 7, 8. The flow velocity on the melt surface was measured from the movie taken at the experiments. The flow pattern and the flow rate almost agreed with the simulation results, as well.

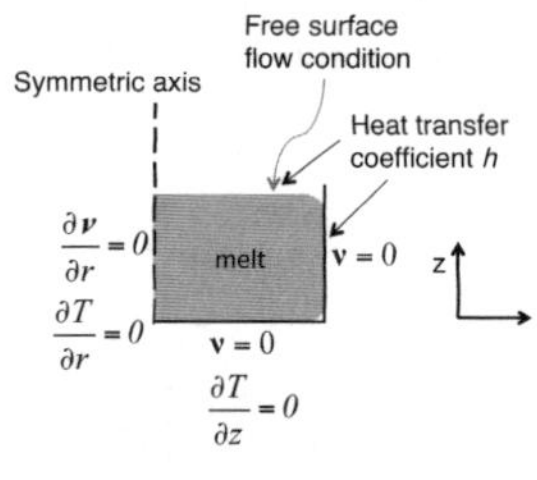

Fig. 9: Boundary conditions used for numerical simulation of flow and heat transfer in NaCl-KCl melt.

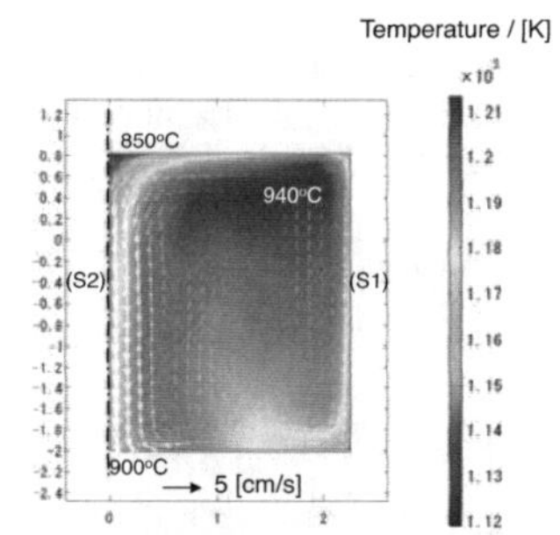

Fig. 10 : Calculated temperature and velocity distribution in NaCl-KCl melt (coil current 300A).

Conclusion

The induction heating and stirring states of NaCl-KCl and vanadium oxide glass melts were observed. Temperature measurement of the molten salt was conducted. Numerical simulation of electromagnetic and heat transfer analysis was performed for the molten salt, and the calculated distributions of magnetic field, temperature and flow velocity were presented. The calculated distributions of temperature, flow pattern and the flow velocity were almost able to explain the measurements and observation of molten salt. The observed flow pattern of vanadium oxide glass (having the similar physical properties) was similar to that of the molten salt.

References

1. S.Gopalakrishnan and A.Tess, Int. J. Therm. Sci., **60** (2012) 142-152.
2. L.Jacout, Y.Fautrelle, A.Gagnound, P.Brun and J.Lacombe, Chem. Eng. Sci., **63** (2008) 2391-2401.
3. B.Nacke, M.Kudryash, T.Behrens, B.Niemann, D.Lopukh, A.Martynov and S.Chepluk, Int. Sci. Colloq. Modelling for Material Processing, Riga, (2006) 209-214.
4. B.Nacke, V.Kichigin, V.Geza and I.Poznyak, Proc. 8th Int. Conf. of EPM, (2015), p.11-14.
5. V.Danek, I.Votava, K.Matiasovsky and J.Balajka, Chem. zvesti, **28** (1974) 728-732.

IOP Conf. Series: Materials Science and Engineering **424** (2018) 012081 doi:10.1088/1757-899X/424/1/012081

Separation of steelmaking slag with mechanical stirring by fluctuated magnetic field

Y. Takaki[1], M. Sasaki[1], K. Ishida[2], Y. Nishina[1], J. Yotsuji[1], J. Tateno[1]

[1] Mechanical Engineering Research Dept., JFE Steel Corporation,
Minamiwatarida-cho 1-1, Kawasaki-ku, Kawasaki 210-0855 Kanagawa, Japan
[2] Research Planning & Administration Dept., JFE Steel Corporation,
Kawasaki-cho 1, Chuo-ku, Chiba 260-0835 Chiba, Japan

Corresponding author : y-takaki@jfe-steel.co.jp

Abstract
Steelmaking slag is by-product of the steelmaking process and has a high content of iron. In order to recover this iron, steelmaking slag is processed with various methods such as crushing, sieving and magnetic separating. Magnetic separators process steelmaking slag in multiple layers for mass production and non-magnetic particles are attracted to a magnet roll by magnetic particles [1]. Therefore, precision of separation is low with conventional pulley magnetic separator. This study proposes an efficient magnetic separating technology. In proposal magnetic separation, a fluctuating magnetic field is generated by rotation of a magnet roll with multiple poles. This quick change of the magnetic field generates strong agitation in the attracted particles and eliminates non-magnetic particles [2]. As the result of separation test with steelmaking slag, the iron content of magnetic side and the overall separation efficiency is higher in proposal method than in conventional method.

Keywords : steelmaking slag; iron; slag recycling; physical separation; magnetic separation

Introduction

In steelmaking process, impurities of molten iron such as silicon (Si), sulfur (S), phosphorous (P) are removed by reaction with CaO and O_2 in the vessel or the converter and discharged as steelmaking slag [3] [4] [5] [6]. To accelerate chemical reaction, CaO and O_2 are injected to molten iron at high blow rate and mixed together. So much iron contaminates into steelmaking slag. For recovering this iron, steelmaking slag is processed with various methods. Fig. 1 shows an example of steelmaking slag processing flow. Firstly, steelmaking slag is cooled in the yard. And then, lumps of slag are crushed into small size by a jaw crusher, an impact mill or a rod mill and classified by a vibrating sieve. Finally, slag is processed by magnetic separator in a particle size for each particle size and iron is separated from slag. Recovered iron is used in iron making or steelmaking process and slag is usually sold to other industries, for example, as a raw material for cement or a roadbed material.

Issue

In steelworks, about hundred thousand tons of steelmaking slag is produced in one year. Therefore, for increasing throughput of magnetic separating process, steelmaking slag is processed in the multiple layers. Fig. 2 shows schema of conventional pulley magnetic separator. Because magnet roll is fixed, magnetic flux density at surface of magnet pulley is constant and attracted magnetic particles don't move. When separating fine particles with pulley magnetic separator, non-magnetic particles are captured to a magnet roll by magnetic particles [1]. As a result, Non-magnetic particles contaminate into magnetic side and precision of magnetic separation decreases.

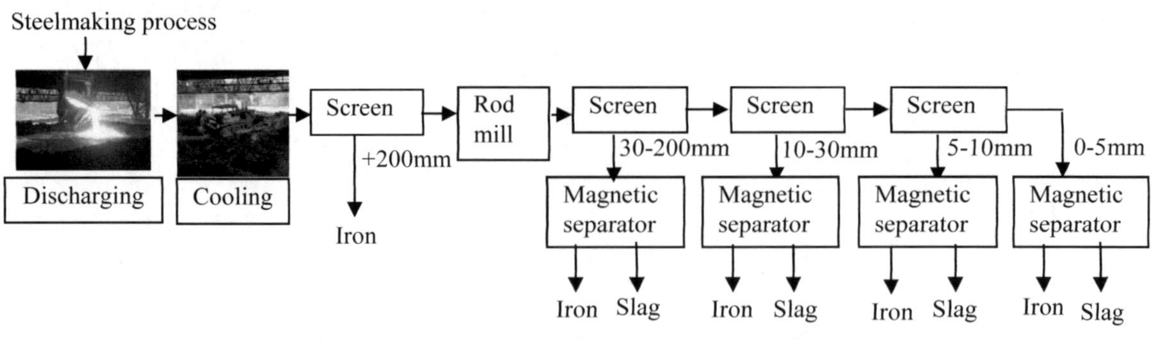

Fig. 1: Example of steelmaking slag processing flow.

IOP Conf. Series: Materials Science and Engineering **424** (2018) 012081 doi:10.1088/1757-899X/424/1/012081

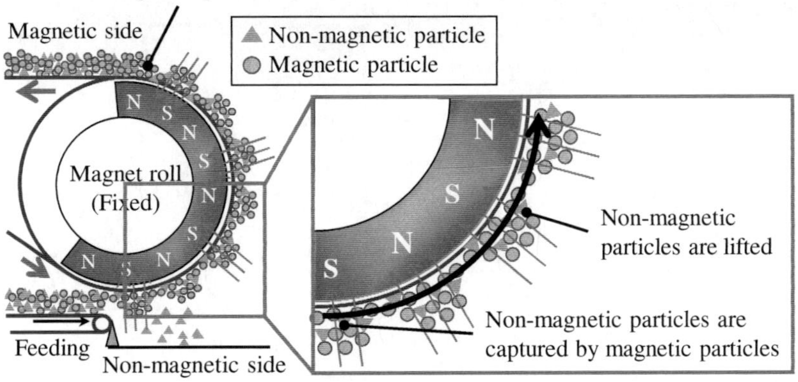

Fig. 2: Schema of conventional pulley magnetic separator.

Principal of proposal method

Fig. 3 shows a schema of fine particles separation with the conventional method. The fine particles can't move independently and non-magnetic particles are captured by attracted magnetic particle. As a result, non-magnetic particles and magnetic particles are collected together in magnetic side. So separation is not enough. On the other hand, Fig. 4 shows a schema of the proposed method using mechanical stirring by a fluctuating magnetic field. First, the magnetic particles are enriched on the surface of the fine particles. Next, the magnetic particles on surface swing under the fluctuating magnetic field unlike the condition with a static magnetic field and non-magnetic particles fall out. Finally, only magnetic particles remain under the magnetic field, and all the non-magnetic particles have fallen out. The fine particles are stirred intensely by changing the direction and strength of the applied magnetic field. As a result, magnetic particles and non-magnetic particles are separated efficiently.

For practical use, quick fluctuation using a mechanical device is necessary. Methods for generating a fluctuating magnetic field can be divided into methods using electromagnets and permanent magnets. The disadvantage of the electromagnets is that an operating device with a very high specification is necessary, as magnetic flux density is inversely proportional to the frequency of fluctuation. On the other hand, permanent magnets made from a rare earth element magnet with a high magnetic flux density are developed recently. Due to its high flux density, the size of the magnet can be reduced. Eddy current separators use this kind of small magnets in magnetic rolls. The permanent magnet method is used in this study.

Fig. 5 shows a schema of proposal magnetic separator with permanent magnets. When a magnetic roll rotates at R [rpm], the frequency F [Hz] of the fluctuation of the magnetic field is given by

$$\mathbf{F} = \frac{\mathbf{N} \times \mathbf{R}}{\mathbf{60}} \quad (1)$$

where, N is the number of magnetic poles. The energy required by this type of device is only that for operation of the magnetic roll, and because permanent magnets do not generate heat, a cooling device is not necessary.

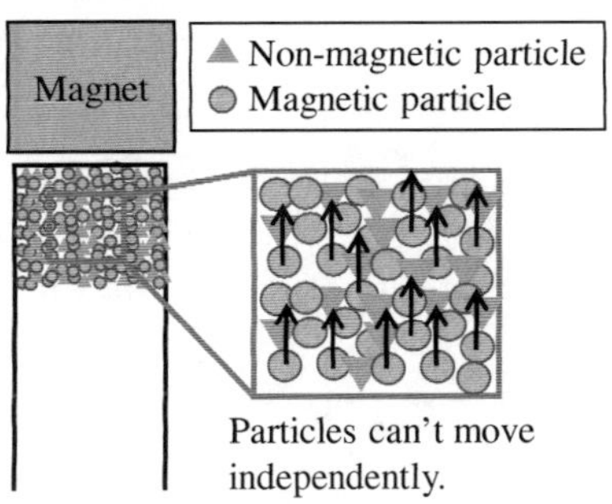

Fig. 3: Schema of conventional static magnetic separation.

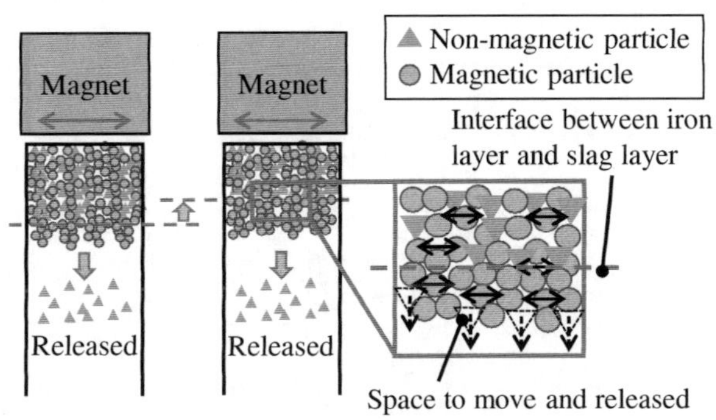

Fig. 4: Schema of proposal fluctuating magnetic field separation.

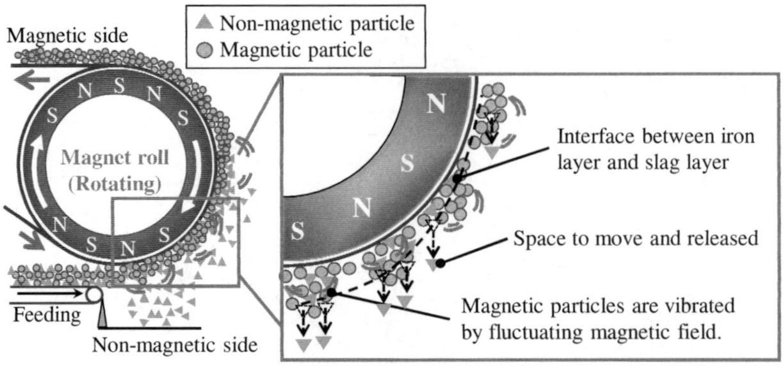

Fig. 5: Schema of proposal magnetic separator.

Separation test of steelmaking slag sample

In order to compare separation precision of conventional method and proposal method, laboratory test with steelmaking slag sample was conducted. Conditions of separation test are shown in Table 1 and properties of slag sample are shown in Table 2. Chemical component of magnetic side and non-magnetic side is analysed.

For evaluation, overall separation efficiency is calculated by Fe content. Overall separation efficiency is used generally for evaluation of separation precision. Definition is following. When mixture of material A and material B is separated, overall separation efficiency E [-] is given by

$$\mathbf{E} = \mathbf{r}_a + \mathbf{r}_\mathrm{b} - \mathbf{1} \tag{2}$$

where, r_a [-] is the recovery rate of material A, r_b [-] is the recovery rate of material B. Recover rate r_a [-] is yield of material A in separation and given by

$$\mathbf{r_a} = \frac{\mathbf{x_{ac}}}{\mathbf{x_a}} \tag{3}$$

where, x_a is the amount of material A in raw material, x_{ac} is the amount of remained material A after separation. Recover rate r_b [-] is calculated in the same way. In this study, material A is Fe and material B is slag (all other components except Fe). Remained material A is Fe in magnetic side. Remained material B is slag in non-magnetic side.

Results of test is shown in Fig. 6 and Fig. 7. As magnetic flux density increases, Fe content of magnetic side decreases. This is because in high magnetic flux density single-edged particles which are low Fe content are attracted to magnet. Anyway, both Fe content and overall separation efficiency is higher with proposal method than with conventional method in all magnetic flux density. From the result of separation test, it is revealed that proposal method performs very good separation for mass product.

However, if frequency of magnetic field is too high, particles will not have enough time to react to fluctuating magnetic force, and fluctuating magnetic field behaves just like static magnetic field. The upper limit of magnetic field frequency is described by Dr. Ishida in theoretical approach [2]. In order to clarify the upper limit of magnetic field frequency, separation test with various magnetic field frequency was conducted. Test conditions are shown in Table 3.

Results of test is shown in Fig. 8 and Fig. 9. At first, as magnetic field frequency increases, both of Fe content of magnetic side and overall separation efficiency increase. On the other hand, in higher frequency than 58 Hz, both of them decrease as increase of magnetic field frequency and finally almost same as conventional method. This result shows that there is upper limit frequency and optimum frequency for magnetic separation.

Table 1: Conditions of separation test.

Magnetic flux density [mT]	36.5, 48, 62, 84.5, 113, 155
Speed of pulley belt [m/min]	15
Thickness of slag sample [mm]	10
Frequency of magnetic field [Hz]	0, 58

Table 2: Properties of steelmaking slag sample.

Diameter of slag [mm]	0-5
T.Fe content of slag [wt%]	51

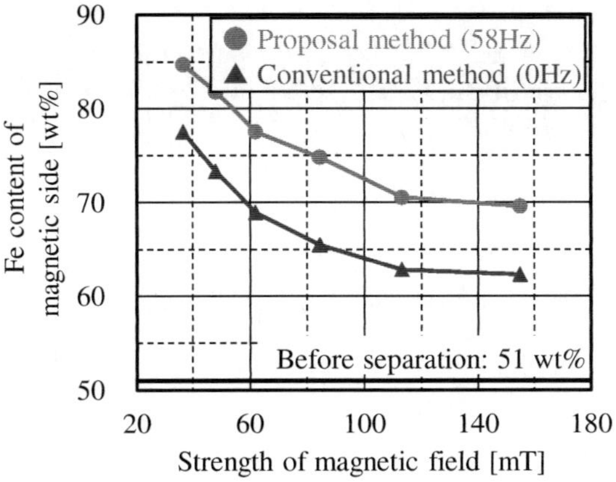

Fig. 6: T.Fe content of magnetic side.
(Magnetic field frequency: 0Hz, 58Hz)

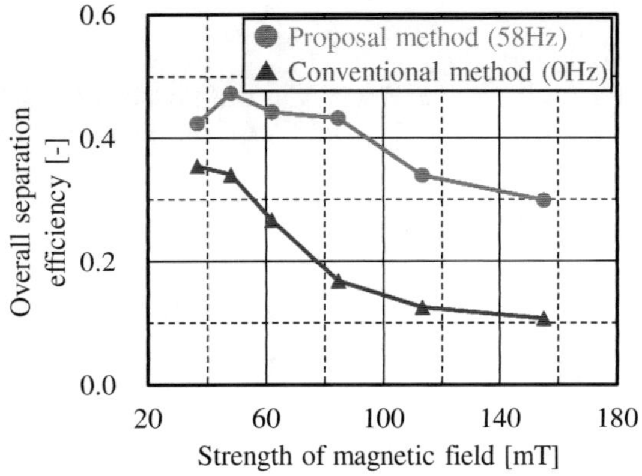

Fig. 7: Overall separation efficiency.
(Magnetic field frequency: 0Hz, 58Hz)

Table 3: Conditions of separation test.

Magnetic flux density [mT]	62
Speed of pulley belt [m/min]	15
Thickness of slag sample [mm]	10
Frequency of magnetic field [Hz]	0, 23, 58, 93, 187, 233

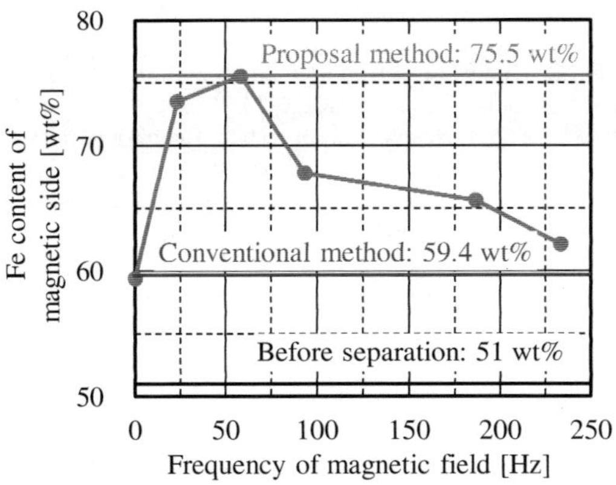

Fig. 8: T.Fe content of magnetic side.
(Magnetic force condition 62 mT)

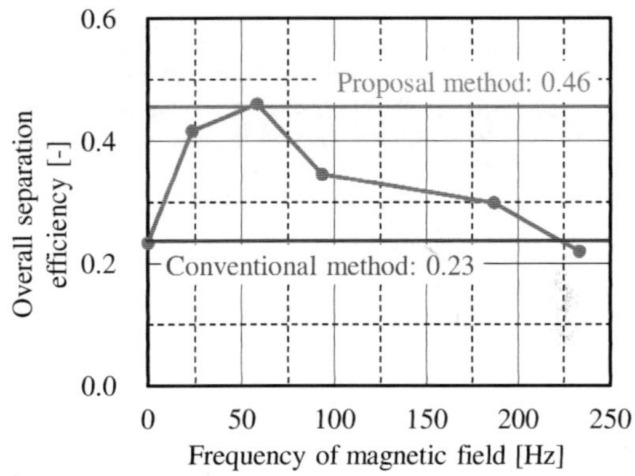

Fig. 9: Overall separation efficiency.
(Magnetic force condition 62 mT)

Conclusion

This study has discussed the new technology of a magnetic separation which can process steelmaking slag in the multiple layers for recycling iron in steelmaking slag. From laboratory test with steelmaking slag sample, it is revealed that new proposal method performs very good separation and there is the upper limit frequency and optimum frequency for magnetic separation.

References

1. J. Svoboda, "Magnetic Techniques of the Treatment of Materials", Kluwer Academic Publishers, (2004).
2. K. Ishida: PhD Thesis of Aachen University, (2014).
3. R. Dippenaar, Ironmaking and Steelmaking, 32, No. 1 (2005), 35-49.
4. R. J. Fruechan, "Overview of Steelmaking Processes and Their Development", The AISE Foundation, (1998), 1-8.
5. P. J. Koros, "Pre-Treatment of Hot Metal", The AISE Foundation, (1998), 413-429.
6. G. J. W. Kor, P. C. Glaws, "Ladle Refining and Vacuum Degassing", The AISE Foundation, (1998), 661-713.

On the Possibility of Melt Stirring in the Ladle by Modulated Traveling Magnetic Field

A. Kapusta[1,2], B. Mikhailovich[2*], M. Havkin[1], B. Tilman[1], K. Smirnov[1], A. Kudriavtcev[1]

[1]Nord Israel Research and Development Ltd, Beer-Sheva, Israel
[2]Department of Mechanical Engineering, Ben-Gurion University of the Negev, Beer-Sheva, Israel

Corresponding author: borismic@bgu.ac.il

Abstract
In the present report, we discuss the possibility of optimization of amplitude- and frequency-modulated traveling magnetic field (AFM TMF) excited in the melt by a standard inductor for the intensification of heat and mass transfer at the out-of-furnace processing in the ladle-furnace. On the basis of theoretical analysis of electrodynamic processes occurring in the melt under the action of traveling magnetic field, algorithms of AFM TMF parameters computation are determined.

Keywords: Ladle furnace, traveling magnetic field, Maxwell equations, amplitude and frequency modulation, resonance frequency.

Introduction

In the widely used technology “ladle-furnace”, vertically travelling harmonic magnetic field excited by a straight stirrer located near the ladle is used. The 3-dimensional induction stirring models presented in [1] are used in induction stirring in the ladle applying the finite-element method. In recent years, some attempts were made to optimize the processes in the melt under the impact of AFM TMF [2].

One of variants of the AFM TMF parameters optimization is given in [3]. Under the action of a harmonic TMF, in the active region of the melt in the ladle, whose thickness is on the order of the depth of magnetic field penetration into the melt, active circulation of the melt is excited covering around 20-30% of the liquid phase volume. The remaining part of the melt is involved into the convective large-scale motion much slower and homogenizes much longer and worse. To intensify melt mixing using the available equipment, it is suggested to apply frequency and amplitude modulation of TMF with the frequency ensuring resonance transfer of the electromagnetic energy from a part of Fourier-harmonics of the magnetic field to the melt.

Below we examine electrodynamic processes in the melt under the action of magnetic field created by a straight TMF stirrer oriented along the ladle height (Fig.1).

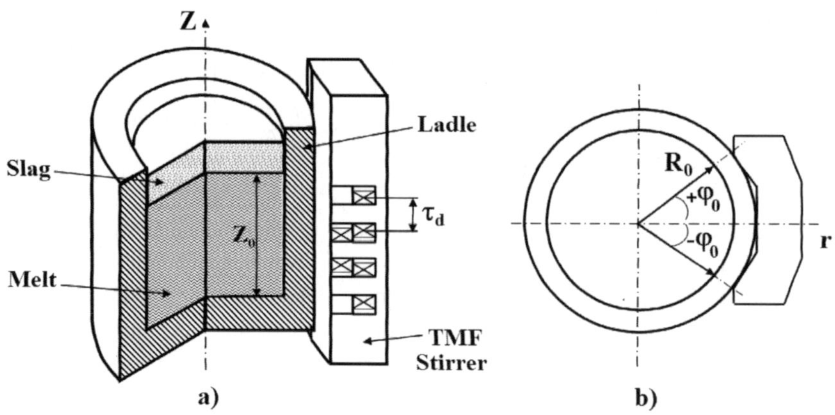

Fig. 1: Schematic common view of the ladle (a), and $r - \varphi$ section of active zone (b)

Problem Formulation and Solution

Electrodynamic processes excited by TMF inductor in the melt in a ladle are described in dimensionless cylindrical coordinates $r, \overline{\varphi}, z$ by the following equation for φ –component of the vectorial potential a_φ:

$$DD_*a_\varphi + \frac{1}{\pi^2 r^2}\frac{\partial^2 a_\varphi}{\partial \overline{\varphi}^2} + \frac{1}{\delta_z^2}\frac{\partial^2 a_\varphi}{\partial z^2} - \frac{\overline{\omega}}{\pi}\left(\frac{\partial a_\varphi}{\partial t} + \frac{V_z}{\delta_z}\frac{\partial a_\varphi}{\partial z}\right) = 0, \quad (1)$$

where $D = \frac{\partial}{\partial r}$, $D_* = \frac{\partial}{\partial r} + \frac{1}{r}$, $r = \frac{r_d}{R_0}$, R_0 – dimensional inner radius of the ladle; $z = \frac{z_d}{R_0}$ – normalized melt height, z_d– dimensional axial coordinate; $\overline{\varphi} = \frac{\varphi}{\pi}$ – normalized angular coordinate; $\overline{\omega} = \mu_0\sigma_m\omega_0 R_0^2$ – dimensionless frequency; ω_0 – carrier angular frequency of TMF; σ_m- specific electrical conductivity; ; μ_0– magnetic permeability of the vacuum; Z_0 - melt level height; $\delta_Z = \frac{Z_0}{R_0}$ - aspect ratio; $V_z = \frac{V_{zd}}{2\tau f_0}$ – dimensionless mean velocity of the melt; V_{zd} – dimensional mean velocity of the melt; τ_d - pole division of the inductor; $\tau = \frac{\tau_d}{R_0}$ (see (Fig. 1).

Boundary conditions on the lateral surface of the melt (Fig. 1b) are of the following form:

Near the inductor

$$D_*a_\varphi\big|_{r=1} = e^{-\frac{i\pi z}{\tau}}\phi(t)W(\overline{\varphi}), \quad (2)$$

On the remaining surface

$$D_*a_\varphi\big|_{r=1} = 0, \quad (3)$$

where, according to [4]

$$\phi(t) = (1 + \chi e^{i\pi\omega_a t})\sum_{n=-\infty}^{\infty} J_n(\xi)\, e^{i\pi\Omega_n t}, \quad (4)$$

and

$$W(\overline{\varphi}) = cos\pi\overline{\varphi}\big|_{-\overline{\varphi}_0}^{\overline{\varphi}_0},$$

where χ – depth of amplitude modulation, Ω_n – dimensionless angular frequency, $J_n(\xi)$ – Bessel function, $\xi = \frac{\Delta\omega}{\omega_f}$ – frequency modulation index, $\Delta\omega$– angular frequency deviation, ω_f–frequency modulation of the frequency.

We transform the vectorial potential a_φ as follows:

$$a_\varphi(r, \overline{\varphi}, z, t) = \alpha_\varphi(r, \overline{\varphi}, t)e^{-\frac{i\pi z}{\tau}} \quad (5)$$

Substituting (5) into (2) and (3) and then (2) and (3) into (1), we obtain the following:

$$DD_*\alpha_\varphi + \frac{1}{\pi^2 r^2}\frac{\partial^2 \alpha_\varphi}{\partial \overline{\varphi}^2} - \frac{\pi^2}{\tau^2\delta_z^2}\alpha_z - \frac{\overline{\omega}}{\pi}\left(\frac{\partial \alpha_z}{\partial t} - i\Omega_0\alpha_z\right) = 0 \quad (6)$$

with boundary conditions in accordance with Fig. 1b

$$D_*\alpha_\varphi\big|_{r=1} = \phi(t)W(\overline{\varphi})\big|_{-\varphi_0}^{\varphi_0}, \quad (7)$$

where Ω_0 has the form

$$\Omega_0 = \frac{\langle V_z\rangle}{\tau\delta_Z}$$

$\langle V_z\rangle$ – mean-integral velocity of z – component of flow velocity in the melt volume calculated by the Reynolds number $Re = 10^5$.

The following transformation transforms semi-uniform Equation (6) with boundary conditions (7) into the following semi-uniform equation with uniform boundary conditions:

$$\alpha_\varphi(r, \overline{\varphi}, t) = \alpha^* - f(r, \overline{\varphi}, t), \quad (8)$$

where

$$f(r, \overline{\varphi}, t) = \frac{r^{\sqrt{1+l^2}}}{1 + \sqrt{1 + l^2}}\cos(\pi\overline{\varphi})\big|_{-\overline{\varphi}_0}^{\overline{\varphi}_0}\phi(t)$$

where $l = 0;\ 1; \dots$.

Equation (6) and boundary conditions (7) can be represented as follows

$$DD_*\alpha^* + \frac{1}{\pi^2 r^2}\frac{\partial^2\alpha^*}{\partial\bar{\varphi}^2} - q\alpha^* - \frac{\bar{\omega}}{\pi}\left(\frac{\partial\alpha^*}{\partial t} - i\Omega_0\alpha^*\right) = f_1(r,\bar{\varphi},t) \quad (9)$$

$$D_*\alpha_\varphi\Big|_{r=1} = 0 \quad (10)$$

where $f_1(r,\bar{\varphi},t) = \left[q + \bar{\omega}\left(\frac{1}{\pi}\frac{\partial\phi}{\partial t} - i\Omega_0\phi\right)\right]\frac{r^{\sqrt{1+l^2}}}{1+\sqrt{1+l^2}}\cos\pi\bar{\varphi}\Big|_{-\bar{\varphi}_0}^{\bar{\varphi}_0}$; $l = 0, 1$; $q = \frac{\pi^2}{\tau^2\delta_z^2}$

Equation (9) with boundary condition (10) is solved by Galerkin's method. We expand the function α^* into series

$$\alpha^* = \sum_{k=1}^{\infty}\sum_{l=0}^{\infty} T_{kl}(t)J_{\chi_{kl}}(\gamma_{kl})\cos l\pi\bar{\varphi} \quad (11)$$

and confine ourselves by the first term of the series (11). Then $\cos l\pi\bar{\varphi} = 1$, and γ_{10} are roots of the equation $D_*J_1(\gamma_{10})|_{r=1} = 0$.

Substituting (11) into (9) and accomplishing procedures of Galerkin's method, we obtain an equation for calculating T_{10}. The first term of the series corresponding to the condition $l = 0$, is determined by the equation

$$DD_*\alpha_{10}^* - q\alpha_{10}^* - \bar{\omega}\left(\frac{1}{\pi}\frac{\partial\alpha_{10}^*}{\partial t} - i\Omega_0\alpha_{10}^*\right) = f_{10}(r,t)W(\bar{\varphi}), \quad (12)$$

where

$$f_{10}(r,t) = 0.5\left[q + \bar{\omega}\left(\frac{1}{\pi}\frac{\partial\phi}{\partial t} - i\Omega_0\phi\right)\right]r$$

The solution of the Equation (12) is as follows

$$\alpha_{10}^*(r,t) = \sum_{k=1}^{\infty} T_{k0}(t)J_1(\gamma_{10}r) \quad (13)$$

Substituting (13) into (12) and applying Galerkin's method, we obtain

$$\frac{1}{\pi}\frac{\partial T_{10}}{\partial t} + (P_{10} + Q - i\Omega_0)T_{10} = 0.5\left(Q + \frac{1}{\pi}\frac{\partial\phi}{\partial t} - i\Omega_0\phi\right)\frac{Int_{10}}{Int_{00}}f(\bar{\varphi}_0) \quad (14)$$

where $P_{10} = \frac{\gamma_{10}{}^2}{\bar{\omega}}$, $Q = \frac{q}{\bar{\omega}}$, is the norm of Bessel's function

$$Int_{00} = \frac{1}{2}J_1^2(\gamma_{10}),\ Int_{10} = \frac{J_2(\gamma_{10})}{\gamma_{10}};\quad f(\bar{\varphi}_0) = \frac{Sin\pi\bar{\varphi}_0}{\pi\bar{\varphi}_0}$$

A complex equation of amplitude oscillations of a harmonic Ω_n (14) has the following solution

$$T_{k0} = (A_{Rk0} + iA_{Ik0})e^{i\pi\Omega_n t}; \quad (15)$$

Substituting (15) into (14), we obtain

$$A_{I10} = \frac{C_{10}P_{10}(\Omega_n - \Omega_0)}{[(P_{10} + Q)^2 + (\Omega_n - \Omega_0)^2]};\quad C_{10} = \frac{Int_{10}f(\bar{\varphi}_0)}{Int_{00}} \quad (16)$$

The investigation of A_{I10} for extremum seeking has shown that A_{I10r} is a resonance amplitude under condition

$$\Omega_{nr} = P_{10} + Q + \Omega_0 \quad (17)$$

Resonance frequencies of frequency modulation ω_{fr} for two Fourier-packs of frequency and amplitude-and-frequency modulations are calculated using the following formula

$$\omega_{(fr)} = \frac{(P_{10} + Q + \Omega_0 - 1)}{n}. \quad (18)$$

Example

By way of example, we present a plot of the dependence of the vectorial potential amplitude A_{Ikl_n} on the Ω_n frequency harmonic (Fig. 2). Parameters for calculation: inner radius $R_0 = 1.58$ m, melt level $Z_0 = 2.66$ m, carrier frequency $f_0 = 1$ Hz, inductor pole division $\tau_d = 0.39$ m, dimensionless resonance frequency $\Omega_{nr} = 4.25$.

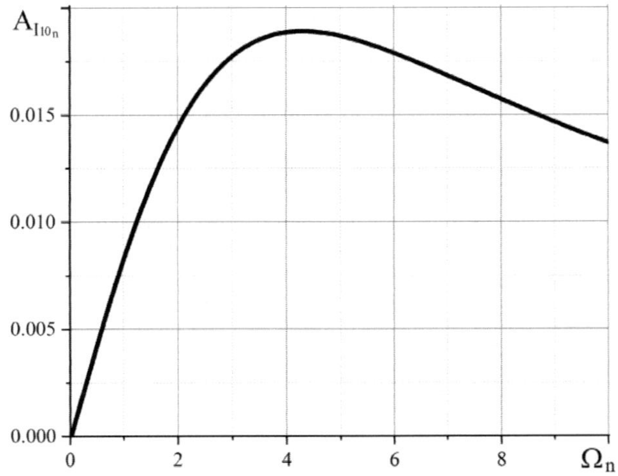

Fig. 2: Dependence of the amplitude of vectorial potential A_{I10_n} on the frequency of Ω_n Fourier-pack harmonics in the ladle.

In view of the fact that the AFM TMF is characterized by a superposition of two Fourier-packs (see Equation (4)), the total amplitude of resonance harmonics of the vector potential is more than twice greater than when using a harmonic field.

Conclusion

As follows from theoretical analysis the use of optimal parameters of AFM TMF leads to an increase of the amplitude of vectorial potential more than twice. Since the electromagnetic body forces are proportional to a squared vectorial potential, the effect of force impact grows more than four times. One can also expect an increase in the intensity of heat and mass transfer into the inactive zone of the melt due to the dispersion of waves generated by the sum of Fourier-harmonics in the vicinity of the resonance area. This point indicates to the possibility of development the highly-efficient out-of-furnace treatment technology using the AFM TMF acting on the entire volume of the ladle.

References

1. M. Pal, R. Eriksson, and P. Jonsson, A CFD model of a 20Kg Induction stirred laboratory scaled ladle. Proceedings of COMSOL Conference, Stockholm, Sweden, Oct-2005.
2. B. Mikhailovich, A. Kapusta, A. Levy, Proc. of EPM 2015, https://hal.archives-ouvertes.fr/hal-01333913
3. M. Havkin, B.Tilman, K. Smirnov, A. Kudriavtcev, Method of optimizing electromagnetic stirring in metallurgical technologies, Foreign License Granted US 62/642,617.
4. A. Angot, Compléments de Mathématiques. Paris, 1957.

Simulation of electrically induced vortical flows

V. Dzelme[1], A. Jakovics[1], A. Chudnovsky[2], E. Baake[3]

[1] Laboratory for Mathematical Modelling of Environmental and Technological Processes, Department of Physics, University of Latvia, 23 Zellu street, Riga, Latvia
[2] BIS Global Ltd., 79/81-15 Kr. Valdemara street, Riga, Latvia
[3] Institute of Electrotechnology, Leibniz University of Hannover, 4 Wilhelm Busch street, D-30167 Hannover, Germany

Corresponding author : valters.dzelme@lu.lv

Abstract
In this work, we study electrically induced flows numerically using open-source software and experimentally. Two systems are considered - single and multiphase (free surface) flows driven by axisymmetric DC and AC current injection. We investigate characteristic velocity and free surface deformation dependence on the injected current and validate it with experimental data. Results show that maximum axial velocity is a linear function of injected current, but free surface deformations are approximately proportional to current squared.

Keywords : magnetohydrodynamics, electrovortex, simulation

Introduction

Electrically induced flows are found in many applications – arc furnaces, welding, etc. [1]. The flow is driven by Lorentz force created by the interaction between injected current and its own magnetic field. Depending on specific electrode configuration different vortical structures can form.

Electrovortical phenomena have been extensively studied theoretically and experimentally [1]. Numerical work usually considers simplified systems, such as hemispherical arc furnaces [2][3] using commercial software. Free and open-source software is not widely used mostly due to lack of user interface and complicated coupling between, for example, electromagnetic and fluid dynamics simulations. The recent development of EOF-Library [4], which is a coupler between Elmer [5] and OpenFOAM [6] and is an open-source library itself, simplifies the task of simulating complex coupled systems. The coupling library has been validated against commercial solutions [7].

The goal of this work is to apply and validate open-source software for electromagnetically induced vortical flows, as well as to study two specific systems with industrial applications in mind. The first (*system A*) is a small cylindrical container with electric current flowing through the melt between a small electrode on top and conducting bottom of container. It is a model system of typical DC arc furnaces. The second system (*system B*) is more complex – it consists of cylindrical container with 50Hz alternating current flowing between a small bottom electrode and conducting side wall. In this case, the current flows through a layer of liquid metal causing free surface deformations near the small electrode at the center. Possible application of this system can be slag removal from the melt in special applications – surface deformation is pushing the slag towards the side wall where it can be easily collected – this is demonstrated experimentally. Another application could be surface wave generation and flow intensification for improved melt purification via evaporation from the surface.

Numerical model

Software used for electromagnetics is Elmer, which solves the Maxwell's equations in potential formulation using the finite element method, but for fluid dynamics – OpenFOAM, which solves the Navier-Stokes, k-ω SST turbulence model equations, and for system B also transport equation for volume fraction, using the finite volume method. The coupling is achieved with the EOF-Library. The schemes of system A and B are shown in Fig. 1 and 2, respectively. Relevant material properties are listed in Tab.1.

Since the 2D axisymmetric solver in Elmer is not designed to solve for azimuthal component of magnetic flux density (or current in meridional plane), quasi 2D approach is used modelling 2° sector of full cylinder. Electromagnetics mesh for system A consists of 133k hexahedral elements (three layers in azimuthal direction), but system B – 180k hexahedral elements (two layers in azimuthal direction).

Fluid flow is also solved in quasi 2D approximation as 2° sector of full cylinder, as OpenFOAM does not have pure 2D capabilities. Fluid dynamics mesh for system A is 22k hexahedral elements (one element thick wedge), but for system B – 250k hexahedral elements (one element thick wedge). The electromagnetic force is computed only once, and the flow simulation is performed with constant momentum source.

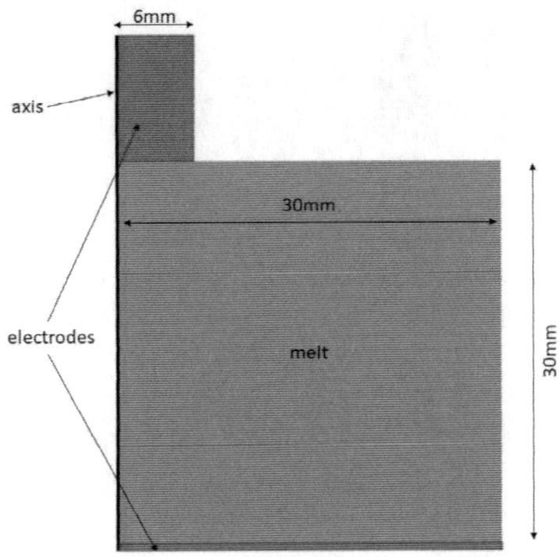

Fig. 1: System A

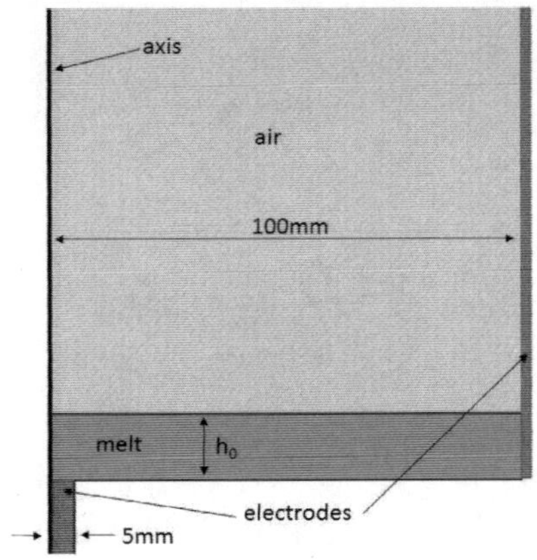

Fig. 2: System B

Table 1. Material properties

	System A	System B
Melt density, kg/m^3	13534	6440
Melt electric cond., S/m	$8.6 \cdot 10^5$	$3.46 \cdot 10^6$
Electrode electric cond., S/m	$5.8 \cdot 10^7$	$5.8 \cdot 10^7$
Melt viscosity, mPa·s	2.20	2.40
Surface tension, N/m	-	0.534

Experiments

Experimental results for system A are taken from [8], where liquid mercury flow velocity was measured using optical fiber sensor. Experiments for system B are performed at the Institute of Electrotechnology of the Leibniz University of Hannover. The picture of the setup is shown in Fig. 3. The two side electrodes are of the same polarity; the other pole is a small electrode at the bottom not seen here; dimensions as shown in Fig.2. Bottom wall of the container is plastic, so that the current is flowing from the bottom electrode to the side wall only through the melt. The current is provided by a 50Hz voltage source. At 50Hz, the skin depth in liquid galinstan is approximately 4cm.

Fig. 3: System B experimental setup

Results

System A

Example of current and velocity distributions are shown in Fig. 4 and 5, respectively. The highest current density is near the edge of the top electrode, and this is also where the Lorentz force concentrates, pushing the melt away from the electrode. Axial velocity distribution on axis for various injected current values is shown in Fig. 6. In Fig. 7, we plot maximum axial velocity (20mm from the top electrode) as a function of injected current. The theoretical analysis provided in [8] says that axial velocity in such system is proportional to the injected current and it is seen in simulation and experimental results.

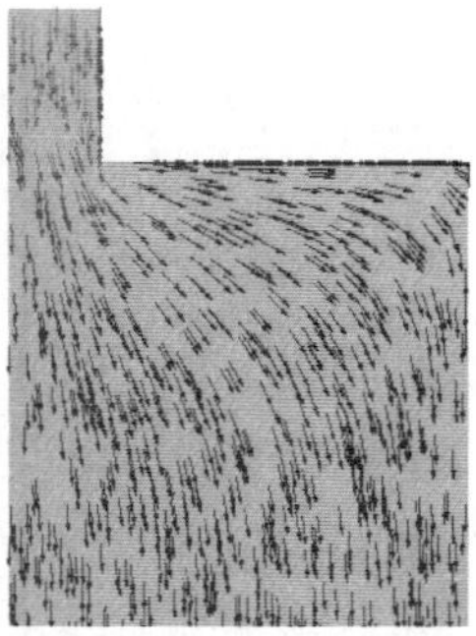

Fig. 4: Example current distribution

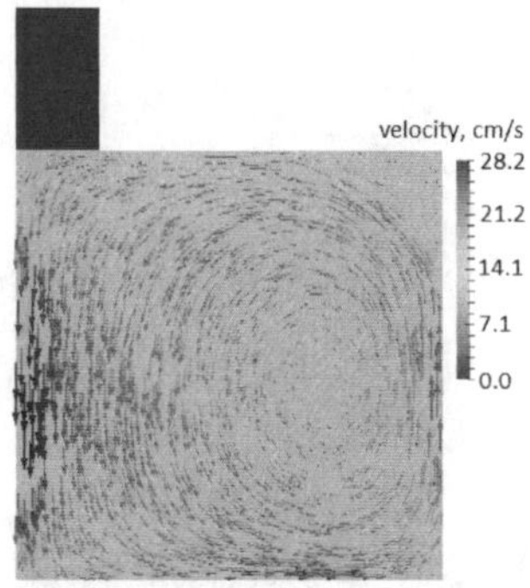

Fig. 5: Example velocity distribution, I=1kA

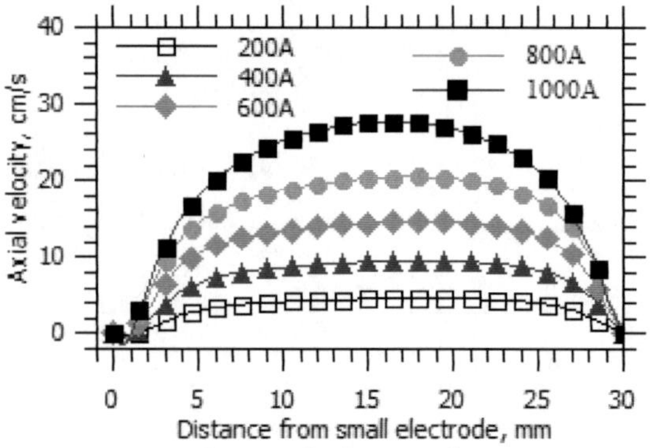

Fig. 6: Velocity on axis

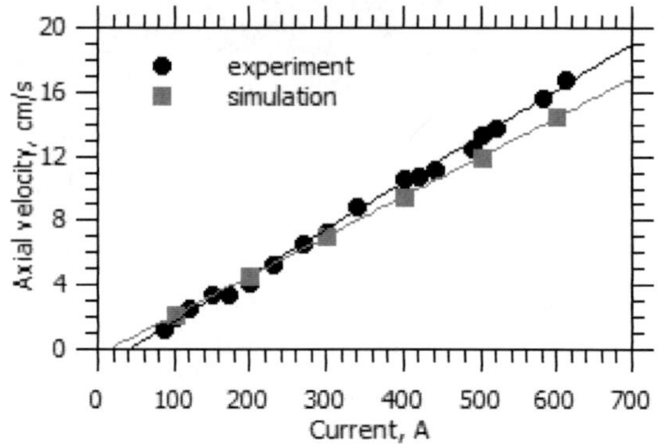

Fig. 7: Max. axial velocity vs current

System B

Example of current distribution (without surface deformation) is shown in Fig. 8 and velocity and surface deformation at I_{rms}=1000A is shown in Fig. 9. Free surface deformation height depending on current squared is shown in Fig. 10. In Fig. 11, we plot maximum axial velocity as a function of injected current.

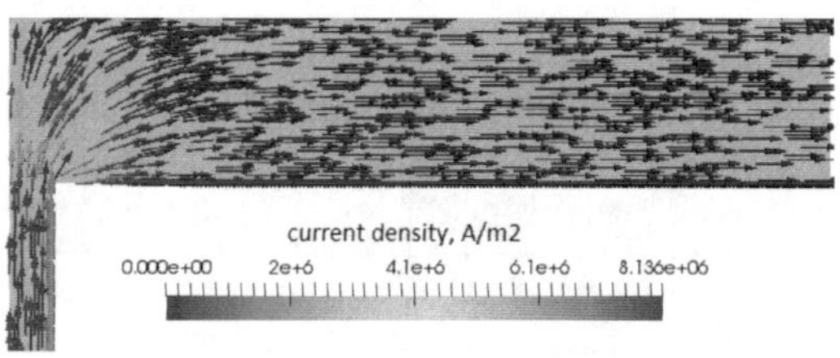

Fig. 8: Example current distribution, h_0=20mm

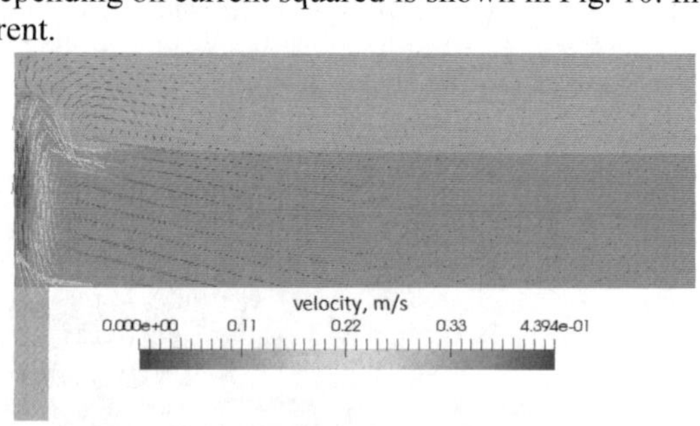

Fig. 9: Example velocity and surface deformation, h_0=20mm, I_{rms}=1kA

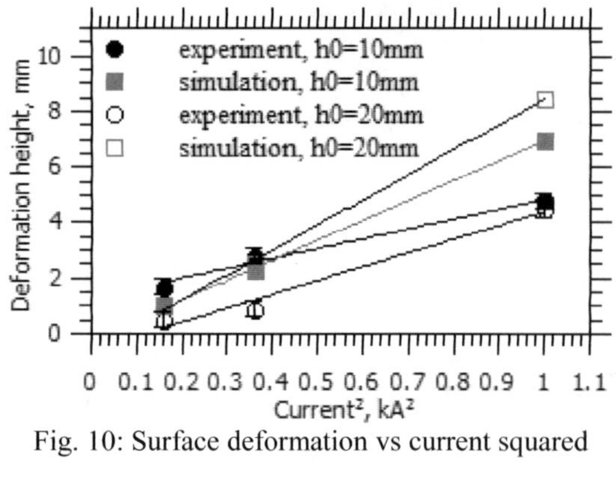

Fig. 10: Surface deformation vs current squared

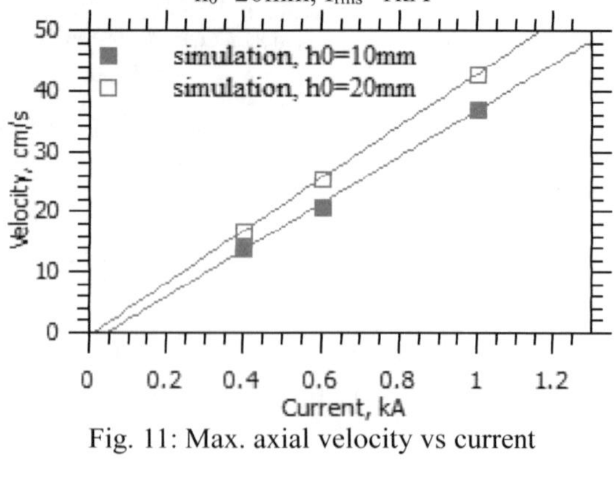

Fig. 11: Max. axial velocity vs current

In this system, no velocity measurements were performed at this stage, but are planned in the future. Deformation height disagreement between experiments and simulations can be due to many reasons. First, the surface height measurements were not done using precise equipment – they were manual measurements using simple tools. Second, the material properties (electric conductivity, viscosity etc) were not exactly known – values taken from literature may not be completely the same as in the experiments. Third, even the slightest imperfections (e.g., electrode centering) can cause different flow patterns and surface deformations. The numerical model, of course, is an idealization, especially being axisymmetric which does not allow the jet movement away from the axis. Nevertheless, the results of both simulation and experiment are the same order of magnitude and show the same tendency – in the considered range of currents surface deformation is approximately proportional to the square of injected current.
Finally, to study possible slag removal towards container walls by free surface deformation, we use small (d=2mm) plastic particles on the surface. Particle distribution without current and with I_{rms}=1000A is shown in Fig. 12 and 13, respectively. It is clearly seen that the surface deformation moves the "slag" away from the zone above the small bottom electrode. To move the particles even closer to the side wall, larger or several electrodes, or higher current could be used.

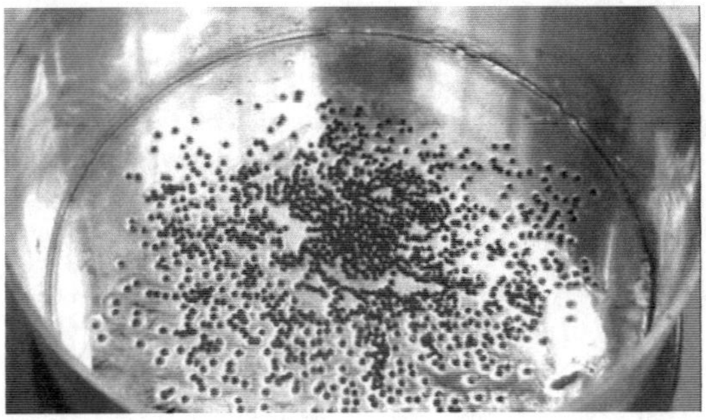

Fig. 12: Particle "slag" distribution; h_0=20mm, no current

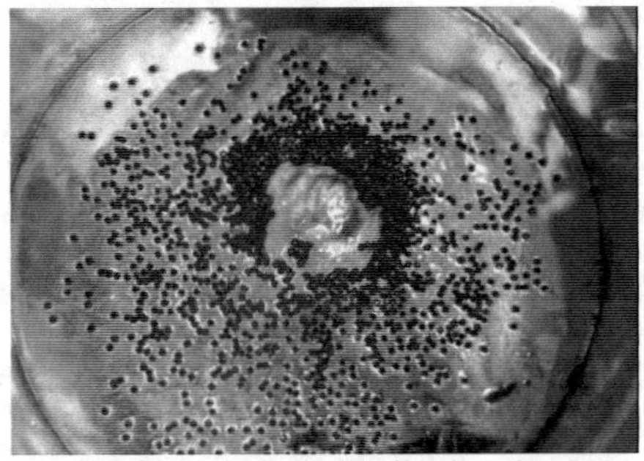

Fig. 13: Particle "slag" distribution; h_0=20mm, I_{rms}=1000A

Conclusions

Firstly, numerical model of single-phase electrically induced flow in a cylindrical container was validated using experimental data – maximum axial velocity is in a good agreement to the measurements. The proportionality of axial velocity to the injected current is also a result agreeing to theoretical considerations [8].
An experimental device was created to study liquid metal free surface deformations induced by current flowing between small bottom electrode and conducting side wall. It was found that the deformation height is approximately proportional to square of the injected current in the considered range. Like in the first system, the maximum axial velocity is proportional to current.
The goal of applying open-source software for electrovortical flow simulations is achieved. The coupling solution EOF-Library is a very efficient interface between Elmer and OpenFOAM, which enables virtually any variable transfer and interpolation between different meshes of electromagnetic and fluid flow simulation.
Further experimental and numerical work will include different bottom electrode configurations to control the free surface deformations, as well as studying the process instabilities, as the deformation height was observed to be quite unstable, which can be of even more importance in industry-scale equipment.

References

1. V. Bojarevics, J. A. Freibergs, E.I. Shilova, E.V. Scherbinin, *Electrically Induced Vortical Flows*, Kluwer Academic Publishers, 1989
2. O. Kazak, Magnetohydrodynamics, Vol. 50 (2014), pp. 193-205
3. Y. Ivochkin et al., Magnetohydrodynamics, Vol. 51 (2015), pp. 337-343
4. EOF-Library, https://eof-library.com
5. ElmerFEM, https://www.csc.fi/web/elmer
6. OpenFOAM, https://openfoam.org
7. J. Vencels et al., Magnetohydrodynamics, Vol. 53 (2017), pp. 643–652
8. V.G. Zhilin et al., Magnetohydrodynamics, Vol. 22 (1986), pp. 110-116

IOP Conf. Series: Materials Science and Engineering **424** (2018) 012084 doi:10.1088/1757-899X/424/1/012084

Electromagnetic flow rate measurement in molten tin circulating in a closed-loop test system

Z. Lyu, Ch. Karcher, Y. Kolesnikov, Th. Boeck

Institute of Thermodynamics and Fluid Mechanics, Technische Universität Ilmenau,
P.O. Box 100565, D-98684 Ilmenau, Germany

Corresponding author: christian.karcher@tu-ilmenau.de

Abstract

Flow rate measurement in molten metals at elevated temperatures is still a challenging task in metallurgic application. Due to the opaqueness and chemical aggressiveness of such melts, non-contact and non-invasive electromagnetic methods are of special interest. In order to develop and to test such methods under controllable laboratory conditions, during the last years at TU Ilmenau we have built up and instrumented the test facility TINTELO (tin test loop). TINTELO is a closed loop facility made of stainless steel in which molten tin at a temperature up to 350°C circulates in the horizontal plane. The flow is driven by an electromagnetic pump based on rotating permanent magnets. The rotation frequency of the pump can be fixed at up to 60 Hz providing a mass flow rate of 17.5 kg/s which corresponds to a Reynolds number of $Re = 2 \cdot 10^5$. TINTELO is instrumented with both a Lorentz force flowmeter and a wall potential flowmeter. The Lorentz force flowmeter measures the force which the electrical conducting moving molten tin exerts on a system of permanent magnets arranged outside of the loop. This force is the counter force to the well-known braking Lorentz force that is proportional to the melt flow rate. Moreover, the wall potential flowmeter records the electric potential difference between two electrodes inserted in the conducting wall of the loop. This potential difference, induced by the moving melt when interacting with an externally applied magnetic field, is proportional to melt velocity. In this paper we present the design and the instrumentation of TINTELO, the calibration methods for the used flowmeters and first results of the flow rate test measurements.

Key words: liquid tin, Lorentz force flowmeter, wall potential flowmeter

Introduction

Electromagnetic flow measurement techniques are promising tools to meet the demand of metallurgic industry of controlling and metering liquid metal melt flow under production conditions. However, field tests are expensive, hard to manage and hard to perform under the very harsh ambient conditions including high temperatures, mechanical vibrations, electromagnetic noise, dust, and wind, among others. Hence, before integrating such methods in the production line, it is favorable to develop and to test such techniques under controlled laboratory conditions. In such laboratory test stands usually low-melting metal alloys are used as test melts including GaInSn, NaK, PbLi, and Wood's metal. However, such substitutes show also some drawbacks like high price, toxicity, or even flammability when handled carelessly. Hence, at TU Ilmenau we have decided to build up a liquid metal test facility operating with tin showing a melting temperature of 232°C. The test facility is named TINTELO (tin test loop). The main goal is to develop and to test the non-contact electromagnetic flow measurement technique called Lorentz force velocimetry (LFV) [1]. This technique has already been applied in several field tests with liquid aluminum and liquid steel [2], [3], [4]. The measurement principle is based on the fact that in electrically conducting fluids eddy currents are induced when the melt flow interacts with an externally applied magnetic field. In turn, these eddy currents interact with the field to generate within the melt Lorentz forces. This force tends to break the flow and is proportional to the flow rate. The scaling relation for the magnitude F_L of this force reads as

$$F_L \sim \sigma Q B^2 L, \tag{1}$$

where σ is the electrical conductivity, Q the flow rate, B the magnetic flux density and L a characteristic length. In LFV the counter force to this force, acting on the externally arranged magnetic field system is measured using a Lorentz force flowmeter (LFF). In this paper we apply this technique to flowrate measurement in liquid tin using the test facility TINTELO. The present study is organized as follows. First, we describe the used test facility TINTELO and its instrumentation. Secondly, we present some details of the calibration of the meters used. Next, we show some first results on flow rate measurement. Finally, we shall give a short summary of the main conclusions.

Test facility TINTELO

A schematic sketch of TINTELO is shown in Fig. 1, while Fig. 2 provides a photograph of the test sections. 1100 kg of pure tin are molten by induction heating in storage tank 1. The typical operation temperature is 350°C. Then the melt is

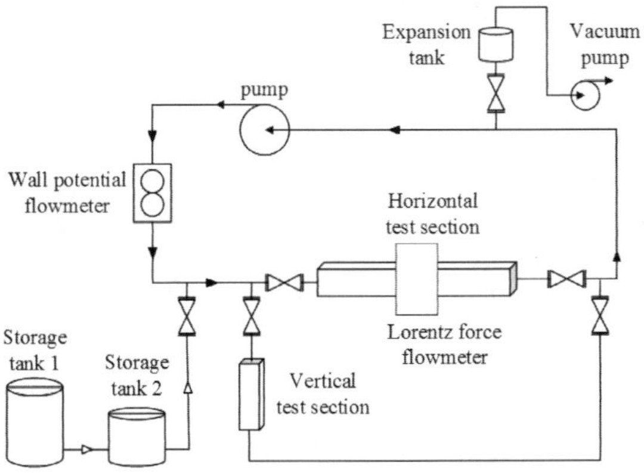

Fig. 1: Schematic sketch of TINTELO.

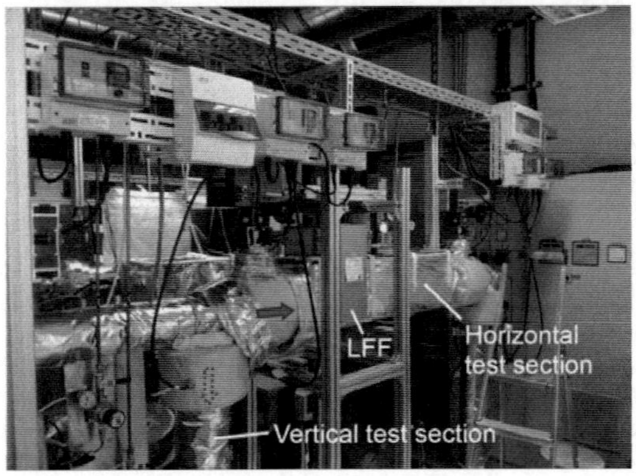

Fig. 2: Photograph of test sections.

partly transferred into storage tank 2 in order to start the filling process. Here, using a vacuum pump, we first evacuate the piping of the test facility from Argon which is used to avoid oxidation. Upon opening of the valve, the molten tin flows into the loop. The test section is completely filled when the melt reaches the expansion tank arranged at the very top of the facility. The melt flow is driven by an electromagnetic pump based on rotating permanent magnets. The frequency of the pump can be fixed at up to 60 Hz providing a mass flow rate of up to 17.5 kg/s. The pipes are made of non-magnetic stainless steel and have an inner diameter of 72.1 mm (DN65). Thus, within TINTELO we can reach flow Reynolds numbers of up to $Re = 2 \cdot 10^5$ which are typical in industrial applications. As it can be seen in Fig. 2, all parts of the facility are thermally well insulated to avoid freezing. Critical parts are even encased by electrical heating jackets,

Depending on the mode of operation, respective valves are set to serve either the horizontal test section or the vertical test section. The horizontal test section is designed as channel with rectangular cross-section and a movable top plate. By that, either closed channel flow or open channel free-surface flow can be established in order to meet flow conditions that are typical in the production of secondary aluminium. The vertical test section is designed as a pipe in order to model flow condition typical in continuous casting of steel, i.e. the flow through the submerged entry nozzle. However, in the present study only the horizontal test section is in operation.

The horizontal test section is equipped with a Lorentz force flowmeter (LFF), see Figs. 2 and 3. A LFF basically consists of two blocks of permanent magnets, connected by a yoke, and an attached force sensor based on digital strain gage. This magnet system provides a magnetic flux density of about 150 mT. The LFF is mounted on the top of the horizontal test section in such a way that a spanwise field in generated that penetrates the entire cross-section of the melt. Thus, as shown in Fig.3, the measured force acting on the magnet system and pointing in the flow direction is the counterforce to the flow-braking Lorentz force.

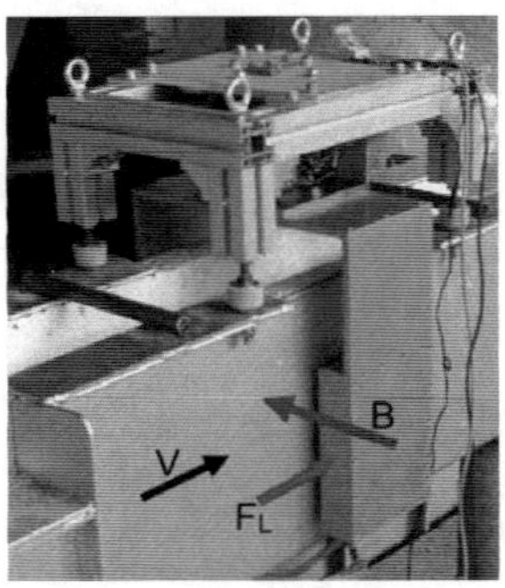

Fig. 3: Lorentz force flowmeter.

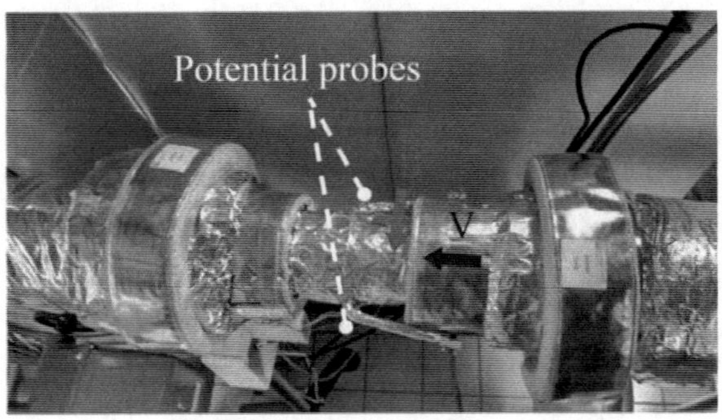

Fig. 4: Wall potential flowmeter.

As a reference measuring device we use a wall potential flowmeter (WPF) which is positioned between pump and test sections, see Figs. 1 and 4. The WPF consists of a pipe made of stainless steel. In the wall of this pipe two potential probes (electrodes) are inserted. The measurement principle is based on the fact that upon the interaction of the

IOP Conf. Series: Materials Science and Engineering **424** (2018) 012084 doi:10.1088/1757-899X/424/1/012084

melt flow and a spanwise magnet field, an electric potential difference $\Delta\varphi$ across the pipe is induced. This potential difference is proportional to the mean melt velocity V according to the relation

$$\Delta\varphi \sim VBD, \qquad (2)$$

where D is the pipe diameter. For proper operation of the WPF it is essential to have very good hydromechanic and electrical wetting conditions between the melt and the inner pipe wall. We solved this problem by soldering the inner pipe wall with a thin layer of tin.

Calibration procedure

The calibration of the LFF is performed by a dry procedure. Here, solid metallic bars of known cross-section and electrical conductivity are pulled at controlled speeds through the LFF. As expected from Eq. 1, results show a strict linear relationship between the measured force and the pulling speed. The obtained data can be transformed to construct the calibration curve under wet conditions. Details of this procedure are given elsewhere [5], [6], [7]. The calibration curves of the WPF and the electromagnetic pump are shown in Figs. 5 and 6. The WPF was calibrated using the test melt GaInSn at known flowrate. However, the data can be directly transferred to liquid tin as the conductivity ratio between wall and melt remains nearly the same and effects of temperature do not matter. As shown in Fig. 5, according to Eq. (2) we obtain a strict linear relationship between potential difference and flowrate. Likewise, the calibration curve of the electromagnetic pump is obtained using GaInSn at known flowrate. The data can be transferred to tin by matching the densities. As it can be seen in Fig. 6, both flowrate and pressure increase linearly with frequency.

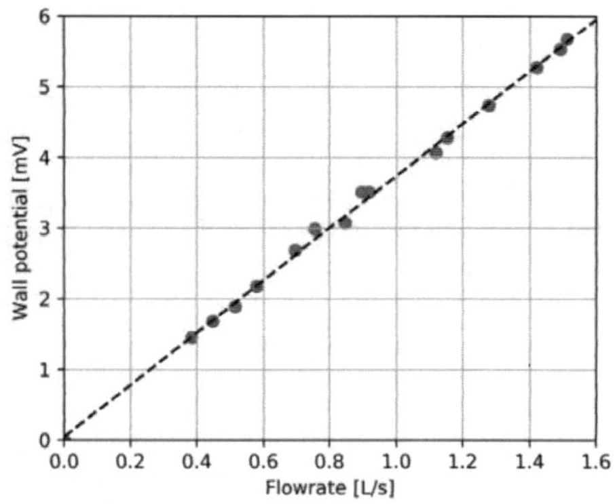

Fig. 5: Calibration curve of wall potential flowmeter.

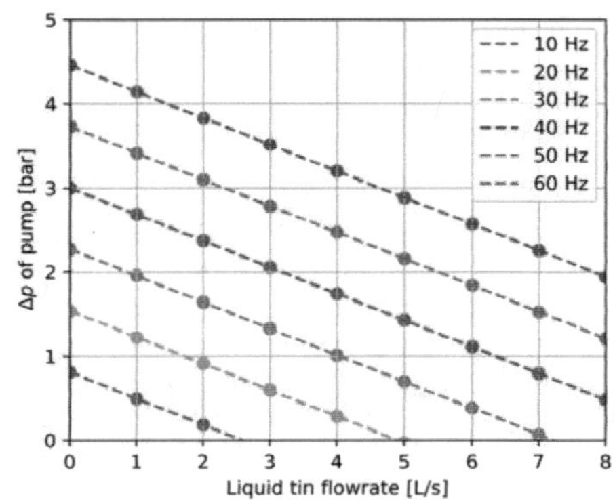

Fig. 6: Calibration curve of electromagnetic pump.

Results and discussion

Some first results of test measurements are shown in Fig. 7. The upper graph (blue curve) shows raw data of a time series of about 25 min of the measured Lorentz force given in mN. The lower graph (black curve) shows the respective raw data of the same time series of wall potential difference given in mV. The sampling rate was 2 Hz. During this run the frequency of the pump was successively changed within the range of 0 – 25 Hz and remained constant for about 30 s. From the graphs it is clearly seen that both electromagnetic flow measurement techniques work well in liquid tin at 350°C. Moreover, it can be seen that the signals obtained by the WPF is more robust than those recorded by the LFF. This reflects the fact that the force sensor is very sensitive to mechanical vibrations caused by the rotating pump and/or the melt flow. It should also be mentioned that we do not observe any hysteresis effects, as upon increasing and decreasing the frequency of the pump, nearly identical measurement values are obtained both LFF and WPF.

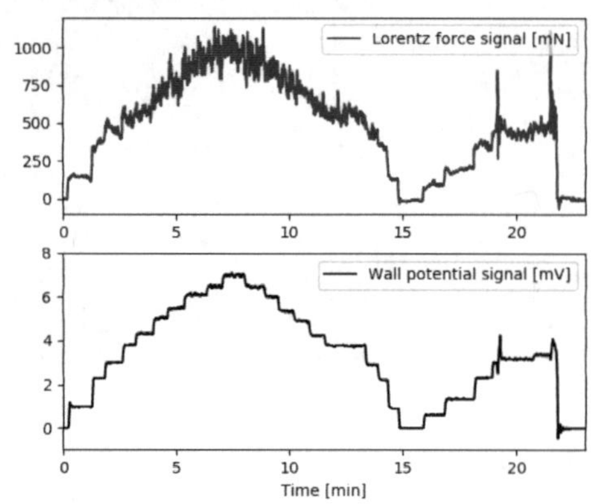

Fig. 7: Time series data of LFV and WPF.

At the end of the run, i.e at about t = 19 min and t = 21 min, the Argon pressure was increased two times in order to compensate back flow of tin into the storage tank due to a slightly leaky ball valve. As it can be seen from Fig. 7, again the LFF is much more sensitive to such sudden changes of pressure and the resulting vibrations. However, in application the somewhat noisy signals of the LFF can be effectively filtered in order to give reliable measurement values.

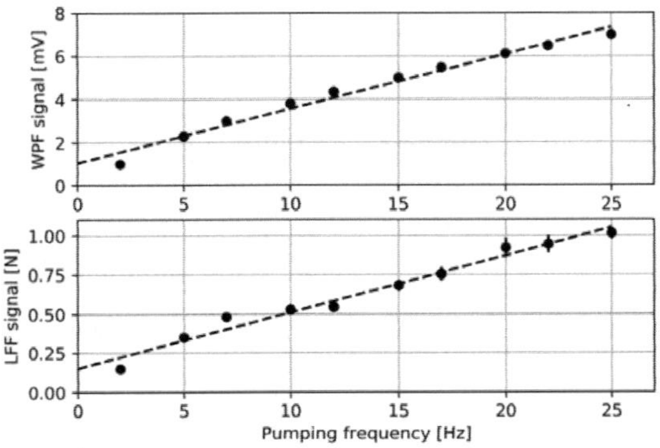

Fig. 8: Data fitting to pumping frequency.

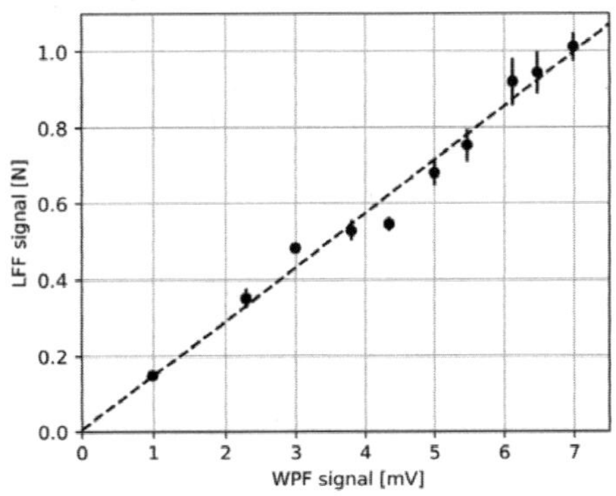

Fig. 9: Correlations between LFF and WPF.

Fig. 8 shows the results of the time-averaged raw data within the frequency range of the pump of 2 – 25 Hz. Here, the upper graph shows the results of the flow measurement using the wall potential flowmeter, while the lower graph refers to the Lorentz force flowmeter. For the averaging, only the raw data of the first 8 min of the time series were used, corresponding to the increase of pump frequency. Again, we can observe the nearly linear dependence of both signals on pumping frequency. We can thus conclude that even at the highest frequency we are still in the regime of low magnetic Reynolds numbers Re_M. A rough estimate shows that Re_M is of order 10^{-1}. Finally, Fig. 9 shows the correlation of the time-averaged signals obtained by LFF and WPF. From this graph we can conclude that there is likewise a linear relationship between the signals. This gives the potential for using both non-contact electromagnetic flow measurement techniques for mutual calibration.

Summary and conclusions

In this study we have investigated experimentally liquid tin flow at a temperature of 350°C. Both a Lorentz force flowmeter and a wall potential flowmeter have been tested in order to demonstrate their feasibility for non-contact electromagnetic flow measurement at such elevated temperatures. Our first test runs clearly show that both techniques operate well. Both methods show an almost linear dependence of their signal amplitudes on the frequency of the flow-driving electromagnetic pump which is related to the flowrate via its calibration curve. Moreover, we observe likewise a linear dependence between both signals.

Acknowledgment

The authors acknowledge financial support by Deutsche Forschungsgemeinschaft (DFG) within the Research Training Group on Lorentz force velocimetry and Lorentz force eddy current testing under grant GRK1567 and within the Research for Proposals of Major Instrumentation on Liquid Tin Channel Ilmenau under grant INST 273/37-1-FUGG. We acknowledge fruitful discussions with Dr. Th. Wondrak from Helmholtz Center Dresden-Rossendorf. Furthermore, we acknowledge scientific support by Dr. I. Bucenieks from University of Lativa in Riga for manufacturing the electromagnetic pump and the wall potential flowmeter and providing the respective calibration data. We also acknowledge the technical support by V. Mitschunas, A. Thieme, S. Buchelt and N. Lukin.

References

1. A. Thess, E. Votjakov, B. Knaepen, O, Zikanov, New journal of Physics, 9 (2007), 299 – 326.
2. Y. Kolesnikov, Ch. Karcher, A. Thess, Metallurgical and materials transactions B, 42 (2012), 441-450.
3. D. Jian, Ch. Karcher, X. Xu, A. Deng, E. Wang, A. Thess, J. Iron Steel Research, 9 (2012), 509 – 513.
4. Ch. Weidermann, PhD dissertation, TU Ilmenau (2013).
5. V. Minchenya, Ch. Karcher, Y. Kolesnikov, A. Thess, Magnetohydrodynamics, 45-4 (2009), 569-578.
6. X, Wang, Y. Kolesinikov, A.Thess, Measuremet Science and Technology, 23 (2012), 0450051-045060.
7. M. Gramß, PhD dissertation, TU Ilmenau (2013).

Institute of Physics
Dirac House, Temple Back
Bristol BS1 6BE UK

ISSN: 1757-8981
ISBN 978-1-5108-7335-3

9 781510 873353

3rd International Conference and Young Scientist School on Magnetic Resonance Imaging in Biomedical Research 2016

Journal of Physics: Conference Series Volume 886

Novosibirsk, Russia
23 – 26 November 2016

IOP
Institute of Physics